Advances in the Interplay Between Quantum and Gravity Physics

NATO Science Series

A Series presenting the results of scientific meetings supported under the NATO Science Programme.

The Series is published by IOS Press, Amsterdam, and Kluwer Academic Publishers in conjunction with the NATO Scientific Affairs Division

Sub-Series

I. Life and Behavioural Sciences	IOS Press
II. Mathematics, Physics and Chemistry	Kluwer Academic Publishers
III. Computer and Systems Science	IOS Press
IV. Earth and Environmental Sciences	Kluwer Academic Publishers
V. Science and Technology Policy	IOS Press

The NATO Science Series continues the series of books published formerly as the NATO ASI Series.

The NATO Science Programme offers support for collaboration in civil science between scientists of countries of the Euro-Atlantic Partnership Council. The types of scientific meeting generally supported are "Advanced Study Institutes" and "Advanced Research Workshops", although other types of meeting are supported from time to time. The NATO Science Series collects together the results of these meetings. The meetings are co-organized bij scientists from NATO countries and scientists from NATO's Partner countries – countries of the CIS and Central and Eastern Europe.

Advanced Study Institutes are high-level tutorial courses offering in-depth study of latest advances in a field.
Advanced Research Workshops are expert meetings aimed at critical assessment of a field, and identification of directions for future action.

As a consequence of the restructuring of the NATO Science Programme in 1999, the NATO Science Series has been re-organised and there are currently Five Sub-series as noted above. Please consult the following web sites for information on previous volumes published in the Series, as well as details of earlier Sub-series.

http://www.nato.int/science
http://www.wkap.nl
http://www.iospress.nl
http://www.wtv-books.de/nato-pco.htm

Series II: Mathematics, Physics and Chemistry – Vol. 60

Advances in the Interplay Between Quantum and Gravity Physics

edited by

Peter G. Bergmann
New York University, U.S.A.

and

Venzo de Sabbata
University of Bologna, Italy

Kluwer Academic Publishers

Dordrecht / Boston / London

Published in cooperation with NATO Scientific Affairs Division

Proceedings of the NATO Advanced Study Institute on
Advances in the Interplay Between Quantum and Gravity Physics
Erice, Italy
30 April–10 May 2001

A C.I.P. Catalogue record for this book is available from the Library of Congress.

ISBN 1-4020-0592-X

Published by Kluwer Academic Publishers,
P.O. Box 17, 3300 AA Dordrecht, The Netherlands.

Sold and distributed in North, Central and South America
by Kluwer Academic Publishers,
101 Philip Drive, Norwell, MA 02061, U.S.A.

In all other countries, sold and distributed
by Kluwer Academic Publishers,
P.O. Box 322, 3300 AH Dordrecht, The Netherlands.

Printed on acid-free paper

Printed in the Netherlands.

CONTENTS

PREFACE

In this XVII Course of the International School of Cosmology and Gravitation devoted to "ADVANCES IN THE INTERPLAY BETWEEN QUANTUM AND GRAVITY PHYSICS" we have considered different aspects of the influence of gravity on quantum systems. In order to achieve this aim, in many lectures, seminars and discussions we have strengthened the interplay between gravity and quantum systems starting from the situation in the early universe based on astrophysical observations, up to the earthly based experiments with atom interferometry for probing the structure of space-time.

Thus we have had timely lectures on the quantum field and horizon of a blaçk hole including reviews of the problem of black holes thermodynamics and entropy, quantum information, quantum black holes, quantum evaporation and Hawking radiation, recent advances in stockastic gravity. We have also discussed quantum fluctuations in inflationary universe, quantum effects and reheating after inflation, and superplanckian energies in Hawking radiation.

In this regard the subject of spinors in purely affine space-time and Dirac matter according to Weyl in the generalized theory of gravitation were developed . The dualism between space-time and matter has been deeply analyzed in order to see why, for general relativity, this is an obstacle for quantization of the theory. Also canonical Gravity and Mach's principle, torsion and curvature as commutator for Quantum Gravity and Dirac Geometry of real space-time were analysed, together with the problem of 5-Dimensional Projective Unified Field theory and Multidimensional Gravity and Cosmology. The quantization of relativistic scalar fields in an Einstein elevator has been considered. Moreover we have examined laboratory experiments and space-based experiments with particular attention to the influence of gravity field of the earth on quantum systems which are represented by atoms and including the quantum mechanical treatement of the atomic motion (accelerated atoms) in the presence of gravitation (i.e.in curved space-time) particularly with regard to their possible application to tests of special and general relativity. With this in mind, the development of gravity gradiomenters and experiments suitable for satellite use were also examines. In addition measurements of the earth's gravity gradient by means of atom interferometry were considered.

Another important subject examined was the introduction of spin in the presence of gravity. From geometric point of view this would require the introduction of torsion. In addition to the problem of the spin we also studied the quantum interpretation of the measurements of Newtonian gravitational constant and laboratory searches for effects beyond classical Gravity. A most fascinating lecture on Majorana's and Bronstein's (both 1906-1938) significant works in matters related to our Corse, some of them unpublished to this day, rounded out the list of presented papers.

We wish to thank all the lecturers and seminar speakers who did so much to make the School successful, and all the participants for contributing to the very stimulating scientific and human atmosphere.

Peter G.Bergmann*

Venzo de Sabbata

* Although Peter Bergmann took part in preparing the School, a serious illness prevented him from actually taking part.

September 2001

QUANTUM INFORMATION AND QUANTUM BLACK HOLES

JACOB D. BEKENSTEIN
The Racah Institute of Physics, Hebrew University of Jerusalem
Givat Ram, Jerusalem 91904, Israel

1. Limitations on Quantum Information from Black Holes

1.1. INTRODUCTION

A naive view of physics—classical or quantum—would have us believe that an infinite amount of information can be contained in a finite 3-D space. After all, matter can have an infinity of classical configurations there, and quantum fields have an infinity of modes in that region. 't Hooft and Susskind's holographic principle [1, 2] has shattered this popular view: it holds that insofar as the information required to describe them is concerned, physical systems are inherently two–dimensional in space. In particular, the information or entropy (see Sec. 1.2 below for the relation) in an isolated system is expected to be bounded from above by one quarter the area of a circumscribing surface expressed in Planck units (holographic bound):

$$S \leq A(4\hbar)^{-1} \tag{1}$$

Except as otherwise stated, I assume units with $G = c = 1$. 't Hooft's original statement of the holographic principle [1] was elucidated by Susskind [2] who showed that the holographic bound is required by the generalized second law (GSL) [3] applied to the wholesale collapse of a physical system into a black hole of its own making. A loophole in Susskind's argument pointed out by Wald [4] can be corrected for quiescent systems by considering either infall of the system into a large black hole, or a tiny auxiliary black hole which devours the system [5]. The holographic bound as above stated can be exceeded by dynamically evolving systems, but Bousso [6] has given a reformulation which works in these cases also.

The attention of the particle theory community has been riveted on the implications of the holographic principle for the structure of the fundamental equations of nature, for example the equivalence between physical

P.G. Bergmann and V. de Sabbata (eds.),
Advances in the Interplay Between Quantum and Gravity Physics, 1–26.

theories formulated in different spacetime dimensions. However, there is no gainsaying the possible practical importance of fundamental restrictions, such as the holographic one on the *information storage capacity* of systems. For one thing, the principles obviously serve as a final arbitrates of the promise of any futuristic information storage technology. In this respect the holographic bound is not an onerous one. For instance, it merely requires that a device with dimensions of order a centimeter hold no more than some 10^{66} bits of information. By contrast, all the books in the Library of Congress hold of the order of 10^{15} bits of information, and no state-of-the-art one-centimeter size memory can hold all that. The holographic bound is thus too liberal for the present and foreseeable technology. So one question to be asked is, can one device a tighter bound on information storage than the holographic one ? In light of the explosive development of fast communications, a further interesting question would be: what fundamental bounds can be set on the *flow* of information ? In this lecture I shall go into such questions wherever they infringe into the realm of gravity, which I here take to be described by standard general relativity.

1.2. INFORMATION AND ENTROPY

It is plain that information can be stored in a system, by man or by nature, only if that system has more than one state accessible for the task. Nothing can be learned from a system that looks the same under all circumstances. With two *distinguishable* states with no bias between them, one can already store and retrieve information, for example, the answer to a "yes or no" question. A two-state system can thus hold a *bit* of information. What about a N state system with no bias between states ? It makes sense to assign it information capacity $\log_2 N$ bits. For one thing, this reduces to the previous case for $N = 2$. In addition, if we have two well separated classical systems, A and B, with $N_{\rm A}$ and $N_{\rm B}$ (unbiased) states, respectively, there are a total of $N_{\rm A}N_{\rm B}$ overall states, so we would assign the joint system information capacity $\log_2(N_{\rm A}N_{\rm B})$. But this is just the sum of information capacities for A and for B; thus is the logarithm function singled out as the relevant one for quantifying information. The states mentioned must be precisely distinguishable. In classical physics they may be, for example, mutually nonoverlapping cells in the phase space of a multiparticle system. In quantum theory they must be mutually orthogonal states, because nonorthogonal states cannot be distinguished with certainty by any measurement [7].

What if the states are biased ? For instance, it might be that before measurement is made, state 1 is twice as probable as state 2. Does this affect the information capacity ? Let us imagine that our system has N

equally probable states, but we divide these in groupings of $N_1, N_2, \cdots$ states with $\sum_i N_i = N$. If our experimental resolution does not permit us to peer into the groupings (or we just do not care to do so), we have to consider them as regular states with probabilities $p_1 = N_1/N, p_2 = N_2/N, \cdots$. It is reasonable to expect that the information capacity of the system, call it $I_{\max}$, should depend only on the set $\{p_i\}$: $I_{\max} = I_{\max}(p_1, p_2, \cdots)$. However, had we insisted and succeeded in peering into the groupings, we could have obtained additional information $\log_2 N_1$ from the first grouping (which turns up with probability p_1), and so on. Therefore, the *total* information capacity of the system can be written in two ways which must agree:

$$I_{\max}(p_1, p_2, \cdots) + \sum_i p_i \log_2 N_i = \log_2 N \tag{2}$$

Using the normalization $\sum_i p_i = 1$ it is easy to solve for $I_{\max}$:

$$I_{\max}(p_1, p_2, \cdots) = -\sum_i p_i \log_2 p_i \tag{3}$$

This is Shannon's famous 1948 formula [8] for the peak information capacity of a system with distinguishable states which occur with *a priori* probabilities $\{p_i\}$. It clearly makes no difference for the final result whether the states $\{i\}$ are composite or "elementary", so long as the probabilities p_i assigned them reflect an operational expectation.

Shannon's $I_{\max}$ looks like Boltzmann's expression for the entropy of a gas, apart from the facts that Boltzmann used natural logarithms, and that he prefaced the expression with the constant k, an historical accident stemming from the use of different units for temperature and energy. Beneath these superficialities the two are the same expression: thermodynamic entropy *is* the information storage capacity of matter. And thermodynamics is derivable from information–theoretic concepts [9, 7]. The endless controversies in the literature around this information–entropy equivalence seem to stem from confusion about the level at which the probabilities $\{p_i\}$ are formulated. The chemist, for instance, determines the entropy S of a piece of iron by methods that reach down to the atomic level; for him the states $\{i\}$ are atomic states. The engineer, by contrast, is interested in storing information in the magnetic domains of the iron in a magnetic tape. He groups atomic states into domain states which give him new probabilities $p_i{}'$. The $I_{\max}$ he calculates from these is much smaller than the chemist's S, but there is no question of conceptual contradiction.

Likewise, confusion has arisen about the relation between von Neumann's entropy in quantum theory, $S = -\mathrm{Tr}\,\rho \ln \rho$ (ρ is the density matrix), and Shannon's $I_{\max}$. Apart from an obvious factor $\ln 2$, the former is identical to $I_{\max}$ when the p_i's in the latter refer to the probabilities to measure values of an observable whose eigenstates coincide with those of ρ (for example energy when ρ stands for a thermal state). But it is possible to design quantum measurements for which the probabilities will give a Shannon $I_{\max}$ which exceeds von Neumann's S. In this case the states are no longer mutually orthogonal, and so are not fully distinguishable, and the maximum information that can be read out of the system is no longer equal to (3), but is bounded by $S \log_2 e$ [7]. At any rate, there is *never* a question of $I_{\max}$ and S being unrelated.

The chemist and engineer in the previous discussion are concerned with two levels of probabilities. In reality there are many: matter is made of atoms which are themselves built of electrons and nuclei. The last are put together from nucleons which are themselves composites of various quarks and gluons. There are states, and so probabilities and an entropy, at every level of this hierarchy. Obviously, the deeper we go, the higher the entropy. In what follows we shall be interested in the entropy (information capacity) S_X calculated at level X, the deepest level of structure. For ordinary matter this means the level of lepton, quark and gluon degrees of freedom, or something deeper if the standard model is up for a big revision. Obviously this S_X bounds from above the information capacity of material media accessible with any actual technology.

1.3. THE ENTROPY BOUND

Consider in asymptotically flat four–dimensional spacetime ($D = n+1 = 4$) a finite system $\mathcal{U}$ with energy E which, for the moment, we think of as spherical with radius R (E and R measured in proper frame). How big can its entropy S be ? The holographic principle limits it to $S \leq 4\pi R^2/\hbar$ (this is in natural logarithm units, nits, rather than base–two logarithm units, bits). Can we do better without knowing more details about $\mathcal{U}$? Indeed, we have the entropy bound

$$S \leq 2\pi E R/\hbar \tag{4}$$

proposed prior to the advent of holography [10]. So long as $\mathcal{U}$'s self-gravity is not strong, $E \ll R$. For example, for laboratory sized systems $E < 10^{-23} R$, while for astronomical systems, barring neutron stars and black holes, $E < 10^{-5} R$. Thus, with few exceptions, the entropy bound is many orders of magnitude tighter than the holographic one. It restricts the information capacity of a one–centimeter device made of ordinary matter to be less than 10^{37} bits, which limit no longer looks unreachable.

In the original derivation of bound (4) I imagined that $\mathcal{U}$ is lowered slowly from far away to the horizon of a stationary black hole, while all the freed potential energy is allowed to do work on a distant agent (a Geroch process [11]). I then applied the GSL to get the bound [10]. This derivation was criticized [12] for not taking into account the buoyancy of $\mathcal{U}$ in the Unruh radiation surrounding it by virtue of its acceleration. A protracted controversy [12, 13] on this issue led to the perception [14] that correction for buoyancy—itself an intricate calculation—merely increases the 2π coefficient in Eq. (4) by a tiny amount provided only that one assumes that $R \geq \hbar/E$. Now elementary particles and composite objects do obey this restriction. (Some solitons in $D = 1+1$ dimensional theories fail to so, but solitons are not expected to bear any entropy anyway.) Although originally meant only for weakly gravitating systems, bound (4) is actually saturated by all Kerr-Newman black holes [10, 15, 16, 17] provided one interprets E as the black hole's mass and R as the Boyer-Lindquist coordinate of the horizon, r_+.

The entropy in bound (4) is the S_X of Sec. 1.2, and not entropy calculated down to a shallower level of structure. This is because gravitation plays a crucial role in many generic ways of deriving the entropy bound [10, 13, 14, 18]. And gravitation is unique among the interactions in that it is aware of all degrees of freedom in its sources (according to the equivalence principle all energy gravitates). How then could bound (4) take into account only entropy corresponding to intermediate degrees of freedom, and so ignore energy–carrying states at the deeper levels ?

Only of late has the ubiquitous role of bound (4) and closely related ones been realized. For instance, Bousso [19] has shown, via the Geroch process argument, that bound (4) also applies *verbatim* in any asymptotically flat spacetime of $D = n+1$ dimensions. Further, he shows that when E is expressed in terms of $\mathcal{U}$'s gravitational radius r_g given by

$$ {r_\mathrm{g}}^{n-2} = \frac{8\Gamma(n/2)E}{(n-1)\pi^{n/2-1}}, \tag{5} $$

the bound takes the form

$$ S \leq \frac{(n-1)\pi^{n/2}\, r_\mathrm{g}^{n-2} R}{4\Gamma(n/2)\, \hbar^{(n-1)/2}}. \tag{6} $$

Bousso has also established a new entropy bound, the D–bound, for systems in $D = n+1$ de Sitter spacetime which occupy a small part of the space inside the cosmological horizon (radius r_c). It is given again by (6) with R's role played by r_c. Spherical *black holes* in $D \geq 4$ spacetime (for which the horizon has a $(n-1)$-D "area") also obey (4) in asymptotically flat spacetimes, and the D–bound in asymptotically de Sitter spacetimes, but no longer saturate these bounds for $D > 4$ [19].

And somewhat earlier E. Verlinde [20] proposed that the entropy S of a complete closed Robertson–Walker universe in $D = n+1$ spacetime dimensions whose contents are described by a conformal field theory (CFT)—the deeper description of a number of massless fields, possibly in interaction—with large central charge (essentially many particle species), is subject to the generic bound

$$S \leq \frac{2\pi R}{n\hbar}[E_{\rm C}(2E - E_{\rm C})]^{1/2} \tag{7}$$

where R is the radius of the S^n space, E the total energy in the fields and $E_{\rm C}$ the Casimir (vacuum) energy. As Verlinde shows, this bound is a straight generalization to n space dimensions of Cardy's famous expression for the entropy of a $D = 1 + 1$ CFT [21]. Verlinde points out that for fixed E the maximum of his bound is $2\pi RE/(n\hbar)$, which never exceeds the original entropy bound (4); indeed Verlinde adopts $S \leq 2\pi RE/(n\hbar)$ as the fiducial form of that bound. A number of recent papers have substantiated Verlinde's bound and so culminate years of protracted efforts by many to make meaningful statements about the entropy (and thus the maximum information) that can be contained in a whole universe [22].

Bounds like (4) are thus of wide applicability. But can one see why a bound like this should be true, at least in some mundane context, without getting embroiled in all the intricacies mentioned ? Indeed, there is an easy *gedanken* experiment which lets us do this. Drop $\mathcal{U}$ into a Schwarzschild black hole of mass $M \gg E$ from a large distance $d \gg M$ away; d is so chosen that the Hawking radiance carries away energy (as measured at infinity) equal to E while $\mathcal{U}$ is falling to the horizon where it is effectively assimilated by the black hole. At the end of the process the black hole is back at mass M and its entropy has not changed. Were the emission reversible, the radiated entropy would be $E/T_{\rm H}$ with $T_H \equiv \hbar(8\pi M)^{-1}$. Curved spacetime makes the entropy emitted a factor ν larger; typical values, depending on particle species, are [23] $\nu = 1.35$–1.64. Thus the overall change in world entropy is

$$\delta S_{\rm ext} = \delta S_{\rm rad} - S = \nu E/T_{\rm H} - S \tag{8}$$

One can certainly choose M larger than R, say, by an order of magnitude so that the system will fall into the hole without being torn up: $M = \zeta R$ with $\zeta =$ a few. Thus by the *ordinary* second law we obtain the bound

$$S < 8\pi\nu\zeta RE/\hbar \tag{9}$$

This bound applies to a rather arbitrary system $\mathcal{U}$ in terms of its total energy E and size R. But in the derivation $\mathcal{U}$ is not allowed to be be strongly gravitating (meaning $R \sim E$) because then M could not be large compared to E, as we have assumed, while ζ is of order a few. We thus have to assume in addition $R \gg E$.

Note that we could not derive (9) by using a heat reservoir in lieu of a black hole. A reservoir which has gained energy E upon $\mathcal{U}$'s assimilation, and has returned to its initial energy by radiating, does not necessarily return to its initial entropy, certainly not until $\mathcal{U}$ equilibrates with the rest of the reservoir. But a (nonrotating uncharged) black hole whose mass has not changed overall, retains its original entropy because that depends only on mass. In addition, for the black hole mass and radius are related in a simple way; this allowed us to replace $T_{\rm H}$ in terms of R. By contrast, for a generic reservoir, size is not simply related to temperature.

Of course, the above derivation ignores the effect of Hawking radiation pressure. How important is this ? Could it blow $\mathcal{U}$ outwards ? If we approximate the radiance as black–body radiance of temperature $T_{\rm H}$ coming from a sphere of radius $2M$, the energy flux at Schwarzschild coordinate r from the hole is

$$F(r) = \frac{\mathcal{N}\hbar}{61,440(\pi M r)^2}, \tag{10}$$

where $\mathcal{N}$ stands for the effective number of massless species radiated (photons contribute 1 to $\mathcal{N}$ and each neutrino species 7/16). This estimate is known to be off by a factor of only a few [24]. This energy (and momentum) flux results in a radiation pressure force $f_{\rm rad}(r) = \pi R^2 F(r)$ on $\mathcal{U}$. More precisely, species which reflect well off $\mathcal{U}$ are approximately twice as effective at exerting force as just stated, while those (neutrinos and gravitons) which go right through $\mathcal{U}$ contribute very little; the $\mathcal{N}$ must thus be reinterpreted accordingly. I have ignored relativistic corrections so that the result, as qualified, is correct mostly for $r \gg M$.

Writing the gravitational force on $\mathcal{U}$ in the Newtonian approximation, $f_{\rm grav}(r) = ME/r^2$, one sees that

$$\frac{f_{\rm rad}(r)}{f_{\rm grav}(r)} = \frac{\mathcal{N}_{\rm eff}\,\hbar R^2}{61,440\pi^2 M^3 E} \tag{11}$$

I have written $\mathcal{N}_{\rm eff}$ here because, as mentioned, some species just pass through $\mathcal{U}$ without exerting force on it. In addition, only those species actually represented in the radiation flowing out during $\mathcal{U}$'s infall have a chance to exert forces. Now an Hawking quantum bears an energy of order $T_{\rm H}$, so the number of quanta radiated together with energy E is approximately $8\pi ME/\hbar$. By our assumption that $\hbar/E < R$ and our stipulation that $M = \zeta R > R$, this is large compared to unity. Since a species can exert pressure only if it is represented by at least one quantum, one obviously has $\mathcal{N}_{\rm eff} < 8\pi ME/\hbar$. Therefore,

$$\frac{f_{\rm rad}(r)}{f_{\rm grav}(r)} < \frac{R^2}{7680\pi M^2} \ll 1 \tag{12}$$

Radiation pressure is thus negligible, and $\mathcal{U}$'s fall is very nearly on a geodesic, at least until $\mathcal{U}$ approaches to within a few Schwarzschild radii. It is intuitively clear that if $d \gg M$, the last (relativistic) stage cannot make any difference, and $\mathcal{U}$ must plunge to the horizon.

Whether d is large enough must be checked. We have taken it such that the infall time equals the time t for the hole to radiate energy E. Newtonially $d \approx 2(t^2M/\pi^2)^{1/3}$, while Eq. (10) gives the estimate $t \approx 5 \times 10^4 EM^2\hbar^{-1}\mathcal{N}^{-1}$ with $\mathcal{N}$ now the full species number. From these equations and $M = \zeta R$ we get that $d \approx 1.2 \times 10^3(\zeta ER/\mathcal{N}\hbar)^{2/3}M$. Thus for $\mathcal{N} < 10^2$ (conservative estimate of *our* world's massless particle content), we have $d > 57M$ for all systems $\mathcal{U}$ satisfying our assumption $R > \hbar/E$. Thus for all these we have established the entropy bound (9).

Our simple argument here leaves the factor $\nu\zeta$ somewhat fuzzy; but it is safe to say that $4\nu\zeta < 10^2$. Thus we recover bound (4) with some overshoot of the coefficient, not a large prize to pay for the simplicity of the derivation. When we come to strongly gravitating systems ($E \sim R$), we cannot derive the bound (4) or even the weaker version (9) by the methods just expounded. Nevertheless, as mentioned, a black hole in $D = 4$ spacetime saturates bound (4) and complies with it for $D > 4$. For strongly gravitating systems in asymptotically flat spacetime, the holographic bound and the entropy bound make very similar predictions for $D = 4$, but for $D > 4$ the holographic bound is the tighter of the two. Unless D is very large, the entropy bound is the tighter bound for weakly gravitating systems, such as those we meet in everyday life.

1.4. BLACK HOLES AS ONE–DIMENSIONAL INFORMATION CONDUITS

The holographic bound (1) is supposed to be telling us that a generic physical system in 4–D spacetime is fundamentally two–dimensional in space. It turns out that viewed as an information absorber or entropy emitter, a black hole in 4–D spacetime is fundamentally one–dimensional in space [30].

To show this one must define a one–dimensional information transmitting system—a channel. In flat spacetime a channel is a complete set of one-way propagating modes of some field, with the modes enumerated by a single parameter. For example, all electromagnetic modes in free space with fixed wave vector *direction* and particular linear polarization constitute a channel; the modes are parametrized solely by frequency. One might implement such a channel with a straight infinitely long coaxial cable (which is well known to transmit all frequencies) capped at its entrance by the analog of a polaroid filter. Acoustic and neutrino channels can also be defined. A fundamental question is: what is the maximum *rate*, in quantum theory, at which information may be transmitted down such a channel for prescribed power P ? The answer was found in the 1960's by several information theory pioneers (see my review with Schiffer[25]), but I want to reproduce here the much later but very simple derivation of Pendry [26], which is of very broad applicability.

Pendry thinks of a possible signal state as corresponding to a particular set of occupation numbers for the various propagating modes. He assumes the channel is uniform in the direction of propagation, which allows him to label the modes by momentum p. But he allows for dispersion, so that a quantum with momentum p has some energy $\varepsilon(p)$. Then the propagation velocity of the quanta is the group velocity $v(p) = d\varepsilon(p)/dp$. Up to a factor $\ln 2$ the information rate capacity is just the maximal one-way entropy current for given P, which obviously occurs for the thermal state, if one excludes the modes moving opposite the direction of interest.

Now the entropy $s(p)$ of any boson mode of momentum p in a thermal state (temperature T) is [27]

$$s(p) = \frac{\varepsilon(p)/T}{e^{\varepsilon(p)/T} - 1} - \ln\left(1 - e^{-\varepsilon(p)/T}\right), \tag{13}$$

so the entropy current in one direction is

$$\dot{S} = \int_0^\infty s(p)\, v(p)\, \frac{dp}{2\pi\hbar} = \int_0^\infty s(p)\, \frac{d\varepsilon}{dp}\, \frac{dp}{2\pi\hbar}, \tag{14}$$

where $dp/2\pi\hbar$ is the number of modes per unit length in the interval dp which propagate in one direction. This factor, when multiplied by the group velocity, gives the one-way current of modes. Suppose $\varepsilon(p)$ is monotonic and extends over the range $[0, \infty)$; we may then cancel dp and integrate over ε. Then after substitution of Eq. (13) and integration by parts we have

$$\dot{S} = \frac{2}{T}\int_0^\infty \frac{\varepsilon}{e^{\varepsilon/T} - 1}\, \frac{d\varepsilon}{2\pi\hbar} = \frac{2}{T}\int_0^\infty \frac{\varepsilon(p)}{e^{\varepsilon(p)/T} - 1}\, v(p)\, \frac{dp}{2\pi\hbar}. \tag{15}$$

The first factor in each integrand is the mean energy per mode, so that the integral represents the one-way power P in the channel. Thus

$$\dot{S} = 2P/T. \tag{16}$$

The integral for P in the first form of Eq. (15) can easily be done:

$$P = \frac{\pi (T)^2}{12\hbar}. \tag{17}$$

Eliminating T between the last two expressions gives Pendry's limit

$$\dot{S} = \left(\frac{\pi P}{3\hbar}\right)^{1/2} \quad \text{or} \quad \dot{I}_{\max} = \left(\frac{\pi P}{3\hbar}\right)^{1/2} \log_2 e. \tag{18}$$

For a fermion channel P in Eq. (17) is a factor 2 smaller, and consequently $\dot{S}$ in Eq. (18) is reduced by a factor $\sqrt{2}$.

The function $\dot{S}(P)$ in Eq. (18) is the so called *noiseless quantum channel capacity*. Surprisingly it is independent, not only of the form of the mode velocity $v(p)$, but also of its scale. Thus the phonon channel capacity is as large as the photon channel capacity despite the difference in speeds. Why? Although phonons convey information at lower speed, the energy of a phonon is proportionately smaller than that of a photon in the equivalent mode. Thus when the capacities of channels harnessing various carriers are expressed in terms of power, they turns out to involve the same constants. Formula. (18) neatly characterizes what we mean by one–dimensional transmission of entropy or information. It refers to transmission by use of a single species of quantum and a specific polarization; different species and alternative polarizations engender separate channels. Although framed in a flat spacetime context, its lack of sensitivity to the dispersion relation of the transmitting *milieu* should make Pendry's limit relevant to curved spacetime also. This because electrodynamics in curved spacetime is equivalent to flat spacetime electrodynamics in a suitable dielectric and paramagnetic medium [28].

By contrast the power and entropy emission rate in a single boson polarization of a closed black body surface with temperature T and area A in flat 4–D spacetime is (half the Stefan–Boltzmann law)

$$P = \frac{\pi^2 T^4 A}{120\hbar^3} \qquad \dot{S} = \frac{4}{3}\frac{P}{T} \tag{19}$$

whereby

$$\dot{S} = \frac{2}{3}\left(\frac{2\pi^2 A P^3}{15\hbar^3}\right)^{1/4} \tag{20}$$

[for fermions P carries an extra factor $7/8$ and formula (20) an extra factor $(8/7)^{1/4}$]. Our manifestly 3–D transmission system deviates from the sleek formula (18) in the exponent of P and in the appearance of the measure A of the system. In emission from a closed curve of length L in two–dimensional space, the factor $(LP^2)^{1/3}$ would replace $(AP^3)^{1/4}$. We may thus gather the dimensionality of the transmission system from the exponent of P in the expression $\dot{S}(P)$ [it is $n/(n+1)$ for $D = n+1$ spacetime dimensions], as well as from the value of the coefficient of P/T in expressions for $\dot{S}$ like (16) or (19) [it is $(n+1)/n$].

Radiation from a Schwarzschild black hole in 4–D spacetime is also given by Eqs. (19) (or their fermion version) with $A = 4\pi(2M)^2$ and $T = T_{\rm H}$, except we must correct the expression for P by a species dependent factor $\bar{\Gamma}$ of order unity [24], and replace the 4/3 in the expression for $\dot{S}$ by the species dependent factor ν already mentioned in Sec. 1.3. Eliminating M between the equations we obtain, in lieu of Eq. (20),

$$\dot{S} = \left(\frac{\nu^2\bar{\Gamma}\pi P}{480\hbar}\right)^{1/2}. \tag{21}$$

(For fermions there is an extra factor 7/8 inside the radical). This looks completely different from the law (20) for the hot closed surface because, unlike for a hot body, a black hole's temperature is related to its mass.

However, (21) *is* of the same form as Pendry's limit (18) for one–channel transmission. From Page [24, 23] we get $\bar{\Gamma} = 1.6267$ and $\nu = 1.5003$ for one photon polarization, so the numerical coefficient of (21) is 15.1 times that in (18). Repeating the above exercise for one species of neutrinos we again find formulae like (21) and (18), this time with $\bar{\Gamma} = 18.045$ and $\nu = 1.6391$; the numerical coefficient of (21) is 48.1 times that of the fermion version of (18).

Evidently in its entropy emission properties a black hole in 4–D spacetime is more like a 1–D channel than like a surface in 3–D space. Why is this ? A formal answer is that, because of the way $T_{\rm H}$ is related to the black hole's radius $2M$, Hawking emission prefers to emerge in the lowest angular momentum mode possible. To exit with angular momentum $j\hbar$, a quantum must have energy (momentum) $\hbar\omega$ of order $j\hbar/2M$. But in the Hawking thermal distribution the dominant $\hbar\omega$ is of order $T_{\rm H} = \hbar(8\pi M)^{-1}$. Thus the emerging j's tend to be small. For example, 97.9% of the photon energy emerges in the $j = 1$ modes ($j = 0$ is forbidden for photons), and 96.3% of the neutrino power is in the $j = \frac{1}{2}$ modes [24]. Thus the black hole emits as close to radially as possible. This means that, crudely speaking, it does so through just one channel.

1.5. INFORMATION PULSES IN CURVED SPACETIME

The discussion in Sec. 1.4 centered on steady state streams of information and energy. What if information is delivered as pulses ? Can one state a bound generalizing (18) ? Further, can one include effects of gravitation on information transfer rate ? To answer these questions let us extend the notion of channel to curved spacetime, at least to stationary curved spacetime. Again, a channel will be a complete set of one-way modes of some field that can be enumerated with a single parameter. Each channel is characterized by species of quanta, polarization (helicity), trajectory, etc. In Sec. 1.4 we characterized the *signal* in a particular channel by power. For a pulse it seems a better idea to use as signal parameters the signal's duration τ and its energy E. Since in curved spacetime a channel is not generally uniform, we choose to measure these parameters in a local Lorentz frame (we shall see presently that it does not matter which Lorentz frame). This precaution allows us to focus on sections of the channel and treat them as if we were working in flat spacetime.

How is $I_{\max}$ related to E and τ ? Since information is dimensionless, $I_{\max}$ must be a function of dimensionless combinations of E, τ, channel parameters and fundamental constants. We exclude channels which transmit massive quanta, e.g. electrons, because rest mass is energy in a form not useful for communication, so that the strictest limits on $I_{\max}$ should emerge for massless signal carriers. Hence Compton lengths do not enter into the argument. Also in order to maximize the information flux, we focus on broadband channels, and exclude any frequency cutoff. Finally, we exclude situations where the signal undergoes dispersion; this has the practical upshot that apart from light's speed c, only one other velocity—the signal velocity c_s—can appear. We consider c_s/c a property of the *channel* because it is common to all signals. If we *temporarily* exclude the gravitational constant, there is thus a single dimensionless combination of *signal* parameters, $\xi = E\tau/\hbar$, at our disposal. Thus

$$I_{\max} = \Im(E\tau/\hbar), \tag{22}$$

where $\Im(\xi)$ is some nonnegative valued function characteristic of the channel, the characteristic information function (CIF) [25].

Let us check formula (22) in flat spacetime in steady state (momentarily return to a long stream of information). Steady state means that the signal can be characterized as statistically stationary in a suitable frame. It should thus be possible to infer the peak information transfer *rate* by focusing on a finite section of the signal bearing information $I_{\max}$ and energy E. It should matter little how long a stretch in τ is used so long as it is not too short. This can only be true if $\dot{I}_{\max} \equiv E\tau^{-1}$ is fully determined by the power $P \equiv E\tau^{-1}$, and this is consistent with Eq. (22) only if $\Im(\xi) \propto \sqrt{\xi}$ for only then does τ cancel out. We have thus recovered Pendry's limit (18), the correct answer for steady state; formula (22) checks out.

The dividing line between steady state information transfer and transfer by means of very long signals is not sharp. This suggests that long pulse signals must also obey a Pendry type formula, albeit approximately [29]. The law $\dot{I}_{\max} \propto (P/\hbar)^{1/2}$ is evidently inapplicable to *brief* information pulses. For such it may be replaced by a linear upper bound [31] which may even transcend some of the limitations we imposed to define $\Im(\xi)$. Consider the information I to be encoded in some material structure $\mathcal{V}$ of radius R and rest energy E which maintains its integrity and dimensions as it flies from emitter to receiver. From Eq. (4) we have the strict inequality $I < 2\pi ER\hbar^{-1}\log_2 e$. The rate at which the information is assimilated by the receiver is obviously restricted by the local time τ it takes for $\mathcal{V}$ to sweep by it. From special relativity $\tau > 2R\gamma^{-1}$ with γ accounting for the Fitzgerald contraction of $\mathcal{V}$ in the frame of the receiver. Thus the peak information reception rate is $I/\tau < \pi E\hbar^{-1}\log_2 e$, or

$$\dot{I}_{\text{rec}} < \pi E_{\text{rec}}\hbar^{-1}\log_2 e \tag{23}$$

where $E_{\text{rec}} \equiv \gamma E$ is $\mathcal{V}$'s energy as measured in the receiver's frame. This replaces the information version of Eq. (18) when it comes to pulses. With $\xi \equiv E_{\text{rec}}\tau\hbar^{-1}$ we thus have the strict linear bound $\Im(\xi) < (\pi\log_2 e)\xi$. There is a lot of evidence [32, 25] that this bound applies even when the signal has no rest frame. One should not be alarmed because the law $\Im(\xi) \propto \sqrt{\xi}$ figuring in (18) exceeds the linear bound for *small* ξ; the law is meant for steady state, which makes sense only in the limit $\xi = E\tau/\hbar \to \infty$.

Let us check the *local* Lorentz invariance of (22). Consider a pair of local Lorentz frames, A and B, encompassing a section of the channel, with B moving to the right with respect to A with speed $V < c_s$, and let $\gamma = (1 - V^2)^{1/2}$. *If* there is a medium, A is taken as its rest frame, and c_s is the signal's speed in this frame. Now let a right moving signal pulse's front (speed c_s) pass the origins of these frames at time $t_{\text{A}} = 0$ when they coincide. At a later time $t_{\text{A}} = t_1$ the pulse's rear has reached A's origin; by then B's origin is at $x_{\text{A}} = Vt_1$. Sometime later, at $t_{\text{A}} = t_2$, the pulse's rear reaches B's origin which is then at $x_{\text{A}} = Vt_2$. Calculating the pulse's length in frame A in two ways gives $(c_s - V)t_2 = c_s t_1$, so that

$$t_1/t_2 = 1 - V/c_s \tag{24}$$

The signal duration is $\tau_{\text{A}} = t_1$ in A; in B, however, it is $\tau_{\text{B}} = t_2\gamma^{-1}$ because t_2 is the time it takes in A for all the pulse to traverse B's origin; thus

$$\tau_{\text{A}} = \gamma\tau_{\text{B}}(1 - V/c_s). \tag{25}$$

Now in A the energy and momentum of a (massless) quantum in the signal stand in the ratio c_s. Thus if interactions are negligible, the energy E_{A} and the corresponding momentum P_{A} of the full pulse stand also in the same ratio. Considering a Lorentz boost to frame B we discover that

$$E_{\text{B}} = \gamma(E_{\text{A}} - VP_{\text{A}}) = \gamma E_{\text{A}}(1 - V/c_s). \tag{26}$$

Therefore, $E_{\text{B}}\tau_{\text{B}} = E_{\text{A}}\tau_{\text{A}}$. We can likewise prove this result for $1 > V > c_s$. In view of this and the fact that information is a Lorentz scalar, the statement (22) is seen to be Lorentz invariant. This has immediate applications. For instance, A can be interpreted as the signal receiver's frame and B as the propagating medium's, or perhaps the transmitter's.

But how is the limit on information transmission rate related at two point along the channel ? In flat spacetime, and in the absence of dispersion, E and τ are evidently conserved with propagation. So is the information, so that Eq. (22) is valid at every point along the channel. Once we are in stationary *curved* spacetime, E and τ are subject to redshift and dilation effects, respectively. However, the effects act in opposite senses on E and τ, and since they depend on the same metric component, $E\tau$ is again conserved throughout the signal's flight. Therefore, Eq. (22) is meaningful throughout the channel. In fact one can use global values (as measured at infinity) of E and τ in Eq. (22). We see that *one and the same formula limits information transmission, propagation and reception rates.*

We have so far excluded appearance of the gravitational constant in the formulae. This means we have been neglecting self–gravity of the signal, because this is measured by the parameter (momentarily restoring G and c) $\varpi \equiv GEc^{-5}\tau^{-1}$, which is on the order of the ratio of the signal's self–potential energy to E, or that of the signal's gravitational radius to its duration. Should we include ϖ as a new argument in Eq. (22) ? One reason for not doing so is that it would obviously spoil the local Lorentz invariance of Eq. (22) because E/τ is not a Lorentz scalar. We cannot allow this for signals propagating in vacuum in a flat background, for this would be tantamount to a violation of special relativity. However, it could be argued that the presence of ϖ in $\Im$ is permissible for signals propagating in a medium ($c_s \neq c$) because it locally defines a preferred Lorentz frame

In a curved background there are further arguments against inclusion of ϖ in $\Im$. In vacuum we can use the requirement of local Lorentz invariance to bar ϖ's appearance, for a sufficiently brief signal should admit being encompassed in its entirety by local Lorentz frames. Further, ϖ evidently decreases as the signal propagates *outward* in the gravitational potential. Thus, $\Im(E\tau/\hbar, \varpi)$ would decrease either outwardly (if $\Im$ increases with ϖ) or inwardly (if it decreases as ϖ increases). If a signal's information saturates the bound $\Im(E\tau/\hbar, \varpi)$ at some point in the potential, then by conservation of information it will exceed the bound once it has propagated somewhat in the direction in which $\Im$ decreases. This leads to a contradiction. One could try to resolve the problem by defining $I_{\max}$ only in terms of the *minimum* value of ϖ in the channel. But it seems strange that, at least for brief signals, one cannot state $I_{\rm nax}$ in terms of local quantities.

It thus seems that for signals propagating in vacuum in curved spacetime, ϖ cannot appear in $\Im$. It is unclear whether this conclusion extends to signal propagation in a medium, because in curved spacetime a medium is never homogeneous, which means, among other things, that c_s varies. This in itself puts in doubt our argument for simplicity of the expression for $I_{\max}$.

1.6. DUMPING INFORMATION INTO A BLACK HOLE

Suppose we have at our disposal a certain power $\mathcal{P}$ to accomplish the task of getting rid of a stream of possibly compromising information by dumping it into a black hole. We may pick the size of black hole which suits us best. By the complementary relation between entropy and information, formula (21) for entropy rate *out* of the black hole suggests a bound $\dot{I}_{\rm max} \sim (\mathcal{P}/\hbar)^{1/2}$ on the dumping rate. Here we verify this new bound.

First we argue that if the signal originates from afar, it is transmitted more or less through a single channel (per field species and polarization). Let us recall the rule for field mode counting. In one space dimension a length L contains $(2\pi)^{-1}L\Delta k$ modes in the wave vector interval Δk. In 3–D space we would have $(2\pi)^{-3}L_x L_y L_z \Delta k_x \Delta k_y \Delta k_z$ modes. From this we may conclude that if a flat 2–surface of area $\mathcal{A}$ radiates into a narrow solid angle $\Delta\Omega$ about its normal, the number of modes out to a distance L from it whose wave vector magnitudes lie between k and $k+\Delta k$ is $(2\pi)^{-3}\mathcal{A}Lk^2\Delta\Omega\Delta k$. The factor $(2\pi)^{-1}L\Delta k$ is obviously the number of modes emitted sequentially in each direction and distinguished by their values of k. One can thus think of $\mathcal{W} = (2\pi)^{-2}\mathcal{A}k^2\Delta\Omega$ as the number of active *channels*.

Now let a transmitter with effective area $\mathcal{A}$ send an information bearing signal towards a Schwarzschild black hole of mass M surrounded by vacuum and situated at distance $d \gg 2M$. Let $\mathcal{A}$ be oriented with its normal towards the black hole. Evidently, as viewed from the transmitter the black hole subtends solid angle $\Delta\Omega = \pi(2M)^2/d^2$, and we must have $\mathcal{A} < 4\pi d^2$. What should we take for k in the formula for $\mathcal{W}$? Being interested in the highest information for given energy (other things being equal), we certainly want to use the smallest k (smallest $\hbar\omega$) possible. But signals composed of too small k's will just be scattered by the black hole. The borderline is $k = 2\pi/\lambda \approx 2\pi/(2M)$. With this we find $\mathcal{W} < 4\pi^2$, which means that information transmission down a black hole is necessarily a few channels' affair (for each field species and polarization, of course) regardless of the scales M and d in the problem.

In light of this we employ formula (22). Further, since $E\tau$ is conserved in Schwarzschild (stationary) spacetime, and closely equals $\mathcal{E}t$, the values being measured at infinity, we have $I_{\rm max} = \Im(\mathcal{E}t/\hbar)$. This for a pulse of duration t as seen from infinity. If we are dealing with a steady state stream of energy and information ($t \to \infty$ and $\mathcal{E} \to \infty$ with $\mathcal{P} \equiv \lim(\mathcal{E}/t)$ finite), we have by the logic of the paragraph following Eq. (22) that the maximum information disposal rate into the black hole is $\dot{I}_{\rm max} \sim (\mathcal{P}/\hbar)^{1/2}$. We have thus checked our guess. The precise proportionality factors for various fields may be worked out from Eq. (21). At any rate, we uncover a "cost" for information disposal into a black hole: the power required grows *quadratically* with the information dumping rate.

2. Black Hole Spectroscopy

2.1. INTRODUCTION

In classical general relativity we think of a black hole's parameters, typically charge Q, angular momentum $\mathbf{J}$ and mass M (but others are possible: see Erick Weinberg's lectures in this collection) as continuous. In reality its angular momentum and charge should have discrete spectra. And what about M ? Is it continuous as in the classical theory, or discrete as for an atom ? Granted that this is a fair question only for quantum gravity, it is still interesting to find out if something can be said about it in the fragmentary state of knowledge today.

I have long argued [33] that certain features of classical black hole physics hint at a discrete mass spectrum of a very definite form. That would make black holes quite like atoms in one sense. Here I would like to summarize the various steps in the logic towards this conclusion, and describe an algebraic quantum formalism [34, 35, 36] designed to deal with this question irrespective of how the final quantum gravity may turn out. I shall use geometric units with $G = c = 1$; $\sqrt{\hbar}$ denotes the Planck mass.

2.2. ADIABATIC INVARIANCE AND BLACK HOLE QUANTIZATION

The present situation in quantum gravity's development, with rival theories (string theory, loop quantum gravity, canonical quantum gravity in several versions, ...) whose elements can still not be set in one to one correspondence is somewhat reminiscent of quantum physics before Schrodinger's equation and Heisenberg's matrix mechanics. People then were very much guided by the rules of the old quantum mechanics. And an important unifying principle of these was Ehrenfest's (Born's excellent book [37] is a convenient source): a quantity which classically is an adiabatic invariant has at the quantum level a discrete spectrum. The Bohr–Sommerfeld quantization rules, which focused on Jacobi actions $\oint p\, dq$ (all of which are adiabatic invariants), and which gave the correct hydrogen spectrum—fine structure included—exemplify the success of Ehrenfest's principle.

In the 1970's the work of Penrose and Floyd [38], Christodoulou [39] and Hawking [40] taught us that transformations of a classical black hole are generically irreversible because the surface area of the event horizon increases in most such changes. While the thermodynamic ramifications of this discovery have been of wide interest to the gravitation and particle communities for three decades, a less visible aspect of this purely mechanical behavior will be the center of attention here.

Christodoulou and Ruffini [39, 41] exhibited a set of black hole transformations which do not involve horizon area increase. The simplest example I can find deals with a point charge ε which is deposited at rest, by whatever means, alongside the horizon of a Reissner–Nordström (RN) black hole with charge Q of the same sign as ε. This means, purely on mechanical grounds, that the energy of the particle as measured at infinity, E, equals $\varepsilon\Phi$ with

$$\Phi = \frac{Q}{M + (M^2 - Q^2)^{1/2}} \tag{27}$$

the electrostatic potential at the horizon. Now differentiation of the expression for horizon area of our Reissner–Nordström black hole, $A = 4\pi(M + (M^2 - Q^2)^{1/2})^2$, gives

$$\begin{aligned} \delta A &= (\delta M - \Phi\delta Q)\Theta^{-1}, &(28)\\ \Theta &\equiv \frac{1}{2}(M^2 - Q^2)^{1/2} A^{-1}. &(29) \end{aligned}$$

By energy and charge conservation, when the particle is captured by the black hole, $\delta M = E$ while $\delta Q = \varepsilon$. But we selected $E = \varepsilon\Phi$, so that $\delta A = 0$ upon capture of the particle. The process of charge assimilation is a slow one since the charge is gently placed on the horizon; and it results in a change of M and Q but with no consequent change in A. Hence A is an adiabatic invariant here. Many other examples have been collected [42, 43, 44] involving the adiabatic addition of angular momentum to the black hole instead of charge, or various wave perturbations. In all these A is unchanged.

The proposed adiabatic principle is a classical one. But if we take into account that by quantum theory the particle to be assimilated has an effective minimal radius (Compton length) which prevents us from placing its center exactly at the horizon without already loosing it, the same sort of calculation [11, 36] shows that the minimal horizon area increase is $(\delta A)_{\rm min} = \xi\hbar$, with ξ a constant of $O(1)$. Adiabatic invariance is thus not literally true, but for super–Planckian black holes the area increase is relatively small:

$$\left(\frac{\delta A}{A}\right)_{\rm min} = \frac{\xi\hbar}{A} < \frac{137\xi e^2}{4\pi Q^2} < \frac{137\xi}{4\pi}\left(\frac{\delta Q}{Q}\right)^2. \tag{30}$$

Here we have used the fact that the elementary charge is of order $\sqrt{137}\,\hbar^{1/2}$, that in all RN black holes $Q \leq (A/4\pi)^{1/2}$, and that the change in Q cannot be less than e. Inequality (30) assures us that the fractional change in horizon area is small compared to that of black hole charge (pressumably $|\delta Q| \ll Q$), and generally also that of black hole mass because

$$\frac{\delta M}{M} = \frac{\Phi\delta Q}{M} > \left(\frac{Q^2}{2M^2}\right)\frac{\delta Q}{Q}. \tag{31}$$

Thus in order for $(\delta A)_{\rm min}/A$ *not* to be small compared to $\delta M/M$, the black hole would have to be virtually neutral. In this sense we still have adiabatic invariance while allowing for quantum mechanics of matter. In fact, adiabatic invariance of A seems to survive into the quantum gravity regime [45].

From the adiabatic invariance of the area of the horizon and Ehrenfest's principle, we should suspect that in quantum gravity the horizon area is replaced by an operator with a discrete spectrum. In the RN example the classical formula for A gives $M = (A/16\pi)^{1/2}[1 + 4\pi Q^2/A]$. If this were true for operators, the discrete spectra of Q and A would imply that M too has a discrete spectrum [35]. Of course, quantum corrections to this last formula might come about [46] without eliminating the discreteness of M. And if the area eigenvalues are themselves degenerate, quantum corrections might split this degeneracy [35, 36]. However, it seems most constructive, in view of the absence of a consensus quantum gravity theory, to consider the pristine situation before all these corrections are effective.

2.3. DYNAMICAL VARIABLES AND CREATION OPERATORS

In quantum theory $\mathbf{J}$ and Q are represented by hermitian operators $\hat{\mathbf{J}}$ and $\hat{Q}$. We assume $[\hat{\mathbf{J}}, \hat{Q}] = 0$ so that the black hole can simultaneously have sharp charge, which we assume to be an integer multiple of e, and angular momentum. In order to get usual spectrum $\mathbf{J}^2 = j(j+1)\hbar^2$; $j = 0, \frac{1}{2}\hbar, \hbar \cdots$ with $J_z \equiv m\hbar = \{-j, -j+1, \cdots, j\}\hbar$, we have to assume that $\hat{\mathbf{J}} \times \hat{\mathbf{J}} = \imath\hbar\hat{\mathbf{J}}$. The argument at the end of Sec. 2.2 predisposes us to expect that there is some observable representing horizon area, $\hat{A}$, which has a simple spectrum, $\{a_1, a_2, \cdots a_n \cdots\}$ with $a_{n+1} > a_n$. Now the horizon area of a black hole is invariant under rotations of its spin; since $\hat{\mathbf{J}}$ is the generator of such rotations in quantum theory, one expects that $[\hat{A}, \hat{\mathbf{J}}] = 0$. Similarly, horizon area is invariant under gauge transformations; in quantum theory their generator is, as usual, the charge $\hat{Q}$. Hence we expect that $[\hat{A}, \hat{Q}] = 0$.

It follows that we can conceive of a basis of one-black hole states of the form $\{|njmqs\rangle\}$ where q is an integer eigenvalue of $\hat{Q}/e$, and s distinguishes between the different black hole states with like a_n, j, m and q (this degeneracy is really a must; see Sec. 2.6). But the above algebra of observables is too simple, and cannot by itself tell us very much new. It has to be extended. Gour [46] has shown how to introduce a "secret" operator corresponding to the quantum number s; his prescription, however, presupposes a uniformly spaced spectrum for $\hat{A}$, which is an idea to be tested here. We, therefore, avoid including something like this in the algebra.

In field theory of particles we would at this point introduce fields and analyze them into creation and anhilation operators; but "field" is an inappropriate concept for black holes which are not even approximately pointlike objects. Nobody stops us, however, from defining the black hole vacuum $|\text{vac}\rangle$ (spacetime with no black holes or particles of any sort), and creation operators for black holes $\hat{R}_{njmqs}$ with the property that $|njmqs\rangle = \hat{R}_{njmqs}|\text{vac}\rangle$. This is essentially a tautology, not a physical assumption, because we introduce as many operators as there are states. In contrast to field theory, we do *not* assume that $\hat{R}_{njmqs}{}^N$ creates a state with N black holes.

By commuting $\hat{A}, \hat{\mathbf{J}}$ and $\hat{Q}$ with the $\hat{R}_{njmqs}$ and iterating we can make more operators. If this process continues indefinitely, no information can be obtained from the algebra unless additional assumptions are made. So let us suppose that the mentioned operators, together with the unit operator $\hat{I}$, form a closed, linear, infinite dimensional algebra. This assumption has two new features: the closure at some low level of commutation, essentially a plea for simplicity, and the linear character of the algebra when formulated in terms of $\hat{A}$. As we shall see in Sec. 2.5, this last implies the additivity of horizon area, which is a reasonable property. By contrast, additivity of mass for two black holes is not reasonable (nonlinearity of gravity), and this is really the reason why one cannot assume linearity of the algebra of $\hat{M}$, $\hat{Q}$, $\hat{\mathbf{J}}$ and $\hat{R}_{njmqs}$. In this sense $\hat{A}$ is special among all functions of the other black hole observables.

By definition $\hat{R}_{njmqs}|\text{vac}\rangle$ is an eigenstate of $\hat{Q}$ with eigenvalue qe, so

$$\exp(\imath\chi\hat{Q})\,\hat{R}_{njmqs}|\text{vac}\rangle = \exp(\imath\chi qe)\,\hat{R}_{njmqs}|\text{vac}\rangle \tag{32}$$

for real χ. Thus, as already mentioned, $\hat{Q}$ is the generator of (global) quantum gauge transformations, and this is equivalent to requiring

$$\exp(\imath\chi\hat{Q})\,\hat{R}_{njmqs}\exp(-\imath\chi\hat{Q}) = \exp(\imath\chi qe)\,\hat{R}_{njmqs} \tag{33}$$

(check by operating with this on $|\text{vac}\rangle$, recalling that $\hat{Q}|\text{vac}\rangle = 0$, and recovering Eq. (32)). Now expansion of Eq. (33) to $O(\chi)$ gives one of our essential commutators,

$$[\hat{Q}, \hat{R}_{njmqs}] = qe\,\hat{R}_{njmqs}. \tag{34}$$

Now obviously

$$\exp(\imath\chi\hat{J}_z/\hbar)\,\hat{R}_{njmqs}|\text{vac}\rangle = \exp(\imath\chi m)\,\hat{R}_{njmqs}|\text{vac}\rangle, \tag{35}$$

Repeating the previous sort of argument we get

$$[\hat{J}_z, \hat{R}_{njmqs}] = m\,\hbar\,\hat{R}_{njmqs}. \tag{36}$$

There is a more significant way to get Eq. (36). Since $\hat{R}_{njmqs}|\text{vac}\rangle$ is defined as a state with spin quantum numbers j and m, the collection of such states with fixed j and all allowed m must transform among themselves under rotations of the black hole like the spherical harmonics $Y_{j\mu}$ (or the corresponding spinorial harmonic when j is half–integer). But $|\text{vac}\rangle$ must obviously be invariant under all rotations, so $\hat{R}_{njmqs}$ may be taken to behave like an irreducible spherical tensor operator of rank j with the usual $2j+1$ components labeled by m [47]. This gives Eq. (36) immediately. In addition, defining the usual raising and lowering operators, $\hat{J}_\pm \equiv J_x \pm \imath J_y$, we infer

$$[\hat{J}_\pm, \hat{R}_{njmqs}] = \sqrt{j(j+1) - m(m\pm 1)}\,\hbar\,\hat{R}_{nj,m\pm 1,qs}. \tag{37}$$

We can use this and the identity [47] $\hat{\mathbf{J}}^2 = (1/2)(\hat{J}_+\hat{J}_- + \hat{J}_-\hat{J}_+) + \hat{J}_z^2$ to work out $[\hat{\mathbf{J}}^2, \hat{R}_{\kappa\, s}]$. It has a rather complicated form; its first term is $j(j+1)\hbar^2\hat{R}_{njmqs}$ which is followed by two terms having on their right hand sides $\hat{J}_+$ and $\hat{J}_-$, respectively. Thus operating with $[\hat{\mathbf{J}}^2, \hat{R}_{njmqs}]$ on $|\text{vac}\rangle$ and taking into account that $\hat{\mathbf{J}}|\text{vac}\rangle = 0$ we get

$$\hat{\mathbf{J}}^2\,\hat{R}_{njmqs}|\text{vac}\rangle = j(j+1)\hbar^2\hat{R}_{njmqs}|\text{vac}\rangle \tag{38}$$

This corresponds to the definition of $\hat{R}_{njmqs}$ as creation operator of a black hole with angular momentum quantum numbers j and m.

2.4. INCLUDING $\hat{A}$ IN THE ALGEBRA

So far all we know about $\hat{A}$ is that it commutes with $\hat{\mathbf{J}}$ and $\hat{Q}$. We can extend the algebra to it by using the Jacobi *identity*

$$[\hat{W}, [\hat{V}, \hat{U}]] + [\hat{V}, [\hat{U}, \hat{W}]] + [\hat{U}, [\hat{W}, \hat{V}]] = 0. \tag{39}$$

Replacing $\hat{W} \to \hat{A}$, $\hat{U} \to \hat{R}_{njmqs}$, and $\hat{V}$ in turn by $\hat{J}_z$, $\hat{J}_\pm$ and $\hat{Q}$, and using Eqs. (34), (36) and (37) as well as the mutual commutativity of all observables, we obtain

$$[\hat{Q}, [\hat{A}, \hat{R}_{njmqs}]] = qe\,[\hat{A}, \hat{R}_{njmqs}]. \tag{40}$$

$$[\hat{J}_z, [\hat{A}, \hat{R}_{njmqs}] = m\,\hbar\,[\hat{A}, \hat{R}_{njmqs}], \tag{41}$$

$$[\hat{J}_\pm, [\hat{A}, \hat{R}_{njmqs}]] = \sqrt{j(j+1) - m(m\pm 1)}\,\hbar\,[\hat{A}, \hat{R}_{nj,m\pm,qs}]. \tag{42}$$

The fact that these commutators mimic those in Eqs. (34), (36) and (37) properly reflects the rotational and gauge invariant status of $\hat{A}$ which forces $[\hat{A}, \hat{R}_{njmqs}]$ to transform exactly like $\hat{R}_{njmqs}$. Note that all the previous commutation relations are invariant under the redefinition $\hat{A} \to \hat{A} + \text{const.}$ We single out the $\hat{A}$ of physical interest by the requirement that $\hat{A}\,|\text{vac}\rangle = 0$.

According to our closure assumption, $[\hat{A}, \hat{R}_{njmqs}]$ has to be a linear combination of some of the operators $\hat{I}, \hat{R}_{n'j'm'q's'}, \hat{A}, \hat{Q}, \hat{J}_z$ and $\hat{J}_{\pm}$. From Eq. (33) and its analog for $\hat{J}_z$ and rotations, it is clear that $\hat{I}, \hat{Q}$ and $\hat{A}$, all of them rotational scalars and gauge invariants, can only show up in the linear combination in the cases $\{nqjms\} = \{n000s\}$. Further, the triplet $\{\hat{J}_{-1}, \hat{J}_0, \hat{J}_{+1}\} \equiv \{\hat{J}_-, \hat{J}_z, \hat{J}_+\}$ is gauge invariant and a spherical irreducible tensor of rank one [47], so $\hat{J}_\nu$ with $\nu = 0, \pm 1$ can only show up in the linear combination in the cases $\{nqjms\} = \{n01\nu s\}$. Furthermore, by rotational invariance all three $\hat{J}_\nu$ must occur with like coefficient. Finally, $\hat{R}_{n'j'm'q's'}$ can only show up if its subscripts q', j' and m' match those of $[\hat{A}, \hat{R}_{njmqs}]$ (we cannot have another j appearing since Eq. (42) makes it clear that $[\hat{A}, \hat{R}_{njmqs}]$ contains a single j). In equations

$$[\hat{A}, \hat{R}_{njmqs}] = \sum_{n's'} h_{ns}^{n's'}\, \hat{R}_{n'jmqs'} + \delta_q{}^0 [\delta_j{}^0\, (C_{ns}\hat{I} + D_{ns}\hat{Q} + E_{ns}\hat{A}) + \delta_j{}^1\, F_{ns}\hat{J}_m] \tag{43}$$

where C_{ns}, D_{ns}, E_{ns} and F_{ns} are c–numbers and $h_{ns}^{n's'}$ is a c–number matrix.

Let us now operate with Eq. (43) on the vacuum; $\hat{Q}, \hat{\mathbf{J}}$ and $\hat{A}$ all anhilate it, so we are left with

$$a_n\, \hat{R}_{njmqs}|\text{vac}\rangle = \sum_{n's'} h_{ns}^{n's'}\, \hat{R}_{n'jmqs'}|\text{vac}\rangle + \delta_q{}^0\, \delta_j{}^0\, C_{ns}|\text{vac}\rangle \tag{44}$$

However, it is clear that the states $\{|nqjms\rangle\}$ are orthogonal to one another (automatically in the space spanned by n, q, j, m and by Schmidt orthogonalization with respect to the s quantum number), and all of them to the vacuum. The previous equation will contradict this unless we demand $C_{ns} = 0$ and $h_{ns}^{n's'} = a_n\, \delta_n{}^{n'}\, \delta_s{}^{s'}$. In that case

$$[\hat{A}, \hat{R}_{njmqs}] = a_n\, \hat{R}_{njmqs} + \delta_q{}^0 [\delta_j{}^0\, (D_{ns}\hat{Q} + E_{ns}\hat{A}) + \delta_j{}^1\; F_{ns}\hat{J}_m\;]. \tag{45}$$

At this stage we exploit the freedom left in $\hat{R}_{njmqs}$ to define a new set of creation operators

$$\hat{\mathcal{R}}_{njmqs} \equiv \hat{R}_{njmqs} + (a_n)^{-1}\delta_q{}^0 [\delta_j{}^0\, (D_{ns}\hat{Q} + E_{ns}\hat{A}) + \delta_j{}^1\, F_{ns}\hat{J}_m\;] \tag{46}$$

After the redefinition the algebra of $\{\hat{I}, \hat{A}, \hat{Q}, \mathbf{J}, \hat{\mathcal{R}}_{njmqs}\}$ is still closed, and the $\hat{\mathcal{R}}_{njmqs}$ create exactly the same states as the $\hat{R}_{njmqs}$. The redefinition transforms Eqs. (45) and (34) into

$$[\hat{A}, \hat{\mathcal{R}}_{njmqs}] = a_n\, \hat{\mathcal{R}}_{njmqs} \tag{47}$$

$$[\hat{Q}, \hat{\mathcal{R}}_{njmqs}] = q\, \hat{\mathcal{R}}_{njmqs}, \tag{48}$$

but changes the forms of (36) and (37) slightly [48].

2.5. THE AREA SPECTRUM

What is the state $\hat{\mathcal{R}}_{njmqs}|n'q'j'm's'\rangle = \hat{\mathcal{R}}_{njmqs}\hat{\mathcal{R}}_{n'q'j'm's'}|\text{vac}\rangle$? In field theory we would unhesitatingly identify it as a two–black hole state. Here it is different. To explain why, it is useful to denote the quantum numbers $\{njmqs\}$ collectively by a Greek index, κ, λ or μ, as the case may be. Likewise, we denote the operators $\hat{Q}$ and $\hat{A}$ by the common symbol $\hat{X}$, and the latter's eigenvalues by ξ_λ. From Eqs. (47)-(48) and the Jacobi identity (39) we discover that

$$[\hat{X}, [\hat{\mathcal{R}}_\kappa, \hat{\mathcal{R}}_\lambda]] = (\xi_\kappa + \xi_\lambda)[\hat{\mathcal{R}}_\kappa, \hat{\mathcal{R}}_\lambda]. \tag{49}$$

Now by the closure condition

$$[\hat{\mathcal{R}}_\kappa, \hat{\mathcal{R}}_\lambda] = \sum_\mu \varepsilon^\mu_{\kappa\lambda}\hat{R}_\mu + \cdots \tag{50}$$

where the ellipsis signifies some linear combination of $\hat{A}$, $\hat{J}_m$ and $\hat{Q}$, and $\varepsilon^\mu_{\kappa\lambda}$ are c-numbers (structure constants). Substituting in Eq. (49) gives

$$\sum_\mu \varepsilon^\mu_{\kappa\lambda}[\hat{X}, \mathcal{R}_\mu] = (\xi_\kappa + \xi_\lambda)(\sum_\mu \varepsilon^\mu_{\kappa\lambda}\hat{\mathcal{R}}_\mu + \ldots), \tag{51}$$

which in view of Eqs. (47)-(48) excludes the $\cdots$ terms, and also tells us that whenever $\varepsilon^\mu_{\kappa\lambda} \neq 0$,

$$\xi_\mu \equiv \xi_\kappa + \xi_\lambda \tag{52}$$

for both types of $\hat{X}$. We shall assume that for given κ and λ at least one of the $\varepsilon^\mu_{\kappa\lambda} \neq 0$ does not vanish (see below for the interpretation).

Operating on $|\text{vac}\rangle$ with Eq. (50) one gets

$$[\hat{\mathcal{R}}_\kappa, \hat{\mathcal{R}}_\lambda]|\text{vac}\rangle = |\bullet\rangle \tag{53}$$

where $|\bullet\rangle$ stands for a *one*–black hole state, a superposition of states $\{njmqs\}$ which by virtue of Eq. (49) all have a common area eigenvalue $a_\kappa + a_\lambda$ and a common charge $q_\kappa + q_\lambda$. This $|\bullet\rangle$ is obviously a physical state (it complies with the charge superselection rule); it may involve superpositions of different j and m.

We conclude that

$$|\Psi\rangle \equiv \hat{\mathcal{R}}_\kappa\hat{\mathcal{R}}_\lambda|\text{vac}\rangle = \frac{1}{2}|\bullet\bullet\rangle + \frac{1}{2}|\bullet\rangle. \tag{54}$$

Here $|\bullet\bullet\rangle = (\mathcal{R}_\kappa\mathcal{R}_\lambda + \mathcal{R}_\lambda\mathcal{R}_\kappa)|\text{vac}\rangle$ is obviously a two-black hole state symmetric under exchange of the κ and λ sets of quantum numbers. Thus trying to add an extra black hole to the state $\{njmqs\}$ actually creates a linear combination of a one- and a two-black hole states. This is reasonable since classically black holes can merge, so there is some amplitude for the new black hole to fuse with the original one. This is the ultimate justification for our assumption that at least one of the $\varepsilon^{\mu}_{\kappa\lambda}$ is nonzero.

Now using Eqs. (47)-(48) twice we find that

$$\hat{X}\,|\Psi\rangle = \hat{\mathcal{R}}_{\kappa}(\hat{X}+\xi_{\kappa})\hat{\mathcal{R}}_{\lambda}|\text{vac}\rangle = (\xi_{\kappa}+\xi_{\lambda})|\Psi\rangle \tag{55}$$

Thus the states $|\Psi\rangle$ and $|\bullet\bullet\rangle$ both have sharp area $a_{\kappa}+a_{\lambda}$ and sharp charge $q_{\kappa}+q_{\lambda}$, just like $|\bullet\rangle$: area, like charge, is an additive quantity for two black holes. This additivity jibes with our geometric notion that areas of separate objects are additive, and serves as further justification for the assumptions leading to our algebra. And because $|\Psi\rangle$ involves $|\bullet\rangle$, we find, in addition, that *the sum of two eigenvalues of $\hat{Q}$ or $\hat{A}$ for a single black hole is also a possible eigenvalue of $\hat{Q}$ or $\hat{A}$, respectively, of a single black hole.*

It turns out that differences of eigenvalues of the $\hat{X}$ are also eigenvalues for one black hole. Consider the hermitian conjugates of Eqs. (47)-(48),

$$[\hat{X},\hat{\mathcal{R}}^{\dagger}_{\kappa}] = -\xi_{\kappa}\hat{\mathcal{R}}^{\dagger}_{\kappa}. \tag{56}$$

What is the meaning of $|\chi\rangle \equiv \hat{\mathcal{R}}^{\dagger}_{\kappa}\,|\text{vac}\rangle$? Operating with Eq. (56) with $\hat{X}=\hat{A}$ on $|\text{vac}\rangle$ and taking the scalar product with $\langle\chi|$ shows that $\hat{A}$ would have a *negative* average in the state $|\chi\rangle$, unless this last state vanishes. Thus, since $\hat{A}$ is a positive definite operator, $\hat{\mathcal{R}}^{\dagger}_{\kappa}$ must anhilate the vacuum. It seems very plausible then that $|\Xi\rangle \equiv \hat{\mathcal{R}}^{\dagger}_{\kappa}\hat{\mathcal{R}}_{\lambda}\,|\text{vac}\rangle$ can only be a one–black hole state, pressumably distinct from $\hat{\mathcal{R}}_{\lambda}\,|\text{vac}\rangle$. Applying Eqs. (47),(48) and (56) gives

$$\hat{X}\,|\Xi\rangle = \left(\hat{\mathcal{R}}^{\dagger}_{\kappa}\hat{X} - \xi_{\kappa}\hat{\mathcal{R}}^{\dagger}_{\kappa}\right)\hat{\mathcal{R}}_{\lambda}|\text{vac}\rangle = (\xi_{\lambda}-\xi_{\kappa})\,|\Xi\rangle. \tag{57}$$

This verifies our claim, with the obvious caveat that if $a_{\lambda}<a_{\kappa}$, $\hat{\mathcal{R}}^{\dagger}_{\kappa}$, must anhilate the state $\hat{\mathcal{R}}_{\lambda}\,|\text{vac}\rangle$ because negative black hole areas are unacceptable (the same must happen if $a_{\lambda}=a_{\kappa}$ with $q_{\lambda}\neq q_{\kappa}$ because a zero area state is necessarily the vacuum, which bears no charge. Hence *the positive difference of two eigenvalues of $\hat{A}$ for a single black hole is also a possible eigenvalue of $\hat{A}$ of a single black hole; the difference of two $\hat{Q}$ eigenvalues is a possible $\hat{Q}$ eigenvalue for one black hole.*

We take it from Sec. 2.2 that the spectrum of $\hat{A}$ is discrete. Now the only discrete set of positive real numbers that is unchanged under addition or absolute value substraction of two members is the set of all the natural numbers multiplied by some common factor. We conclude that the one-black hole area spectrum is just $\{na_1 | n = 1, 2, \cdots\}$, where a_1 is some positive scale of area (here we make no attempt to determine which eigenvalues correspond to which charges and spin; see [48]). It is also clear that the two rules (in italics) lead to a one-black hole charge spectrum composed of all integers multiplied by a common factor, e. We have thus formally obtained from the algebra the kind of area spectra predicted long ago [33], and summarized in Sec. 2.2. For nonrotating neutral black holes, the corresponding mass eigenvalues are proportional to $\{\sqrt{n} \,|\, n = 1, 2, \ldots\}$.

Based on a variety of interpretations of the nonrotating neutral black hole in canonical quantum gravity, Schiffer (whose paper's title is used as title of the present lecture), Peleg, Kastrup, Louko and Mäkelä, Barvinsky and Kunstatter, Berezin, Vaz and Witten [49] have all obtained a mass spectrum of the mentioned form, but with no consensus as to the exact numerical coefficient. Some such calculations give a not uniformly spaced spectrum [50]. Regarding RN black holes, Mäkelä and Repo[51] find the *sum* of areas of outer and inner horizons to scale like an integer, and a recent paper of theirs and coworkers extends this rule to charged rotating black holes [52]. Vaz and Witten [53] find a law of this form rather for the *difference* of these areas. And Barvinsky, Das and Kunstatter [45] find an *external* horizon area with a spectrum precisely equally spaced, but with its zero point shifted by a charge dependent quantity. Overall, the predictions are thus similar to those following from the algebra.

2.6. DEGENERACY OF AREA EIGENVALUES

According to Mukhanov [54] degeneracy plays a central role in any discussion of area levels. What can we say about it from our approach ? By rotational invariance neither area eigenvalues nor degeneracy factors can depend on the quantum number m. Let us assume that the spectrum $a_n = na_1$ for $n = 1, 2, \cdots$ is common to every combination of quantum numbers j and q (there is one alternative to this [48]). The degeneracy factor will be of the form $g_n = g_n(j, q)$. Now for fixed $\{n_\kappa, j_\kappa, m_\kappa, q_\kappa\}$ with not all of j_κ, m_κ and q_κ vanishing, there are $g_{n_\kappa}(j_\kappa, q_\kappa)$ independent one–black hole states $\hat{R}_\kappa|\mathrm{vac}\rangle$ distinguished by the values of s. Analogously, the set $\{n_\lambda = 1, j_\lambda = 0, m_\lambda = 0, q_\lambda = 0\}$ specifies $g_1(0, 0)$ independent states $\hat{R}_\lambda|\mathrm{vac}\rangle$, all different from the previous ones because not all quantum numbers agree. One can thus form $g_1(0, 0) \cdot g_{n_\kappa}(j_\kappa, q_\kappa)$ one–black hole states, $[\hat{R}_\kappa, \hat{R}_\lambda]|\mathrm{vac}\rangle$, with area eigenvalues $(n_\kappa+1)a_1$, and charge and spin just like the states $\hat{R}_\kappa|\mathrm{vac}\rangle$. Let us assume these new states are independent. Then their number cannot exceed the total number of states with area $(n_\kappa+1)a_1$ and quantum numbers j_κ and q_κ: $g_{n_\kappa+1}(j_\kappa, q_\kappa) \geq g_1(0, 0) \cdot g_{n_\kappa}(j_\kappa, q_\kappa)$. Iterating this inequality starting from $n_\kappa = 1$ gives (we drop κ)

$$g_n(j, q) \geq g_1(j, q) \cdot g_1(0, 0)^{n-1}. \tag{58}$$

If $g_1(0, 0) \neq 1$, Eq. (58) tells us that the degeneracy rises at least exponentially with area. Since the area spectrum is rather sparse, black hole entropy must receive its principal contribution from the logarithm of the degeneracy of area levels, so what we have just found is that black hole entropy must grow at least as fast as the horizon area. Thus we have a pleasant microscopic explanation of the black hole entropy–area relation. Further, as first emphasized by Mukhanov [54], one can calibrate the area spectrum by use of the degeneracy $\leftrightarrow$ entropy correspondence relation. For $g_1(0, 0) = 2$ this gives $a_1 = 4\hbar \ln 2$ with the corresponding mass spectrum $\{(\hbar \ln 2/4\pi)^{1/2}\sqrt{n} \,|\, n = 1, 2, \cdots\}$ [55].

Acknowledgments This research is supported by grant No. 129/00-1 of the Israel Science Foundation.

References

1. 't Hooft, G. (1993) in Aly, A., Ellis, J. and Randjbar–Daemi, S. (eds.), *Salam–festschrifft*, World Scientific, Singapore, gr–qc/9310026.
2. Susskind, L. (1995) *J. Math. Phys.*, **36**, 6377.
3. Bekenstein, J.D. (1974) *Lett. Nuovo Cimento*, **4**, 737 and *Phys. Rev.* D, **9**, 3292.
4. Wald, R.M. (2001) *Living Reviews in Relativity*, **2001-6**.
5. Bekenstein, J.D. (2000) *Phys. Letters*B, **481**, 339; (2001) in Ruffini, R. (ed.), *Proceedings of the Ninth Marcel Grossmann Meeting on General Relativity*, World Scientific, Singapore.
6. Bousso, R. (1999) *J. High Energy Phys.*, **9906**, 028.
7. Peres, A. (1993) *Quantum Theory: Concepts and Methods*, Kluwer, Dordrecht.
8. Shannon, C. and Weaver, W. (1949) *The Mathematical Theory of Communication*, Univ. of Illinois Press, Urbana.
9. Katz, A. (1967) *Principles of Statistical Mechanics*, Freeman, San Francisco.
10. Bekenstein, J.D. (1981) *Phys. Rev.* D, **23**, 287.
11. Bekenstein, J.D. (1973) *Phys. Rev.* D, **7**, 2333.
12. Unruh, W.G. and Wald, R.M. (1982) *Phys. Rev.* D, **25**, 942 and (1983) *Phys. Rev.* D, **27**, 2271; Pelath, M.A. and Wald, R.M. (1999) *Phys. Rev.* D, **60**, 104009.
13. Bekenstein, J.D. (1982) *Phys. Rev.* D, **26**, 950; (1983) *Phys. Rev.* D, **27**, 2262.
14. Bekenstein, J.D. (1994) *Phys. Rev.* D, **49**, 1912; (1999) *Phys. Rev.* D, **60**, 124010.
15. Zaslavskii, O. (1991) *Phys. Lett.* A, **160**, 339.
16. Hod, S. (1999) *Phys. Rev.* D, **60**, 104031; (2000) *Phys. Rev.* D, **61**, 024018 and **61**, 024023.
17. Mayo, A.E. (1999) *Phys. Rev.* D, **60**, 104044; Bekenstein, J.D. and Mayo, A.E. (2000) *Phys. Rev.* D, **61**, 024022.
18. Bekenstein, J.D. (1996) in Jantzen, R.T. and Keiser, G.M. (eds.), *Proceedings of the VII Marcel Grossmann Meeting on General Relativity*, World Scientific, Singapore, gr–qc/9409015.
19. Bousso, R. (2001) *J. High Energy Phys.*, **0104**, 035.
20. Verlinde, E. (2000) preprint hep–th/0008140.
21. Cardy, J. L. (1986) *Nucl. Phys.* B, **270**, 317.
22. Bekenstein, J.D. (1989) *Int. J. Theor. Phys.*, **28**, 967; Schiffer, M. (1991) *Int. J. Theor. Phys.*, **30**, 419; Fischler, W. and Susskind, L. (1998) preprint hep-

th/9806039; Bousso, R. (1999) *J. High Energy Phys.*, **9907**, 004; Easther R. and Lowe, D. (1999) *Phys. Rev. Letters*, **82**, 4967; Veneziano, G. (1999) *Phys. Lett.* B, **454**, 22; Kaloper, N. and Linde, A. (1999) *Phys. Rev.* D, **60**, 103509; Bak D. and Rey, S.-J. (2000) *Class. Quant. Grav.*, **L83**, 17; Brustein, R. and Veneziano, G. (2000) *Phys. Rev. Letters*, **84**, 5695.
23. Page, D.N. (1983) *Phys. Rev. Letters*, **50**, 1013.
24. Page, D.N. (1976) *Phys. Rev.* D, **13**, 198.
25. Bekenstein, J.D. and Schiffer, M. (1990) *Int. J. Mod. Physics C*, **1**, 355.
26. Pendry, J.B. (1983) *J. Phys. A*, **16**, 2161.
27. Landau, L.D. and Lifshitz, E.M. (1980) *Statistical Physics, Part I*, Third Edition, Pergamon, Oxford.
28. Volkov, A.M., Izmest'ev, A.A. and Skrotskii, G.V. (1971) *Sov. Phys. JETP*, **32**, 686.
29. Marko, H. (1965) *Kybernetik*, **2**, 274.
30. Bekenstein, J.D. and Mayo, A.E. (2001) preprint gr-qc/0105055, to appear in General Relativity and Gravitation.
31. Bekenstein, J.D. (1981) *Phys. Rev. Letters*, **46**, 623.
32. Bekenstein, J.D. (1988) *Phys. Rev.* A, **37**, 3434.
33. Bekenstein, J.D. (1974) *Lett. Nuovo Cimento*, **11**, 467.
34. Bekenstein, J.D. (1996) in da Silva, A. J. et. al (eds.), *XVII Brazilian National Meeting on Particles and Fields*, Brazilian Physical Society, São Paulo.
35. Bekenstein, J.D. (1998) in *Proceedings of the VIII Marcel Grossmann Meeting on General Relativity*, Piran, T. and Ruffini, R. eds, World Scientific, Singapore, gr-qc/9710076.
36. Bekenstein, J.D. (1999) in Novello, M. (ed.), *IX Brazilian School of Cosmology and Gravitation*, Atlantiscience, Paris, gr-qc/9808028.
37. Born, M. (1969) *Atomic Physics*, Eight Edition, Blackie, London.
38. Penrose, R. and Floyd, R.M. (1971) *Nature*, **229**, 177.
39. Christodoulou, D. (1970) *Phys. Rev. Letters*, **25**, 1596
40. Hawking, S.W. (1971) *Phys. Rev. Letters*, **26**, 1344.
41. Christodoulou, D. and and Ruffini, R. (1971) *Phys. Rev.* D, **4**, 3552.
42. Bekenstein, J.D. (1998) in Iyer, B.R. and Bhawal, B. (eds.), *Black Holes, Gravitational Radiation and the Universe*, Kluwer, Dordrecht.
43. Mayo, A.E. (1998), *Phys. Rev.* D **58**, 104007.
44. Duez, M.W. et. al (1999) *Phys. Rev.* D, **60**, 104024.
45. Barvinsky, A, Das, S. and Kunstatter, G. (2000) preprint gr-qc/0012066 and (2001) preprint hep-th/0102061.
46. Gour, G. (2000) *Phys. Rev.* D, **61**, 124007.
47. Merzbacher E., 1970, *Quantum Mechanics.*, Second Edition, Wiley, New York.
48. Bekenstein, J.D. (2002) in Duff, M. and Liu, J. T. (eds.), *2001: A Spacetime Odyssey*, World Scientific Publishing, Singapore, hep-th/0107045.
49. Schiffer, M. (1989) "Black hole spectroscopy", preprint IFT/P-38/89, São Paulo; Peleg, Y. (1995) *Phys. Lett.* B, **356**, 462; Kastrup, H. (1996) *Phys. Lett.* B, **385**, 75; Louko, J. and Mäkelä, J. (1996) *Phys. Rev.* D, **54**, 4982; Barvinsky, A. and Kunstatter, G. (1996) *Phys. Lett.* B, **329**, 231; Berezin, V.A. (1997) *Phys. Rev.* D, **55**, 2139; Vaz, C. and Witten, L. (1999) *Phys. Rev.* D, **60**, 024009; Vaz, C. (2000) *Phys. Rev.* D, **61** 064017.
50. Dolgov, A.D. and Khriplovich, I B. (1997) *Phys. Lett.* B, **400**, 12; Berezin, V.A. (1997) *Phys. Rev.* D, **55**, 2139; Berezin, V.A., Boyarsky, A.M. and Neronov, A.Yu. (1998) *Phys. Rev.* D, **57** 1118.
51. Mäkelä, J. and Repo, P. 1997 *Phys. Rev.* D, **57**, 4899.
52. Mäkelä, J., Repo, P., Luomajoki, M. and Piilonen, J. (2000) *Phys. Rev.* D, **64**, 024018 (2001)
53. Vaz, C. and Witten, L. (2001) *Phys. Rev.* D, **63**, 024008.
54. Mukhanov, V.F. (1986) *JETP Letters*, **44**, 63.

ATOMIC CLOCKS AND ATOM INTERFEROMETRY

CHRISTIAN J. BORDÉ
Laboratoire de Physique des Lasers
UMR 7538 CNRS, Université Paris-Nord
99, Avenue J.-B. Clément, 93430, Villetaneuse, France
and
ERGA/LERMA, UMR 8540 CNRS, Université Paris VI
4 Place Jussieu, 75005 Paris, France

Abstract. We show that the language of atom interferometry [1] provides a unified picture for microwave and optical atomic clocks as well as for gravito-inertial sensors. The sensitivity and accuracy of these devices, is now such that a new theoretical framework [2] common to all these interferometers, is required and which includes: 1 - A fully quantum mechanical treatment of the atomic motion in free space and in the presence of a gravitational field (most cold atom interferometric devices use atoms in "free fall" in a fountain geometry), 2 - An account of simultaneous actions of gravitational and electromagnetic fields in the interaction zones, 3 - A second quantization of the matter fields to take into account their fermionic or bosonic character in order to discuss the role of coherent sources and their noise properties, 4 - A covariant treatment including spin to evaluate general relativistic effects. A theoretical description of atomic clocks revisited along these lines, is presented, using both an exact propagator of atom waves in gravito-inertial fields [3] and a covariant Dirac equation in the presence of weak gravitational fields [4]. Using this framework, recoil effects, spin-related effects, beam curvature effects, the sensitivity to gravito-inertial fields and the influence of the coherence of the atom source can be discussed in the context of present and future atomic clocks and gravito-inertial sensors.

1. Introduction

The first purpose of these few pages is to clarify the link between atomic clocks and the recent field of atom interferometry [1] and to show that

P.G. Bergmann and V. de Sabbata (eds.),
Advances in the Interplay Between Quantum and Gravity Physics, 27–55.

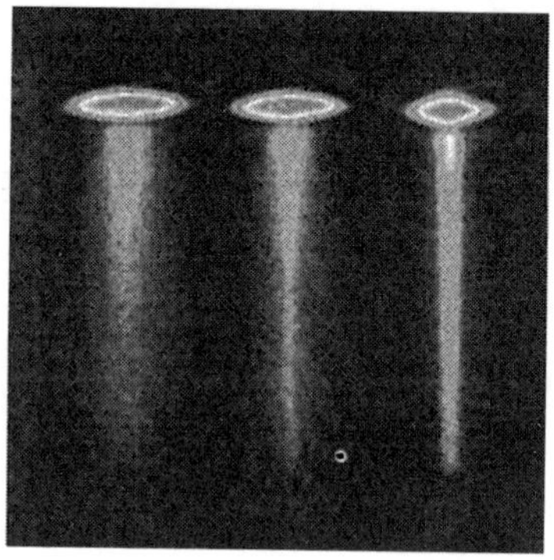

Figure 1. Rubidium atoms are extracted from a cold Rubidium gas (left) and from a Bose-Einstein condensate (right). *(Courtesy of the University of Munich [7])*

indeed, microwave and optical atomic clocks are genuine atom interferometers [2]. The wave character of atoms is getting more and more manifest in these devices: the recoil energy $\hbar\delta = \hbar^2k^2/2M$ is not negligible any more in Cesium clocks ($\delta/2\pi\nu \simeq 1.5\ 10^{-16}$). Atom sources may now be coherent sources of matter-waves (Bose-Einstein condensates [6, 7], atom lasers or atomasers [8]) as illustrated in Fig. 1. We have to deal with a very different picture from that of small clocks carried by classical point particles. The atomic frame of reference may not be well-defined. In modern microwave atomic clocks (see Fig. 2) atoms interact twice with an electromagnetic field (this is the method of separated fields introduced by N. F. Ramsey around 1950) giving rise to interference fringes (Fig. 3), which can now be reinterpreted as an interference between the de Broglie waves associated with the external motion of the atoms. Atomic clocks are thus now fully quantum devices in which both the internal and external degrees of freedom must be quantized.

Gravitation and inertia play a key role in slow atom clocks. The Einstein red shift and the second-order Doppler shift may become important and thus atomic clocks have to be treated also as relativistic devices.

Finally, we recall how the idea of separated e.m. fields in space or time has been extended to the optical domain in order to build atom interferometers, which can be used as optical clocks, but also as very sensitive gravito-inertial sensors.

2. Atom waves

The wave properties of atoms are fully described by a dispersion law relating the de Broglie frequency to the de Broglie wave vector, which is obtained

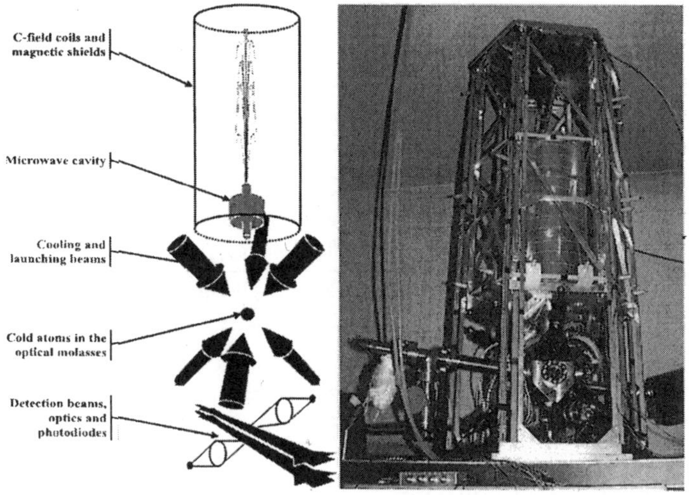

Figure 2. Principle and realization of a fountain clock (*courtesy of BNM-LPTF[5]*)

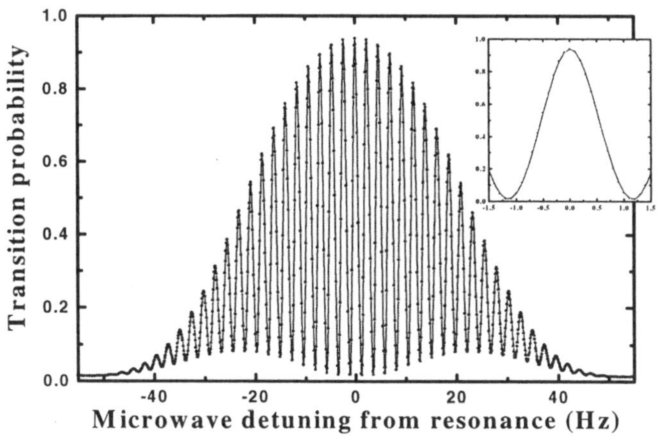

Figure 3. Ramsey fringes obtained with the Cesium fountain clock (*Courtesy of BNM-LPTF[5]*)

from the law connecting the energy $E(\overrightarrow{p})$ to the momentum $\overrightarrow{p}$ by the introduction of Planck constant. In free space (Fig. 4) the corresponding

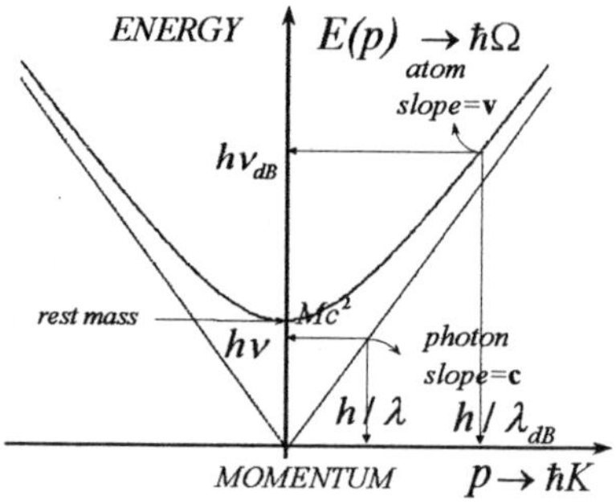

Figure 4. Energy-momentum curves: (1D cut of the mass shell) hyperbola for a massive particle e.g. an atom in a given internal energy state and straight lines for photons. The slope is the group velocity of the de Broglie wave.

curve is the hyperbola of equation:

$$E(\overrightarrow{p}) = \sqrt{M^2c^4 + p^2c^2} \tag{1}$$

The amplitude of an atom wave may therefore be written generally as:

$$\begin{aligned} a(\overrightarrow{r}, t) &= \int dE \frac{d^3p}{(2\pi\hbar)^{3/2}} e^{i[\overrightarrow{p}\cdot(\overrightarrow{r}-\overrightarrow{r}_0)-E(t-t_0)]/\hbar} \\ &\quad \delta\left(E^2 - E^2(\overrightarrow{p})\right)\theta(E)\,a(\overrightarrow{p}, E) \\ &= \int \frac{dE}{2E(\overrightarrow{p})} \frac{d^3p}{(2\pi\hbar)^{3/2}} e^{i[\overrightarrow{p}\cdot(\overrightarrow{r}-\overrightarrow{r}_0)-E(t-t_0)]/\hbar} \\ &\quad \delta\left(E - E(\overrightarrow{p})\right) a(\overrightarrow{p}, E) \end{aligned} \tag{2}$$

In the non-relativistic limit:

$$a(\overrightarrow{r}, t) = \int \frac{d^3p}{(2\pi\hbar)^{3/2}} e^{i\left[\overrightarrow{p}\cdot(\overrightarrow{r}-\overrightarrow{r}_0)-\left(Mc^2+p^2/2M\right)(t-t_0)\right]/\hbar} a(\overrightarrow{p}) \tag{3}$$

If $a(\overrightarrow{p})$ is expanded, for example, in Hermite-Gauss functions,

$$\begin{aligned} a(p_x, p_y, p_z) &= \frac{1}{(2\pi)^{3/2}\sqrt{\Delta_x\Delta_y\Delta_z}} \sum_{lmn} H_l\left(\frac{p_x}{\Delta_x}\right) H_m\left(\frac{p_y}{\Delta_y}\right) H_n\left(\frac{p_z}{\Delta_z}\right) \\ &\quad \exp\left[-\frac{p_x^2}{2\Delta_x^2}\right] \exp\left[-\frac{p_y^2}{2\Delta_y^2}\right] \exp\left[-\frac{p_z^2}{2\Delta_z^2}\right] \end{aligned} \tag{4}$$

we obtain a complete orthogonal set of free-propagation modes. The lowest order modes correspond to minimum uncertainty wave packets. These free-space propagation modes transform with the same ABCD law as in Gaussian laser optics. The ABCD matrices are functions of time in this case and we shall see below how the ABCD law can be generalized in the presence of gravito-inertial fields. For a travelling wave the δ function can be written $\delta\left(p_x - \sqrt{2ME_{kin} - p_y^2 - p_z^2}\right)$ and we may keep the kinetic energy $E_{kin} = E - Mc^2$ instead of p_x in the expression of the modes. This is a good choice if the atom wave is monochromatic, in which case the integral over energy combines with an amplitude proportional to $\delta\left(E_{kin} - E_0\right)$. In the paraxial approximation $\delta\left(p_x - \sqrt{2ME_{kin} - p_y^2 - p_z^2}\right)$ goes into $\delta\left(p_x - \sqrt{2ME_{kin}}\left(1 - \left(p_y^2 + p_z^2\right)/\left(4ME_{kin}\right)\right)\right)$ and

$$
\begin{aligned}
a(\overrightarrow{r},t) \quad \propto \quad & \int dp_y dp_z a(p_y,p_z) \exp\left[-\frac{i}{\hbar}\frac{p_y^2+p_z^2}{2\sqrt{2ME_0}}\left(x-x_0\right)+\frac{i}{\hbar}\left(p_y y+p_z z\right)\right] \\
& \exp\left[\frac{i}{\hbar}\sqrt{2ME_0}\left(x-x_0\right)-\frac{i}{\hbar}\left(Mc^2+E_0\right)\left(t-t_0\right)\right] \qquad (5)
\end{aligned}
$$

The same Hermite-Gauss expansion as above may be used for $a(p_y, p_z)$. Again the propagation is described by ABCD matrices, which are in this case functions of x.

3. Interaction of two-level atoms with electromagnetic waves

We shall make a systematic use of energy-momentum diagrams to discuss the problem of interaction of two-level atoms with two separated field zones in a Ramsey excitation scheme (Figure 5).

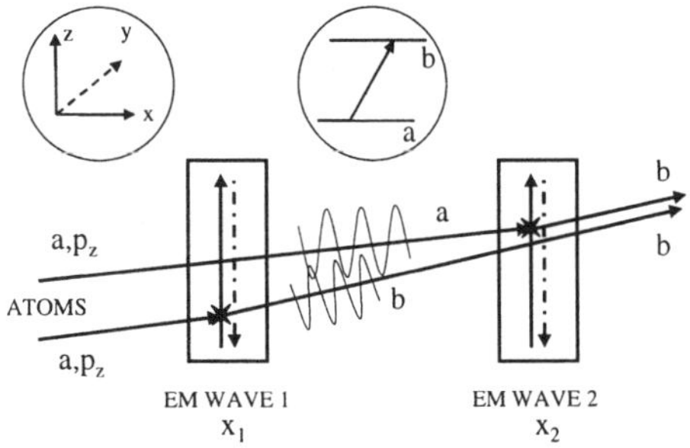

Figure 5. Illustration of the reinterpretation of the separated field method as the interference between de Broglie waves. Case of successive interactions with copropagating waves.

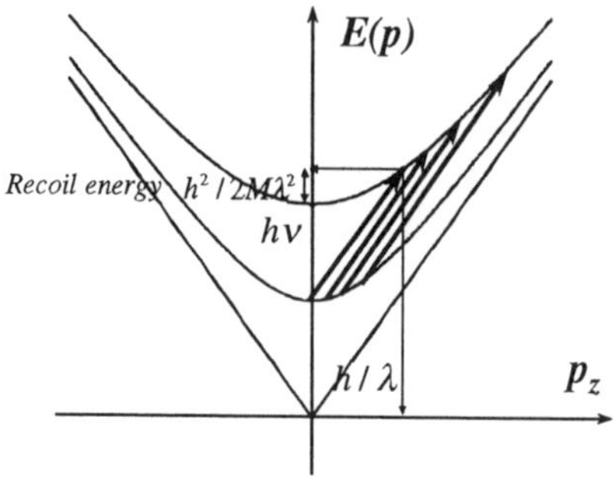

Figure 6. Energy vs transverse momentum in the absorption of a photon by a two-level atom. A distribution of absorbed frequencies correspond to a distribution of momenta.

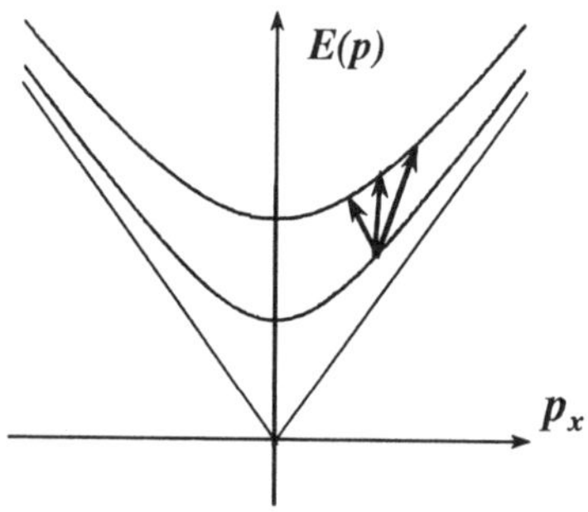

Figure 7. Energy vs longitudinal momentum in the absorption of a photon by a two-level atom. When the e.m. is confined in the longitudinal direction, there is a corresponding distribution of wave vector components in that direction, which allows for a spread of absorbed frequencies (transit-time broadening or Rabi pedestal) and a change of the longitudinal momentum of the atom.

Figures 6 and 7 illustrate the energy and momentum conservation between this two-level atom and effective photons from each travelling wave in the transverse and longitudinal directions and display the recoil energy, the first and second-order Doppler shifts and the transit broadening. It is clear from Fig. 7 that, out of resonance, an additional longitudinal momentum is transferred to the atoms in the excited state. This velocity change along the forward direction is the basis for the so-called mechanical reinterpretation of Ramsey fringes [9, 10, 11, 12, 13]. We will now illustrate this point in more detail through a simple first-order theory of Ramsey fringes. state.

Let us consider a beam of two-level atoms with $E_a < E_b$ initially in state a which interacts successively with two field zones respectively centered at x_1 and x_2 as in Fig. 5 and let us calculate the excited state amplitude to first order in each field zone [2, 3]. One can check that the following expression is indeed a first-order solution of Schroedinger equation:

$$\begin{aligned} b^{(1)}(\vec{r},t) &= \frac{1}{i\hbar}\int_{-\infty}^{t} dt' \int \frac{d^3p}{(2\pi\hbar)^{3/2}} \int \frac{d^3k}{(2\pi)^{3/2}} V_{ba}(\vec{k},t') e^{i\vec{k}\cdot(\vec{r}-\vec{r_1})} \\ & e^{i\left[E_b(\vec{p}+\hbar\vec{k})-E_a(\vec{p})\right](t'-t)/\hbar} \\ & e^{i[\vec{p}\cdot(\vec{r}-\vec{r}_0)-E_a(\vec{p})(t-t_0)]/\hbar}\langle a|\langle\vec{p}|\Psi^{(0)}\rangle \end{aligned} \tag{6}$$

where the energy is given by the dispersion relation 1 and can be expanded in a Taylor series:

$$E(\vec{p}+\hbar\vec{k}) = E(\vec{p})+\frac{\hbar\vec{k}\cdot\vec{p}c^2}{E(\vec{p})}+\frac{(\hbar k)^2c^2}{2E(\vec{p})}+... = E(\vec{p})+\hbar\vec{k}\cdot\vec{v}+\frac{(\hbar k)^2}{2M}+... \tag{7}$$

The matrix element of the Hamiltonian of interaction with the e.m. waves is $V_{ba}(\vec{r},t) = -\Omega_{ba}e^{i(\pm kz-\omega t+\varphi)}U(\vec{r})$, where Ω_{ba} is a Rabi frequency. Let us introduce a monochromatic electromagnetic wave with a Gaussian distribution of k_x (example for the illustration) and, for simplicity, let us ignore the dimension y :

$$\begin{aligned} b_{\pm}^{(1)}(\vec{r},t) &= i\Omega_{ba}e^{i(\pm kz-\omega t+\varphi)}\frac{w}{2\sqrt{\pi}}\int\frac{d^3p}{(2\pi\hbar)^{3/2}} \\ & \int_{-\infty}^{+\infty} dk_x e^{-\frac{w^2k_x^2}{4}} e^{ik_x(x-x_1)} e^{i[\omega-\omega_{ba}\mp kv_z-k_xv_x-\delta](t-t_1)} \\ & \int_{-\infty}^{t} dt' e^{-i[\omega-\omega_{ba}\mp kv_z-k_xv_x-\delta](t'-t_1)} \\ & e^{i[\vec{p}\cdot(\vec{r}-\vec{r}_0)-E_a(\vec{p})(t-t_0)]/\hbar}a^{(0)}(\vec{p}) \end{aligned} \tag{8}$$

with $\hbar\omega_{ba}(\vec{p}) = E_b(\vec{p}) - E_a(\vec{p}) \approx \hbar\omega_{ba}\sqrt{1-v^2/c^2} = \hbar\omega_{ba}(v)$ and $a^{(0)}(\vec{p}) = \langle a|\langle\vec{p}|\Psi^{(0)}\rangle$. In the time integral the upper bound t may be extended to infinity if the considered wave packet has left the interaction zone (this is justified in the footnote [1], where the exact calculation of reference

[1]The exact calculation gives:

$$\begin{aligned} b_{\pm}^{(1)}(\vec{r},t) &= i\frac{\sqrt{\pi}w\Omega_{ba}}{2}e^{i(\pm kz-\omega t+\varphi)}\int\frac{d^3p}{(2\pi\hbar)^{3/2}}\frac{1}{v_x} \\ & \Big[2e^{-w^2(\omega-\omega_{ba}\mp kv_z-\delta)^2/4v_x^2}e^{i(\omega-\omega_{ba}\mp kv_z-\delta)(x-x_1)/v_x} \\ & - w(i\rho)\Big]e^{i[\vec{p}\cdot(\vec{r}-\vec{r}_0)-E_a(\vec{p})(t-t_0)]/\hbar}a^{(0)}(\vec{p}) \end{aligned} \tag{9}$$

with $\rho = (x-x_1)/w_0 + i(\omega-\omega_{ba}\mp kv_z-\delta)w_0/2v_x$ and where $w(z)$ is the error function of complex arguments. The second term vanishes with the distance $(x-x_1)/w_0$ leaving the accelerated or decelerated first contribution as the dominant one.

[12] is recalled). We obtain a δ function expressing energy conservation as expected from the S-matrix:

$$2\pi\delta(\omega - \omega_{ba}(v) \mp kv_z - k_x v_x - \delta)$$

and corresponding to Figures 6 and 7. If the resonance condition $\omega - \omega_{ba}(v) \mp kv_z - \delta = 0$ is satisfied in Fig. 6 this implies $k_x = 0$ in Fig. 7, otherwise the effect of energy conservation is to select a particular component $k_x = (\omega - \omega_{ba}(v) - kv_z - \delta)/v_x$.

We obtain the first-order excited state transition amplitude:

$$\begin{aligned} b_{\pm}^{(1)}(\vec{r},t) &= i\sqrt{\pi}e^{i(\pm kz-\omega t+\varphi)} \int \frac{d^3p}{(2\pi\hbar)^{3/2}} \frac{w\Omega_{ba}}{v_x} e^{-w^2(\omega-\omega_{ba}\mp kv_z-\delta)^2/4v_x^2} \quad (10) \\ & e^{i(\omega-\omega_{ba}\mp kv_z-\delta)(x-x_1)/v_x} e^{i[\vec{p}\cdot(\vec{r}-\vec{r}_0)-E_a(\vec{p})(t-t_0)]/\hbar} a^{(0)}(\vec{p}) \end{aligned}$$

as the product of the e.m. carrier times a Rabi frequency and a Rabi envelope, times an additional momentum phase factor for each initial wave packet Fourier component. This additional longitudinal momentum is proportional to the detuning and is responsible for the Ramsey fringes, since de Broglie waves associated with each path have a different wavelength in the dark zone (Fig. 5) and the transition probability integrated over the detection volume is:

$$\begin{aligned} \int d^3r\, b_{1\pm}^{(1)}(\vec{r},t) b_{2\pm}^{(1)*}(\vec{r},t) &\propto \int dp_z e^{-w^2(\omega-\omega_{ba}\mp kv_z-\delta)^2/2v_x^2} \quad (11) \\ & e^{i(\omega-\omega_{ba}\mp kv_z-\delta)(x_2-x_1)/v_x} a^{(0)}(p_z) a^{(0)*}(p_z) \end{aligned}$$

This Ramsey interference pattern has a blue recoil shift δ and is the superposition of fringe sub-systems corresponding to each velocity class, shifted by the first-order Doppler effect. If the transverse velocity distribution is too broad (absence of diaphragm) or in the optical domain, this will blur out the fringes. To make the connection with atom optics, this superposition can be rewritten as a correlation function involving the degree of coherence of the atom source [2]:

$$\int dz a^{(0)}(z \mp \frac{\hbar k}{M}(x_2 - x_1)/v_x, t) a^{(0)*}(z,t) \quad (12)$$

[2]The degree of transverse coherence of a thermal atom source is given by the density matrix element: $\langle z,t|\,\rho\,|z',t'\rangle$ which is simply the free propagator for a "complex time" argument: $t - t' - i\hbar/(k_B T)$. Its width is the thermal de Broglie wavelength h/Mu and it gives rise to the Doppler width $k_B u$. Incidentally, an accurate value for the Boltzman constant could be obtained through the accurate frequency measurement of a Doppler width.

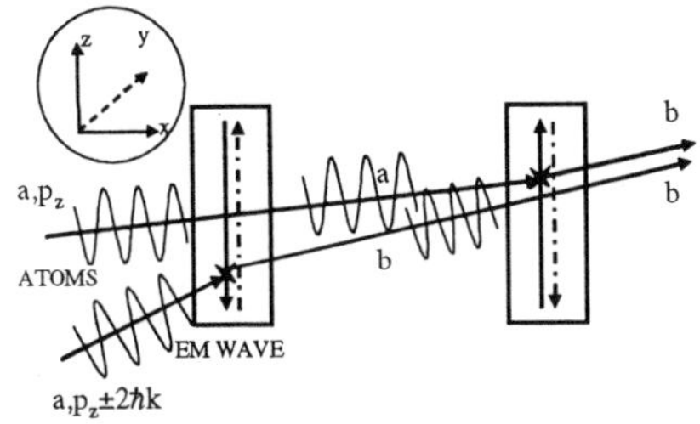

Figure 8. Illustration of the reinterpretation of the separated field method as the interference between de Broglie waves. Case of successive interactions with counterpropagating waves.

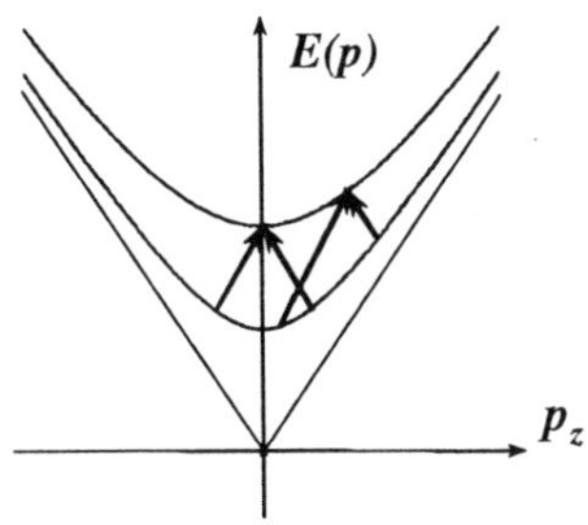

Figure 9. Energy vs transverse momentum exchanges in the case of successive interactions with counterpropagating waves.

Fringes will be obtained as long as $\hbar k\,(x_2 - x_1)\,/Mv_x$ is smaller than the coherence width of the atom source. A second mechanism, which was considered only recently, is the interaction with oppositely travelling waves in each zone as in Figs 8 and 9.

This is possible only if the initial wave packet has Fourier components which differ by $\pm 2\hbar k$ (size of atomic cloud $< \lambda_{em}$) since:

$$\int dz b_{1\pm}^{(1)}(\overrightarrow{r},t) b_{2\mp}^{(1)*}(\overrightarrow{r},t) \propto \int dz e^{\pm 2ikz} \int dp_z dp_z' \exp\left[i\left(p_z - p_z'\right) z/\hbar\right] \ldots$$
$$\ldots a^{(0)}(p_z) a^{(0)*}(p_z') \Rightarrow \delta\left(p_z - p_z' \pm 2\hbar k\right) \qquad (13)$$

The resulting signal exhibit fringes with an opposite recoil shift $-\delta$. Unlike the previous one, this signal depends upon the propagation characteristics

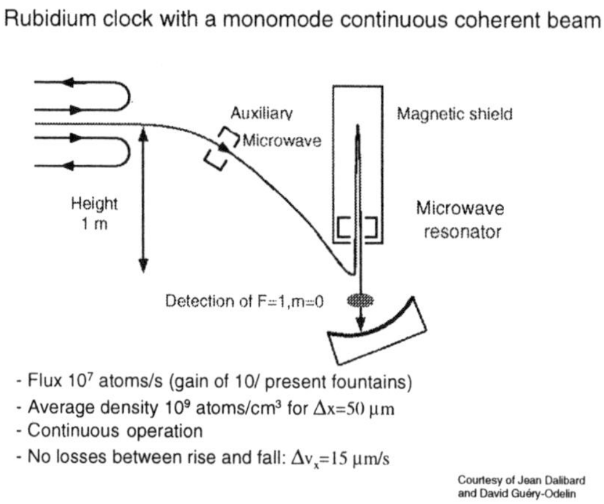

Figure 10. Schematic view of a possible future fountain clock using the continuous coherent atom wave source which is being developed at LKB (Paris). The magnetic mirror is in the strong field regime, in which four magnetic sublevels bounce upwards (including one connected with F=2, m=0), and the four others are attracted (including one connected with F=1, m=0). This guarantees a detection on a dark background.

of the incident atom wave:

$$\int dz b^{(1)}_{1\pm}(\overrightarrow{r},t) b^{(1)*}_{2\mp}(\overrightarrow{r},t) \propto e^{i(\omega-\omega_{ba}+\delta)(x_2-x_1)/v_x}$$

$$\cdot \int dp_z e^{-i(\pm kv_z+2\delta)(2x_0-x_2-x_1)/v_x} \quad a^{(0)}(p_z)a^{(0)*}(p_z \pm 2\hbar k) \quad (14)$$

This integral is easily calculated for Gaussian wave packets and statistical mixtures. If the waist position x_0 of the atom wave is not well-defined (e.g. in a thermal beam), energy conservation requires $kv_z = \mp 2\delta$ and will not be satisfied for most velocities (Fig. 9) and this signal will tend to average out. For a coherent atom wave, if the waist is located at x_1, this second contribution will have the same magnitude as the first one and the overall recoil shift will cancel [14]. If it is focussed at the midpoint $x_0 = (x_1+x_2)/2$ (perfect time reversal), this signal will be free of Doppler effect and will tend to dominate and impose its opposite recoil shift. This would be the case, for example, of a coherent atom wave with a waist at the top of a fountain clock, which could be achieved in a clock design such as that represented on Fig. 10.

4. Introduction of gravito-inertial fields

The well-known stationary solution of Schroedinger equation in presence of a gravitational field involving the Airy function can be applied directly to the fountain clock. However, calculations with the Airy function are not so easy and here, we prefer to take a more general time-dependent approach, which is mathematically simpler and more powerful. To take gravito-inertial fields into account in the treatment of fountain clocks and other atom interferometers, we shall consider quite generally the non-relativistic Schroedinger equation obtained as the non-relativistic limit of a general relativistic equation described in the second part of this course:

$$\begin{aligned} i\hbar\frac{\partial|\Psi(t)\rangle}{\partial t} &= \Big[H_0 + \frac{1}{2M}\vec{p}_{op}\cdot \overset{\Rightarrow}{g}(t)\cdot\vec{p}_{op} - \vec{\Omega}(t)\cdot(\vec{L}_{op}+\vec{S}_{op}) \\ &\quad - M\vec{g}(t)\cdot\vec{r}_{op} - \frac{\dot{M}}{2}\vec{r}_{op}\cdot\overset{\Rightarrow}{\gamma}(t)\cdot\vec{r}_{op} + V(\vec{r}_{op},t)\Big]|\Psi(t)\rangle \end{aligned} \qquad (15)$$

where H_0 is an internal atomic Hamiltonian and $V(\vec{r}_{op},t)$ some general interaction Hamiltonian with an external field. Gravito-inertial fields are represented by the tensors $\overset{\Rightarrow}{g}(t)$ and $\overset{\Rightarrow}{\gamma}(t)$ and by the vectors $\vec{\Omega}(t)$ and $\vec{g}(t)$. The same terms can also be used to represent the effect of various external electromagnetic fields. The operators $\vec{L}_{op} = \vec{r}_{op}\times\vec{p}_{op}$ and $\vec{S}_{op}$ are respectively the orbital and spin angular momentum operators. Apart from in $V(\vec{r}_{op},t)$ we have limited the dependence of the Hamiltonian to second-order in the operators $\vec{p}_{op}$ and $\vec{r}_{op}$.

The rotation terms are easily removed with a unitary transformation, which rotates all quantities [3, 15]. The exact propagator of this equation in the absence of $V(\vec{r}_{op},t)$ has been derived by introducing a vector $\vec{\xi}$ such that[3]:

$$\ddot{\vec{\xi}} - \overset{\Rightarrow}{\gamma}(t)\cdot\vec{\xi} - \vec{g} = 0 \qquad (16)$$

and the ABCD matrices of Gaussian optics[3, 15]. As an example, for one space dimension z, the following result (corrected from reference [2]) is obtained for the wave packet at (z,t) :

$$\begin{aligned} &\exp\left[\frac{iM}{\hbar}\dot{\xi}(z-\xi)\right]\exp\left[\frac{iM}{\hbar}\int_{t'}^{t}(\dot{\xi}^2/2+\gamma\xi^2/2-g\xi)dt_1\right] \\ &\int_{-\infty}^{+\infty}dz'\left(\frac{M}{2\pi i\hbar B}\right)^{1/2}\exp\left[(iM/2\hbar B)(D(z-\xi)^2-2(z-\xi)z'+Az'^2)\right] \\ &\exp\left[iMv_0(z'-z_0)/\hbar\right]F(z'-z_0,X_0,Y_0) \\ = \ &\exp\left[\frac{iS(t,t_0)}{\hbar}\right]\exp\left[\frac{iM}{\hbar}v(t)\,(z-z(t))\right]F(z-z(t),X(t),Y(t)) \end{aligned} \qquad (17)$$

[3]For simplicity, we have left out the tensor $\overset{\Rightarrow}{g}(t)$ in what follows. This tensor is useful to represent the effect of gravitational waves. The reader interested in keeping this term is invited to use the more general equation derived in [3].

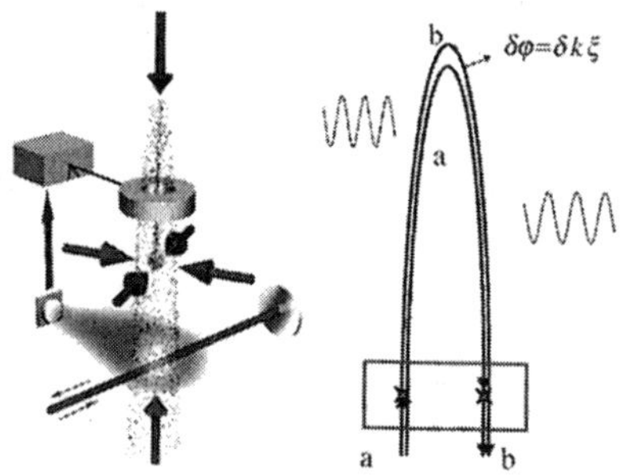

Figure 11. Phase shift in a fountain clock

$$\begin{aligned}\text{where } S(t,t_0) &= M\dot{\xi}(Az_0+Bv_0+\xi)-\frac{M}{2}\int_{t_0}^{t}(\dot{\xi}^2+\gamma\xi^2)dt_1\\ &+\frac{M}{2}(ACz_0^2+DBv_0^2+2BCz_0v_0)\\ &= \frac{M}{\gamma^{3/2}}\left[\frac{g^2}{4}(\sinh 2x-2x)+\frac{\gamma}{4}\left(v_0^2-2gz_0+\gamma z_0^2\right)\sinh 2x\right.\\ &\left.+\sqrt{\gamma}v_0\left(-g+\gamma z_0\right)\sinh^2 x\right] \qquad (18)\end{aligned}$$

is the classical action, with $x=\sqrt{\gamma}(t-t_0)$ and where

$$F\left(z-z_0,X_0,Y_0\right)=\frac{1}{\sqrt{X_0}}\exp\left[\frac{iM}{2\hbar}\frac{Y_0}{X_0}(z-z_0)^2\right] \qquad (19)$$

is a Gaussian (more generally Hermite-Gaussian) wave packet at the initial time t_0 in which the central position z_0, the initial velocity v_0 and the initial complex width parameters X_0, Y_0 in phase space, have to be replaced by their values at time t given by the $ABCD\xi$ transformation law:

$$\begin{aligned} z(t) &= Az_0+Bv_0+\xi \quad ; \quad X(t)=AX_0+BY_0\\ v(t) &= Cz_0+Dv_0+\dot{\xi} \quad ; \quad Y(t)=CX_0+DY_0 \qquad (20)\end{aligned}$$

In the limit where $\gamma\to 0$, $A=D\to 1$, $B\to t-t_0$, $C\to 0$, $\xi\to -(1/2)g(t-t_0)^2$ and $S\to M(t-t_0)\left(v_0^2-2gz_0-2gv_0(t-t_0)+2g^2(t-t_0)^2/3\right)/2$.

Let us apply this result to fountain clocks (Fig.11).

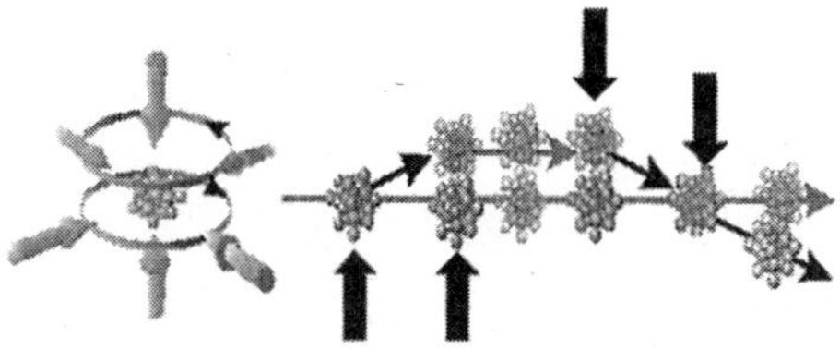

Figure 12. Magneto-optical trap (MOT on the left) and sequence of four laser beam spatial zones or time pulses to generate a closed atom interferometer in space or space-time (on the right).

We have seen that the total phase factor acquired by the atomic wave packet is: $\exp\left[\frac{iS(t,t_0)}{\hbar}\right]\exp\left[ip(t)\left(z-z(t)\right)/\hbar\right]$ [4]. On one arm the additional momentum communicated after the first interaction:

$$\hbar\delta k = \hbar\left(\omega - \omega_{ba}(v) \mp kv_z - \delta\right)/v_x$$

combined with $\xi = -(1/2)gT^2$, in the second phase factor, gives the phase shift responsible for the Ramsey fringes $\delta k\xi$. Note that this phase shift is indeed the same as in the atom gravimeter (see below) and an atom fountain clock is essentially a gravimeter with a recoil momentum communicated longitudinally proportional to the detuning. The phase factor which comes from the action gives the proper combination of gravitational phase shift and second-order Doppler effect (analogy with the Langevin twin paradox):

$$\left(S_b - S_a\right)/\hbar = -\omega_{ba}\left[1 + \frac{1}{6}\frac{v_0^2}{c^2}\right]\left(\frac{2v_0}{g}\right) \tag{21}$$

5. Optical atomic clocks

In the optical domain, more interaction zones are necessary to close the interferometer[16, 17, 18] and cancel the transverse phase shift (see figure 12).

[4]Quite generally, the phase shift along each arm $\left(S\left(t,t_0\right) - \left[\vec{p}\left(t\right).\vec{r}\left(t\right)\right]_{t_0}^{t}\right)/\hbar$ is equal to minus the time integral of the kinetic energy.

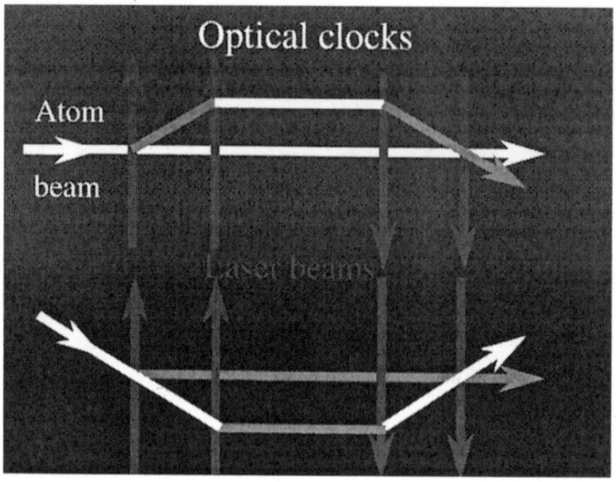

Figure 13. Pair of interferometers generated by two counterpropagating pairs of two copropagating laser beams. In one case the atoms are in their ground state in the central gap while they are in the excited state in the second interferometer central drift zone. The four interaction regions may be separated in space or in the time domain. They may involve single photon or multiphoton (Raman or cascade) transitions.

This closed circuit may have different shapes [18]: parallelogram (case of three or four copropagating laser beams) or trapezoid (two counterpropagating pairs of copropagating laser beams) or more complicated geometries with larger deviations [21, 19, 20]. The choice depends on the type of phase shift that one wishes to measure: symmetric configurations are sensitive only to inertial effects while asymmetric ones depend also on the laser detuning. A variety of transition processes and effective fields may be used: single photon, two-photon transitions of Raman or cascade type [17, 18]. These interactions may be separated in space to generate a spatial interferometer [10, 25] or in time to generate a space-time interferometer, as in recent realizations of optical clocks [9, 22, 23], which use magneto-optical traps of Ca, Sr or Mg. Cold atoms are released from a magneto-optical trap and submitted to the four pulses required to generate the interferometer. An interference pattern such as that displayed in Fig. 14 is obtained. High performances of stability and accuracy have been achieved in these various experiments.

In the case of the trapezoid geometry, there are two different interferometers created by two counterpropagating pairs of copropagating laser beams (Figure 15) with opposite recoil shifts. The two fringe systems are separated by 2δ. This splitting is clearly resolved on Fig. 16 where the velocity distribution in a thermal beam of Magnesium averages out the side fringes. In current realizations of optical clocks the fringe spacing is set precisely

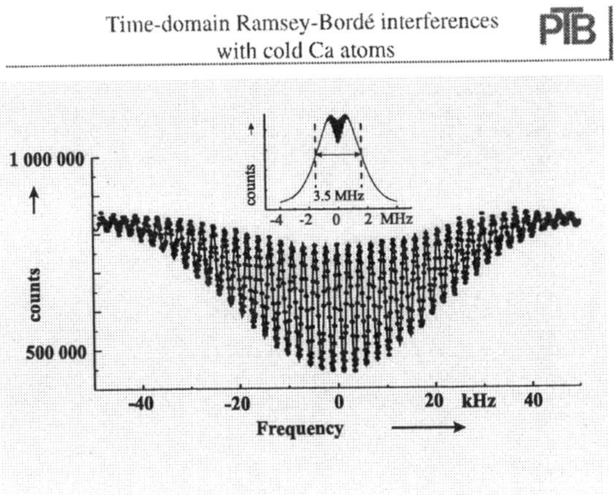

Figure 14. Interference pattern of an optical Calcium clock (*Courtesy of F. Riehle, P.T.B.*)

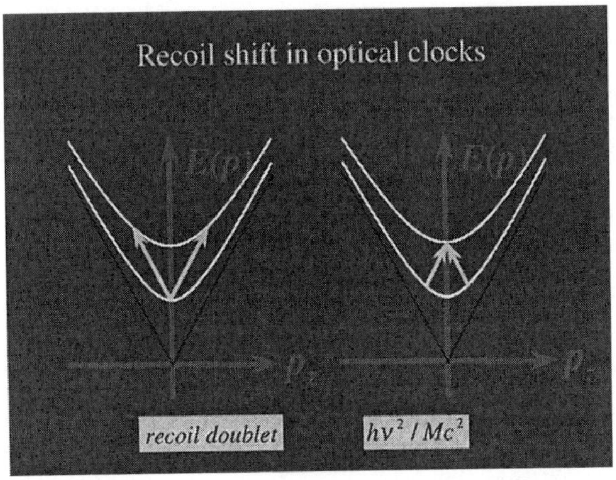

Figure 15. Energy-momentum diagrams illustrating exchanges in the two interferometers obtained with two counterpropagating pairs of two copropagating laser beams. The recoil shifts of central fringes are opposite. The first manifestation of this splitting was observed in the saturation spectrum of the methane molecule [24] and was the first quantitative demonstration of the exchange of momentum $\hbar k$ between light and an atomic species.

equal to this recoil splitting, in order to have two superimposed interference patterns as in Fig. 14.

In other experiments this recoil splitting is used to perform an accurate frequency measurement of $\hbar/m_{atom}$ and consequently of the fine structure

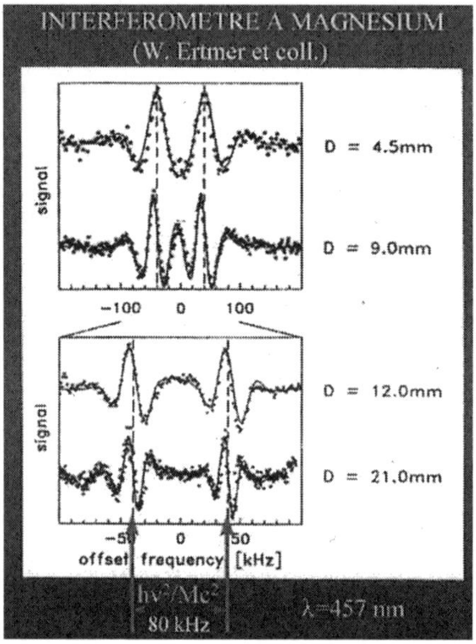

Figure 16. Interference patterns obtained with a Magnesium interferometer by the group of W. Ertmer [9]. Because of the thermal velocity distribution, only the central fringes dominate and one can see clearly the recoil splitting between the two fringe systems, with a resolution (reciprocal fringe width) increasing with the distance between the beam splitters.

constant α thanks to the relation:

$$\alpha^2 = \frac{2R_\infty}{c}\frac{m_p}{m_e}\frac{m_{atom}}{m_p}\frac{h}{m_{atom}}$$

where all quantities (Rydberg constant, ratio of proton mass m_p to the electron mass m_e ..) may be known with an accuracy equal or better than 10^{-9}, which means that α may be determined directly with an accuracy at the 10^{-9} level [20]. Since the recoil shift is proportional to the difference in kinetic energies in both arms of the interferometer, it can be increased very significantly (quadratically) by an accumulation of momentum quanta $\hbar k$ on one arm with respect to the other arm [21, 19, 20]. Recent determinations on Earth by the group of S. Chu have been pushed as far as the 10^{-8} level, which is of great importance as a test of QED, given the discrepancies between various other determinations and independently of any QED calculation.

The theory of optical clocks begins with perturbative and numerical approaches around 1977 [16]. A more sophisticated theory, which is still used to describe experimental results, introduces 2x2 ABCD matrices in the internal spinor space of the two-level system and free propagation between pulses/field zones and was first published in 1982 [26, 27]. In 1990, the ABCDξ formalism for atom wave propagation in gravito-inertial fields has been presented, for the first time, in Les Houches [15]. The strong field S-matrix treatment of the electromagnetic field zones was then published in 1994 [13, 10]. In 1995, the problem of Rabi oscillations in a gravitational field has been treated in analogy/complementarity with the frequency chirp in curved wave-fronts [28]. Finally the dispersive properties of the group velocity of atom waves in strong e.m. fields have been described as a generalization of the dynamical neutron diffraction theory [30] in neutron beam splitters [29, 31, 32]. To-day we combine all these elements in a new sophisticated and realistic quantum description of optical clocks. This effort is also underway for atomic inertial sensors and is essential to develop strategies to eliminate the inertial field sensitivity of optical clocks [22]. In the next paragraphs, we outline the complete general relativistic derivation of phase shifts which was published in 1999 [4].

6. General relativistic framework for atom interferometry:

It is possible to include all possible effects of inertial fields, as well as all the general relativistic effects of gravitation in a consistent and synthetic framework[33, 29, 4], in which the atomic fields are second-quantized. The starting point is the use of coupled field equations for atomic fields of a given spin in curved space-time: e.g. coupled Klein-Gordon, Dirac or Proca equations. Gravitation is described by the metric tensor $g_{\mu\nu}$ and by tetrads, which enter in these equations. Several strategies can then be adopted: one can perform Foldy-Wouthuysen transformations [34], but conceptual difficulties arise in the case of arbitrary $g_{\mu\nu}$; one can go to the weak-field limit $g_{\mu\nu} = \eta_{\mu\nu} + h_{\mu\nu}$ with $|h_{\mu\nu}| \ll 1$ and use renormalized spinors and finally one can consider $h_{\mu\nu}$ as a spin-two tensor-field in flat space-time [36, 37, 38] and use ordinary relativistic quantum field theory. Using this last approach, it has been possible to derive field equations that display all interesting terms, coupling Dirac atomic fields, gravitational and electromagnetic fields and simple expressions of the corresponding relativistic phase shifts in atom interferometers [4].

The evolution equation of the state vector $|\Psi(t)\rangle$ in the interaction picture is

$$i\hbar\frac{d}{dt}\,|\Psi(t)\rangle = \int d^3x\ \theta^{\dagger}(x)\mathcal{V}_G(x)\theta(x)\ |\Psi(t)\rangle\,, \tag{22}$$

where the operator $\mathcal{V}_G(x)$, acting on the field operator $\theta(x)$, is given in compact form by:

$$\mathcal{V}_G = \frac{c}{4}\alpha^\mu h_{\mu\nu} p^\nu + h.c. = \frac{c}{4}\left\{\alpha^\mu h_{\mu\nu}, p^\nu\right\}_+ \tag{23}$$

$$\text{with } p^0 = -\alpha^j p_j + \gamma^0 mc \text{ and } p_j = i\hbar\partial_j \tag{24}$$

The free field operator θ is written as:
$\theta(x) = \sum_{r=1}^{2}\int(d^3p)\left[c_r(\overrightarrow{p})\chi^{(+)}_{\overrightarrow{p},r}(x) + d_r^\dagger(\overrightarrow{p})\chi^{(-)}_{\overrightarrow{p},r}(x)\right]$, where $c_r(\overrightarrow{p})$ and $d_r(\overrightarrow{p})$ are the annihilation operators for the particles or antiparticles, respectively, and $\chi^{(\pm)}_{\overrightarrow{p},r}$ are the positive or negative energy solutions of the free Dirac equation:

$$\chi^{(\pm)}_{\overrightarrow{p},r}(x) = \frac{1}{(2\pi\hbar)^{3/2}}\sqrt{\frac{Mc^2}{E(\overrightarrow{p})}}\, u^{(r)}_{(\pm)}(\overrightarrow{p}) e^{\mp i(E(\overrightarrow{p})t - \overrightarrow{p}.\overrightarrow{r})/\hbar} \tag{25}$$

We are interested in the output spinor corresponding to one-particle (antiparticle) states: e.g. $\psi(x) = \langle 0|\,\theta(x)\,|\Psi(t)\rangle$ for atoms. The evolution of this spinor is governed by the equation:

$$i\hbar\partial_t\psi = -i\hbar c\gamma^0\gamma^j\partial_j\psi + Mc^2\gamma^0\psi + \mathcal{V}_G(x)\psi \tag{26}$$

to which we may add terms corresponding to diagonal magnetic dipole and off-diagonal electric dipole interactions [33, 29]. This equation has been used in reference [29, 4] to discuss all the terms that lead to a phase shift in an interferometer.

For the phase shift, the general result is:

$$\begin{aligned}
\delta\varphi = & -\frac{1}{\hbar}\int_{t_0}^{t} dt' \left\{ \frac{c^2}{2E(\vec{p})} p^\mu h_{\mu\nu}(\vec{x}_0 + \vec{v}t', t') p^\nu \right. \\
& + \frac{\gamma}{m(\gamma+1)}\left[\frac{c^2 p^\mu \vec{\nabla} h_{\mu\nu}(\vec{x}_0 + \vec{v}t', t') p^\nu}{2E^2(\vec{p})} \times \vec{p}\right]\cdot\vec{s} \\
& \left. - \frac{c}{2}\left[\vec{\nabla}\times\left(\vec{h}(\vec{x}_0 + \vec{v}t', t') - \overrightarrow{\overrightarrow{h}}(\vec{x}_0 + \vec{v}t', t')\cdot\frac{\vec{p}c}{E(\vec{p})}\right)\right]\cdot\vec{s}\right\}
\end{aligned} \tag{27}$$

where $\vec{s}$ is the mean spin vector

$$\vec{s} = \sum_{r,r'}\beta^*_{r,i}\beta_{r',i}\hbar w^{(r)\dagger}\vec{a}w^{(r')}/2\gamma \tag{28}$$

where $\vec{a} = (\vec{\sigma}_\perp + \gamma\vec{\sigma}_\parallel)$ is the spatial part of the Thomas-Pauli-Lubanski 4-vector operator [39].

Expression (27) displays all the terms which may lead to a gravitational phase shift in a matter-wave interferometer:

- the terms involving h_{00} lead to the gravitational shift ($h_{00} = -2 \vec{g} \cdot \vec{r}/c^2$), to shifts involving higher derivatives of the gravitational potential and to the analog of the Thomas precession (spin-orbit coupling corrected by the Thomas factor).
- the terms which involve $\vec{h} = \{h^{0k}\}$, give the Sagnac effect in a rotating frame ($\vec{h} = \vec{\Omega} \times \vec{\mathbf{r}}/c$), the spin-rotation coupling and a relativistic correction (analogous to the Thomas term for h_{00}). They describe also the Lense-Thirring effects coming from inertial frame-dragging by a massive rotating body, which is a source for $\vec{h}$.
- the other terms, which involve the tensor $\overset{\Rightarrow}{h} = \{h^{ij}\}$ describe genuine General Relativity effects such as the effect of gravitational waves and de Sitter geodetic precession (which also includes the Thomas term for h_{00}[5]).

In fact the phase calculation is usually more involved since formula (27) applies only to the case of straight unperturbed trajectories. In practice, however, one cannot ignore the fact that, when calculating the phase to first-order for a given term of the Hamiltonian, the motion of the particles is affected by other terms. One example, mentioned above, is the calculation of the gravitational shift within the atom beam splitters, in which one cannot ignore the important effects of the diffracting electromagnetic field on the trajectories of the particles [29, 31, 28, 32]. Gravitational phase shifts have to be calculated along these trajectories. Another example is the gravity field itself, which, on earth, gives parabolic trajectories for atoms. The phase shift for the other terms in equation (27) has to be calculated along these parabolas. A convenient way to achieve these calculations is to replace $\vec{x}_0 + \vec{v}t'$ and $\vec{v}$ in equation (27) by the classical trajectory $\{\vec{x}(t'), \vec{v}(t')\}$ obtained in the *ABCD* formalism. In the non-relativistic limit, one is brought back to the Schroedinger equation and, up to second degree in position and momentum operators, the best approach is to take the full benefit of the *ABCD* formalism developed above, which gives exact results. Higher-order terms can be treated as perturbations along unperturbed trajectories.

The reader will find calculations of the phases corresponding to the various terms in references [18, 17, 29, 40, 41]. In these calculations, one should never forget that the external field $h_{\mu\nu}$ acts not only on the atoms but also on other components of the experiments, such as mirrors and laser beams and that, depending on the chosen gauge, additional contributions may enter in the final expression of the phase which should, of course, be gauge independent. As an example, the Sagnac phase which can be removed

[5]Some authors reserve the name "Thomas precession" for the contribution coming specifically from an acceleration $\vec{a}$ (which has been included here in h_{00}) and separate it from de Sitter precession.

from the above formula by a simple coordinate transformation will reappear in the beam splitter phases.

The expressions valid for spins 0 and 1/2 may be conjectured to be valid for arbitrary spin if $\vec{\sigma}/2$ is replaced by the corresponding spin operator $\vec{S}$. The extension of the formulas is presently underway using higher spin formalisms.

Formula (27) also displays the analogy with electromagnetic interactions: $h_{\mu\nu}p^{\nu}/2$ plays the role of the 4-potential A_{μ} and $(E(\vec{p})/2c)\vec{\nabla} \times \left(\vec{h} - \overset{\Rightarrow}{h} \cdot \vec{p}c/E(\vec{p})\right)$ plays the role of a gravitomagnetic field $\vec{\nabla} \times \vec{A}$. This new correspondence between the gravitational interaction and the electromagnetic interaction generalizes the so-called gravitoelectric and gravitomagnetic interactions introduced by de Witt [42] and Papini [43].

The spin-independent part of this phase shift (Linet-Tourrenc [44] term) $(c^2/2E(\vec{p}))\, p_{\mu}h^{\mu\nu}p_{\nu}$, (combination which appears also in the generalized Thomas precession) corresponds to $u_{\mu}A^{\mu}/\gamma$ where u_{μ} is the 4-velocity p_{μ}/M and the corresponding circulation of potential takes the form of the Aharonov-Bohm phase formula $\oint A^{\mu}dx_{\mu}$. Using Stokes theorem in four dimensions, this integral gives the phase shift as the ratio of the flux of gravitoelectromagnetic forces through the interferometer space or spacetime area to a quantum of flux of force $\hbar$ or $\hbar c$:

$$\begin{aligned} -\hbar\delta\varphi &= \oint \frac{c^2}{2E(\vec{p})} p^{\mu}h_{\mu\nu}p^{\nu}dt = \oint \frac{1}{2}p^{\mu}h_{\mu\nu}dx^{\nu} \\ &= \frac{1}{2}\iint d\sigma^{\mu\nu}\left(\partial_{\mu}A_{\nu} - \partial_{\nu}A_{\mu}\right) \text{ with } A_{\nu} = \frac{1}{2}p^{\mu}h_{\mu\nu} \end{aligned} \tag{29}$$

in analogy with electromagnetism. This formula gives:

$$\delta\varphi = -\vec{k}.\vec{g}\, T^2 \tag{30}$$

for the gravitational phase shift[17, 18] as the flux of a gravitoelectric field $-c^2\vec{\nabla}h_{00}/2 = \vec{g}$ through a space-time area (which is the case above for the fountain clock), whereas the Sagnac phase shift[6] is the flux of a gravitomagnetic field $c^2\vec{\nabla}\times\vec{h} = 2c\vec{\Omega}$ through an area $\vec{A}$ in space [17, 18], which atomic clocks usually do not have:

$$\delta\varphi = \frac{2c\vec{\Omega}.\vec{A}}{\hbar c/M} \tag{31}$$

[6]This Sagnac phase shift (in units of 2π) can be written as the projection of the orbital angular momentum (in units of $\hbar$) of the interfering particles. An example of nuclear Sagnac interferometer is provided by rotating molecules, for which this phase shift is naturally quantized.

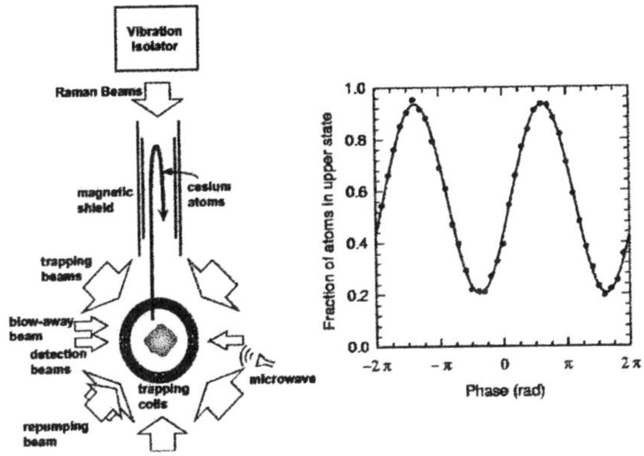

Figure 17. Atom-wave gravimeter developed at Stanford University by S. Chu and coworkers [47]. 5.10^8 Cesium atoms extracted from a low pressure background vapor are loaded in a MOT in 600 ms and launched upwards. A sequence of microwave, velocity sensitive Raman and state selective blow away pulses, places 3.10^6 atoms in the 6 $S_{1/2}$ F=3 $M_F = 0$ state with an effective vertical temperature of $\sim$ 10 nK. The atomic cloud enters a magnetically shielded region and is illuminated by the sequence $\pi/2 - \pi - \pi/2$ of interferometer Raman pulses which enter from below. The retromirror is vibration isolated with an actively stabilized system. The right picture displays typical fringes for T=160 ms. Each point corresponds to a single launch of the atoms separated by 1.3 s and taken over one minute.

The spin-rotation term is discussed in more detail in references [29, 4]. Its effect on atomic clocks needs to be carefully evaluated, since, unlike the magnetic field, the gravitomagnetic field cannot be shielded.

7. State-of-the-art for gravito-inertial sensors

To emphasize the sensitivity of atom interferometers to inertial and gravitational fields, a short overview of realizations of gravito-inertial sensors, including gravimeters, gradiometers and atomic gyros is proposed to the reader in this last section. A first very successful application of atom interferometry is gravimetry and was developed by S. Chu and his collaborators (Fig. 17). This is an extension of the celebrated COW experiments for neutrons [45] to the atom world. In 1991, in one of the early experiments of atom interferometry [46], this group demonstrated a resolution of 3.10^{-8} for g in 40 minutes integration time. In recent earth gravity measurements, the relative sensitivity is $\delta g/g \simeq 3.10^{-9}$ after 60 seconds and the absolute accuracy 5.10^{-9}[20, 47]. This resolution is sufficient to see clearly the effect of ocean loading on the earth tides. Also, the agreement with a conventional

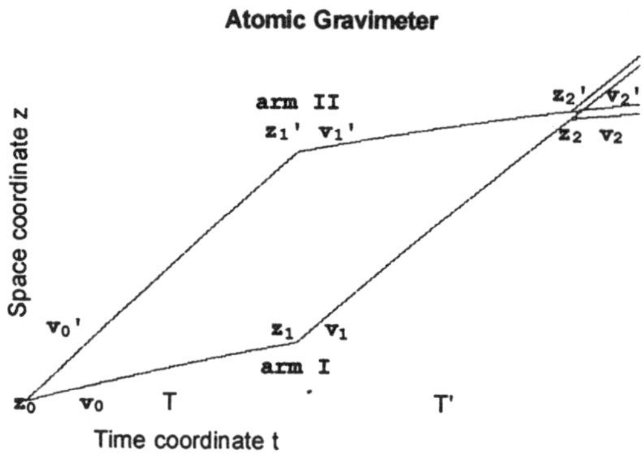

Figure 18. Space-time diagram of the atomic gravimeter.

corner cube gravimeter (FG-5) is at the $7.10^{-9}g$ level, which constitutes by itself, a test of the equivalence principle between an atomic species and a macroscopic object at that level. The corresponding space-time diagram is given in Fig.18. From which the following phase shift is calculated with the ABCDξ formalism [3]:

$$\begin{aligned} \delta\varphi &= -k((z_2 + z_2')/2 - z_1 - z_1' + z_0) \\ &= -\frac{k}{\sqrt{\gamma}}\left[\left[\sinh\left(\sqrt{\gamma}(T+T')\right) - 2\sinh\left(\sqrt{\gamma}T\right)\right](v_0 + \frac{\hbar k}{2M}) \right. \\ &\quad \left. +\sqrt{\gamma}\left[1 + \cosh\left(\sqrt{\gamma}(T+T')\right) - 2\cosh\left(\sqrt{\gamma}T\right)\right](z_0 - \frac{g}{\gamma})\right] \end{aligned} \quad (32)$$

which, to first-order in γ and for $T = T'$ reduces to:

$$kgT^2 + k\gamma T^2\left[\frac{7}{12}gT^2 - (v_0 + \frac{\hbar k}{2M})T - z_0\right] \quad (33)$$

where the first term, which is precisely expression (30), gives huge phase shifts for matter-waves compared to what could be obtained with light rays, since the times T can be of the order of one second, and gT^2 being of the order of ten meters is compared to an optical wavelength. The next term is a significant correction due to the gravitational field gradient γ. One can also measure directly these field gradients, with two gravimeters using two clouds of cold atoms and sharing the same vertical laser beam splitters. It is then no more necessary to have a very sophisticated inertial platform for the reference mirror and it is possible to measure directly

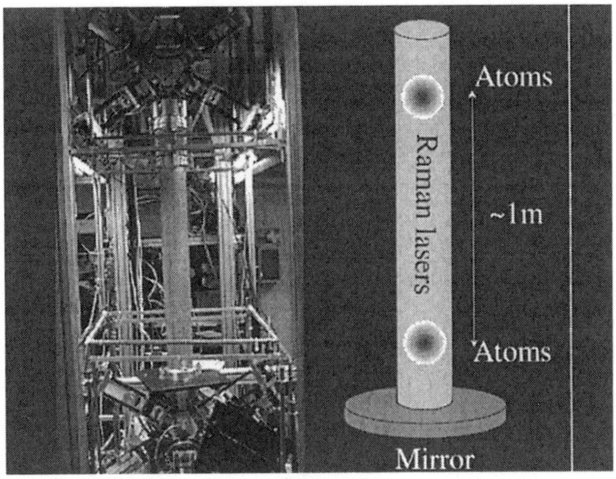

Figure 19. Atom wave gradiometer of Yale University. The two clouds of atoms share the same Raman beams, which generate two atom interferometers separated vertically by one meter.

the differential acceleration between these two clouds. This is the principle, illustrated in Fig.19, of the gradiometers developed first in Stanford then in Yale by M. Kasevich and his coworkers [20, 49]. The present sensitivity is $4.10^{-9}s^{-2}/\sqrt{Hz}$ or 4 E/$\sqrt{Hz}$ and the accuracy 1E/$\sqrt{Hz}$ for an extrapolated 10 m separation between accelerometers. For the future, one may think of using this principle for gravitational wave detection in space [48].

The first demonstration of a gyro using atom waves was performed at the P.T.B. in Braunschweig in a joint collaboration with the author [50]. The photograph of this experiment is given in Fig. 20. The Sagnac phase shift formula (31) given above reveals the considerable gain in sensitivity brought by matter-waves, since, in this formula, the rest mass energy Mc^2 has to be replaced by the photon energy $h\nu$ in the case of light waves. This expression can be derived by a number of equivalent approaches and the simplest one is to use the rotation operator in Schroedinger equation, which rotates the wave vectors of the beam splitters. This formula applies to the trapezoid geometry used in the previous experiment and also to the parallelogram geometry, as suggested in [18], analog of the Mach-Zehnder optical interferometer, which has been used in more recent experiments and which has the advantage of being insensitive to laser detuning. For rotations, the best sensitivity achieved up to now is 6.10^{-10}rad.s^{-1}Hz$^{-1/2}$ [51] with the set-up of Fig. 21. Clearly, if the atomic motion is reversed h_{00} is unchanged, while $\vec{h}.\vec{p}$ is reversed. This property is used to separate rotations and accelerations through the use of counterpropagating atomic

Figure 20. Photograph of the first atom wave gyro [50]. The calcium atomic beam originating from the oven on the left crosses two counterpropagating pairs of copropagating laser beams generated be the pair of cat's eyes facing each other on the right platform. These four beams act as beam splitters, deflectors and recombiner for the atom waves and the excited state output of the interferometer is monitored via the fluorescence light by the photomultiplier on the extreme right of the apparatus. The whole set-up is mounted on a rotating table.

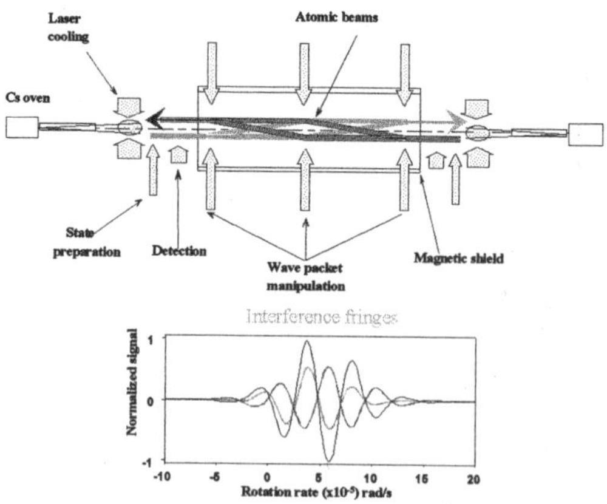

Figure 21. Atomic gyro developed at Yale in the group of Mark Kasevich. The Sagnac phases corresponding to the two opposite thermal atom beams have opposite signs and are substracted. The Earth rotation is responsible for the offset.

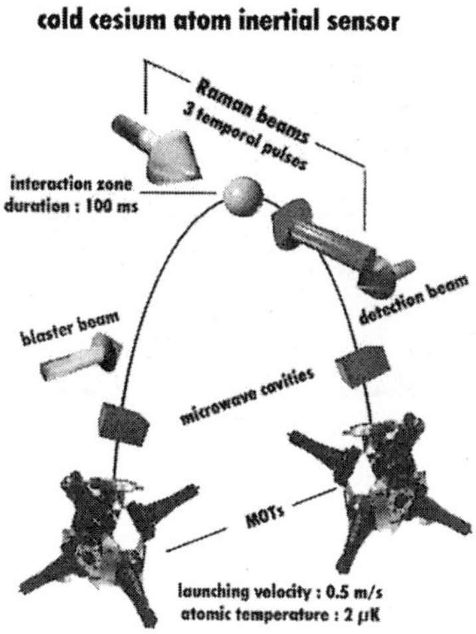

Figure 22. Principle of the cold Cesium gyro-accelerometer developed at the Observatory in Paris in collaboration between D. Holleville, J. Fils, A. Landragin, N. Dimarcq , A. Clairon (LHA and BNM-LPTF), Ph. Bouyer (IOTA), Ch. Salomon (LKB) and Ch. Bordé (LPL-ERGA).

beams. A compact device (30 cm height) using cold atoms and the Cesium clock technology (PHARAO) is under development in Paris, in a joint collaboration between several laboratories [52], and is shown in Figs.22 and 23. Here again, cold atoms are launched along counterpropagating parabolic trajectories, in order to separate the various components of the acceleration and of the rotation fields.

The sensitivity numbers quoted here are expected to improve rapidly in the near future, especially in space experiments, in which general relativistic effects should become detectable. The space project called HYPER [54] aims precisely at the detection of such effects thanks to the possibility to have long drift times in space (Fig.24). This will increase considerably the sensivity of these devices. The technology of trapping and manipulating cold atoms developed for the project ACES (Atomic Clock Ensemble in Space) will be directly applicable to inertial sensors for many applications in deep space navigation of space probes. Among the goals of HYPER, there is a very accurate measurement of the fine structure constant α, a test of the equivalence principle at the atomic level, using two different atomic species in the interferometer, a detection of the periodic signal coming from the latitudinal dependence of the Lense-Thirring effect in polar orbit, decoherence studies ....In fact, atom interferometers are so sensitive

Figure 23. Cold atom gyro-accelerometer under construction at the Observatory in Paris.

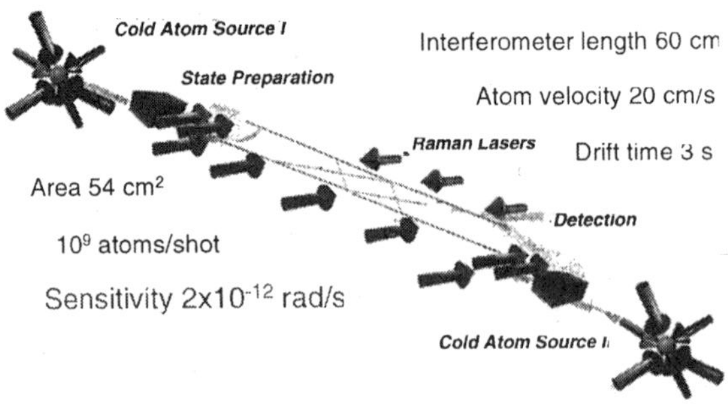

Figure 24. Schematic picture of an atom interferometer in space providing long drift times and a large interferometer area.

to gravito-inertial fields and to their gradients that, it will be necessary during the next years to develop many new techniques and tricks in the field of atom optics (e.g. phase conjugation of atom waves ...) to isolate the specific signature of investigated phenomena.

An accurate measurement of the effect of gravitation and inertia on antimatter also appears as a possibility already discussed in reference [53] with a transmission-grating interferometer, although we believe, for obvious reasons, that an antiatom interferometer using laser beams for the antihydrogen beam splitters (so-called Ramsey-Bordé interferometers) would be better suited for such an experiment. coherent beams of antihydrogen will be produced either by Bose-Einstein condensation and/or by stimulated bosonic amplification[7] [8].

References

1. *Atom Interferometry*, ed. P. Berman, Academic Press (1997).
2. Ch.J. Bordé, *Quantum theory of clocks and of gravitational sensors using atom interferometry, in:* Laser Spectroscopy, Eds. R. Blatt, J. Eschner, D. Leibfried and F. Schmidt-Kaler, World Scientific (1999) pp. 160-169.
3. Ch.J. Bordé, *Theoretical tools for atom optics and interferometry*, C.R. Acad. Sci. Paris, t.2, Série IV, 509-530 (2001).
4. Ch.J. Bordé, J.-C. Houard and A. Karasiewicz, *Relativistic phase shifts for Dirac particles interacting with weak gravitational fields in matter-wave interferometers*, in: Gyros, Clocks and Interferometers: Testing Relativistic Gravity in Space, Eds C. Lämmerzahl, C.W.F. Everitt and F.W. Hehl, Springer-Verlag, (2001) pp. 403-438 and gr-qc/0008033.
5. P. Lemonde et al., Cold atom clocks on earth and in space, in A.N. Luiten (Ed.), Frequency measurement and control, Topics Appl. Phys. **79**, 131-152 (2001). Springer-Verlag.
6. I. Bloch, T.W. Hänsch and T. Esslinger, *Measurement of the spatial coherence of a trapped Bose gas at the phase transition*, Nature **403**,166-170 (2000) and references therein.
7. I. Bloch, T.W. Hänsch and T. Esslinger, *Atom Laser with a cw Output Coupler*, Phys. Rev. Lett. **82**, 3008-3001 (1999).
8. Ch.J. Bordé, *Amplification of atomic fields by stimulated emission of atoms*, Physics Letters **A204**, 217-222 (1995); Ch.J. Bordé, *Amplification of atomic waves by stimulated emission of atoms*, in Laser Spectroscopy, edited by M. Inguscio, M. Allegrini and A. Sasso, World Scientific, 303-307 (1996); Ch.J. Bordé, *Amplification de champs atomiques par émission stimulée d'atomes*, Annales de Physique, **20**, 477-485 (1995).
9. U. Sterr, K. Sengstock, W. Ertmer, F. Riehle and J. Helmcke, *Atom interferometry based on separated light fields*, in [1].
10. Ch.J. Bordé, N. Courtier, F. du Burck, A.N. Goncharov and M. Gorlicki, *Molecular interferometry experiments*, Phys. Lett. A **188**, 187-197 (1994).
11. U. Sterr, K. Sengstock, J. H. Müller, D. Bettermann, W. Ertmer, *The magnesium Ramsey interferometer: applications and prospects*, Appl. Phys. B **54**, 341 (1992).
12. Ch.J. Bordé, J.L. Hall, C.V. Kunasz and D.G. Hummer, *Saturated absorption line shape: calculation of the transit-time broadening by a perturbation approach*, Phys. Rev. **14**, 236-263 (1976).
13. J. Ishikawa, F. Riehle, J. Helmcke and Ch. J. Bordé, *Strong-field effects in coherent saturation spectroscopy of atomic beams*, Phys. Rev. **A49**, 4794-4825 (1994).
14. P. Wolf et al., *Recoil effects in microwave atomic frequency standards*, Proceedings of the 2001 IEEE Frequency Control Symposium, Seattle, USA, (2001), in press.
15. Ch. J. Bordé, *Propagation of Laser beams and of atomic systems*, Les Houches Lectures, Session LIII, 1990, *Fundamental Systems in Quantum Optics*, J. Dalibard, J.-M. Raimond and J. Zinn-Justin eds, Elsevier Science Publishers (1991) p.287-380.

[7]using for example a reaction like: antiproton +positronium→antihydrogen+electron

16. Ch.J. Bordé, *Sur les franges de Ramsey en spectroscopie sans élargissement Doppler*, C. R. Acad. Sc. Paris, **284B**, 101-104 (1977), *Progress in understanding sub-Doppler line shapes*, in *Laser Spectroscopy III*, Springer-Verlag (1977).
17. Ch.J. Bordé, *Atomic interferometry with internal state labelling*, Phys. Lett., **A140**, 10-12 (1989).
18. Ch.J. Bordé, *Atomic interferometry and laser spectroscopy*, in: Laser Spectroscopy X, World Scientific (1991) pp. 239-245.
19. D.S.Weiss, B.C. Young and S. Chu, *Precision measurement of $\hbar/m_{Cs}$ based on photon recoil using laser-cooled atoms and atomic interferometry*, Applied Physics **B 59**, 217-256 (1994).
20. B.C. Young, M. Kasevich and S. Chu, *Precision atom interferometry with light pulses*, in [1] and references therein.
21. Ch.J. Bordé, M. Weitz and T.W. Hänsch, *New optical interferometers for precise measurements of recoil shifts. Application to atomic hydrogen*, in Laser Spectroscopy, L. Bloomfield, T. Gallagher and D. Larson eds, American Institute of Physics (1994) pp. 76-78.
22. T. Trebst, T. Binnewies, J. Helmcke and F. Riehle, *Suppression of spurious phase shifts in an optical frequency standard*, IEEE Trans. on Instr. and Meas. **50**, 535-538 (2001).
23. Th. Udem et al., *Absolute frequency measurements of the Hg^+ and Ca optical clock transitions with a femtosecond laser*, Phys. Rev. Lett. **86**, 4996 (2001).
24. J.L. Hall, Ch. J. Bordé and K. Uehara, *Direct optical resolution of the recoil effect using saturated absorption spectroscopy*, Phys. Rev. Lett. **37**, 1339 (1976).
25. Ch. J. Bordé, S. Avrillier, A. van Lerberghe, Ch. Salomon, Ch. Bréant, D. Bassi, and G. Scoles, *Observation of optical Ramsey fringes in the 10 μm spectral region using a supersonic beam of SF_6*, Appl. Phys. **B28**, 82 (1982) and Journal de Physique Colloques, **42**, C8-15-C8-19 (1981).
26. Ch.J. Bordé, *Développements récents en spectroscopie infrarouge à ultra-haute résolution*, Revue du Cethedec - Ondes et Signal, **NS83-1**, 1-118 (1983); Erratum n° 76, 191 (1983).
27. Ch. J. Bordé, Ch. Salomon, S. Avrillier, A. van Lerberghe, Ch. Bréant, D. Bassi, and G. Scoles, *Optical Ramsey fringes with traveling waves*, Phys. Rev. **A30**, 1836-1848 (1984).
28. C. Lämmerzahl and Ch.J. Bordé, *Rabi oscillations in gravitational fields: exact solution*, Physics Letters **A203**, 59-67 (1995).
29. Ch.J. Bordé, *Matter-wave interferometers: a synthetic approach*, in [1].
30. H. Rauch and D. Petrascheck, *Dynamical Neutron Diffraction and its Application*, in Neutron Diffraction, ed. by H. Dachs (Springer, Berlin, 1978) pp. 303-351 and references therein.
31. Ch.J. Bordé and C. Lämmerzahl, *Atom beam interferometry as two-level particle scattering by a periodic potential*, Ann. Physik (Leipzig) **8**, 83-110 (1999).
32. C. Lämmerzahl and Ch.J. Bordé, *Atom interferometry in gravitational fields: influence of gravitation on the beam splitter*, General Relativity and Gravitation, **31**, 635 (1999).
33. Ch.J. Bordé, A. Karasiewicz and Ph. Tourrenc, *General relativistic framework for atomic interferometry*, Int. J. of Mod. Phys.**D 3**, 157-161 (1994).
34. F.W. Hehl and Wei-Tou Ni, *Inertial effects of a Dirac particle*, Phys Rev. **D 42**, 2045-2048 (1990).
35. K.-P. Marzlin and J. Audretsch, Phys.Rev. A**53**, 1004 (1996).
36. S.N. Gupta, *Quantization of Einstein's gravitational field: linear approximation*, Proc. Phys. Soc. **A 65**, 161-169 (1952) and *Quantization of Einstein's gravitational field: general treatment*, Proc. Phys. Soc. **A 65**, 608-619 (1952).
37. R.P. Feynman, F.B. Morinigo and W.G. Wagner, *Feynman Lectures on Gravitation*, edited by B. Hatfield (Addison-Wesley, Reading MA 1995).
38. B.M. Barker, S.N. Gupta and R.D. Haracz, *One-graviton exchange interaction of*

elementary particles, Phys. Rev. **149**, 1027 (1966) and references therein.

39. L.H. Thomas, *The Kinematics of an Electron with an Axis*, Phil. Mag. **3**, 1-22 (1927).
40. J. Audretsch and K.-P. Marzlin, *Atom interferometry with arbitrary laser configurations: exact phase shift for potentials including inertia and gravitation*, J. Phys. II France **4**, 2073-2087 (1994).
41. J. Audretsch and K.-P. Marzlin, *Ramsey fringes in atomic interferometry: measurability of the influence of space-time curvature*, Phys. Rev. **A 50**, 2080 (1994).
42. B.S. DeWitt, *Superconductors and gravitational drag*, Phys. Rev. Lett. **16**, 1092 (1966).
43. G. Papini, *Particle wave functions in weak gravitational fields*, Nuovo Cimento **52B**, 136-140 (1967).
44. B. Linet et P. Tourrenc, *Changement de phase dans un champ de gravitation: possibilité de détection interférentielle*, Can. J. Phys. **54**, 1129-1133 (1976).
45. R. Collela, A.W. Overhauser and S.A.Werner, *Observation of Gravitationally Induced Quantum Interference*, Phys. Rev. Lett. **34**, 1472 (1975).
46. M. Kasevich and S. Chu, *Atomic interferometry using stimulated Raman transitions*, Phys. Rev. Lett. **67**, 181-184 (1991).
47. A. Peters, K. Y. Chung and S. Chu, *A measurement of gravitational acceleration by dropping atoms*, Nature **400**, 849 (1999) and *High precision gravity measurements using atom interferometry*, Metrologia **38**, 25-61 (2001).
48. For an early treatment of a gravitational wave detector using an atom interferometric gradiometer see: Ch.J. Bordé, J. Sharma, Ph. Tourrenc and Th. Damour, *Theoretical approaches to laser spectroscopy in the presence of gravitational fields*, J. Physique Lettres **44**, L-983-990 (1983).
49. M.J. Snadden, J.M. McGuirk, P. Bouyer, K.G. Haritos and M.A. Kasevich, *Measurement of the Earth's gravity gradient with an atom interferometer-based gravity gradiometer*, Phys. Rev. Lett. **81**, 971-974 (1998).
50. F. Riehle, Th. Kisters, A. Witte, J. Helmcke and Ch.J. Bordé, *Optical Ramsey Spectroscopy in a Rotating Frame: Sagnac Effect in a Matter-Wave Interferometer*, Phys. Rev. Lett. **67**, 177-180 (1991).
51. A. Landragin, T. L. Gustavson and M. A. Kasevich, *Precision atomic gyroscope*, in Laser Spectroscopy, Proceedings of the 14th International Conference on Laser Spectroscopy, Eds. R. Blatt, J. Eschner, D. Leibfried and F. Schmidt-Kaler, World Scientific, Singapore (1999) pp. 170-176.
52. D. Holleville, J. Fils, P. Petit, N. Dimarcq, A. Clairon, P. Bouyer, Ch. Bordé et Ch. Salomon, *Réalisation d'un gyromètre à atomes froids*, J. Phys IV (France) **10**, Pr8-171 (2000).
53. T.J. Phillips, *Measuring the gravitational acceleration of antimatter with an antihydrogen interferometer*, Hyperfine Interactions **100**, 163-172 (1996).
54. R. Bingham et al., HYPER, *Hyper-Precision Cold Atom Interferometry in Space*, Assessment Study Report, ESA-SCI (2000) 10.

CANONICAL GRAVITY AND MACH'S PRINCIPLE: KINEMATIC AND DYNAMIC SOLUTIONS OF THE SPACE PROBLEM

HORST-HEINO VON BORZESZKOWSKI
Technical University Berlin, Institute for Theoretical Physics
Hardenbergstrasse 36, D-10623 Berlin, Germany
e-mail: borzeszk@itp.physik.tu-berlin.de

HANS-JÜRGEN TREDER
Rosa-Luxemburg-Str. 17a
D-14482 Potsdam, Germany

Abstract

From the viewpoint of the matter and space problems, we compare different theories of gravitation to the purely affine theory.

1. Introduction

There are arguments in favor of the program to consider "generalized theories of general relativity" working in geometric spaces that are different from the Riemannian one. Two of these reasons are as follows:
(i) There are certain conceptual problems of the theory of general relativity (GRT) that seem demand its generalization. First of all, there exists the problem of the non-localizability of gravitational energy. As a consequence, in GRT one arrives at a "degenerated" canonical formalism leading to problems with quantum gravity.
(ii) For Einstein's general-relativistic field equations of GRT are covariant with respect to the inhomogeneous and the homogeneous parts of the Poincaré group they are not compatible either with Mach's principle of the "once given world system of reference" or the Mach-Einstein doctrine of complete induction of inertia by the cosmic masses. This is obvious from Einstein's "hole argument" [18] and Weyl's inverse hole argument of the "predominance of ether" [66] in GRT.

But, in generalizing GRT, one must not forget about the fact that the space-time manifold, primarily or secondarily, has to be endowed with a Lorentz metric. This is even true for purely affine theories as the Einstein-Schrödinger theory [25, 60], where the geometry is specified by a connection as basic variable, while the metric is only of secondary importance. In the final analysis, a metric structure is always needed in order to arrive at a theory that is physically interpretable. For, only this structure allows one to introduce the notions of spatial distance, time interval, angle, and relativistic velocity.

P.G. Bergmann and V. de Sabbata (eds.),
Advances in the Interplay Between Quantum and Gravity Physics, 57–76.

As a consequence of the principle of equivalence, however, the metric has to be related to the gravitational field. Therefore, in contrast to the special-relativistic approach, it cannot be introduced *a priori*, but has to be specified in accordance with the other unrenouncable geometric structure, the connection. In other words, a theory satisfying the principle of equivalence has to be established in such away that it solves the so-called space problem (Weyl [67], Cartan [17]; cf. also [30]). First of all, this means to found a gravitational dynamics that does not clash with the "kinematic" conditions ensuring the compatibility between metric and connection.

As far as the kinematic part of the Weyl-Cartan space problem is concerned, it was shown by Schrödinger [60] that the relationship

$$\nabla_l g_{ik} = 0 \tag{1}$$

between the metric g_{ik} and the connection Γ^i_{kl} defined by ∇_l is a sufficient condition for their compatibility. For physical reasons, Einstein [26] considered (1) even as a necessary condition.

In the present lecture, we compare metric theories to bi-metric theories, theories with teleparallelism, metric-affine (Einstein-Cartan) theories, and purely affine theories of gravitation in order to discuss the implications arising from the equivalence principle, first, for the energy problem and the related question as to a canonical theory of gravitation and, second, for the space problem and the related question of the realization of Mach's principle. (The lecture is based on the two papers [12, 13]; for more details on the space problem, cf. in particular [12] .)

2. Metric, Bi-Metric, and Metric-Affine Theories

2.1 METRIC THEORIES

In metric theories, the space problem is solved *a priori* by assuming a Riemannian space-time, i.e., by assuming a four-dimensional manifold with a Lorentz metric, where the connection given by the metric and their first derivatives satisfies identically. In this case, the field equations for the metric cannot conflict with Eq. (1). Then the free gravitational Lagrange density is formed from g_{ik} and $\partial_l g_{ik}$

$$\boldsymbol{L}_f = \boldsymbol{L}_f(g_{ik}, \partial_l g_{ik}, \partial_l \partial_m g_{ik}), \tag{2}$$

while the matter Lagrange density depends on both the gravitational and matter variables, g_{ik} and $\phi^{(A)}$. Imposing, as an implication of the principle of equivalence, the principle of minimal coupling one generally has

$$\boldsymbol{L}_m = \boldsymbol{L}_m(g_{ik}, \partial_l g_{ik}, \phi^{(A)}, \partial_l \phi^{(A)}). \tag{3}$$

The Euler variation of the action integral

$$I = \int (\boldsymbol{L}_f + 2\kappa \boldsymbol{L}_m) d^4x \tag{4}$$

by the matter field provides the matter field equations

$$\frac{\delta L_m}{\delta \phi^{(A)}} = 0 \tag{5}$$

and by the metric the gravitational equations

$$G_{ik} = -\kappa P_{ik}, \tag{6}$$

where the Einstein tensor G_{ik} and the metric energy-momentum tensor P_{ik} of matter are defined by

$$G_{ik} := \frac{\delta L_f}{\delta g^{ik}} \quad \text{and} \quad P_{ik} = \frac{2}{\sqrt{-g}} \frac{\delta L_m}{\delta g^{ik}}. \tag{7}$$

Furthermore, assuming that all field equations are satisfied, one obtains [5]

$$\delta^* L + (L \xi^m)_{,m} = 0 \tag{8}$$

with $\delta x^i := \xi^i$, the Lie differential $\delta^* g^{ik}$ of g^{ik},

$$\delta^* g^{ik} = g^{sk} \xi^i{}_{,s} + g^{is} \xi^k{}_{,s} \tag{9}$$

and the corresponding Lie differential $\delta^* \phi^{(A)}$ of $\phi^{(A)}$. For $\xi^i = \alpha^i = const.$ Eq. (8) yields the Noether identity

$$(P_i^k + \frac{1}{2\kappa} t_i^k)_{,k} = \nabla_k P_i^k = 0, \tag{10}$$

where t_i^k is the gravitational energy-momentum pseudo-tensor). Via the field equations, (15) can shown to be equal to the contracted Bianchi identities.

Following Lorentz [49], one can read the field equations (6) as the statement that the total metric energy-momentum density of the of matter and gravity is equal to zero. That such a reading of (6) has a physical meaning was shown for Einstein's GRT, where $L_f = \sqrt{-g} R$, and for fourth-order gravitational equations, where the Lagrangian reads $L_f = \sqrt{-g}(R + \alpha R_{ik} R^{ik} + \beta R^2)$ [16, 69].

Purely metric theories are insofar physically completely satisfying as (i) they solve the space problem, (ii) they are in agreement with the principles of equivalence and general relativity, and (iii) their "conservation" laws given by Noether's theorem follow, via the field equations, automatically from the Bianchi identities. Otherwise, and that is their disadvantage, as a matter of these principles, they do not allow for the existence of a genuine law of energy-momentum conservation. This provides problems for the quantization of gravity (in the case of GRT, this has been discussed in detail, e.g., in [2] and [6]).

Further more, as a matter of fact, this theory does not satisfy Mach's principle. (For a discussion of different versions of Mach's principle and their relation to GRT, see [4].)

2.2. BI-METRIC THEORIES

Bi-metric theories have in addition to the Riemann-Einstein metric g_{ik}, a second metric $\bar{g}_{ik}$. Such theories go back to ideas of Rosen [57], Band [3], and Papapetrou [54], but their satisfying elaboration was only given by Kohler [43, 44].[1] In this theory, the Weyl-Cartan space problem is solved in the same manner as in the uni-metric theories discussed above (both metrics g_{ik} and $\bar{g}_{ik}$ are presupposed to satisfy Eq. (1)). And, even more, since the second metric $\bar{g}_{ik}$ is assumed to be a pseudo-Euclidean one, one has 10 Killing vectors such that also the Helmholtz- Lie space problem is solved.

Corresponding to the two metrics there exist the two (Riemann-Christoffel) connections $\Gamma^i{}_{kl}$ and $\bar{\Gamma}^i{}_{kl}$, where the latter vanishes in pseudo-Cartesian coordinates. Their difference is a tensor $\rho^i{}_{kl}$ describing the gravitational field such that the connection $\Gamma^i{}_{kl}$ is the sum of the inertial field $\bar{\Gamma}^i{}_{kl}$ and the gravitational field $\rho^i{}_{kl}$:

$$\Gamma^i{}_{kl} = \bar{\Gamma}^i{}_{kl} + \rho^i{}_{kl} \tag{11}$$

Kohler specified the free Lagrangian,

$$\boldsymbol{L}_K = \boldsymbol{L}_f(g_{ik}, \partial_l g_{ik}, \bar{g}_{ik}, \partial_l \bar{g}_{ik}) + \boldsymbol{L}_m(g_{ik}, \partial_l g_{ik}, \phi^{(A)}, \partial_l \phi^{(A)}) \tag{12}$$

such that, as in the above uni-metric theories, the metric energy-momentum tensor fulfills again the dynamical equation with respect to $\Gamma^i{}_{kl}$

$$\nabla_k P_i^k = 0. \tag{13}$$

But now, due to the existence of the second metric, this can be rewritten as

$$\bar{\nabla}_k \left(\sqrt{\frac{g}{\bar{g}}} (P_i^k + t_i{}^k) \right) = 0 \tag{14}$$

where $\bar{\nabla}_k$ is the covariant derivative with respect to $\bar{\Gamma}^i{}_{kl}$, g is the determinant of g_{ik}, and $\bar{g}$ the determinant of $\bar{g}_{ik}$. In contrast to $t_i{}^k$ in Eq. (10), the expression $t_i{}^k$ arising in Eq. (14) is a tensor under general coordinate transformations.

According to Einstein's principle of equivalence, in bi-metric theories of gravitation the matter field equations are formulated in the space V_4 whose metric is identified with the gravitational potential g_{ik}. The equations for this field g_{ik}, however, are referred to a flat Minkowski space $\bar{V}_4$ with the metric $\bar{g}_{ik}$. With respect to this flat background of the gravitational field, then (the covariance of the gravitational equations being provided) integral laws of conservation may be derived, and the phase vectors of the quantum theory can be defined as Killing vectors of the $\bar{V}_4$.

The variation of the Lagrangian (12) by g_{ik} provides 10 field equations, while its variation by $\bar{g}_{ik}$ leads to four additional equations. The latter ones can be interpreted as coordinate conditions soldering the two metrics; they are equivalent to Eq. (13). Thus, in this theory the dynamical equations Eq. (13) (representing, according to Eq. (14), a conservation law) are no implication of the Bianchi identities and the field equations. This is the price one has to pay for the formulation of differential laws of energy-momentum conservation [47]. Except for this point, one finds the same situation as in the uni-metric theories considered above.

2.3. THEORIES WITH TELEPARALLELISM (EINSTEIN-MAYER-TYPE THEORIES)

Theories with teleparallelism consider the tetrad field $h^A{}_i(x^l)$ given by the coframe field $h^A = h^A{}_i dx^i$ as basic quantity; the metric is a secondary concept defined as

$$g_{ik} = \eta_{AB} h^A{}_i h^B{}_k \tag{15}$$

with $\eta = diag(1.-1,-1,-1)$. For the $h^A{}_i$ are assumed to be fixed by the gravitational field equations they define a teleparallelism in the Riemannian space. Thus the Weyl-Cartan space problem is solved in the same way as in GRT.

Let us start now from a Weitzenboeck Lagrange density which is a scalar density with respect to coordinate transformations but not invariant under local Lorentz transformations. One finds here a restricted invariance, namely an invariance with respect to global Lorentz transformations. To discuss the points in this paper under consideration we shall confine ourselves to the Einstein-Mayer class of Lagrangians [21-24, 28, 29]

$$\boldsymbol{L}_{EM} = \sqrt{-g}R + ahF_{Aik}F^{Aik} + bh\phi_A\phi^A, \tag{16}$$

where $h = \det(h^A{}_i) = \sqrt{-g}$, $R = g^{ik}R_{ik}$ is the Ricci scalar, a and b are numerical constants, the ϕ_A are defined as $\phi_A = h^i{}_A \gamma^m{}_{im}$, and the F_{Aik} are Cartan's anholonomy objects

$$F_{Aik} = h_{Ai,k} - h_{Ak,i} = h_A{}^l(\gamma_{lik} - \gamma_{lki}) \tag{17}$$

with the Ricci rotation coefficients $\gamma_{ikl} = h^A{}_i h_{Ak;l}$. The Lagrangian (16) lies on the basis of Møller's tetrad theory of gravity [50].[2] (From the viewpoint of a unified gravito-electromagnetic theory in the sense of Einstein's program of 1929, it is also discussed in [62]; however, in contrast to Einstein's approach, there, like in Møller's theory an additional matter Lagrangian is introduced.) For historical reasons, in the following we call the Lagrangian (16) "Einstein-Mayer Lagrangian", while the tetrad equivalent of the Einstein-Hilbert Lagrangian will be called "Møller Lagrangian". (For other metric-teleparallel theories, see, e.g., Kopczyński [45], Hayashi and Shirafuji [34]).

Now, in the anholonomic (Einstein-Cartan) representation of the space-time structure the reference systems, i.e., the tetrads $h^A{}_i(x^l)$, are the gravitational field variables. From the standpoint of the theories here under consideration, Einstein's GRT is a degenerate case, now following from Eq. (16) for $a = b = 0$.

As it was shown in [52, 56], assuming for matter a Lagrange density depending only on the tetrads via the algebraic combination (15) one is led to the field equations

$$\frac{1}{h}\frac{\delta L_{EM}}{\delta h^{Ai}} h^A{}_k = -\kappa T_{ik} \tag{18a}$$

with

$$\frac{\delta L_{EM}}{\delta h^{Ai}} h^A{}_k - \frac{\delta L_{EM}}{\delta h^{Ak}} h^A{}_i = 0. \tag{18b}$$

The 10+6 equations (18a, b) determine the 16 components of $h^A{}_i$ up to constant Lorentz rotations of the tetrads. Together with the Noether identities

$$H^i{}_{k,i} = F^i{}_{k;i} - F^{il}\gamma_{ilk} \quad \text{where} \quad H_{ik} \equiv E_{ik} + \theta_{(ik)} \quad \text{and} \quad F_{ik} = \theta_{[ik]}, \tag{19}$$

and equation (18b) this leads again to the dynamical equations (10).

As argued in [8-10], in a Riemannian theory with Einstein-Cartan teleparallelism one finds quite a satisfying situation for the gravitational energy. One has a Lagrangian of canonical structure that is coordinate-covariant but, like Møller's Lagrangian, not covariant with respect to local Lorentz transformations. Thus the Lagrangian leads to field equations which do not satisfy the general principle of relativity. They represent equations fixing the 16 components of the tetrads $h^A{}_i$ instead of their 10 combinations $g_{ik} = h^A{}_i h_{Ak}$, and the Hamiltonian $\boldsymbol{H}$ is given by the *00*-component of the energy-momentum complex (which is a tensor with respect to the group of global Lorentz transformations lying on the basis of this theory). In [31, 32] such theories were considered as models realizing that local version of the Mach principle, according to which the cosmic matter distribution fixes the reference system.

Similarly to the Kohler case, the dynamical equations of the theory with teleparallelism are not simply an implication of the Bianchi identity and the field equations (now given by (18a). They result from the additional conditions, where now these conditions (18b) specify, instead of the coordinates, the reference tetrads. Again, this is the price that one has to pay for having differential laws of conservation. Or better, this is a part of the price, for, the other part is that one has even to sacrifice the local Lorentz covariance and, thus, the general principle of relativity.

2.4 EINSTEIN-CARTAN THEORIES

In [53] it was shown that in the more general case, where the matter Lagrange density depends independently on the 16 tetrad and the 64 connection components, a theory with teleparallelism can be considered as a constrained Einstein-Cartan theory. Therefore, Einstein-Cartan theories can be regarded as straightforward generalization of Einstein-Mayer-type theories. Thus, let us turn to Einstein-Cartan theories [36-43]. (Due to the fact that these theories generally work in Riemann-Cartan spaces without teleparallelism, they, of course, do not satisfy the Mach-Einstein doctrine.)

As to the general principle of relativity, such theories resemble the purely metric theories. They are covariant under local Lorentz and general coordinate transformations. A great similarity is also found as to the energy-momentum problem. Indeed, starting from a Lagrange density of the form

$$\boldsymbol{L}_{EC} = \boldsymbol{L}_f(h^A{}_i, \Lambda^{AB}{}_i) + \kappa \boldsymbol{L}_m(h^A{}_i, \Lambda^{AB}{}_i, \phi^{(A)}, \partial_l \phi^{(A)}), \tag{20}$$

where the tetrad and the connection fields, h and Γ, are regarded as independent variables, the Euler variation of $\phi^{(A)}$, $h^A{}_i$, and $\Gamma^{AB}{}_i$ yields the matter field equations and the following gravitational equations (h denotes the determinant of $h^A{}_i$),

$$-\frac{\delta L_f}{\delta h^A{}_i} = \kappa h T_A{}^i := \kappa \frac{\delta L_m}{\delta h^A{}_i} \tag{21}$$

and

$$-\frac{\delta L_f}{\delta \Lambda^{AB}{}_i} = \kappa h \tau_{AB}{}^i := \kappa \frac{\delta L_m}{\delta \Lambda^{AB}{}_i} \ . \tag{22}$$

The source term in Eq. (21), $T_A{}^i$, is the canonical energy-momentum tensor, and the source term in Eq. (22), $\tau_{AB}{}^i$, represents the canonical spin-angular momentum tensor of the matter field. Therefore, as in GRT and its uni-metric generalizations, Eq. (21) can be read as the statement that the total energy-momentum density is equal to zero. (Now, however, this statement does not concern the metric, but the canonical tensor.)

As to the fact that one has the same symmetries as in GRT, there does not exist a differential law of conservation, either. There exist dynamical equations which are a generalization of Eq. (13), and, for the class of theories under consideration in [31-39], it turns out that, provided the field equations are fulfilled, the generalized dynamical equations are again an automatic by-product of the Bianchi identities which hold in Riemann-Cartan space [38].

For spinless matter these dynamical equations reduce to (13). This implies that spinless point particles move along geodesics of the Riemann-Christoffel connection, but not along autoparallels. In [41] this fact is considered as an objection to the proposed class of Einstein-Cartan theories. Otherwise, as was shown in [48] that, starting from the Dirac equation generalized to metric-affine spaces with non-vanishing curvature, non-metricity, and torsion, one is automatically led to the geodesic motion of pole particles.

After mentioning a series of similarities between GRT and Einstein-Cartan theories, finally it should be stressed that, as to the Weyl-Cartan space problem, there is a great difference. The point is that in Einstein-Cartan theories the validity of Eq. (1) cannot be kinematically guaranteed. To solve the space problem now means to solve it by using an appropriate dynamic starting point. That is, one has to look for a Lagrangian such that the solutions of the corresponding gravitational field equations (21) and (22) satisfy Eq. (1), i.e., they have to yield solutions satisfying the constraints $\Lambda_{ABl} = -\Lambda_{BAl}$ (or in the spinorial case, $\Lambda_{\alpha\beta l} = \Lambda_{\beta\alpha l}$). From this point of view, for example, the Einstein-Hilbert Lagrangian, $\sqrt{-g}R$, viewed as a functional of the metric and an arbitrarily generalized connection, has to be excluded from consideration [7]; it leads to

$$g_{ik//l} = -2 g_{is} \Gamma^s{}_{[kl]} + \frac{2}{3}(\Gamma_i g_{kl} + \Gamma_k g_{il}) \tag{23}$$

where torsion vector $\Gamma_i = \Gamma^s_{is}$ and // denotes the derivative with respect to the generalized connection.

In purely affine theories, one does not encounter this problem because generally there is no metric present, and in the special case of an Riemann-Cartan geometry the validity of the Ricci and Weyl lemmata are satisfied *per definitionem.*

[In parentheses, it should be mentioned that this problem of consistency occurs also in Ashtekar's approach to canonical GRT [1, 16]. Starting from the Lagrangian (51) in its self-dual spinor representation, the authors seem to feel that, in the case of a Dirac matter field, one is automatically led to a connection which does not satisfy the Weyl lemma. In order to avoid its explicit appearance on the base manifold as second connection competing with that one for tensors, they try to avoid dotted spinors. This forces them to change the Dirac theory because the standard theory requires undotted and dotted spinors.]

3. Affine Theory of Gravity

3.1 AFFINE GEOMETRY

First, let us make some general remarks on the linear connection field from the point of view of the tangent bundle.[3] We consider the connection defined in the tangent bundle $T(V_4)$ over V_4. Then to the transformation $(x^i, x^A{}_i) \to (x^{i'}, x^{A'}{}_{i'})$ of the bundle coordinates there corresponds the following transformation of the local components $\Lambda^A{}_{Bi}$ of the Lorentz connection Γ

$$\Lambda^{A'}{}_{B'i'} = \frac{\partial x^k}{\partial x^{i'}} \omega^{A'}{}_C \omega_{B'}{}^D \Lambda^C{}_{Dk} + \omega^{A'}{}_C \omega_{B'}{}^C{}_{,i'} . \tag{24}$$

Here $\partial x^k/\partial x^{i'}$ corresponds to the transformation $x^i \to x^{i'}$ and $\omega^{A'}{}_B$ (with $\omega^{A'}{}_C \omega_{B'}{}^C = \delta^{A'}{}_{B'}$) to the transformation $x^A{}_i \to x^{A'}{}_{i'}$.

Using in $T(V_4)$ natural bundle coordinates $x^k{}_i$ one has

$$\omega^{A'}{}_{A'} = \omega^{a'}{}_a = \partial x^{a'}/\partial x^a , \tag{25}$$

and Eq. (24) reduces to the (Riemann-Christoffel) transformation law of a linear connection under coordinate transformations (it acts only on space-time indices):

$$\Gamma^{a'}{}_{b'i'} = \frac{\partial x^k}{\partial x^{i'}} \frac{\partial x^{a'}}{\partial x^m} \frac{\partial x^n}{\partial x^{b'}} \Gamma^m_{nk} + \frac{\partial x^{a'}}{\partial x^m} \frac{\partial^2 x^m}{\partial x^{b'} \partial x^{i'}} . \tag{26}$$

On the other hand, for $x^i \to \delta^i{}_l x^l$ (i.e., $\partial x^i/\partial x^{i'} = \delta^i_{i'}$, from Eq. (24) one obtains the (Ricci-Levi-Civita) transformation law of the connection under Lorentz transformations (it acts only on the anholonomic indices)

$$\Lambda^{A'}{}_{B'i'} = \omega^{A'}{}_C \omega_{B'}{}^D \Lambda^C{}_{Di} + \omega^{A'}{}_C \omega_{B'}{}^C{}_{,i} . \tag{27}$$

Considering now the transformation $x^A{}_i = h^A{}_k x^k{}_i$ (with $h^A{}_i \in GL(4, \boldsymbol{R})$) from anholonomic to holonomic bundle coordinates, then, according to Eq. (24) the following (Einstein-Cartan) transformation $\Gamma^i_{kl} \to \Gamma^A_{Bl}$ of the connection Γ is associated with it:

$$\Lambda^A{}_{Bi} = h^A{}_m h_B{}^k \Gamma^m{}_{ki} + h^A{}_m h_B{}^m{}_{,i} = h^A{}_m h_B{}^m{}_{;i} . \tag{28}$$

This relation is equivalent to the so-called Einstein lemma (// denotes here the covariant derivative with respect to the connection Γ^i_{kl} and /// the derivative acting simultaneously on holonomic and anholonomic indices)

$$h^B{}_{i////l} := h^B{}_{i,l} - h^B{}_r \Gamma^r{}_{il} + h^C{}_i \Lambda^B{}_{Cl} = h^B{}_{i//l} + h^C{}_i \Lambda^B{}_{Cl} = 0 . \tag{29}$$

The local Lorentz transformations and the coordinate transformations which are above considered as different transformations of the coordinates in a Lorentz bundle can also be regarded as the homogeneous and inhomogeneous parts of the local Poincaré group[4]

$$\Delta x^{i'} = \omega^{i'}{}_k (x^l) dx^k + \frac{\partial x^{i'}}{\partial x^k} dx^k \tag{30}$$

acting in an affine space. Then, relation (29) can be considered as an implication of the transition from one item of the Poincaré transformation to the other.[5]

The mathematically equivalent relations (26)-(30) can also be justified by the physical requirement that a physical theory has to enable one to define measurement values. That this requirement is really satisfied by (29), and thus by the related formulas (26-28), can be seen as follows. The Einstein coefficients $A^i{}_{kl} := h_B{}^i h^B{}_{k,l} = -h^i{}_{B,l} h^B{}_k$ are the components of the connection defining the teleparallelism of Einstein and Cartan: The measurement values $V_B = h_B{}^i V_i$ and $V^B = h^B{}_i V^i$ of vectors V_i and covectors V^i do not change when they are transported by means of this connection:

$$V_{i////l} = V_{i,l} - V_r A^r{}_{il} = 0 \quad \leftrightarrow \quad V_{B,i} = 0 \tag{31a}$$
$$V^i{}_{////l} = V^i{}_{,l} + V^r A^i{}_{rl} = 0 \quad \leftrightarrow \quad V^B{}_{,i} = 0 . \tag{31b}$$

That the coefficients $A^i{}_{kl}$ really provide a teleparallelism can be seen from the fact that the transport is transitive and integrable and thus independent of the way along which it is done. This is reflected by the vanishing of the curvature tensor formed from the Einstein coefficients:

$$A^i{}_{klm} = -A^i{}_{kl,m} + A^i{}_{km,l} - A^i{}_{rm} A^r{}_{kl} + A^i{}_{rl} A^r{}_{km} = 0 . \tag{32}$$

With the *Ansatz* given in Eq. (28), this provides the Cartan-Schouten representation of the Riemann tensor in anholonomic coordinates,

$$\begin{aligned} R^i{}_{klm} &= -\Gamma^i{}_{kl,m} + \Gamma^i{}_{km,l} - \Gamma^i{}_{rm} \Gamma^r{}_{kl} + \Gamma^i{}_{rl} \Gamma^r{}_{km} \\ &= \Lambda^i{}_{kl,m} - \Lambda^i{}_{km,l} + \Lambda^i{}_{rm} \Lambda^r{}_{kl} - \Lambda^i{}_{rl} \Lambda^r{}_{km} + \Lambda^r{}_{kl} A^i{}_{rm} - \Lambda^r{}_{km} A^i{}_{rl} + \Lambda^i{}_{rm} A^r{}_{kl} - \Lambda^i{}_{rl} A^r{}_{km} . \end{aligned} \tag{33}$$

In order to couple gravity to matter one has to introduce spinorial quantities in affine spaces. To do so, it is helpful to remind of the fact that spinors φ^ν, $\varphi^{\dot\nu}$ (with ν, $\dot\nu$ = 1, 2) are "scalars that satisfy a particular rule of differentiation" [29, 68]). In other words, we employ the one-to-two correspondence of the Lorentz group O(3,1) to the unimodular group SL(2, **C**) and require that the differentiation has to be covariant with respect to unimodular transformations $\alpha_\nu{}^\mu$:

$$\varphi'_\nu = \alpha_\nu{}^\mu \varphi_\mu , \quad \varphi'^\nu = \alpha^\nu{}_\mu \varphi^\mu , \quad \varphi'_{\dot\nu} = \alpha_{\dot\nu}{}^{\dot\mu} \varphi_{\dot\mu} , \quad \varphi'^{\dot\nu} = \alpha^{\dot\nu}{}_{\dot\mu} \varphi^{\dot\mu} \tag{34}$$

with

$$\det \alpha_{\nu}{}^{\mu} = \det \alpha_{\dot{\nu}}{}^{\dot{\mu}} = 1, \tag{35}$$

and the relations

$$\alpha^{\mu}{}_{\nu} = \frac{\partial(\det \alpha_{\kappa}{}^{\lambda})}{\partial \alpha_{\mu}{}^{\nu}}, \qquad \alpha^{\dot{\mu}}{}_{\dot{\nu}} = \frac{\partial(\det \alpha_{\dot{\kappa}}{}^{\dot{\lambda}})}{\partial \alpha_{\dot{\mu}}{}^{\dot{\nu}}}. \tag{36}$$

The coefficients of the spinor connection, $\Lambda^{\mu}{}_{\nu l}$ and $\Lambda^{\dot{\mu}}{}_{\dot{\nu} l}$,which allow to built the covariant derivatives,

$$\varphi_{\nu//l} = \varphi_{\nu,l} - \Lambda^{\mu}{}_{\nu l}\varphi_{\mu}, \qquad \varphi^{\nu}{}_{//l} = \varphi^{\nu}{}_{,l} + \Lambda^{\nu}{}_{\mu l}\varphi^{\mu} \tag{37}$$

are specified by demanding Einstein's rule of teleparallelism (6) which now takes the form
of "Weyl's lemma" demanding the covariant constancy of the spin-vectors[6] $\sigma^{\mu\dot{\nu}}{}_{k}$:

$$\sigma^{\mu\dot{\nu}}{}_{k///l} = \sigma^{\mu\dot{\nu}}{}_{k,l} - \sigma^{\mu\dot{\nu}}{}_{r}\Gamma^{r}_{kl} + \sigma^{\alpha\dot{\nu}}{}_{k}\Lambda^{\mu}{}_{\alpha l} + \sigma^{\mu\dot{\beta}}{}_{k}\Lambda^{\dot{\nu}}_{\dot{\beta} l} = \sigma^{\mu\dot{\nu}}{}_{k,l} - \sigma^{\mu\dot{\nu}}{}_{r}A^{r}{}_{kl} = 0. \tag{38}$$

This provides the following expressions for the coefficients of the connection,

$$A^{i}{}_{kl} = \sigma^{i}{}_{\mu\dot{\nu}}\sigma_{k}{}^{\mu\dot{\nu}}{}_{,l} \tag{39a}$$

$$\Lambda^{i}{}_{kl} = \sigma^{i}{}_{\mu\dot{\alpha}}\sigma_{k}{}^{\mu\dot{\beta}}\Lambda^{\dot{\alpha}}{}_{\dot{\beta} l} + \sigma^{i}{}_{\alpha\dot{\mu}}\sigma_{k}{}^{\beta\dot{\mu}}\Lambda^{\alpha}{}_{\beta l} \tag{39b}$$

$$\Lambda^{\alpha}{}_{\beta l} = \frac{1}{2}\sigma^{\alpha\dot{\mu}}{}_{k}\sigma_{\beta\dot{\mu}}{}^{k}{}_{//l} \tag{39c}$$

$$\Lambda^{\dot{\alpha}}{}_{\dot{\beta} l} = \frac{1}{2}\sigma^{\mu\dot{\alpha}}{}_{k}\sigma_{\mu\dot{\beta}}{}^{k}{}_{//l}. \tag{39d}$$

3.2 RIEMANN-CARTAN GEOMETRY AS A SPECIAL CASE OF AFFINE GEOMETRY

In special cases, this purely affine framework can turn into a *metric-affine framework*, namely when the field equations $(\delta L/\delta\Gamma)\delta\Gamma = 0$ have a solution $\Gamma^{i}{}_{kl}$ that allows one to solve the Weyl-Cartan space problem by introducing a covariantly constant (symmetric) metric tensor of Minkowski signature satisfying the "Ricci lemma"

$$g_{ik//l} = 0 \tag{40}$$

such that

$$\begin{aligned} \Gamma^{i}_{kl} &= \left\{{}^{\,i}_{kl}\right\} + \Gamma^{i}_{[kl]} + g^{im}g_{kr}\Gamma^{r}_{[lm]} + g^{im}g_{lr}\Gamma^{r}_{[km]} \\ &= \left\{{}^{\,i}_{kl}\right\} + K^{i}_{kl} \end{aligned} \tag{41}$$

where $\left\{{}^{\,i}_{kl}\right\}$is the Christoffel connection and K^{i}_{kl} the contorsion which is anti-symmetric in the first two indices, $K_{ikl} = -K_{kil}$.[7]

Otherwise, the Weyl lemma (38) translating the connection from holonomic (space-time) to anholnomic (internal) coordinates provides for the anholonomic representation of the connection Γ the expression

$$\Lambda^A{}_{Bl} = \gamma^A{}_{Bl} + K^A{}_{Bl} \tag{42}$$

$$\gamma_{ikl} = -\gamma_{kil}, \quad K_{ikl} = -K_{kil}, \tag{43}$$

where $\gamma^A{}_{Bl}$ are again the Ricci rotation coefficients, $\gamma^A{}_{Bl} = h^A{}_k h^k_{B//l}$, and the last expression in (42) is defined as $K^A{}_{Bl} = h^A{}_i h_B{}^k K^i_{kl}$; both are antisymmetric in the first two indices. As a consequence, the anholonomic components of the internal connection is antisymmetric in the first two indices, too:

$$\Lambda_{ikl} = -\Lambda_{kil}. \tag{44}$$

Exploiting the validity of the lemma of Ricci (40) and introducing the relation $g_{ik} = h^A{}_i h_{Ak}$, finally one finds

$$g_{ik//l} = g_{ik////l} = (h^A{}_i h^B{}_k \eta_{AB})_{////l} = 0 \tag{45}$$

such that

$$\eta_{AB,l} + \Lambda_{ABl} + \Lambda_{BAl} = 0 \tag{46}$$

and, thus,

$$\eta_{AB,l} = 0 \tag{47}$$

is satisfied. This recovers the usual spinor formalism where $\gamma_{\alpha\beta}$ and $\gamma_{\dot{\alpha}\dot{\beta}}$ are the "metrics" in the spin spaces:[8]

$$\begin{aligned} &\gamma_{\alpha\beta} = -\gamma_{\beta\alpha}, && \gamma_{\dot{\alpha}\dot{\beta}} = -\gamma_{\dot{\beta}\dot{\alpha}}, \\ &\gamma_{\alpha\beta,l} = \gamma_{\dot{\alpha}\dot{\beta},l} = 0, && \\ &\Lambda_{\alpha\beta l} = \Lambda_{\beta\alpha l}, && \Lambda_{\dot{\alpha}\dot{\beta}l} = \Lambda_{\dot{\beta}\dot{\alpha}l}. \end{aligned} \tag{48}$$

3.3 VACUUM EQUATIONS (LAGRANGIAN AND HAMILTONIAN FORMS)

In the *affine framework* based on the symmetries given by the Poincaré group, geometry is completely determined by the affine connection $\Gamma^i{}_{kl}$. Thus, a field theory has to start from a purely affine Lagrangian

$$\boldsymbol{L} = \boldsymbol{L}(\Gamma, \psi). \tag{49}$$

Its variation with respect to $\Gamma^i{}_{kl}$ and the matter field ψ provides the following *purely affine field equations*

$$\frac{\delta L}{\delta \Gamma}\delta\Gamma = 0 \qquad \frac{\delta L}{\delta \psi}\delta\psi = 0\,. \tag{50}$$

A classical example of an affine theory is given by the Einstein-Schrödinger theory stemming from the Lagrangian density [20, 60] (for its canonical structure see, in particular [64])

$$\boldsymbol{H} = \frac{2}{\lambda}(-\det(R_{mn}))^{1/2}. \tag{51}$$

According to the intention of their founders, it should represent a unified geometric description of gravitational and matter fields so that an additional matter Lagrangian need and must not be introduced. For we, however, want to regard the expression (51) as the Lagrangian of vacuum gravity, here a matter Lagrangian has to be supplemented (see § 3.4).

It should be stressed that, as long as we aim at field equations of second order, the Lagrangian (51) could only be supplemented by the square root of the determinant of the homothetic curvature. The most general Lagrangian then reads [20]

$$\boldsymbol{H}^* = \alpha\boldsymbol{H} + \beta\frac{1}{\lambda}(-\det(R^s_{smn})^{1/2}, \tag{52}$$

where the tensor of the homothetic curvature is given as $R^s_{smn} = -\Gamma^s_{sm,n} + \Gamma^s_{sn,m}$. But, for in our framework we shall be interested in special solutions of the field equations which enable us to introduce a metric we can omit a second term on the right-hand side of (52). Indeed, this term is equal to zero because the vanishing of the homothetic curvature is a necessary condition for the solution of the Weyl-Cartan space problem.

Using the Einstein affine tensor

$$U^i{}_{kl} = \Gamma^i_{kl} - \delta^i_l\Gamma^r_{kr} \quad (\leftrightarrow \Gamma^i_{kl} = U^i_{kl} - \frac{1}{3}\delta^i_l\Gamma^r_{kr}) \tag{53}$$

as field coordinates the field equations stemming from the Lagrangian (51) read [64]

$$\mathfrak{R}^{ik}{}_l := \frac{\delta \boldsymbol{H}}{\delta U^l{}_{ik}} = 0, \tag{54a}$$

$$\mathfrak{R}^{ik}{}_l = \left(\frac{\delta \boldsymbol{H}}{\delta R_{ik}}\right)_{,l} + \frac{\delta \boldsymbol{H}}{\delta R_{sk}}\left(U^i_{sl} - \frac{1}{3}\delta^i_l U^t_{st}\right) + \frac{\delta \boldsymbol{H}}{\delta R_{is}}\left(U^k_{ls} - \frac{1}{3}\delta^k_l U^t_{ts}\right), \tag{54b}$$

where

$$\mathfrak{R}^{[ik]}{}_k = \left(\frac{\delta \boldsymbol{H}}{\delta R_{[ik]}}\right)_{,k} = 0; \tag{55}$$

the Ricci tensor is given as

$$R^m{}_{klm} = R_{kl} = -\Gamma^t_{kl,t} + \Gamma^t_{kt,l} - \Gamma^t_{rt}\Gamma^r_{kl} + \Gamma^t_{rl}\Gamma^r_{kt}$$
$$= -U^t_{kl,t} + U^r{}_{kt}U^t{}_{rl} - \frac{1}{3}U^r_{kr}U^t_{tl}. \tag{56}$$

These equations satisfy the differential identity

$$\mathfrak{R}^{[ik]}{}_{k,i} = 0 \tag{57}$$

which first time was derived by Einstein and Kaufmann (see [1]). Eq. (57) holds due to the invariance of the Lagrangian (51) under the transformation,

$$\Gamma^{*i}_{kl} = \Gamma^i_{kl} + \delta^i_k \phi_{,l}, \qquad U^{*i}_{kl} = U^i_{kl} + \delta^i_{[k}\phi_{,l]}; \tag{58}$$

it holds:

$$R^{*i}_{klm} = R^i{}_{klm}, \qquad R^*{}_{kl} = R_{kl}, \quad R^* = R. \tag{59}$$

Indeed, from the relations (58) and (59) one obtains the following:

$$\begin{aligned}\delta\int \boldsymbol{H} d^4x &= \int \mathfrak{R}^{ik}{}_l \delta U^l{}_{ik} d^4x\\ &= 2\int \mathfrak{R}^{ik}{}_l \delta^l{}_{[i}\phi_{,k]} d^4x\\ &= 2\int \mathfrak{R}^{[ik]}{}_i \phi_{,k} d^4x\\ &= 2\int (\mathfrak{R}^{[ik]}{}_{i,k}\phi) d^4x = 0;\end{aligned} \tag{60}$$

that means, one has relation (57).

As was shown by Bergmann [5], the principles of general relativity (i.e., of coordinate covariance) and of equivalence (i.e., of minimal coupling), implying a total Lagrangian of the form[9]

$$\boldsymbol{L} = \boldsymbol{L}_G[g,\Gamma] + \boldsymbol{L}_{mat}[g,\Gamma,\psi], \tag{61}$$

always lead to the following identity

$$-(\boldsymbol{L}\xi^i)_{,t} = \frac{\partial \boldsymbol{L}}{\partial g}\bar{\delta}g + \frac{\partial \boldsymbol{L}}{\partial \Gamma}\bar{\delta}\Gamma + \frac{\partial \boldsymbol{L}}{\partial \Gamma_{,i}}\bar{\delta}\Gamma_{,i} + \frac{\partial \boldsymbol{L}}{\partial \psi}\bar{\delta}\psi + \frac{\partial \boldsymbol{L}}{\partial \psi_{,i}}\bar{\delta}\psi_{,i} \tag{62}$$

where $\bar{\delta}$ is the change of a quantity with fixed values of the coordinates, $\delta x^i = \xi^i$ is the infinitesimal change of the coordinates coming in via the relation $\bar{\delta}\mathfrak{B} = -(\mathfrak{B}\xi^i)_{,i}$ satisfied for a scalar density $\mathfrak{B}$, and $\delta\boldsymbol{L}/\delta\Gamma$ and $\delta\boldsymbol{L}/\delta\psi$, respectively, are the Euler variations of the Lagrangian with respect to the connection and the matter fields. If the gravitational and matter field equations are satisfied there exists a covariant vector density

$$\vartheta^i = \boldsymbol{L}\xi^i - \frac{\delta \boldsymbol{L}}{\delta \Gamma^l{}_{mn,i}}\bar{\delta}\Gamma^l{}_{mn} - \frac{\delta \boldsymbol{L}}{\delta \psi_{A,i}}\bar{\delta}\psi_A \tag{63}$$

satisfying the equation (Noether identity)

$$\vartheta^i{}_{,i} = 0 . \tag{64}$$

The divergence (64) is a scalar density, too. In [5], it is also shown that, if we consider the case of a rigid displacement along one coordinate direction, $\xi^i = \alpha^i = \text{const.}$, then the expressions (63) and (64) reduce to the following relations,

$$\vartheta^i = \vartheta^i{}_k \alpha^k , \qquad \vartheta^i{}_{k,i} = 0 , \tag{65a}$$

$$\vartheta^i{}_k = t^{il}{}_{k,l} + \left(\psi_{,k} \frac{\partial L_{mat}}{\partial \psi_{,i}} - \delta^i{}_k L_{mat} + \frac{2}{\sqrt{-g}} \frac{\delta L_{mat}}{\delta g_{li}} g^{lk} \right). \tag{65b}$$

The coefficients $\vartheta^i{}_k$ are no covariant tensors, but only affine tensors and can be considered to be the components of the energy-momentum complex of the total system including gravitational and matter parts.[10]

In our case of the pure vacuum Lagrangian $\boldsymbol{H}$, the relation (62) takes the form

$$\frac{\delta \boldsymbol{H}}{\delta U^i{}_{kl}} \bar{\delta} U^i_{kl} - \left(\frac{\partial \boldsymbol{H}}{\partial R_{ik}} \bar{\delta} U^s_{ik} \right)_{,s} = 0 , \tag{66}$$

where

$$\bar{\delta} U^l{}_{ik} = U^s{}_{ik} \xi^l_{,s} - U^l{}_{sk} \xi^s_{,i} - U^l{}_{is} \xi^s_{,k} - \delta^l_i \xi^s_{,sk} + \xi^l_{,ik} - U^l_{ik,s} \xi^s \tag{67}$$

is the Lie derivative of $U^i{}_{kl}$ so that, for rigid translations $\xi^i = \text{const.}$, in the case that the field equations (54) are satisfied Einstein's conservation law results:

$$\left(\frac{\partial \boldsymbol{H}}{\partial R_{ik}} \bar{\delta} U^s_{ik} \right)_{,s} = 0 . \tag{68}$$

Thus, the corresponding Noether identity reads

$$t^i_{k,i} = 0 , \tag{69}$$

where

$$t^i_k = \frac{\partial \boldsymbol{H}}{\partial R_{mn}} U^i_{mn,k} . \tag{70}$$

Therefore, the Hamiltonian density corresponding to the Lagrangian $\boldsymbol{H}$ is given by the expression

$$t^0_0 = \frac{\partial \boldsymbol{H}}{\partial R_{mn}} U^0_{mn,0} . \tag{71}$$

This shows that the Einstein-Schrödinger field equations (54) can be rewritten in a canonical form, where the field coordinates are given by the components of the Einstein affine tensor U^0_{mn} and the field momenta by the components of the Ricci tensor R_{mn}. Interestingly, as relation (36) shows, the momenta are invariant with respect to the gauge condition (58).

As to the formula (14), it makes sense for the coordinate x^0 is locally well-defined. Indeed, the requirement that the matter field is a SL(2, **C**) spinor garantees that the 4-dimensional manifold has locally a causal structure which enables one to distinguish timelike- from spacelike-related events and, thus, the time coordinate from the three spatial coordinates. This topological structure defines Einstein's notion of simultaneity, according to which events can only be simultaneous if there cannot exist a causal nexus between them.

As far as the canonical form is concerned, one finds here the same situation as in Maxwell's electrodynamics. Starting in the Maxwell case from the Lagrangian

$$\boldsymbol{L}_M = \frac{\sqrt{-g}}{4} F_{ik} F^{ik} \tag{72}$$

one obtains the canonical energy-momentum tensor

$$t_i{}^k = \frac{\partial \boldsymbol{L}_M}{\partial A_{r,k}} A^r{}_{,i} - \delta_i^k \boldsymbol{L}_M = 2 \frac{\partial \boldsymbol{L}_M}{\partial F_{rk}} A^r{}_{,i} - \delta_i^k \boldsymbol{L}_M \tag{73}$$

revealing that the field momenta that are canonically conjugate to the field coordinates are given by the field tensor which is an invariant of the gauge group of the electromagnetic theory, $\overline{A}_i = A_i + \phi_{,i}$.

Concluding this paragraph, let us remember the situation in Einstein's theory of general relativity in order to explicate the advantage of the purely affine theory over this theory. The coordinate-covariant and Lorentz-covariant form of the Lagrangian is given by the Einstein-Hilbert scalar density

$$\boldsymbol{L}_{EH} = \sqrt{-g} g^{ik} \delta^m{}_l R^l{}_{ikm} = \sqrt{-g} g^{ik} R_{ik}. \tag{74}$$

The density $\boldsymbol{L}_{EH}$ is not of a canonical form, but there are two canonical densities which are Euler-equivalent to $\boldsymbol{L}_{EH}$. In the holonomic representation, this is Einstein's Lagrange density [19].

$$\boldsymbol{L}_E = \sqrt{-g} g^{ik} \left(\left\{ {l \atop ik} \right\} \left\{ {m \atop lm} \right\} - \left\{ {l \atop im} \right\} \left\{ {m \atop kl} \right\} \right) \tag{75}$$

(here the Γ are given by the Christoffel symbols $\{\ \}$) and, in the anholonomic representation, Møller's Lagrange density [50].

$$\boldsymbol{L}_M = h(\gamma^i{}_{lm} \gamma^{ml}{}_i - \gamma^a{}_{la} \gamma^{bl}{}_b). \tag{76}$$

$\boldsymbol{L}_E$ is Lorentz-invariant, but not coordinate-covariant, and $\boldsymbol{L}_M$ is coordinate-covariant, but not Lorentz-covariant. If, in the holonomic representation, we regard the components of the metric as the field coordinates then the canonically conjugate momenta are algebraic concomitants of the Christoffel connection. In the anholonomic representation, one finds a similar situation: If one chooses the tetrad components as field coordinates then the algebraic concomitants of the Ricci rotation coefficients are the canonically conjugate momenta.

Due to the missing covariances, in GRT the corresponding Hamilton-Jacobi formalisms (and thus the canonical quantum formalisms) are not covariant, in the first case not coordinate-covariant and in the second case not Lorentz-covariant. Therefore, in order to give the formalisms a definite physical sense one has to specify the coordinates or the tetrads by adding either four coordinate conditions or six conditions that, together with the field equations, fix the tetrads uniquely.

In case of purely affine theory, one finds insofar an improved situation as (i) one arises at a canonical formulation without destroying the covariant form of the Lagrangian (51) and (ii) the resulting theory is of the same structure as Maxwell's electrodynamics. Thus one has to expect first-class constraints four of them correspond to the diffeomorphism group and one to the gauge group (58).

3.4 MATTER COUPLINGS

Coupling now matter fields to gravity, we have to start from the Lagrangian (49) having the form (with the coupling constant α)

$$\boldsymbol{L} = \boldsymbol{H}[\Gamma] + \alpha \boldsymbol{L}_{mat}[\Gamma, \psi] \tag{77}$$

leading to the field equations

$$\frac{\delta \boldsymbol{H}}{\delta \Gamma} = -\alpha \frac{\delta \boldsymbol{L}_{mat}}{\delta \Gamma}, \qquad \frac{\delta \boldsymbol{L}_{mat}}{\delta \psi} = 0. \tag{78}$$

The coupling constant α has the dimension of κ, $[\alpha] = [\kappa] := [8\pi G/c^4]$, since for index symmetry, i.e., for $\Gamma^i_{kl} = \Gamma^i_{lk}$ and $R_{ik} = R_{ki}$, the Lagrangian (51) reduces to the expression $\boldsymbol{H}' = \sqrt{-g}(R - 2\lambda)$ [63, 64]. Thus, to remain in accordance to the theory of general relativity, we set $\alpha = 2\kappa$. Equations (78) reveal that the purely affine theory couples only to spinorial matter Indeed, the scalar and Yang-Mills matter Lagrangians do not depend on the connection, and tensorial fields of higher spin than one are no candidate for matter sources of gravity.

Thus let us consider the coupling of the spin-1/2 Dirac field given by the Dirac Lagrangian

$$\boldsymbol{L}_D = \frac{i\hbar c}{2} |\sigma| \left\{ (\varphi^{\dot{\nu}} \sigma^l{}_{\mu\dot{\nu}} \varphi^\mu{}_{///} - \chi^{\dot{\nu}} \sigma^l{}_{\mu\dot{\nu}} \chi^\mu{}_{///}) \sqrt{2} + \frac{imc}{\hbar} (\varphi_\mu \chi^\mu - \varphi_{\dot{\mu}} \chi^{\dot{\mu}}) \right\}. \tag{79}$$

Due to the validity of Weyl's lemma (38), one can also start with the complex conjugate Lagrangian $\boldsymbol{L}_D^*$ or from the Hermitean symmetric Lagrangian $1/2(\boldsymbol{L}_D + \boldsymbol{L}_D^*)$, they are Euler-equivalent (for a discussion of this point, see [61, 65]). These Lagrangians depend on $\sigma^l{}_{\mu\dot{\nu}}$ and thus on the tetrads $h^A{}_l$ which in the purely affine framework are not assumed to be gravitational potentials but anholonomic coordinates.

In complete analogy to electrodynamics, the Lagrangian (77) with $\boldsymbol{L}_{mat} = \boldsymbol{L}_D[\Gamma, \varphi, \chi]$ is invariant with respect to gauge transformations of second kind,

$$\Gamma^i_{kl} \rightarrow \Gamma^i_{kl} + \delta^i_k \phi_{,l}, \tag{80a}$$
$$\Gamma^i_{kl} \rightarrow \Gamma^i_{kl} + \delta^i_k \phi_{,l}, \tag{80b}$$

where (80b) corresponds to the transformation $\psi \rightarrow e^{i\phi}\psi$ of the Dirac spinor ψ. In virtue of the Noether theorem, this invariance leads to the vanishing of the covariant divergence of the current density. [The Lagrangian $1/2(\boldsymbol{L}_D + \boldsymbol{L}_D^*)$ is also invariant under the gauge transformation of first kind (80b).]

It should be mentioned here that, requiring the index-symmetry $\Gamma^i{}_{kl} = \Gamma^i{}_{lk}$ and $R_{ik} = R_{ki}$, the Lagrangian (77) provides the vacuum field equations (54) which lead to a Riemann-Einstein space, $R_{ik} = \lambda g_{ik}$ with $g_{ik//l} = 0$ [20, 63, 64]. Therefore, such a space V_4 is incompatible with the existence of matter iff we start with the Lagrangian (77). Each metrical space stemming from (77) and containing matter must be an Einstein-Cartan space with non-vanishing torsion.

Without meeting any problems, one can of course consider special solutions or approximate cases of Eq. (78) satisfying relation (40). Then one arrives at the above-described special case of a Riemann-Cartan geometry.

Finally, let us mention that, due to this fact, the field equations (78) imply the possibility to realize that version of the Mach-Einstein doctrine, according to which the cosmic matter distribution generates (or induces) the metric field in the sense that it depends on the matter whether a metric exists at all. Indeed, provided there are no solutions of the equations with matter source defining via Eq. (40) a metric g, then the matter can be considered as inducing g.

4. Conclusion

Physically interpretable theories need a metric. However, due to the principle of equivalence, the metric cannot be presupposed *a priori*. Taking this fact into consideration and rejecting any restriction on the principle of general relativity one finds the following situation.

For theories assuming as gravitational field quantities the metric or the metric supplemented by further geometric structures, one is necessarily lead to purely metric theories like GRT and Einstein-Cartan theories. Both, however, suffer from the fact that the only energetic statement one can make consists in that the (metric and canonical, respectively) total energy-momentum density vanishes. In contrast to Einstein-Cartan theories, in GRT this implies an degenerated canonical formalism and related problems for quantum GRT. Otherwise, as to the space problem, metric theories seem to be less problematic since for them it can be solved kinematically, while for Einstein-Cartan theories it has be solved dynamically.

In the case of purely affine theory, one has the same energy problem as in metric and Einstein-Cartan theories. Nevertheless, one arrives at an comparatively simple canonical formalism and does not encounter the space problem in the same way as in Einstein-Cartan theories. In addition, this theory possibly allows to realize the Mach-Einstein doctrine by appropriate Riemann-Cartan solutions of the affine gravitational equations.

5. Notes

1 See also [47] and [14].

2 In order to remove singularities from the theory of gravitation, later Møller [52] also introduced Lagrangians that are quadratic in the Weitzenboeck invariants.

3 For the fibre bundle theory see, for example, [42]; for the above discussion, see also [65].

4 In general, instead of the local Poincaré group (semiproduct of translations and pseudo-orthogonal transformations), one can consider the affine group (semiproduct of translations and general linear transformations) [cf., e.g., [39, 46]]. Here we confine ourselves to the Poincaré group since this group can be interpreted most simply as an requirement following from the principle of general relativity.

5 Following Schouten [59], we here consider the two parts of the Poincaré transformation as associated to anholonomic and holonomic transformations. For another understanding, see the Poincaré gauge field theory ([39] and the literature cited therein). In this approach working in the framework of metric-affine space, the homogeneous (linear) item of the Poincaré group is ascribed to a tensorial connection part Γ^L which has to be added to the inhomogeneous (translational) part Γ^T. This procedure exploits the fact that the sum of a connection and a tensor is again a connection. Apart from the fact that this tensor can be chosen arbitrarily, to our mind, the generalized connection "$\Gamma^L + \Gamma^T$" can conflict with the Weyl lemma (38) establishing the connection between the holonomic and anholonomic representations (see below).

6 This is the notation introduced by Infeld and van der Waerden [40]. In modern literature, sometimes it is also called "soldering form".

7 For a given metric g, (40) has to be considered as a differential equation for Γ whose solution is given by (18). In our case, there is a connection Γ given (as solution of the affine field equations) so that Eq. (40) must be considered as differential equation for g. It is of course a complicated matter to formulate the general conditions under which Eq. (40) can be solved. For a symmetric connection, the conditions on this connection to be a metric connection were given in [58].

8 By means of another line of arguments, this was also shown by Hayashi [31]. Even more, it was demonstrated there that this formalism can only be introduced iff the nonmetricity is of the Weyl form or vanishes.

9 For the sake of generality, in this passage, we shall assume that there exists also a metric which is independent of the connection, and we include the matter term which comes in § 3.4 into the game.

10 To be more precise, the points summarized in this passage are in [8] partly formulated for the case of GRT. But they are evidently true for all Lagrangians of the form (61).

6. References

1. Ashtekar, A., Romano, J. D., and Tate, R. S. (1989) *Phys. Rev.* **D40**, 2572.
2. Ashtekar, A., and Stachel, J. (1991) Conceptual Problems of Quantum Gravity, Birkhäuser, Boston.
3. Band, W. (1942) *Phys. Rev.* **61**, 698.
4. Barbour, J., and Pfister, H. (1995) *Machs Principle: From Newton's Bucket to Quantum Gravity*, Birkhäuser, Boston.
5. Bergmann, P. G. (1962) General relativity of theory, in S. Flügge (ed.), *Encyclopedia of Physics*, Springer, Berlin et. al., vol. II.
6. v. Borzeszkowski, H.-H., and Treder, H.-J. (1993) *Gen. Rel. Grav.* **25**, 291.
7. v. Borzeszkowski, H.-H., and Treder, H.-J. (1996) Mixed field theories and Weyl-Cartan geometry, in P. G. Bergmann, V. de Sabbata, H.-J. Treder (eds.),*Quantum Gravity*, World Scientific, Singapore et al.
8. v. Borzeszkowski, H.-H., and Treder, H.-J. (1996) *Found. Phys.* **26**, 929.
9. v. Borzeszkowski, H.-H., and Treder, H.-J. (1997) *Found. Phys.* **27**, 595;
10. v. Borzeszkowski, H.-H., and Treder, H.-J. (1998) *Found. Phys.* **28**, 273.
11. v. Borzeszkowski, H.-H., and Treder, H.-J. (2000) Bohr's and Mach's conceptions of non-locality in gravitation", in P. G. Bergmann, V. de Sabbata, and J. N. Goldberg (eds.), *Classical and Quantum Nonlocality*, World Scientific, Singapore.

12. v. Borzeszkowski, H.-H., and Treder, H.-J. (2001) *Gen. Rel. Grav.* **33**, No. 8 (to be published).
13. v. Borzeszkowski, H.-H., and Treder, H.-J. (2001) Metric and connection: kinematic and dynamic solutions of the space problem, in W. Jahnke, A. Pelster, H.-J. Schmidt, M. Bachmann (eds.), *Fluctuating Paths and Fields*, World Scientific, Singapore.
14. v. Borzeszkowski, H.-H., Kasper, U., Kreisel, E., Liebscher, D.-E., and Treder, H.-J. (1971) *Gravitationstheorie und Äquivalenzprinzip*, H.-J. Treder (ed.), Akademie Verlag, Berlin.
15. v. Borzeszkowski, H.-H., Treder, H.-J., and Yourgrau, W. (1978) *Ann. Physik* (Leipzig) **35**, 471.
16. Capovilla, R., Dell, J., Jacobson, T., and Mason, L. (1991) *Class. Quantum Grav.* **8**, 41.
17. Cartan, E. (1923) Sur un théorème fundamental de H. Weyl, *Jour. d. math. p. et a.* **2**, 167 (Cf. also: Cartan, E. (1955)*Œuvres complétes*, Gauthier-Villars, Paris, p. 633.)
18. Einstein, A. (1914) Bemerkungen, Appendix to the reprint of the paper by M. Grossmann and A. Einstein, *Math. Phys.* **62**, 260.
19. Einstein, A. (1916) Hamiltonsches Prinzip und allgemeine Relativitätstheorie, *Sitz. Ber. Preuss. Akad. Wiss. Berlin*, 111.
20. Einstein, A. (1923) Zur allgemeinen Relativitätstheorie, *Sitz. Ber. Preuss. Akad. Wiss. Berlin*, 32.
21. Einstein, A. (1928) Neue Möglichkeit für eine einheitliche Feldtheorie von Gravitation und Elektrizität, *Berliner Ber.*, p. 224.
22. Einstein, A. (1928) Riemann-Geometrie unter Aufrechterhaltung des Begriffs des Fernparallelismus, *Berichte Preuss. Akad. Wiss. Berlin*, p.219.
23. Einstein, A. (1929) Zur einheitlichen Feldtheorie, *Berliner Ber.*, p. 2.
24. Einstein, A. (1929) Einheitliche Feldtheorie und Hamiltonsches Prinzip, *Berliner Ber.*, p. 124.
25. Einstein, A. (1950, 1955) *The Meaning of Relativity* (4th and 5th edn.), Princeton University Press, Princeton.
26. Einstein, A. (1968) Remark to Weyl, in H. Weyl, Gravitation und Elektrizität, in *Hermann Weyl, Gesammelte Abhandlungen*, Springer, Berlin et al., vol. 2, p. 40.
27. Einstein, A. (1969) *Über spezielle und allgemeine Relativitätstheorie*, Akademie Verlag, Berlin; Pergamon Press, Oxford; Vieweg, Braunschweig; (21th ed.), App. 5: Relativität und Raumproblem.
28. Einstein, A., and Mayer, W. (1931) Systematische Untersuchung über kompatible Feldgleichungen, welche von einem Riemannschen Raum mit Fernparallelismus gesetzt werden können, *Berliner Ber.*, p. 3.
29. Einstein, A., and Mayer, W. (1932) Semivektoren und Spinoren, *Sitz. Ber. Preuss. Akad. Wiss. Berlin*, 522-550.
30. Freudenthal, H. (1956) *Math. Z.* **63**, 374.
31. Hayashi, K. (1976) *Phys. Lett.* **65B**, 437.
32. Hayashi, K., and Nakano, T. (1967) *Progr. Theor. Phys.* **38**, 49.
33. Hayashi, K., and Shirafuji, T. (1977) *Prog. Theor. Phys.* **57**, 302.
34. Hayashi, K., and Shirafuji, T. (1979) *Phys. Rev.* **D 19**, 3524.
35. Hayashi, K., and Shirafuji, T. (1980) *Progr. Theor. Phys.* **64**, 866, 883, 1435.
36. Hehl. F.W. (1980) Four lectures on Poincaré gauge theory, in P. G. Bergmann and V. de Sabbata (eds.), *Cosmology and Gravitation,* Plenum, New York.

37. Hehl, F.W., van der Heyde, P., Kerlick, G. D., and Nesfer, J. M. (1976). *Rev. Mod. Phys.* **48**, 393.
38. Hehl, F. W., and McCrea, J. D. (1986) *Found. Phys.* **16**, 267.
39. Hehl, F.W., McCrea, D., Mielke, E. W., and Nee'man, Y. (1995) Metric-affine theory of gravity: Field equations, Noether identities, world spinors, and breaking of dilation invariance *Phys. Reports* **258**, 1-171.
40. Infeld, L., and van der Waerden, B. L. (1933) Die Wellengleichung des Elektrons in der allgemeinen Relativitätstheorie, *Sitz. Ber. Preuss. Akad. Wiss. Berlin*, 380-401.
41. Kleinert, H. (2000) *Gen. Rel. Grav.* **32**, 769; Kleinert, H. (2000) *Gen. Rel. Grav.* **32**, 1271.
42. Kobayashi, L., and Nomizu, K. (1963) *Foundations of Differential Geometry* , Interscience, London.
43. Kohler, M. (1952) *Z. Physik* **131**, 571.
44. Kohler, M. (1953) *Z. Physik* **134**, 286, 306.
45. Kopczyński, W. (1982) *J. Phys. A* **15**, 493.
46. Kopczyński, W., and Trautmann, A. (1992) *Spacetime and Gravitation*, Wiley, Chisester and PWN Polish Scientific Publishers, Warsaw.
47. v. Laue, M. (1965) *Die Relativitätstheorie, Zweiter Band. Die Allgemeine Relativitätstheorie*, Friedr. Vieweg & Sohn, Braunschweig, (5th ed.).
48. Liebscher, D.-E. (1973) *Ann. Physik* (Leipzig) **30**, 309 and 321.
49. Lorentz, H. A. (1917) On Hamilton's principle in Einstein's theory of gavitation, *Proc. K. Akad. Wet. Amsterdam* **19**, 751.
50. Møller, C. (1961) *Math.-Fys. Skr. Dansk. Vidensk. Selsk.* **1**, No. 10.
51. Møller, C. (1978) *Math.-Fys. Skr. Dan. Vid. Selskab* **39**, No. 13.
52. Møller, C. (1979) Are the singularities in the theory of gravitation inevitable?, in H.-J. Treder (ed.), *Einstein-Centenarium,* Akademie Verlag, Berlin.
53. Müller-Hoissen, F., and Nitsch, J. (1983) *Phys. Rev. D* **28**, 718.
54. Papapetrou, A. (1948) *Proc. Irish Acad.* **52**, 11.
55. Pauli, W., and Fierz, M. (1939) *Helv. Phys. Acta* **XII**, 297.
56. Pellegrini, C., and Pleba°ski, J. (1963) Tetrad fields and gravitational fields, *Math.-Fys. Skr. Dan. Vid. Selskab* **2**, No. 4.
57. Rosen, N (1940) *Phys. Rev.* **57**, 147, 150.
58. Schmidt, B. G. (1973) *Commun. Math. Phys.* **29**, 55.
59. Schouten, J. A. (1954) *Ricci-Calculus.*, Springer, Berlin.
60. Schrödinger, E. (1950) *Space – Time – Structure*, Cambridge University Press, Cambridge.
61. Treder, H.-J. (1971) *Gen. Rel. Grav.* **2**, 313.
62. Treder, H.-J.: (1978) *Ann. Phys. Leipzig* **35**, 371.
63. Treder, H.-J. (1992) *Found. Phys* **22**, 301.
64. Treder, H.-J. (1994) *Astron. Nachr.* **315**, 1.
65. Treder, H.-J., and v. Borzeszkowski, H.-H. (1973) *Int. J. Theor. Phys.* **8**, 219.
66. Weyl, H. (1924) *Die Naturwissenschaften* **12**, 197-204.
67. Weyl, H. (1923) *Mathematische Analyse des Raumproblems*, Springer, Berlin.
68. Weyl, H. (1929) *Zs. Physik* **56**, 530.
69. Yourgrau, W., v. Borzeszkowski, H.-H., and Treder, H.-J. (1979) *Astron. Nachr.* **300**, 57.

RADIATION AND GRAVITATION

M J CLARK*
Department of Physics and Astronomy, 401 A H Neilsen Physics Building, University of Tennessee, Knoxville, Tennessee 37996-1200, USA

ABSTRACT

The standard physical model for the interaction of radiation with matter, widely used to calculate the transfer of energy, can also be used to model the exchange process for virtual gravitons. Using this approach, it is possible to derive Newton's law of gravitation and to show that the gravitational constant G ($\mathrm{kg}^{-1}\,\mathrm{m}^{3}\,\mathrm{s}^{-2}$) is comprised of other physical constants, including the atomic mass unit, u (kg), the speed of light, c ($\mathrm{m\,s}^{-1}$) and Planck's constant $\hbar$ ($\mathrm{kg\,m}^{2}\,\mathrm{s}^{-1}$). The model is shown to be compliant with the weak equivalence principle and is therefore consistent with experimental observation, apart from the well known exceptions that are explained by General Relativity. An alternative physical model for graviton exchange is also presented which gives an identical formulation for G. The models lead to a prediction of an increase in G with temperature.

PACS Codes: 03.65.-w, 04.60.-m, 12.20.-m.

1. Introduction

The qualitative aspects of quantum field theory can be applied to gravity, particularly the concept of the graviton being a massless, force carrying, virtual particle. Gravitons are assumed to mediate the gravitational force between two particles in exactly the same way that virtual photons mediate the electrostatic force between two charged particles. Both the electrostatic and the gravitational force are long range because photons and gravitons are massless, and they both obey the inverse square law, because the forces depend on the continual fluence of virtual particles. In this paper, it will be shown that the concept of a fluence of virtual gravitons emanating from a mass and exchanging with nucleons in other masses can be applied in a physical model for gravitational interactions. The model used is based on the standard physical model for the interaction of quantum radiation with matter, widely used in atomic and nuclear physics, and in radiological protection and dosimetry.

The concept of fluence (m^{-2}) of virtual gravitons will be developed here and used to derive Newton's law of gravitation. Fluence concepts are used widely in physical models for quantum interactions of real particles and the transfer of energy[1, 2]. Real particles are free and are directly observable independently from their source. Virtual particles are fundamentally different. They are bound, messenger particles which mediate forces between real particles, and they are not directly observable. Despite being non-observable, virtual particles produce real observable effects, and the tremendous success of Quantum Electrodynamics is testament to the utility of the concept of virtual particle exchange processes[3, 4, 5]. In order to account for the differences between real and virtual particles, the concept of a fluence of virtual particles must have constraints so that their distinct properties are incorporated. The primary constraint is that it should not be possible to distinguish between a virtual particle either being emitted or being absorbed, and therefore any physical model must be completely reversible in space and time. Virtual particles are simply

* M J Clark is at the National Radiological Protection Board, Chilton, Didcot, OX11 0RQ, UK and this address should be used for correspondence (e-mail *mike.clark@nrpb.org.uk*).

P.G. Bergmann and V. de Sabbata (eds.),
Advances in the Interplay Between Quantum and Gravity Physics, 77–84.

exchanged between mass-energy, as shown in Fig.1. for atomic mass units u (kg). The models described here comply with this constraint.

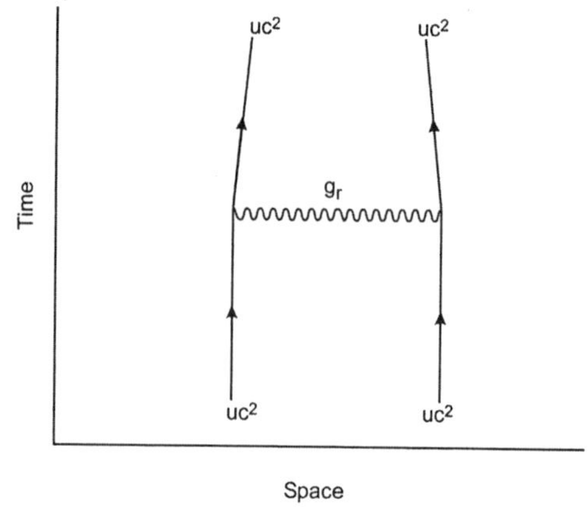

Fig. 1. Feynman diagram for graviton exchange (g_r) between atomic mass units (amu, u).

2. Basic postulates for a virtual graviton fluence model

Assume a mass, M (kg), is emitting virtual gravitons whose fluence at the surface of the mass is Φ_o (m^{-2}). It is postulated that the gravitational force F_g (kg m s^{-2}) acting on another mass m (kg), positioned at a distance X (m) from M, is proportional to the product of the mass-energy, $M\,c^2$, and the fluence of virtual gravitons at its surface, Φ_o. Hence, $F_g \propto M\,c^2\,\Phi_o$, where c (m s^{-1}) is the velocity of light. For masses experiencing gravitational forces, M is the relativistic mass and therefore $M\,c^2$ is equal to the total energy, a quantity conserved in all frames of reference. If mass M is emitting gravitons in all directions in space, only a certain fraction of these gravitons will exchange with nucleons in the other mass m. Hence, $F_g \propto f_X M\,c^2\,\Phi_o$, where f_X is a geometrically derived factor used to calculate the fraction of virtual graviton fluence incident on m at any arbitrary distance from the surface, X (m). In general, f_X can be derived using the divergence theorem (Gauss's Law). Assuming that both M and m are spherical objects with radii R_M and r_m respectively, then f_X will be the ratio of the surface cross-sectional area of m to that of an imaginary sphere of radius X, giving $f_X = \pi\,r_m^{\,2} / 4\,\pi\,X^2 = r_m^{\,2} / 4\,X^2$. Although the fraction of graviton fluence incident on mass m can be calculated, not all gravitons are necessarily exchanged with nucleons in m. Hence another factor, f_a, is required to allow for the absorption of gravitons via the exchange process, giving $F_g = k\,f_a\,f_X M\,c^2\Phi_o$, where k (m) is a constant of proportionality which for the present will be called the exchange constant. The nature of k will be examined later in this paper.

3. Bethe's formula for quantum interactions

Photon radiation obeys a simple exponential law of absorption[6], $I = I_0 \exp(-\mu r)$, where I is the intensity of a monochromatic parallel beam attenuated through a thickness r in a mass, I_0 is the intensity of the beam incident at the surface of the mass and μ *is* a constant. Using atomic theory and nuclear physics, Bethe[7] derived a formula, $\mu\,r = N_A\,\sigma\,\rho_m\,d_r\,/A$', where N_A is Avogadro's number ($mole^{-1}$), ρ_m is the density (kg m^{-3}) of the material in mass m, σ is the cross-sectional area (m^2) of atoms in m, d_r is the mean distance (m) travelled by particles in m and A' is the atomic or molar mass (kg mole^{-1}) of the material in m. In terms of particle fluence, this process of absorption or attenuation is given by $\Phi = \Phi_0 \exp(-N_A\, \sigma\, \rho_m\, d_r / A')$, where Φ_0 is the incident fluence and Φ is the fluence at a depth d_r. It is assumed that an exponential absorption law applies similarly to the exchange of virtual gravitons, and the assumption of a plane parallel beam is a reasonable geometry for gravitons exchanging at large distances. The fraction of the gravitons interacting with the body in space will therefore be, $f_a = 1 - \exp(-N_A \sigma\, \rho_m\, d_r / A')$. Substituting this and rearranging, the expression for F_g becomes,

$$F_g = k\left[1 - \exp\left(\frac{-N_A \sigma \rho_m 4r_m/3}{A'}\right)\right]\frac{r_m^2}{4X^2} M c^2 \Phi_0 \qquad (1)$$

This assumes that both masses are spherical and the mean distance in a broad parallel geometry is $4r_m/3$ for a sphere of radius r_m.

In Bethe's formula the number density (m^{-3}) of atoms or molecules in a given material is $N_A\, \rho_m / A'$, where Avogadro's number N_A is the number of atoms or molecules per gram mole, while the density ρ_m is kg m^{-3}. In SI units, A' must be the atomic or molar mass in kilograms of one gram mole (kg mole^{-1}). If the graviton exchange cross sections for nucleons are assumed to be the same (σ), then Bethe's formula simplifies to $N_A\,(A\, \sigma / A')\rho_m\, d_r$, where A is the atomic mass number. For any individual isotope, A/A' is simply 1/1 10^{-3} kg mole^{-1} and therefore $N_k = N_A A/A' = 6.022\ 10^{26}$ kg^{-1}, the number of nucleons in a kilogram of any material. Strictly, $N_k = 1/u$ where u is the atomic mass unit (kg) and σ is the cross section for an atomic mass unit (amu) and this aspect will be examined in more detail later. To ensure consistency in SI units and to maintain a link with Bethe's original formulation, the adapted form of Avogadro's number, N_k (kg^{-1}), will also be used in this paper.

4. Interaction cross sections and Newton's law

The interaction cross section for individual nucleons, σ (m^2), can be defined as an effective area that the nucleon presents to the virtual graviton fluence for the exchange process. In terms of quantum mechanics and a probability of interaction, P_i, an interaction cross section is more precisely defined[1] as $\sigma(m^2) = P_i / \Phi\ (m^{-2})$. In the case here, $P_i = \alpha_g$, the coupling constant for virtual graviton exchange between nucleons in mass M, and nucleons in other masses, and Φ (m^{-2}) is the graviton fluence. Substituting this in the expression for F_g and assuming α_g to be very small, thereby making the exponential term very small, then because $1 - \exp(-\delta x) = \delta x$, the expression for F_g becomes;

$$F_g = kN_K \frac{\alpha_g}{\Phi_0}\frac{m}{4\pi r_m^3/3}\frac{4r_m^3/3}{4X^2} M c^2 \Phi_0 = kN_K \alpha_g \frac{1}{4\pi} c^2 \frac{Mm}{X^2} \qquad (2)$$

Clearly equation (2) is a direct analogue of Newton's law of gravitation $F_g = GMm/X^2$ and exactly the same equation is obtained if M or m are irregular, non-spherical masses. Furthermore, exactly the same equation is obtained if the fluence of gravitons from mass m is assumed to be incident on mass M at a distance of X, which is consistent with the interaction being a virtual exchange process. It should not be possible to distinguish between virtual gravitons emanating from M and those coming from m.

The graviton exchange model shows that the gravitational constant G ($kg^{-1}\,m^3\,s^{-2}$) can be expressed in terms of other physical constants: the exchange constant k (m); Avogadro's number (adapted) N_k (kg^{-1}) = $1/u$, where u is the atomic mass unit (amu, kg); the coupling constant for virtual graviton exchange α_g; the speed of light c ($m\,s^{-1}$); and π.

$$G = \frac{k}{4\pi} N_K \alpha_g c^2 = \frac{k}{4\pi}\frac{1}{u}\alpha_g c^2 \qquad (3)$$

At this stage, the precise nature of the exchange constant k is unknown. It will be examined further in sections 6 and 7.

5. Compliance with the weak equivalence principle

The use of Bethe's formula above assumed a single cross section for graviton exchange, σ, for all nucleons. If instead mass-energy as "charge" is assumed for the graviton exchange process, could there be small differences in exchange probability due to the different mass energies of nucleons and electrons? Assuming different interaction cross sections for graviton exchange between protons, σ_p, electrons, σ_e and neutrons, σ_n, a more exact expression of Bethe's formula is,

$$\mu r = N_A \left[\frac{Z(\sigma_p + \sigma_e) + (A - Z)\sigma_n}{A'} \right] \rho d_r \qquad (4)$$

Here Z is the atomic number (number of protons), A is the atomic mass number (number of protons and neutrons) and A' is the atomic or molar mass. If σ_p, σ_e and σ_n are assumed to be directly proportional to the mass of each particle and due account is taken of nuclear binding energy[8, 9] by subtracting a notional cross section σ_{be}, the formula simplifies to $(\sigma_{amu}/u)\,\rho\, d_r$ exactly for any chosen atom or isotope. This implies that α_g, and therefore also G, should be a constant irrespective of the composition of the masses involved, because α_g is the probability of virtual graviton exchange between a defined quantity of mass-energy. Hence all matter and mass-energy (including anti-matter) will experience the same acceleration in a gravitational field. This is consistent with many laboratory and cosmological observations of the principle of equivalence which have set extremely low upper limits on the extent to which gravitational forces are influenced by the composition of the attracting masses[10, 11, 12, 13, 14]. There is a continuing experimental program in this area, including proposals to measure G using masses in orbital satellites[15].

6. Summation of all virtual exchange paths

A formula for G can also be obtained by considering the total number of possible graviton exchanges between atomic mass units (amus) in the masses M and m. The atomic mass unit has been chosen because it is an established standard, and it is assumed that the force of gravity is proportional to the mass-energy of two amus exchanging virtual gravitons, $F_g \propto 2uc^2$. The total number of possible graviton exchanges between amus in M and m is the product $(M/u)(m/u)$ and the coupling constant for exchange is α_g. Incorporating the divergence theorem and assuming spherical symmetry, the factor $1/4\pi X^2$ is obtained. Combining these terms and incorporating a constant of proportionality λ_g (m), the equation derived for F_g is similar to that given in equation (2),

$$F_g = \lambda_g \frac{1}{4\pi X^2} \alpha_g \left(\frac{M}{u}\right)\left(\frac{m}{u}\right) 2uc^2 = \lambda_g \frac{1}{2\pi} \alpha_g \frac{1}{u} c^2 \frac{Mm}{X^2} \qquad (5)$$

bearing in mind that $1/u = N_k$. Equations (2) and (5) would exactly the same if $k = 2\lambda_g$. A schematic diagram of the components of this physical model is shown in Fig. 2. This model is intrinsically reversible in space and time because it is based on a straightforward summation of all possible exchange pathways. The concepts of emission or absorption of the graviton fluence are not required here.

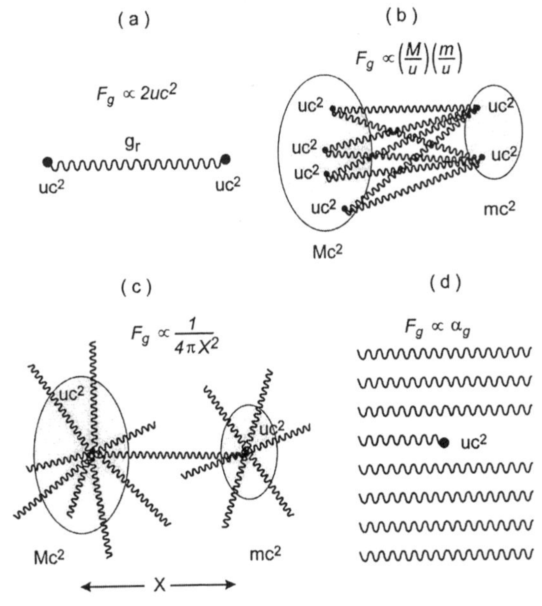

Fig. 2. Schematic diagram of assumptions in the pathways model for virtual graviton exchange:

(a) The force of gravity F_g is proportional to the mass-energy of two atomic mass units (amu, u).
(b) F_g is proportional to the total number of possible virtual graviton exchanges between atomic mass units (amu, u) in masses M and m, which is the product $(M/u)(m/u)$.
(c) Apply the divergence theorem (Gauss's law) to allow for the effect of distance X between M and m, giving $F_g \propto 1/4\pi X^2$.
(d) F_g is proportional to the gravitational coupling constant, α_g.

The formulation in equation (5) has the advantage of clarifying how any relativistic corrections should be applied in the model. The atomic mass unit u is a precisely defined rest mass quantity and is therefore not subject to any corrections. In contrast, M and m are relativistic masses and therefore corrections should appear in the $(M/u)(m/u)$ terms, and will appear as a product of the two appropriate terms. Such corrections mean that measurements of G will increase with temperature. In principle this should be measurable, but a significant increase is only likely to be observed at extremes of temperature, i.e., above 1000K and experimental verification is therefore likely to be very difficult. At these temperatures radiation energy becomes significant, and above 3000K it becomes comparable with matter energy[9]. As pointed out by Gillies[11], virtually all published experimental investigations of variations of G with temperature were carried out before 1925 and showed null results. Because of the practical difficulties, only a relatively narrow temperature range below 100 C has been investigated. The increase in gravitational forces with increasing temperature is discussed in chapter 3 of reference 4, and others have considered the prediction of an increase in G with temperature[16, 17, 18].

7. The Compton wavelength

It is shown elsewhere[19] that, by applying the second model described in Sec. 6. to virtual photon exchange between charged particles, the λ_g term is the Compton wavelength of the particles. Substituting the Compton wavelength for the atomic mass unit u into equation (3) derived above, the following formula for G is obtained,

$$G = \lambda_g \frac{1}{2\pi} \alpha_g \frac{1}{u} c^2 = \frac{h}{uc} \frac{1}{2\pi} \alpha_g \frac{1}{u} c^2 = \frac{\hbar \alpha_g c}{u^2} \qquad (6)$$

A similar formula can be derived[9] using dimensional analysis to calculate a value for α_g, but this usually includes the mass of the proton m_p or a nucleon rather than u. Here a formula for G has been derived instead on the basis of two separate physical models for virtual graviton exchange between mass-energy.

The formula provides guidance to where experimental studies on G should focus. As mentioned above, there is a clear prediction that G will increase with temperature because the formulation is based on relativistic masses, and there are also cosmological arguments that α_g will increase with temperature. In contrast, because $\hbar$, c, and u are fundamental constants, laboratory experiments attempting to discover a composition dependence of G are likely always to produce null results. Moreover, the value for α_g is so small ($5.82\ 10^{-39}$) the effects of gravitational shielding will also be null for all practical purposes. Calculations elsewhere show[19] that the half value layer (HVL, the thickness[2] of material required to reduce a fluence of particles to half its incidence level) for gravitational shielding is in excess of $1\ 10^{29}$ light years of lead. This is a vast thickness of material, the distance being greater than the scale of the observable universe. Hence gravitational shielding is unlikely to be observed in the cosmos, let alone in laboratory experiments.

8. Planck quantities

The formula for G can provide useful clarification of concepts that incorporate G when they are derived on purely dimensional grounds. For example, it offers a new insight into the Planck length l_P, showing that it incorporates the Compton wavelength for u,

$$l_P = \left(\frac{\hbar G}{c^3}\right)^{1/2} = \left(\frac{\hbar^2 \alpha_g c}{c^3 u^2}\right)^{1/2} = \frac{\hbar}{uc}\sqrt{\alpha_g} \qquad (7)$$

Bearing in mind that a probability of interaction P_i is the square of the sum of probability amplitudes A_n, then $\alpha_g = (A_1 + A_2 + A_3 + A_4 + \ldots A_N)^2 = A_g^2$, we can write $l_P = 1/2\pi\,(\lambda_C\, A_g)$. As noted above, $\alpha_g = 5.82\ 10^{-39}$ and therefore $A_g = 7.63\ 10^{-20}$ which means that $l_P = 1/2\pi(1.33\ 10^{-15} \times 7.63\ 10^{-20}) = 1.62\ 10^{-35}$ m. The substitution of the formula for G into the formula for the Planck mass is similarly illuminating,

$$m_P = \left(\frac{\hbar c}{G}\right)^{1/2} = \left(\frac{\hbar c u^2}{\hbar \alpha_g c}\right)^{1/2} = \left(\frac{u^2}{\alpha_g}\right)^{1/2} = \frac{u}{A_g} \tag{8}$$

Using the same value for A_g, this gives $m_P = 2.18\ 10^{-8}$ kg. Hence the numerical parameter that connects Planck quantities to other constants of length, mass and time, is simply the sum of probability amplitudes for virtual graviton exchange between the defined units of mass-energy.

9. Conclusions

The use of physical models for virtual graviton exchange between mass-energy gives results that are entirely consistent with Newtonian gravitation. The gravitational constant G is shown to be composite of other physical constants, $\hbar$, c, u and α_g, the coupling constant for virtual graviton exchange. The model predicts that measurements of G will increase with temperature of the masses involved, and the formula for G gives a new interpretation of dimensionally derived quantities such as the Planck length.

Acknowledgements

I would like to thank the Directors of the International School of Cosmology and Gravitation for the invitation to present this paper to the 17th course at the Ettore Majorana Centre for Scientific Culture in Erice, Sicily. I would also like to thank colleagues at NRPB and elsewhere who have commented on the ideas and encouraged me to publish a paper. In particular, I am very grateful to Professor G T Gillies (University of Virginia) and Professor A Sanders (University of Tennessee) for their interest and practical help to do this.

References

[1] International Commission on Radiation Units and Measurements, in *Fundamental Quantities and Units for Ionising Radiation, ICRU Report 60* (ICRU, Bethesda MD, 1998).
[2] A. B. Chilton, J. K. Shultis & R. E. Faw, in *Principles of Radiation Shielding* (Prentice-Hall, NJ, 1984).
[3] P. A. M. Dirac, in *Quantum Mechanics* (4th edition, Oxford: Clarendon Press, 1958).
[4] R. P. Feynman, F. B. Marinigo and W. G. Wagner, in *Feynman Lectures on Gravitation*, edited by B. Hatfield (Addison-Wesley, Reading MA, 1995).
[5] P. C. W. Davies, in *The Forces of Nature* (2nd Edition Cambridge University Press, Cambridge UK, 1987).
[6] R. B. Leighton, in *Principles of Modern Physics* (McGraw-Hill, N Y, 1959).
[7] H. A. Bethe and J. Ashkin, in *Experimental Nuclear Physics*, edited by E. Segre (J. Wiley & Sons, NY, 1953), **Vol. 1,** Chap. Passage of radiations through matter.
[8] W. S. C. Williams, in *Nuclear and Particle Physics* (Clarendon Press, Oxford UK, 1996).
[9] J. W. Rohlf, in *Modern Physics from α to Z^0* (J Wiley & Sons,Inc., New York, 1994).
[10] A. Cook, Rep. *Prog. Phys*. **51**, 707-757 (1988).
[11] G. T. Gillies, *Rep. Prog. Phys.* **60**, 151-225 (1997).
[12] R. v. Eotvos, D. Pekar and E. Fekete, *Ann. Phys*.(Leipzig) **68**, 11 (1922).
[13] J. H. Gundlach, G. L. Smith, E. G. Adelberger, B. R. Heckel and H. E. Swanson, *Phys. Rev. Lett.* **78**, 2523 (1997).
[14] G. L. Smith, C. D. Hoyle, J. H. Gundlach, E. G. Adelberger, B. R. Heckel and H. E. Swanson, *Phys. Rev. D* **61**, 022001 (1999).
[15] A. L. Sanders et alia, *Meas. Sci. Tech.* **10**(6), 514-524 (1999).
[16] C. Massa , *Helv. Phys. Act.* **62**, 420-423 (1989).
[17] C Massa, *Helv. Phys. Act.* **62**, 424-426 (1989).
[18] A. K. T. Assis and R. A. Clemente, *Nuovo Cimento B* **108**(6), 713-6 (1993).
[19] M J Clark, submitted for publication.

Erice 2001

International School of Cosmology and Gravitation

Torsion and Curvature as commutator for Quantum Gravity

Venzo de Sabbata(*) Luca Ronchetti(**)

and

Alfred Yu(*)

(*) Dept.of Physics, University of Bologna, Italy

(**) Dept.of Physics, University of Ferrara, Italy

Introduction

The problem that will be discussed is that of quantum gravity; first of all we will explain which road we follow in this paper and, after that, we will see if it is possible to quantize general relativity in a real space-time. For that reasons we begin by considering the Einstein - Cartan theory [1] i.e. general relativity plus torsion, that from physical point of view means the introduction of spin in the theory of gravity.

We know [1,2,3] that one of the most important geometrical properties of torsion is that a closed contour in U_4 manifold becomes non-closed contour in the flat space-time V_4 i.e.the integral

$$1^{\alpha} = \oint Q^{\alpha}_{.\beta\gamma} dx^{\beta} \wedge dx^{\gamma} = \oint Q^{\alpha}_{.\beta\gamma} dS^{\beta\gamma} \neq 0 \qquad (1)$$

over a closed infinitesimal contour is different from zero

P.G. Bergmann and V. de Sabbata (eds.),
Advances in the Interplay Between Quantum and Gravity Physics, 85–101.

and can be treated as defects in space-time. But we know also that torsion is linked to the intrinsic spin ħ and, as the spin is quantized, we can say that the defects in space-time topology should occur in multiples of the Planck length $(\hbar G/c^3)^{1/2}$ i.e.

$$\oint Q^{\alpha}_{\cdot\beta\gamma} dx^{\beta} \wedge dx^{\gamma} = n\ (\hbar G/c^3)^{1/2} n^{\alpha} \qquad (2)$$

(n is an integer and n^{α} a unit vector).
Eq. (2) is analogous to the well known relation $\oint p\,dq = n\hbar$, i.e. the Bohr-Sommerfeld relation and would define a minimal fundamental length and a minimal time, i.e.the Planck length entering through the minimal unit of spin or action ħ. So ħ has to deal with the intrinsic defect built into torsion structure of space time.

We can also observe (see ref.[1]) that from relation (2), considering the fourth component, we can define time at the quantum geometric level through torsion as

$$t = (1/c) \oint Q\ dA = n(\hbar G/c^5)^{1/2} \qquad (3)$$

so torsion is essential to have a minimum unit of time ≠ 0! This in fact would give us the smallest definable unit of time as $(\hbar G/c^5)^{1/2} \cong 10^{-43}$s. In the limit of $\hbar \Longrightarrow 0$ (classical geometry of general relativity) or $c \Longrightarrow \infty$ (Newtonian case), we would recover the unphysical $t \Longrightarrow 0$ of classical cosmology or physics. So both ħ and c must be finite to give a geometric unit for time (i.e.$\hbar \Rightarrow 0$ and $c \Rightarrow \infty$ are equivalent). The fact that ħ is related to a quantized timelike vector, discretize time. This quantum of time or minimal unit of time has also an experimental confirmation when considering a limiting frequancy $f_{max} \approx (c^5/\hbar G)^{1/2}$ in perturbative QED, in estimating self energies of electrons and other particles, i.e. the self energy integral (in momentum space) taken over the momenta of all virtual photons.

To make the integral converge Feynman in his paper on QED [4], multiplied the photon propagator, k^{-2}, by the ad hoc factor: $-f^2/(k^2- f^2)$, where k is the frequancy (momentum) of the virtual photon. This convergence factor although it preserves relativistic invariance, is objectionable because of its ad hoc character without any theoretical justification. Feynman considers f to be arbitrarily large without definite theoretical basis.

Here the presence of space-time defects associated with the torsion due to the intrinsic spin would give a natural basis for the maximal value for $f^2_{max} \approx c^5/G\hbar \approx 10^{86}$ which appears as a high frequency cut off which is not arbitrary.

Returning back to eq.(2), we can say that in this way we have quantized the torsion field; but what about the curvature? We believe that there is the possibility of including also the curvature. In fact, following ref.[1] we can show that torsion Q and curvature R can be conjugate variables for which we can write commutation relations. We know from the equation of geodesic deviation that curvature causes relative acceleration between neighbouring test particles so that we have momentum uncertainty related to curvature as:

$$ma^{\mu}ds = \Delta p^{\mu} = m\, R^{\mu}{}_{\alpha\beta\gamma} \frac{dx^{\alpha}}{ds} dx^{\beta}\eta^{\gamma} = mcR^{\mu}\, dS \qquad (4)$$

(where η^{γ} is the separation vector between neighboring geodesic). So as position fluctuations are given by torsion, momentum fluctuation are due to curvature and we can interpret quantum effects (and then uncertainty principle) as consequences of space-time deformation i.e.

$$\Delta p^{\mu}\cdot\Delta x_{\mu} \geq \hbar \qquad (5)$$

where

$$\Delta x^{\mu} = Q^{\mu}dS \qquad (6)$$

and

$$\Delta p^{\mu} = mcR^{\mu}dS \qquad (7)$$

We see that Q (torsion) and curvature (R) play the role of conjugate variables of the geometry (gravitational field), thus enabling us to write commutation relations between curvature and torsion (analogous to $[x,p] = i\hbar$) as:

$$\left[Q\ ,\ R\right] = i(\hbar G/c^{3})^{-3/2} = iL_{Pl}^{-3} \qquad (8)$$

where L_{Pl} is the Planck length.

At this point we like to remember the important works of Treder and Borzeszkowski [5,6,7,8] who consider the Einstein-Schrödinger affine theory where the field coordinates are Einstein affine tensors $U^i_{kl} = \Gamma^i_{kl} - \Gamma^r_{kr}\delta^i_l$ and the conjugate momenta are given by the tensor $R^{kl}\delta^0_i$ (where $\Gamma^i_{[kl]} \neq 0$ and $R_{kl} \neq R_{lk}$. So in purely affine theory of Einstein and Schrödinger we have (see [7,8]): $[U^a_{bc}, R^{kl}\delta^0_i] = -i\ \delta^{akl}_{bci}\delta^3(x)$ where $R_{kl} \neq R_{lk}$ is again the Einstein tensor and $U^i_{kl} = \Gamma^i_{kl} - \Gamma^r_{kr}\delta^i_l$ Einstein's affine tensor with $U^i_{[kl]} \neq 0$.

We stress the fact that in the papers [1,3] [7,8] we are working with "non- canonical" commutation rules between ordinary space- coordinates x^i and the torsion Q^i_{kl} respectively the curvature R^i_{kml} where these coordinates are not conjugate to Q and R and then these commutation rules are not founded canonically but only by geometrical hypothesis, in analogy to the (semi-classical) quantization of Bohr and Sommerfeld.

Now in a first attempt to consider torsion and curvature as conjugate variables enabling us to write commutation relations we like to examine torsion and curvature in the context of geometrical algebra in the four dimensional space-time [9] in order to have both, spinor and tensor, in the real space-time. As we know that in Dirac theory we have a totally antisymmetric spin density, we are in position to introduce the torsion tri-vector [3]

$$Q = Q^{\alpha\beta\gamma}\gamma_\alpha \wedge \gamma_\beta \wedge \gamma_\gamma \qquad (9)$$

as element of space-time algebra, where $\{\gamma_\alpha\}$ are the base vectors for which we have

$$\gamma_\alpha \gamma_\beta = g_{\alpha\beta} + \gamma_\alpha \wedge \gamma_\beta . \qquad (10)$$

Moreover, given the curvature bivector

$$\Omega^{\alpha\beta} = (1/2) R^{\alpha\beta\mu\nu} \gamma_\mu \wedge \gamma_\nu \qquad (11)$$

one can form the curvature tri-vectors

$$R^\alpha = \Omega^{\alpha\beta} \wedge \gamma_\beta \ . \qquad (12)$$

Now we have the trivectors Q and R^α. Consider now the antisymmetric part of the geometric product

$$[Q, R^\alpha] = (1/2)(QR^\alpha - R^\alpha Q) \qquad (13)$$

This type of product between two tri-vectors gives a bivector (for example it is easy to verify

$$[\gamma_0 \wedge \gamma_1 \wedge \gamma_2 \ , \ \gamma_0 \wedge \gamma_1 \wedge \gamma_3] = \gamma_2 \wedge \gamma_3 \qquad (14)$$

remembering that $\gamma_0 \gamma_1 \gamma_2 = -i\gamma_3$ where 'i' indicates the pseudoscalar unit. In the language of geometric algebra, the imaginary unit of complex numbers is substituted by a bivector; then we can have commutation relations of canonic type from eqs.(9,11,12):

$$[Q\ ,\ R^{a}]\ \equiv\ (1/2)(QR^{\alpha} - R^{\alpha}Q) =$$

$$= (1/2)\left[Q^{\mu\nu\beta}\gamma_{\mu}\wedge\gamma_{\nu}\wedge\gamma_{\beta}\ ,\ R^{\alpha\tau\nu\beta}\gamma_{\tau}\wedge\gamma_{\nu}\wedge\gamma_{\beta}\right] =$$

$$= (1/2)(Q^{\mu\nu\beta}R^{\alpha\tau}{}_{\nu\beta} - R^{\alpha\mu}{}_{\nu\beta}Q^{\tau\nu\beta})\gamma_{\mu}\wedge\gamma_{\tau} = \gamma_{2}\wedge\gamma_{1}\ L_{Pl}^{-3} \qquad (15)$$

for every α, where, being the first member a bivector, also the second member is a bivector, coherently with the fact that the imaginary unit in Dirac equation is substitute by the bivector $\gamma_2\wedge\gamma_1$ (see, for instance [2]).

2. A quadratic Hamiltonian

In order to give a stronger basis to the consideration of torsion and curvature as canonical conjugate variables, we try to see if is possible to improve the theory by introducing a quadratic Hamiltonian function of torsion and curvature. We use as conjugate variable the torsion and the curvature trivectors because the first Bianchi identity (see [2]) guarantees that it is different from zero only in the presence of torsion. Since the commutation relations between Q and R^a are defined in the geometric algebra we need to modify the usual lagrangian field theory.

Whitout going into details (see ref.[3]) we can introduce the following quadratic Lagrangian in torsion and curvature in order to have the usual expression for conjugate variables in a quantum theory:

$$\mathcal{L} = (\hbar c/2) R^a R_a - (c^4/2G) Q^a Q_a \tag{16}$$

$$\mathcal{L} = (\hbar c/2) \left(R^a R_a - Q^a Q_a / L_{Pl}^2 \right) \tag{17}$$

In agreement with the Bianchi identities, the momenta conjugate to the torsion bivectors are the trivectors:

$$\Pi^a = \partial\mathcal{L}/\partial(DQ^a) = \hbar c R^a \tag{18}$$

(in natural unit $\hbar=c=1$ we can write Π^a or R^a without distinction) The Hamiltonian results:

$$H = DQ^a \Pi_a - \mathcal{L} = (\hbar c/2) \left(R^a R_a + Q^a Q_a / L_{Pl}^2 \right) \tag{19}$$

Then we have the Hamilton equations:

$$\begin{cases} DQ^a = \partial H/\partial R^a = R^a \\ DR^a = - \partial H/\partial Q^a = - Q^a/L_{Pl}^2 \end{cases} \tag{20}$$

where D is a covariant operator in presence of torsion (see eq.(51) of ref,[3]).

The starting point are the results of ref.[10]: the field equation for the multivector ψ can be written in the manifestly invariant way:

$$\nabla \left(\frac{\partial\mathcal{L}}{\partial(\nabla\psi)} \right) = \frac{\partial\mathcal{L}}{\partial\psi} \tag{21}$$

where the gradient operator $\nabla = \gamma_\mu \partial^\mu$ acts as the geometric product.

The presence of ∇ or, in other hand, the peculiarity of the intrinsic geometric calculus, suggests that we can define the conjugate momentum field as

$$\Pi = \frac{\partial \mathcal{L}}{\partial (\nabla \psi)} \tag{22}$$

and the Hamiltonian

$$H = (\nabla \psi) \Pi - \mathcal{L} \tag{23}$$

This is different from the usual $H = \Pi \dot{\psi} - \mathcal{L}$ formula, but, following eq.(21) for the Lagrangian, we propose also a modified Hamiltonian that is manifestly invariant like eq.(21) so we cannot work with $\partial \psi / \partial x^0$ because it depends from the choice of the time-coordianate x^0. So $\Pi = \frac{\partial \mathcal{L}}{\partial (\nabla \psi)}$ is the natural choice; in fact eq.(21) generalizes the Lagrangian eq. $(d/dt)(\partial \mathcal{L}/\partial \dot{q}) - \partial \mathcal{L}/\partial q) = 0$ simply substituting d/dt with ∇.

Notice that eq.(21) allows for vectors, tensors and spinors variables, to be handled in a single equation a considerable unification (see ref.[10]).

So we have the Hamilton equations

$$\begin{cases} \nabla \psi = \partial H / \partial \Pi \\ \nabla \Pi = - \partial H / \partial \psi \end{cases} \tag{24}$$

For example we have the Maxwell equations in vacuum taking $\mathcal{L} = F^2/8\pi$ where the vector potential $A = A^{\mu}\gamma_{\mu}$ and the bivector $F = \nabla \wedge A$ are conjugate variables (remembering that the geometric product gives $F = \nabla A = \nabla \cdot A + \nabla \wedge A$ and Lorentz condition $\nabla \cdot A = 0$).

Moreover since we use the curved U_4 manifold in agreement with the minimal coupling principle, we substitute the covariant operator D for ∇ defining

$$D \equiv \nabla + \gamma_a [\omega^a, \quad] \tag{25}$$

where the $\{\gamma_a\}$ is a orthonormal frame of tetrads, (for which

$$\gamma_a \gamma_b = \eta_{ab} + \gamma_a \wedge \gamma_b), \tag{26}$$

(related with the coordinate system $\gamma_{\mu} = e^a_{\mu}\gamma_a$) (see (see ref.[11]) and $\omega^a = (1/2)\omega^{abc}\gamma_b \wedge \gamma_c$ is the connection bivectors. If in the eq. (14) we put D instead of ∇, we have

$$D\left(\frac{\partial \mathcal{L}}{\partial (D\psi)}\right) = \frac{\partial \mathcal{L}}{\partial \psi} \tag{27}$$

In that way introducing the quadratic Lagrangian (17), we find the Hamilton equations (20).

In that way we have shown that torsion bivector and curvature trivector are canonical conjugate variables.

3. Commutator and anticommutator

Now we must modify the commutation relations found in the

section 1, because in the previous works we have considered torsion and curvature trivectors, while here we have to do with a bivector Q^{α} (torsion) and a trivector R^{α} (curvature) which are our canonical conjugate variables.

In other words we try to improve the theory and we find important developments about the possibility to have commutation relations between torsion and curvature with some consequences regarding considerations about supersymmetry.

As we have seen, the commutators in eq.(15) allow a simple interpretation in agreement with the geometric content of Dirac equation, but they are not entirely satisfactory because in the Hamiltonian (19) the conjugate variables are the torsion bivector Q^{α} and the curvature trivector R^{α}, while in eq.(15) the commutation is made between trivectors.

In fact working with the geometric algebra it is possible to have commutation relations between multivectors of different grade and not only between geometric objects of the same type. Yet eq.(15) has meaning and it is not to be rejected, because we have started with a totally antisymmetric torsion and then the torsion trivector Q is the complete and single object of geometric algebra which contains the torsion field.

Now the geometric product between a grade-r multivector A_r and a grade-s multivector B_s can be decomposed as (see ref.[12] page 10, eq.1.36)

$$A_r B_s = (A_r B_s)_{r+s} + (A_r B_s)_{r+s-2} \quad .....+ \quad (AB)_{|r-s|} \qquad (28)$$

If r = 2 and s = 3, being in a four dimensional space-time, we are left with two terms: $(AB)_1$ and $(AB)_3$. In other words the geometric product BT, where B is a generic bivector and T a generic trivector, can be easily calculated remembering that every trivector T is dual of some vector A (see ref.[13] page 107 eqs.(B1,B2)) i.e.

$$T = iA \qquad (29)$$

where i = $\gamma_0\gamma_1\gamma_2\gamma_3$ is the unit pseudoscalar and then A is the vector dual of T (remember that the dual application, the multiplication by i transform an r-vector in (4-r) vector). Moreover

$$BA = B\cdot A + B \wedge A \qquad (30)$$

where the inner product

$$B\cdot A = (1/2)/(BA - AB) \equiv [B,A] \qquad (31)$$

has grade 1 and the outer product

$$B \wedge A = (1/2)/(BA + AB) \equiv \{B,A\} \qquad (32)$$

has grade 3.

Then, using also iB = Bi, one finds

$$BT = BiA = iBA = i[B,A] + i\{B,A\} = [B,T] + \{B,T\} \qquad (33)$$

where the commutator between B and T has grade 3 (trivector) and the anticommutator has grade 1 (vector).

Finally we can write

$$Q^{\alpha}R_{\alpha} = [Q^{\alpha},R_{\alpha}] + \{Q^{\alpha},R_{\alpha}\} \tag{34}$$

where

$$[Q^{\alpha},R_{\alpha}] = (1/2)Q^{\alpha\mu\nu}R_{\alpha\mu}{}^{\sigma\rho}\ \gamma_{\nu}\wedge\gamma_{\sigma}\wedge\gamma_{\rho} \tag{35}$$

and

$$\{Q^{\alpha},R_{\alpha}\} = (1/2)Q^{\alpha\mu\nu}R_{\alpha\mu\nu}{}^{\rho}\ \gamma_{\rho} \tag{36}$$

As in eq.(15), in agreement with uncertainty relation $\Delta Q\Delta R \geq L_{Pl}^{-3}$, we can put

$$[Q^{\alpha},R_{\alpha}] = L_{Pl}^{-3}\ iu \tag{37}$$

$$\{Q^{\alpha},R_{\alpha}\} = L_{Pl}^{-3}\ v \tag{38}$$

where u and v are unit vectors.

Therefore, given the conjugate variables Q^{α} and R^{α} , we have both commutator and anticommutator; we believe that this fact can be related to supersymmetry in the sense that one can treat simultaneously fermionic fields and bosonic fields if one consider, as in the second quantization procedure, the development of the fields in terms of creation and annihilation operators which present analogies with the relations (37,38) between torsion and curvature. But this is an argument for future works.

4. Comments and Conclusions

The foregoing analysis is based on the following observations:

(A) the tidal force between two neighbouring particles is proportional to the curvature tensor and hence the change in momentum vector may be made proportional to it as given by equation (4) and the change in displacement after one loop is proportional to torsion as given by (6).

Hence it is tempting to regard torsion and curvature as playing an analogous role corresponding to displacement and momentum respectively. Thus various commutator relations involving torsion and curvature may be defined.

(B) The above observation is further enhanced by the possibility of defining a Hamiltonian in equation (19) so that torsion and curvature are conjugate to each other. Hence further commutators as well as anticommutators may be defined via the Geometric Algebra.

Consequent to the above propositions, however, a new problem arises. The foregoing is supposed to be based on the standard U_4 theory where there are already established the Einstein-Cartan equations: one relating the generalized Einstein tensor to the generalized energy momentum tensor and the other relating the contorsion tensor to the spin density tensor. These equations are supposed to be self-consistent and self-contained. And yet the commutator relations such as equation (15) constitutes additional equations. In fact (15) should, strictly speaking, be a tensor equation with 24 components. So the question arises as to whether the Einstein-Cartan equations and the additional commutator relations do not constitute an over determined system.

Alternatively, are the additional commutator relations compatible with the Einstein-Cartan equations? This problem awaits further clarification.

(C) There is, however, a natural way to define a commutator that avoids the above problem. It has been shown in [10] and [14] that with the real representation of the unitary group as the structure group that defines the Dirac geometry of real spacetime, a spinor φ may be represented by a tangent vector A^α and we have

$$(\nabla_\lambda \nabla_\mu - \nabla_\mu \nabla_\lambda) A^\alpha = R_{\lambda\mu\nu}^{\cdot\cdot\cdot\alpha} A^\nu - Q_{\lambda\mu}^{\cdot\cdot\nu} \nabla_\nu A^\alpha \tag{39}$$

where the momentum operator $\hat{P}_\lambda = \nabla_\lambda$ has been replaced by the G-connection. For flat spacetime RHS of (39) vanishes and the usual relation in flat spacetime quantum theory is obtained. On putting $\alpha = \lambda$ in (39) we obtaion

$$(\nabla_\lambda \nabla_\mu - \nabla_\mu \nabla_\lambda) A^\lambda = R_{[\mu\nu]} A^\nu - Q_{\lambda\mu}^{\cdot\cdot\nu} \nabla_\nu A^\lambda \tag{40}$$

With this definition of the commutator, spacetime can be said already quantized by the presence of curvature and torsion. Now, as shown in [14],

$$R_{[\mu\nu]} = \nabla_\alpha Q_{\mu\nu}^{\cdot\cdot\alpha} \tag{41}$$

and hence for $Q_{\mu\nu}^{\cdot\cdot\alpha} = 0$ the RHS of [40] vanishes.

References

[1] V.de Sabbata - Il Nuovo Cimento 107A, 363 (1994)

[2] B.K.Datta, V.de Sabbata and L.Ronchetti - Il Nuovo Cimento 113B, 711 (1998)

[3] V.de Sabbata and L.Ronchetti - Found.Phys. 29, 1099 (1999)

[4] P.A.M.Dirac - "New direction in Physics", J.Whiley, N.Y.(1978)

[5] H.-J.Treder - Ann.der Physik 48, 497 (1991)

[6] H.-J.Treder - Astr.Nachr.315,1 (1994)

[7] H.-H.v.Borzeszkowski and H.-J.Treder - "Classical Gravity and Quantum Matter Field" in "Quantum Gravity" ed.by P.G.Bergmann, V.de Sabbata and H.-J.Treder, World Sci.Singapore, (1996) pp.32-42

[8] H.-H.v.Borzeszkowski and H.-J.Treder - "Torsion and the Weyl-Cartan Space Problem in Purely Affine Theory" in "Spin in Gravity" ed.by P.G.Bergmann, V.de Sabbata, G.T.Gillies, and P.I.Pronin World Sci.Singapore, (1998) pp.9-32

[9] Yu Xin (Alfred Yu) - "General Relativity on Spinor-Tensor manifold" in "Quantum gravity" ed.by P.G.Bergmann, V.de Sabbata and H.-J.Treder, World Sci.Singapore, (1996) pp.382-311

[10] A.Lasenby, C.Doran and S.Gull - Found.Phys. 23,1295 (1993)

[11] B.K.Datta and V.de Sabbata - "Hestenes' geometric algebra and real spinor field" in "Spin in Gravity" ed.by P.G.Bergmann, V.de Sabbata G.T.Gillies and P.I.Pronin, World Sci.Singapore (1996) pp.33-50

[12] D.Hestenes - "Clifford Algebra to Geometric calculus" D.Reidel Publ.Co. Dordrecht, Holland 1982

[13] V.de Sabbata, L.Ronchetti and B.K.Datta - "Non-commuting geometry and spin fluctuations" in "Classical and quantum nonlocality" ed.by P.G.Bergmann, V.de Sabbata and G.Goldberg, World Sci.Singapore, (2000) pp.85-110

[14] Alfred Yu - "Dirac geometry of real spacetime", to be published

EXPERIMENTAL PERSPECTIVES ON THE INTERPLAY BETWEEN QUANTUM AND GRAVITY PHYSICS

G. T. GILLIES
Department of Physics, University of Virginia
Charlottesville, VA 22904-4714, U.S.A.
E-mail: gtg@virginia.edu

C. S. UNNIKRISHNAN
Gravitation Group, Tata Institute of Fundamental Research
Mumbai – 400 005, India
and
NAPP Group, Indian Institute of Astrophysics
Bangalore – 560 034, India
E-mail: unni@tifr.res.in

ABSTRACT

There is a rich history of experimental effort that has been aimed at exploring the effects of gravitational fields on the elementary building blocks of matter: atoms, nuclei and subatomic particles, including the photon and neutrino. Much of this work was originally empirically motivated, with many of the investigations driven simply by curiosity over how the gravitational field might affect either various attributes of thequantized states of matter or the behavior of fundamental particles. More recently, searches for novel or anomalous aspects of gravity (torsion, screening effects, etc.) have employed "exotic" matter (eg., spin-polarized substances, superconductors, etc.) the properties of which can be understood only in terms of quantum mechanics. Substantial theoretical developments have often either preceded or accompanied that class of experiments. All of this reflects the desire of experimental physics to contribute insight into the problem of the interplay between quantum and gravity physics. An overview of this broad class of experimental studies is presented here, with some emphasis on searches for (1) new gravitational effects on single particles and particle beams, (2) certain types of potential violation of the weak equivalence principle, and (3)gravitomagnetic anomalies.

1. Introduction

There are no experiments in gravitational physics that are easy to do at significant levels of precision. All of them are plagued by the weakness of the gravitational force and the inability to screen the interaction between two test masses from all other gravitational effects. These two characteristics of the gravitational force make such experiments vulnerable to a host of competing effects and confounding influences. The very slow progress in advancing the accuracy of measurements of the absolute value of the Newtonian gravitational constant (approximately 1 order of magnitude per century since Cavendish!) speaks to this point. In spite of these difficulties, the past few decades have seen many decisive tests of the predictions of general relativity, verifications of the inverse square law, searches for new weak forces, and so

P.G. Bergmann and V. de Sabbata (eds.),
Advances in the Interplay Between Quantum and Gravity Physics, 103–121.

on. The weak equivalence principle (WEP) and the universality of free fall have provided an important, unifying focal point around which much of the data from these experimental efforts has been interpreted and ultimately understood.

Reliance on, or testing of, the WEP has carried over to many classes of experiments that are seen as investigations of quantum gravity or, more exactly, experiments where quantum mechanical and gravitational effects both play a legitimate role. If the latter is taken as the definition of a "quantum gravity" experiment, then a very wide range of phenomenon fall into this classification. Equally diverse are the motivations for doing such experiments. They have ranged from simple empirical curiosity over which way fundamental particles move in a gravitational field to very complex arguments centering on the possible spin-dependence of the gravitational force. In the purest form, those quantum gravity predictions concerning single gravitons and processes that take place at the Planck scales of length and time might never be amenable to direct testing, but that does not make the doable experiments any less interesting (or difficult!) or their results any less important in framing the direction of research in this area.

It is our purpose here to provide a succinct overview of this field and to discuss some of the proposed new experiments that have been described in the literature. We do not present a thorough review or a critical assessment of the field, but rather we offer the article as a starting point for those interested in learning more on the topic. Among the more comprehensive treatments is that prepared by Lämmerzahl (1998), who characterizes the various experiments within the structures of Local Lorentz Invariance and Local Position Invariance. We open with a section on free fall measurements on fundamental particles and continue with a brief discussion of neutron and atom interferometry experiments. Next, we provide a description of certain quantum-limit tests of the WEP and list and assess the various searches for gravitomagnetic anomalies. We close with some commentary on various proposed tests of quantum-gravitational phenomenon and some additional inferences for future work. In a companion paper (these Proceedings) we discuss the specific role played by the Newtonian gravitational constant as a testing ground for a variety of concepts important to the advancement of experimental quantum gravity.

2. Free Fall of Particles and Atoms in a Gravitational Field, and Other Non-Interferometric Phenomena

Prior to the late 1940s there was little, if any, discussion in the literature of the gravitational effects on individual sub-atomic particles and on free atoms and ions. The gravitational acceleration of bulk matter had been studied carefully by then, but neutron beams from nuclear reactors and atomic beams from collimated sources were still relatively new and were only then finding use in investigating gravitational effects on particles and beams. In its simplest form, the question was one of asking whether or not an elementary particle or an atom obeyed the equation $F = mg$, where m was the mass of the particle or atom in question, and g was the gravitational acceleration of the Earth acting on it. The contemporary thinking viewed such experiments as tests of the extrapolation of gravitational acceleration over the scale of masses from ponderable bodies down to microscopic ones. Eventually, free-fall tests on particles and atoms were analyzed in terms of the WEP. An interesting corollary to these approaches is to examine such experiments in terms of the correspondence principle: the quantum mechanical description of physical phenomenon must not violate the established Newtonian principles; on the other hand how far down in the scales of length and mass can the classical laws of gravity be applied before they become incomplete?

A number of the experiments that fall in this category are discussed briefly here, as are some additional ones that explore the behavior of quantum particles in a gravitational field. Studies employing matter-wave inteferometry will be discussed in the next section. Table 1 provides a succinct listing of the experiments in this category.

The earliest such test of free fall was done by Estermann, Simpson and Stern (1947) who observed the gravitational deflection of beams of K and Cs atoms emitted from an oven. The gravitational deflection was measured by using an inhomogeneous magnetic field to compensate for it. From the Cs atom's magnetic moment, a gradient of $\approx$ 25 gauss/cm was estimated as being the value needed to null the deflection of the beam in the experimental arrangement that was used. An agreement of 1% between the predicted and observed values confirmed that indeed, atomic masses responded to gravitational acceleration in the same way that bulk matter did.

A few years later McReynolds (1951) did the first free-fall experiment on a beam of neutrons, observing their fall over a 12 m path. He obtained a value for the gravitational acceleration acting on them, of $g = (935 \pm 70)$ cm/s^2, which matched the nominal value of g to within roughly 7%. Dabbs et al. (1965) extended this result by using the Oak Ridge Research Reactor to observe the gravitational deflection of beams over an evacuated pathway 180 m long. Two independent measurements of g made in this way yielded $g = (975.4 \pm 3.1)$ cm/s^2 and $g = (973.1 \pm 7.4)$ cm/s^2, with the two measurements corresponding to data taken using the <100> and <200> lattice spacings of the beryllium filter used in the experimental arrangement, respectively. The results were within 0.6% of the local value of g. They found no evidence for a splitting in g due to the $\pm 1/2$ spin projections of the neutrons. At this level of sensitivity one does not expect to find any spin dependent quantum effects, even in nonstandard theories. Even so, this was probably the earliest experiment in which a quantum number served as a variable in a gravitational physics experiment.

At nearly the same time, Witteborn and Fairbank (1967) were attempting the very difficult task of studying the gravitational effects on free charged particles, in this case, allowed to fall inside of an evacuated drift tube. The total force that they measured was < 0.09mg where m was the mas of the electron, thus confirming the Schiff-Barnhill conjecture of a gravity-induced electric field in such experiments that would offset the gravitational force on the individual electrons. The principle of this experiment was to have been used on positrons as well, for the purpose of seeing if there was any difference in the gravitational accelerations of matter and antimatter, but that study was never done. In more recent times, a very significant effort was mounted by groups from Los Alamos and CERN, who sought to measure the gravitational acceleration on antiprotons using the LEAR beam and a sophisticated form of Penning-style trap. Electronic cooling would be used to slow the antiprotons to essentially room temperature, for subsequent use in the experiment. A detailed discussion of the motivations for and the technical plans of this experiment has been prepared by Nieto and Goldman (1991). This worked reached the point of being able to store approximately 10^6 antiprotons before the decommissioning of LEAR, Holzscheiter et al. (1996). Future plans call for continuation of the effort using antihydrogen.

In a very interesting investigation, Jain, Lukens and Tsai (1987) made the first gravitational measurements on charged particles done in such a way as to reveal the equivalent of a gravitational red shift for them. They used superconducting electronics, specifically two

Josephson junctions that were separated vertically by a height of 7.2 cm, with both phase-locked to a common external source of microwaves. The total potential difference around the loop was found to be less than 1 x 10^{-22} V. However, there was a voltage difference of 2.35 x 10^{-21} driven between the junctions by the gravitational red shift. The red shift modified the electrochemical potential, μ, such that it became $\mu' = \mu(1 + \lambda)$ where $\lambda = zg/c^2$ and thus depends on the gravitational potential through the separation distance z.

For purposes of comparison, we include in Table 1 a reference to one of the early experiments at Stanford with the "atomic fountain" technique of atom interferometry. This approach, which uses sophisticated laser-cooling methodologies, was already making it possible to measure the gravitational acceleration on sodium atoms to a resolution of 3 x 10^{-8}g after 2,000 s integration times almost a decade ago, Kasevich and Chu (1992). This marked a turning point on quantum-limit measurements of g: rather than being essentially only of scientific interest, it then possible to use this approach to create new precision-measurement technologies for engineering applications. Moreover, the Newtonian ("weak-field") limit of general relativity is thus established, with high sensitivity, as being correct in the quantum limit, thus constituting a test of a kind of inverse correspondence principle.

The general conclusions are that neutrons and atoms accelerate in the Earth's gravitational field the same way that bulk matter does. Electrons, even in superconducting materials also have a passive gravitational mass that behaves similarly to that of bulk matter; the question of the active gravitational mass of electrons is discussed in our companion paper (these proceedings). The question of universality of free fall of antiparticles is still open, but may be resolved by experiments now in planning and progress.

3. Gravitational and Inertial Effects in Neutron and Atom Interferometry

Perhaps the most well known of all quantum gravity experiments are those involving matter wave interferometry, and neutron and atom interferometry in particular. Beginning with the original investigations of Colella, Overhauser and Werner (COW, 1975), this general approach has been used successfully to study how the quantum mechanical phase, eg., of the neutron is affected by a variety of external influences. In that first paper, COW discussed an apparatus and approach in which gravity and quantum mechanics simultaneously play a role. The interferometer itself is of the Bonse-Hart design, in which a silicon rod about 5 cm in diameter is cut in such a way that there are three planes protruding from the flat surface of a cylinder cut in half along its axis (see the excellent illustrated tutorial article by Greenberger and Overhauser (1980) for photos and detailed discussions of the apparatus). The neutron beams (wavelength = 1.445 Å) scatter from these three planes in such a way that the resulting intensity patterns upon recombination demonstrate interference at external detectors. When the interferometer crystal is rotated through an angle ϕ about an off-axis line, the gravitational field of the Earth modulates the phase, β, at the recombination point such that $\beta = q_{grav}\sin\phi$, where q_{grav} is a function of the gravitational acceleration, Planck's constant, and the geometry of the apparatus and the rotation angles. The modulated pattern that was observed came within $\approx$ 10% of being exactly that which was predicted. Various correction factors have been applied in subsequent experiments, such that this discripency has been virtually eliminated.

In a later experiment, Werner et al. (1979) used a similar approach to observe the effect of the Earth's rotation on the neutron's phase. They found that an analog to the Sagnac effect for photons exists for neutrons, and confirmed the presence of it in the overall expression for the quantum-mechanical phase of the neutron. Bonse and Wroblewski (1983) extended these lines of work to show that neutron interferometry in accelerating (but gravity-free) frames can be used to verify the classical transformation laws for noninertial frames. They did this by oscillating the interferometer in such a way that the gravitational effects on the interfering beams very nearly cancelled each other, and ultimately found that gravitational effects were equivalent to acceleration effects in such a measurement, to within approximately 4%.

A number of experiments with atom interferometers have demonstrated a variety of similar effects. For instance, Riehle et al. (1991) correlated a shift in the Ramsey fringes of Ca atomic beam with the rotation frequency of the apparatus, thus demonstrating the Sagnac effect. Borde and colleagues (1994) extended such technologies to the use of I_2 molecules with velocity tuning in each arm of the interferometer. Lämmerzahl and Bordé (1999) described laser-induced atomic-beam splitting in gravitational fields, and discussed the outcome of an atom interferometry experiment that is corrected for gravitational effects. Meanwhile, Oberthaler et al. (1996) introduced the concept of Moiré deflectometry, in which rotational, acceleration and gravitational effects are sensed in atomic or molecular beams traversing a set of three sequential gratings that are in relative motion. In this approach, no collimation is needed, so high intensities can be used and significant gains in sensitivity can thus be obtained.

The work on quantum gravity effects in matter-wave interferometry has been well motivated over the years from a theoretical perspective. The main goal has been to develop expressions for the quantum mechanical phase of, eg., the interfering neutron beams as a function of gravitational and mechanical states of the apparatus. A few of the many substantial efforts in this area are the following. Anandan (1977) developed a special relativistic treatment of neutrons in a gravitational field, which produced an expression for the phase that had four terms, two of which are still too small be detected. Mashhoon (1988) introduced the need for a spin-rotation term, which would arise if the neutron spins are flipped in one of the arms of the interferometer. Mannheim (1998) derived the phase shift from completely classical arguments based on the kinetic energy of the particles in the gravitational potential that was acting on them. A general relativistic treatment of the quantum mechanical phase has been presented by Varjú and Ryder (2000), who use the Dirac equation in the Schwarzschild field of the Earth as the starting point. Their result has five terms in it, including the Mashhoon spin-rotation term, and two small terms (one proportional to the acceleration and the other proportional to a spin-orbit type of coupling) that are beyond the limits of detection. Their article serves as a useful starting point for those interested in this topic, as they compare the results of most of the well-known calculations of the phase. Bordé et al. (1994) have also developed a general relativisitic framework for interpreting the results of atom interferometry experiments, and their description includes spin-gravity couplings.

The whole field of matter wave interferometry is growing, and new tools, such as the improved beam-lines and the S18 thermal neutron interferometer at Institute Laue-Langevin, are drawing several new workers to it. The exquisite work at Stanford by Chu and colleagues (eg., Kasevich and Chu, 1992; Peters et al., 1999) using atomic fountain atom interferometers has reached unprecedented levels of precision. There are a number of proposals for new and interesting experiments that have appeared in the literature in recent years. For instance, de Sabbata et al. (1991) have suggested the use of neutron interferometry to search for new weak forces and as an experimental test of torsion in the gravitational field. Unnikrishnan (1999) suggested the possibility of observing the gravitational analogue of the scalar Aharonov-Bohm effect in atom interferometers. Adunas and colleagues (2000) have proposed a WEP violation due to a breakdown of superposition within quantum gravity. They suggest that free fall experiments on Ce atoms might reveal a spread in their accelerations, depending on the eigenstates. Camacho (2001) has described a variation of that concept, in which he discusses the possibilities of detecting Yukawa-type add-on contributions to the Newtonian potential and testing restricted path integral formalisms via COW experiments. A dated but still very useful review is that of Staudenmann et al. (1980). A more recent and very extensive summary of the field has been written by Wagh and Rakhecha (1996). Stedman (1985, 1997) has written interesting and useful summaries of the use of ring interferometers of all types in tests of classical and quantum gravity. Table 2 provides a succinct listing of several of the results mentioned above.

4. Tests of the Equivalence Principle in the Quantum Limit: Results and Proposals

There has been great interest in searching for violations of the WEP in the quantum limit. What is typically meant by this, is somehow examining the equivalence of the passive gravitational and inertial masses of fundamental particles and atoms. Most of the experiments described in the previous sections have viewed themselves as tests of WEP at the same time as they've also attempted to explore some related aspect of gravity at the microscopic scale.

However, some workers have very explicitly designed their experiments to evaluate WEP and reported their results uniquely within that context. Some representative examples of efforts in this area include the following. Koester (1976) reported the verification of the WEP for the neutron via an experiment that used a gravity refractometer to measure the scattering lengths dependently and independently of gravity. His result was expressed as the ratio $\gamma = (m_i/m)(g_f/g_o)$, where m_i and m are the inertial and passive gravitational mass of the neutron, and g_f and g_o are the gravitational accelerations of the free neutron and of bulk matter, respectively. His final result had an uncertainty of 250 ppm, and was stated as $\gamma = 1.000\ 16 \pm 0.000\ 25$.

The CPLEAR collaboration at CERN reported a test of WEP with neutral kaons, Apostolakis et al. (1999). Importantly, this was a test of WEP for both particles and antiparticles, as they observed the kaon and antikaon decays into pions and antipions. Part of their motivation was to see if any WEP violation that might be detected between matter and antimatter could signal the presence of the graviphoton, a gravitational strength light vector particle of interest in quantum theories of gravity. Assuming exact CPT symmetry, they searched for mass differences in the particle-antiparticle pairs before and after decay, and they were able to establish limits at the parts-per-billion level for WEP violations within the scalar, vector and tensor potentials of the Sun. In contrast to many other experiments discussed in this review, this experiment has the flavor of a true quantum gravity experiment.

An interesting line of work presently underway is associated with the Brookhaven muon (g-2) experiment. Mann (1997) and Alvarez and Mann (1997) have pointed out that anomalous magnetic moments of the elementary leptons might form the basis for evaluating the Einstein Equivalence Principle (EEP). They use the Th$\epsilon\mu$ formalism to explore the theoretical basis for possible new constraints on the EEP.

Unnikrishnan (2001) has discussed the equivalence principle within the context of quantum mechanics. He explored the possibility of formulating an independent equivalence principle for spin and noted that precision spectroscopy of spin-orbit doublets in atoms can provide indirect high precision tests of gravito-magnetic fields.

A very novel proposal has been offered by Ross (1999), who has suggested that the electromagnetic zero-point energy inside an aluminum shpere might be subject to a modification large enough to permit detection by Eötvös-style laboratory experiments. The resulting WEP test would be unique in that it would examine this purely quantum-mechanical form of energy. The modification of the zero-point energy arises from a photon cut-off associated with the Casimir force between the hemispheres of the sphere. An explicit calculation of the zero-point energy leads to a ratio of missing zero-point energy to total rest mass energy of 1.6×10^{-14}. If the missing energy violates the WEP, it should be detectable with future improvements to existing torsion balance devices.

A substantial number of phenomonological tests of the WEP have been developed for neutrinos. One of the earliest was by Krauss and Tremaine (1988), who pointed out that the nearly simultaneous observations of photons and neutrinos from Supernova 1987A permited evaluation of the PPN parameter γ (which is related to the Shapiro time delay) for both types of particles. The result was that the values are the same to within 0.5% of each other, thus confirming WEP for them to within that accuracy, the first time it had been tested for relativistic particles. Pakvasa, Simmons and Weiler (1989) later used the SN 1987A data to establish a limit of 1×10^{-6} on the size of EEP violations for electron neutrinos and their antiparticles. Gasparini (1988) noted that if WEP is violated, then gravity does not couple equally to all leptonic flavors and this may be associated with neutrino oscillations. He used the existing limits on neutrino oscillations to put an upper limit on a WEP violation, ϵ, in the gravitational potential of the Earth as being $\Delta\epsilon \lesssim 10^{-11}$ (assuming massless neutrinos). Some of the others who have studied WEP violations using neutrinos include Minakata and Nunokawa (1995), Bahcall et al. (1995), and Halprin et al. (1996), among several others.

Table 3 contains a listing of selected WEP tests that have been carried out for fundamental particles.

5. Superconductivity and Quantum Gravity

The superconducting state of matter can be understood only in terms of quantum mechanics. Superconductivity has long been a tool of gravitational physics and precision measurement: the SQUID detector ("Superconducting Quantum Interference Device") is used to monitor the strains induced in superconductively levitated gravity wave detectors, superconducting gravity gradiometers have been used in null tests of the inverse square law, and the precession of the London moment will be used as the signal in the superconducting relativity gyro in Gravity Probe B. Over the last 10 years many people have tried, in a sense, to invert this situation and use the tools of precision measurement to search for novel gravity-related effects in superconductors. Such phenomenon could fall under the rubric of "quantum gravity" in the same broad sense that the neutron interferometry experiments do, viz., any experiments of that type would inherently be such that both quantum mechanics and gravitational acceleration play a role.

The reason why this has been a topic of interest has to do with the concept of the gravitomagnetic field. In analogy with electromagnetism, the gravitomagnetic field would arise from the motion of matter, similar to the magnetic field arising from the motion of charge. The measurements to be made by Gravity Probe B of the general relativistic frame-dragging effects on gyroscopes in polar orbit about the Earth represent the most thoroughly studied experimental scenario in this area to date, and the hope is that they will be sent into space within the next year.

However, many believe that the ground is quite fertile for exploration of other potential gravitomagnetic effects inside of superconductors, where the Meissner effect, persistant currents, and other exotic phenomena occur.

As an example, Peng (1990) and Li and Torr (1991, 1992) looked for a theoretical basis for new phenomena in superconductors, reasoning that although gravitational effects are very weak, manifistations of new phenomena might be found in superconductors where other forces are essentially nulled out. In particular, they presented arguments contending that the tiny residual imperfections in the Meissner effect might be due not only to impurities in the superconducting material, but also to an interaction between gravity and matter in the superconducting state. Moreover, they suggested that external gravitomagnetic fields might drive small but potentially measurable magnetic fields inside superconducting bodies. Kowitt (1994) argued that the proposed effects were over-estimated due to flaws in the models.

These lines of thought began receiving considerable scrutiny shortly after attention became focused on a previously little known paper by Podkletnov and Nieminen (1992) that claimed to see a surprisingly large gravitational shielding effect associated with a high temperature superconducting disk. A small mass (5.48 g) weighed above the disk was reportedly found to loose 0.05 % to 0.3 % of its weight depending on the speed of rotation of the disk. In examining this result, de Podesta and Bull (1995) noticed that the weight change could be explained by a lack of correction for buoyancy effects. Unnikrishnan (1996) pointed out that the screening interpretation could not be possible when the height of the sample above the disk was taken into account versus the area of the Earth's surface screened by the disk. (He also carried out a preliminary experiment in which no such effect was seen.) Li and coworkers (1997) did a decisive experiment, in which a LoCoste-Romberg gravimeter was positioned above a bulk YBCO superconductor and high precision measurements of the local gravitational acceleration were made. Their results put a stringent upper limit of 8 parts in 10^8 on any deviations from normal gravitational acceleration, thus on the proposed superconductor/gravity couplings. Buzea and Agop (2000) have sought to resurrect interest in screening phenomenon, though, by suggesting that electromagnetic fields can produce a type of gravitational Meissner effect, and they propose a new experiment using ultracold neutrons (speed = 0.01 m/s) moving through a superconducting cylinder as a way of checking this. We note only that very stringent limits have been placed on gravitational shielding/screening effects in general. The University of Zürich measurement of G has been re-analyzed by Unnikrishnan and Gillies (2000), with the result that the Majorana extinction coefficient has the very small value $h \leq 4.3 \times 10^{-14}$ cm^2/g. Moreover, gravimetry data obtained during a recent total solar eclipse in China, while originally thought to support a hypothesis of gravitational screening, were in fact shown by Unnikrishnan, Mohapatra and Gillies (2001) to establish an even more stringent limit on the extinction coefficient, $h \leq 2 \times 10^{-17}$ cm^2/g.

Other possible gravitomagnetic couplings have been suggested, though, and experiments have been proposed to detect them. Ljubičić and Logan (1992), for instance, have proposed a quantum physics-based experiment for testing Mach's principle by measuring the magnetic field in a conductor produced by the gravitomagnetism of the Earth's rotation. Ahmedov (1999) suggests that a new galvan-gravitomagnetic effect arising from the gravitational analog of the Hall voltage might be observed in superconducting test bodies in counter-revolving orbits around the Earth.

A listing of experiments and proposals germane to the quantum gravity of superconducting matter is presented in Table 4.

6. Other Experimental or Observational Connections to Quantum Gravity

One of the most important aspects of any quantum theory of gravity would the role and properties of the graviton in it. The graviton would be a very weakly interacting particle, and there are no presently conceivable notions for how one might detect it experimentally. Perhaps ironically, however, there are ways to put upper limits on the possible rest mass of the graviton, and several workers have done so over the past quarter century; some of them are the following and Will (1998) reviews others. Goldhaber and Nieto (1974) noted that by taking the maximum distance over which one observes gravitational binding of galaxies (i.e., the maximum range over which gravitation is known to act), one can derive a bound on the graviton mass, μ_g, from the Yukawa exponential cutoff, having the value $\mu_g \lesssim 2 \times 10^{-62}$ g $\equiv 1.1 \times 10^{-29}$ eV. From observational limits on the parameters for expansion of the universe, Gershtein, Logunov and Mestvirishvili (1998) found an even tighter limit of $\mu_g \lesssim 4.5 \times 10^{-66}$ g $\equiv 2.5 \times 10^{-33}$ eV. Will (1998) estimates that bounds of $\mu_g \lesssim 4.4 \times 10^{-59}$ g $\equiv 2.5 \times 10^{-26}$ eV might be obtainable from eventual space-based observations of gravitational waves from inspiralling compact binary star systems. Larson and Hiscock (2000) have noted that space-based detection of the gravitational radiation from white dwarf binary systems might lead to a limit of $\mu_g \lesssim 1.7 \times 10^{-57}$ g $\equiv 1 \times 10^{-24}$ eV. Deployment of long-baseline interferometers in space for detecting gravitational radiation, such as might be done by the LISA system, are still many years away, so it will be some time before such predictions can be checked by that method. Table 5 summarizes these results.

Morris and De Leon (1995) asked if it were possible to detect the local expansion rate of the universe via high precision experiments on atomic systems, thus using quantum physics to infer something of interest on the cosmological scale. After a careful analysis of candidate quantum mechanical systems, they found that any predictable effects would be too small to observe, but they did conclude that the physical size of atomic or molecular systems would remain constant in spite of the large scale expansion. The conclusion is reminiscent of those that can be drawn from two-metric hypothesis of Dirac. Lastly, Pinto (2000) has proposed that the Planck mass might be measured directly by a laboratory experiment in which a cold atomic beam is passed over a conducting spherical mass. The Planck mass, $M_P = (\hbar c/G)^{1/2}$, appears as an identifiable quantity in the ratio of the gravitational to Casimir forces for such an experiment, thus opening the door to a class of studies in which gravitation and quantum electrodynamics are both central to the outcome.

There is also the intriguing possibility of exploring analogies between hydrodynamical systems and quantum black holes. Such a process might lead to a conceptual basis for new experiments that will help physicists understand quantum gravitational effects of black holes. Unruh (1981) observed that the motion of sound waves in a convergent fluid flow is analogous to the behavior of a quantum field in a classical gravitational field. He obtained the result that the sound wave modes behave like those of the normal modes of a scalar field in the Schwarzschild metric, and concluded that the sonic "black hole" will emit sound waves with a thermal spectrum. This provides a notional possibility of studying black hole evaporation in a laboratory setting (although it would be quite difficult to do so from a practical perspective). Another recent proposal is that of Garay et al. (2000) who realized that there are dynamically stable and unstable configurations in dilute Bose-Einstein condensates which, when the mean free path is small compared to the interparticle separation, exhibit a behavior resembling that of black holes. The dynamical instabilities involve creation of quasiparticle pairs in positive and negative energy states as occurs in black hole evaporation. They concluded that similar proposals can be offered for other model systems, as well. We note that this is in principle a very rich field where quantum gravitational effects can be studied either in simulations or in actual laboratory experiment, by analogy.

7. Conclusions

Many experiments have been done over the last few decades that can claim to provide us with information relevant to developing a deeper understanding of how gravitational physics and quantum mechanics interact. Genuine tests of quantized theories of gravity, such as the detection of the graviton, are perhaps far outside of technological possibility, but very useful inferences about the rest mass of such a particle, about the behavior of free quantum particles in curved space-time, and about the role played by gravitation in quantum mechanical processes such as superconductivity have been and will continue to be investigated. There are other significant areas where similarly interesting studies either have been or are being carried out, for instance in the search for spin-dependence in the gravitational force and for torsion in gravity. Very useful reviews and reports of experiments were presented by Ritter et al. (1998) and Pronin (1998) at the XVth Course of the School of Cosmology and Gravitation, and the interested reader is referred to those Proceedings and also to the interesting papers of Mashhoon (2000) and Obukhov (2001) for further details.

Acknowledgements

The hospitality of the E. Majorana Center for Scientific Culture is gratefully acknowledged, during the XVIIth Course of the School of Cosmology and Gravitation. Prof. Venzo de Sabbata and the staff of the Center have done much to make these Schools a significant venue for the presentation and discussion of fundamental topics.

References

Adunas, G. Z., Rodriguez_Milla, E., and Ahluwalia, D. V., *Phys. Lett. B* **458** (2000) 215.

Ahmedov, B. J., *Phys. Lett. A* **256** (1999) 9.

Alvarez, C. and Mann, R. B., *Phys. Lett. B* **409** (1997) 83.

Anandan, J., *Phys. Rev. D* **15** (1977) 1448.

Apostolakis, A., Aslanides, E., Backenstoss, G. et al., *Phys. Lett. B* **452** (1999) 425.

Bahcall, J. N., Krastev, P. I., and Leung, C. N., *Phys. Rev. D* **52** (1995) 1770.

Bonse, U. and Wroblewski, T., *Phys. Rev. Lett.* **51** (1983) 1401.

Bordé, C. J., Courtier, N., Duburck, F., Goncharov, A. N., and Gorlicki, M., *Phys. Lett. A* **188** (1994) 187.

Bordé, C. J., Karasiewicz, A., and Tourrenc, P., *Int. J. Mod. Phys. D* **3** (1994) 157.

Braginsky, V. B., Polnarev, A. G., and Thorne, K. S., *Phys. Rev. Lett.* **53** (1984) 863.

Buzea, C. G. and Agop, M., *Physica C* **341** (2000) 307.

Camacho, A., *Int. J. Mod. Phys. A* **16** (2001) 83.

Colella R., Overhauser, A. W., and Werner, S. A., *Phys. Rev. Lett.* **34** (1975) 1472.

Dabbs, J. W. T., Harvey, J. A., Paya, D., and Horstmann, H., *Phys. Rev. B* **139** (1965) 756.

de Podesta, M. and Bull, M., *Physica C* **253** (1995) 199.

de Sabbata, V., Pronin, P. I., and Sivaram, C., *Int. J. Theor. Phys.* **30** (1991) 1671.

Estermann, I., Simpson, O. C., and Stern, O., *Phys. Rev.* **71** (1947) 238.

Garay, L. J., Anglin, J. R., Cirac, J. I., and Zoller, P., *Phys. Rev. Lett*. **85** (2000) 4643.

Gasparini, M., *Phys. Rev. D* **38** (1988) 2635.

Gershtein, S. S., Logunov, A. A., and Mestvirishvili, M. A., *Doklady Akademii Nauk* **360** (1998) 332 (translated as *Doklady Physics* **43** (1998) 293).

Goldhaber, A. S. and Nieto, M. M., *Phys. Rev. D* **9** (1974) 1119.

Greenberger, D. M. and Overhauser, A. W., *Scientif. Am.* **242**, 66 (1980).

Halprin, A., Leung, C. N., and Pantaleone, J., *Phys. Rev. D* **53** (1996) 5365.

Holzscheiter, M. H., Feng, X., Goldman, T., King, N. S. P., Lewis, R. A., Nieto, M. M., and Smith, G. A., *Phys. Lett. A* **214** (1996) 279.

Jain, A. K., Lukens, J. E., and Tsai, J.-S., *Phys. Rev. Lett.* **58** (1987) 1165.

Kasevich, M. and Chu, S., *Appl. Phys. B* **54** (1992) 321.

Koester, L., *Phys. Rev. D* **14** (1976) 907.

Kowitt, M., *Phys. Rev. B* **49** (1994) 704.

Krauss, L., M. and Tremaine, S., *Phys. Rev. Lett.* **60** (1988) 176.

Lämmerzahl, C., in *Spin in Gravity*, P. G. Bergmann et al., eds., (World Scientific, Singapore, 1998), pp. 91 - 117.

Lämmerzahl, C. and Bordé, C. J., *Gen. Rel. Gravit.* **31** (1999) 635.

Larson, S. L. and Hiscock, W. A., *Phys. Rev. D* **61** (2000) 104008.

Li, N. and Torr, D. G., *Phys. Rev. D* **43** (1991) 457.

Li, N. and Torr, D. G., *Phys. Rev. B* **46** (1992) 5489.

Li, N., Noever, D., Robertson, T., Koczor, R., and Brantley, W., *Physica C* **281** (1997) 260.

Ljubičić, A. and Logan, B. A., *Phys. Lett. A* **172** (1992) 3.

Mann, R. B., Mod. *Phys. Lett. A* **12** (1997) 1209.

Mannheim, P. D., *Phys. Rev. A* **57** (1998) 1260.

Mashhoon, B., *Phys. Rev. Lett.* **61** (1988) 2639.

Mashhoon, B., *Class. Quant. Grav.* **17** (2000) 2399.

McReynolds, A. W., *Phys. Rev.* **83** (1951) 172.

Minikata, H. and Nunokawa, H., *Phys. Rev. D* **51** (1995) 6625.

Morris, J. R. and De Leon, N., *Il Nuov. Cim.* **110 B** (1995) 419.

Nieto, M. M. and Goldman, T., *Phys. Rep.* **205** (1991) 221.

Oberthaler, M. K., Bernet, S., Rasel, E. M., Schmiedmayer, J., and Zeilinger, A., *Phys. Rev. A* **54** (1996) 3165.

Obukhov, Y. N., *Phys. Rev. Lett.* **86** (2001) 192.

Pakvasa, S., Simmons, W. A., and Weiler, T. J., *Phys. Rev. D* **39** (1989) 1761.

Peng, H., *Gen. Rel. Gravit.* **22** (1990) 609.

Peters, A., Chung, K. Y., and Chu, S., *Nature* **400** (1999) 849.

Pinto, F., *Mod. Phys. Lett. A* **15** (2000) 2229.

Podkletnov, E. and Nieminen, R., *Physica C* **203** (1992) 441.

Pronin, P. I., in *Spin in Gravity*, P. G. Bergmann et al., eds., (World Scientific, Singapore, 1998), pp. 173 - 198.

Riehle, F., Kisters, T., Witte, A., Helmcke, J., and Bordé, C. J., *Phys. Rev. Lett.* **67** (1991) 177.

Ritter, R. C., Gillies, G. T., and Winkler, L. I., in *Spin in Gravity*, P. G. Bergmann et al., eds., (World Scientific, Singapore, 1998), pp. 199 - 212.

Ross, D. K., *Il Nuov. Cim.* **114B** (1999) 1073.

Staudenmann, J.-L., Werner, S. A., Colella, R., and Overhauser, A. W., *Phys. Rev. A* **21** (1980) 1419.

Stedman, G. E., *Contemp. Phys.* **26** (1985) 311.

Stedman, G. E., *Rep. Prog. Phys.* **60** (1997) 615.

Unnikrishnan, C. S., *Physica C* **266** (1996) 133.

Unnikrishnan, C. S., in *Proceedings of the Eighth Marcel Grossmann Meeting on Recent Developments in Theoretical and Experimental General Relativity, Gravitation and Relativistic Field Theories*, T. Piran and R. Ruffini, eds., (World Scientific, Singapore, 1999), 1203 - 1205.

Unnikrishnan, C. S. and Gillies, G. T., *Phys. Rev. D* **61** (2000) 101101.

Unnikrishnan, C. S., Mohapatra, A. K., and Gillies, G. T., *Phys. Rev. D* **63** (2001) 062002.

Unnikrishnan, C. S., *Mod. Phys. Lett. A* **16** (2001) 429.

Unruh, W. G., *Phys. Rev. Lett.* **46** (1981) 1351.

Varjú, K. and Ryder, L. H., *Am. J. Phys.* **68** (2000) 404.

Velyukhov, G. E., *Zh. Eksperim. Teor. Fiz.* **8** (1968) 372 (translation: *JETP Lett.* **8** (1968) 229).

Wagh, A. G. and Rakhecha, V. C., *Prog. Particle Nucl. Phys.* **37** (1996) 485.

Werner, S. A., Staudenmann, J.-L., and Colella, R., *Phys. Rev. Lett.* **42** (1979) 1103.

Will, C. M., *Phys. Rev. D* **57** (1998) 2061.

Wineland, D. J. and Ramsey, N. F., *Phys. Rev. A* **5** (1972) 821.

Witteborn, F. C. and Fairbank, W. M., *Phys. Rev. Lett.* **19** (1967) 1049.

Young, B. A., *Phys. Rev. Lett.* **22** (1969) 1445.

Table 1. Listing of selected experiments on free-fall of fundamental particles and atoms.

Date	Authors	Particle(s)/atoms	Observations	Results
1947	Estermann, Simpson and Stern	Cs and K atoms	magnetically-compensated free fall of atomic beams	mg compensated to $\approx 1\%$
1951	McReynolds	neutrons	gravitational deflection of thermal neutron beam	$g = (935 \pm 70)$ cm/s^2
1965	Dabbs et al.	neutrons	gravitational deflection of slow versus fast neutrons	$973 \leq g \leq 975$ cm/s^2, same for $\pm½$ spin proj.
1967	Witteborn and Fairbank	electrons	measured forces during free fall in a drift tube	net force $\leq 0.09\ m_e g$ (Schiff-Barnhill effect)
1968	Velyukhov	protons	gyromagnetic precession rate when $\parallel$, anti-$\parallel$, and $\perp$ to g	31 Hz faster when $\parallel$or anti-$\parallel$to g
1969	Young	protons	gravity-induced shift in the Larmor frequency	≤ 0.3 Hz
1972	Wineland and Ramsey	deuterons	gravity-induced shift in the Larmor frequency	$\leq 10^{-4}$ Hz
1987	Jain, Lukens and Tsai	electrons	gravitational redshift on charge carriers	2.35×10^{-21} V found and sustained in circuit
1992	Kasevich and Chu	Na atoms	laser cooled atomic fountain	$\delta g \leq 3 \times 10^{-8}$ cm/s^2
1996	Holzscheiter et al.	antiprotons	goal was to measure gravitational acceleration of antiprotons	10^6 antiprotons trapped, g was not measured

Table 2. Listing of selected experiments on neutron and atom interferometric tests of gravitational effects on matter waves.

Date	Authors	Particle(s)/atoms	Observations	Results
1975	Colella, Overhauser and Werner	neutrons	gravitational potential changes phase of the neutron wave function	measured intensities matched predictions to within 10%
1977	Anandan	neutrons	special relativistic treatment of neutron in gravitational field	4 components to the phase shift predicted
1979	Werner, Staudenmann and Colella	neutrons	detected effect of Earth's rotation on phase of the neutron	$\Delta\Phi_{SAGNAC} = 2 \times 10^{-4}\ \Delta\Phi$
1983	Bonse and Wroblewski	neutrons	measurement of neutron quantum interference in noninertial frames	$\Phi_g = \Phi_a$ to within ≈ 4%
1988	Mashhoon	neutrons	prediction of a spin-rotation term when spins are flipped in one arm	not measured yet, but also predicted by others
1991	de Sabbata et al.	neutrons	measurement of rotation of plane of polarization of neutron beams	proposed search for 5th force and torsion in gravity
1991	Riehle et al.	Ca atoms	measurement of the Sagnac effect using optical Ramsey spectroscopy	Ramsey fringes shifted by ± 8 kHz, direction dependent
1996	Oberthaler et al.	A atoms	Moiré deflectometer (analog of a Mach-Zehnder interferometer)	sensitivity = $3.6\ \Omega_E\ s^{1/2}$ (Earth rate for 10 s averages)
1999	Peters, Chung and Chu	Ce atoms	stimulated Raman transitions in atomic fountain	$\delta g/g \leq 3 \times 10^{-9}$
2001	Camacho	neutrons	predicts Yukawa terms in phase	not measured yet

Table 3. Selected tests of the weak equivalence principle using quantum particles.

Date	Authors	Particle(s)/atoms	Observations	Results
1975	Colella, Overhauser and Werner	neutrons	neutron interferometry in Earth's field	WEP would be verified for non-grav. accelerations
1976	Koester	neutrons	neutron gravity refractometer used to for scattering lengths	$\gamma = 1.000\,16 \pm 0.000\,25$
1988	Krauss and Tremaine	neutrinos, photons	time-of-arrival measurements from SU 1987A	violation of PPN parameter γ is less than 0.5%
1988	Gasparini	neutrinos (assumed massless)	analysis of neutrino oscillations in the Earth's potential	$\Delta\epsilon \leq 10^{-11}$
1989	Pakvasa, Simmons,and Weiler	electron neutrinos and antineutrinos	analysis of SN 1987A data	EEP verified to 1×10^{-6}
1997	Mann, and Alvarez and Mann	muons	proposed (g-2)	measurements in progress
1999	CPLEAR Collaboration	kaons, antikaons	test of WEP for particles vs. antiparticles via deacy to pions in gravitational potential of Sun	6.5×10^{-9} for scalar potential 4.3×10^{-9} for vector potential 1.8×10^{-9} for tensor potential
1999	Ross	zero-point energy	proposed Eötvös test of aluminum sphere	predicted violation would be 1.6×10^{-14}
1999	Peters, Chung and Chu	Caesium atoms	comparison of free fall of atoms falling corner cube gravimeter	accelerations equal to within 7 parts in 10^9

Table 4. Various experiments and proposed tests of gravitational effects in superconductors, and of gravitomagnetism.

Date	Authors	Particles/Materials	Observations	Results
1984	Braginsky et al.	Tunsten test mass	monitor the motions of a Foucault pendulum at the South Pole	gravitomagnetic field would precess pendulum ≈ 0.3"/year
1991/92	Li and Torr	Cooper pairs	interaction of superconductor with an external gravitomagnetic field	Meissner effect violation, internal magnetic field arises
1992	Ljubičić and Logan	Cu	proposed test of Mach's principle via gravitomagnetism of the Earth	magnetic field induced in a test mass would be $\approx 10^{-14}$ T
1992	Podkletnov and Nieminen	$YBa_2Cu_3O_{7-x}$	test mass weighed above a rotating high temperature superconductor	reported mass loss was 0.05 % to 0.3%
1995	de Podesta and Bull	SiO_2	consideration of buoyancy of SiO_2 test mass in Podkletnov experiment	buoyancy correction nulls experimental effect
1996	Unnikrishnan	$YBa_2Cu_3O_{7-x}$	modeling of shielding geometry in Podkletnov experiment	observed effect not consistent with geometric restrictions
1997	Ni et al.	Type II YBCO	gravimeter-based measurement of Earth's field above sample	$\delta g/g \leq 2 \times 10^{-8}$
1999	Ahmedov	semiconductor	galvano-gravitomagnetic analog of the Hall effect (proposal)	analog of Hall coefficient $= 0.8 \times 10^{-22}$
2000	Buzea and Agop	neutrons in HTS	proposed gravitataional Meissner effect	g-field "focusing length" would be ~ 1 cm

Table 5. Various proposed methods or observational limits on the rest mass of the graviton.

Date	Authors	Particles/Materials	Observations	Results
1974	Goldhaber and Nieto	graviton	rest mass limit derived from scale over which galaxies are bound	$\mu_g \leq 2 \times 10^{-62}$ g $\equiv 1.1 \times 10^{-29}$ eV
1998	Gershtein et al.	graviton	rest mass limit derived from the observations of Hubble expansion	$\mu_g \leq 4.5 \times 10^{-66}$ g $\equiv 2.5 \times 10^{-33}$ eV
1998	Will	graviton	rest mass limit derived from the inspiralling of binary systems	$\mu_g \leq 4.4 \times 10^{-59}$ g $\equiv 2.5 \times 10^{-26}$ eV
2000	Larson and Hiscock	graviton	rest mass limit derived from the gravitational radiation of binaries	$\mu_g \leq 1.7 \times 10^{-57}$ g $\equiv 1 \times 10^{-24}$ eV

QUANTUM PHYSICS-MOTIVATED MEASUREMENT AND INTERPRETATION OF THE NEWTONIAN GRAVITATIONAL CONSTANT

G. T. GILLIES
Department of Physics, University of Virginia
Charlottesville, VA 22904-4714, U.S.A.
E-mail: gtg@virginia.edu

C. S. UNNIKRISHNAN
Gravitation Group, Tata Institute of Fundamental Research
Mumbai – 400 005, India
and
NAPP Group, Indian Institute of Astrophysics
Bangalore – 560 034, India
E-mail: unni@tifr.res.in

ABSTRACT

One of the most easily recognized yet quantitatively elusive constants in all of physics is the Newtonian gravitational constant, G. Because the gravitational force is so weak and it cannot be shielded, the determination of the absolute value of G is very difficult and its various measured values are in poor agreement. This situation has often led to suggestions that G might be varying or that some new type of quantum-based physics might be revealing itself. Certain experiments have been designed specifically to explore these possibilities, and various of them will be discussed here. Also, because the experiments aimed at re-determining G are typically done with high precision, it is often possible to reliably analyze the results of a particular experiment in novel ways to see if new limits can be placed on other gravitational effects (eg., shielding, mass equivalence, etc.). Some examples of that class of work will be considered here, as well. Lastly, and from a much broader perspective, the presence of G in the expressions for the Planck mass, length and time, the implications of spatial and temporal variation arising in multi-dimensional theories, and its ubiquity in cosmological models suggest some interesting physical scenarios in which the laws of quantum mechanics are applicable. A temperature-dependent G is one example, and this and others will be explored.

1. Introduction

The year 1998 marked the 200th anniversary of the measurement of the "mean density of the Earth" by Henry Cavendish (1798), and his use of the torsion balance to do so marked a very significant milestone in the then fledgling science of experimental gravitational physics. This was the first time that the gravitational interaction between two test bodies had been measured in a way which minimized the influence of the gravitational field of the Earth. From his early effort came the long-standing quest to determine the absolute value of the constant of proportionality in Newton's inverse square law of gravitation, $F = GmM/r^2$, where F is the force of gravitational attraction acting between the masses m and M, the centers of which are a distance r apart. The constant G, typically referred

P.G. Bergmann and V. de Sabbata (eds.),
Advances in the Interplay Between Quantum and Gravity Physics, 123–131.

to as the universal or Newtonian gravitational constant, has been the subject of much controversy in the last 10 years, largely because the many attempts to determine its absolute value have not been in agreement with each other. The situation is well known, and the early history behind it has been recorded most recently by Gillies (1987), who also reviewed the modern measurements (1997). An excellent source for detailed descriptions of the present-day experiments is the special commemorative issue of *Measurement Science and Technology* that was published by the Institute of Physics (1999) and which served to document the proceedings of the Cavendish Bicentennial Conference.

Partly because there is uncertainty in what the actual value of G should be, an aspect of the role played by G in modern physics that is often somewhat overlooked is the ability that experimental determinations of it have to provide insight into the general field of quantum gravity. It is our goal here to discuss some of those possibilities and call attention to potentially decisive experiements that will test theories of interest in this area. We begin with a brief synopsis of the experimental situation with respect to the absolute value of G, and then proceed to assess how certain measurements of it might be used to test an important question within quantum gravity, "Do leptons generate gravity?" We then go on to address the topic of variations of G with time, and point out that such an effect has been suggested as playing a role in placing a limit on a possible quintessence coupling. Next we address the predictions of a temperature-dependent value of G, and describe the experiments that might be used for such investigations. Lastly, we list a number of proposed and actual experiments directly related to measuring G or the gravitational force in the quantum limit. We close with a few remarks about the directions of future research.

2. The Presently Accepted Value of G and Related Measurements

Until recently, the accepted value of G was $G = (6.672\,59 \pm 0.000\,85) \times 10^{-11}\ m^3s^{-2}kg^{-1}$ based on the adjustment of the fundamental constants carried out by Cohen and Taylor (1987) under the auspices of CODATA. In the latest adjustment of the fundamental constants, Mohr and Taylor (2000), the value of G was found to be in the rather anomalous situation of having its uncertainty increased, with the result that $G = (6.673 \pm 0.010) \times 10^{-11}\ m^3s^{-2}kg^{-1}$, a value virtually identical to that which had been recommended prior to the 1986 adjustment. This reversal of fortunes was due in considerable part to the significant disagreement of the result obtained by workers at the Physikalisch-Technische Bundesanstalt in Germany, which was over 0.5% larger than virtually all of the others. That experiment was done so carefully, though, that no cause of the discrepancy has yet been identified. For a thorough discussion of all of the data, see the review by Gillies (1997). See also the very recent commentary by Quinn (2000).

At the time of the latest re-assignment of the value of G, several high-precision experiments were in progress, with the goal being to resolve this situation. Two of them have since been completed. That of Gundlach and Merkowitz (2000) found the value of G to be $G = (6.674\,215 \pm 0.000\,092) \times 10^{-11}\ m^3s^{-2}kg^{-1}$ yielding an uncertainty of about 14 ppm. That of Quinn et al. at the International Bureau of Weights and Measures (2001, submitted for publication) found the value $G = (6.675\,59 \pm 0.000\,27) \times 10^{-11}\ m^3s^{-2}kg^{-1}$ with the uncertainty thus being about 40 ppm. The two numbers are roughly 200 ppm different and thus exclude each other within their uncertainties, but the level of agreement is nevertheless substantially better than that which guided the 2000 CODATA adjustment. The experiment at the Universität Zürich has also produced results at an interesting level of uncertainty, but it will be discussed within other contexts below.

During this time, progress has also been made on the very difficult problem of making a bona fide theoretical prediction of the value of G. Gillies (1997) and Damour (1999) review the status of the efforts in this area. Kazakov (2001) has cautioned, though, that the Standard Model includes some 19 free parameters that represent the coupling constants of the interactions, and hence it is not clear how one could ultimately arrive at a predicted value of G without additional symmetrization and input from experiments. Much remains to be done in that arena.

Some areas where theory has made significant inroads are in the discussions of different types of variation of G, for instance with time and in space. Extra dimensions are implicated in both cases. The compactification of extra dimensions might be evidenced by a slow temporal variation of G or by a significant deviation from the laboratory value of G at submillimeter distance scales. A positive result in either type of measurement would have significant impacts on the design of quantized theories of gravity.

3. Active Mass of Leptons

The Universität Zürich experiment for the measurement of G, reported by Schurr et al. (1998a, 1998b) is a particularly interesting one from the standpoint of quantum gravity. The reason for this has to do with the materials that were used. The beam-balance method for detecting differential gravitational forces on test masses first used tanks of water and later tanks of mercury as moveable attracting masses. The significance of this harks back to a conjecture by Jacobson (1983) that the question is open as to whether or not the leptons in matter act as sources of gravitational force, i.e., whether or not leptons have an active gravitational mass. This is also related to the specific question of point particles behave gravitationally in the same way that extended particles do. He noted that experiments aimed at making absolute measurements of G could in principle answer this question. Unfortunately, at that time no such experiments had sufficient precision to permit such an analysis to be carried out successfully.

Unnikrishnan and Gillies (2001a, 2001b, 2001c, all submitted for publication) observed that the Universität Zürich apparatus, however, afforded this possibility. The difference in the ratios of active to passive mass for water vs. mercury in the case where leptons did not contribute to gravity would be large enough to produce a difference in the values of G roughly 30 ppm larger than the nominal disagreement between the two actual measured values, thus providing the direct evidence that electrons do generate gravity, i.e., of the equivalence of active and passive mass of electrons ("EAPM"). The only previous experiment aimed specifically at determining a difference between active and passive mass was the well known Kreuzer experiment at Princeton, but its sensitivity was not sufficient to address the question of leptonic gravity. Unnikrishnan and Gillies (2001a) likewise noted that the absolute value of G measuements of Pontikis (1972) also provided a weak constraint on lepton gravity, but that a repetition of the work of Gundlach and Merkowitz (2000) using lead attracting masses could possibly settle the question in a definitive manner.

By coincidence, Unnikrishnan and Gillies (2000) had also noted that the Universität Zürich experiment could be used to set the most stringent laboratory value obtained until then on the Majorana extinction coefficient, h, that governs the size of gravitational shielding effects. They found $h < 4.3 \times 10^{-14}$ cm^2g^{-1}. This limit (and a tighter one of $h < 2 \times 10^{-17}$ cm^2g^{-1}, Unnikrishnan et al. (2001)) stand in contradistinction to the claims of Dedov et al. (1999) that signs of gravitational screening can be found in various measurements of the absolute value of G.

4. An Implication of the Time Variation of G

There have been roughly 40 measurements or observational inferences of the time variation of G that have been made in the past 15 or so years; see Gillies (1997). All of them put very tight limits on the size of any such effect, roughly $\lesssim 1 \times 10^{-11}$ per year or less. There are many implications of a non- zero $\dot{G}/G$ and several authors, beginning with Dirac, have discussed them over the last 75 years.

A fairly recent discussion focusing on the "quintessence" (i.e., a dynamical vacuum energy) in the universe has brought the question of $\dot{G}/G$ back to a level of some significance, however. The coupling of the quintessence field to the gravitational field has been considered by Chiba (1999), who suggests that the size of such a coupling might be constrained by the experimental tests of the time variation of G. If the size of the inter-field coupling is given by ξ, Chiba finds the relationship between $\dot{G}/G$ and ξ to be such that the experimental and observational data on $\dot{G}/G$ tightly constrain negative values of ξ while only loosely constraining the positive values. Specifically, he finds $|\xi| \lesssim 10^{-2}$, a much less stringent limit than the size of the coupling of the quintessence to the electromagnetic field, which is $\lesssim 10^{-6}$, but a value consistent with the existence of such a field.

From the perspective of quantum gravitation, the presence of a quintessence field would be a significant signal of the need for a very small cosmological constant, and this would be consistent with new cosmologies, eg., that of Rooprai and Lohiya (2001), in which a possible violation of active vs. passive mass for electrons would again be an issue. (Therefore, measurements of both G and $\dot{G}/G$ could be important in examining quintessence.)

5. The Temperature Dependence of G

Gauge theories, such as those pursued by Linde (1980), can predict a temperature dependence in the G. Following the discussion of Massa (1989), the general form of such dependencies should be $G = G_0(1 - \alpha G_0 T^2)^{-1}$ where G_0 is the "zero temperature" value of G (i.e., the laboratory value). The constant α will be unique to the particular theory under consideration, but is usually thought to be on the order of the ratio of the electromagnetic to gravitational coupling constants, which is roughly $\approx 10^{-20}$ $s^2kgK^{-2}cm^{-3}$. Massa has noted that the expression for G(T) suggests that there will be an upper bound on the temperature of any physical process given by $T = T_c = (\alpha G_0)^{-1/2}$ where T_c is called the "critical temperature." Using the cited value of α, we find $T_c \approx 10^{12}$ K, which is a value some 20 orders of magnitude below the Planck temperature, which is given by the expression $T_P = (hc^5/Gk^2)^{1/2}$ where k is Boltzmann's constant and G is the value of the gravitational constant at the present epoch, from which $T_P = 1.4 \times 10^{32}$ K.

The meaning of this discrepancy between the values of T_c and T_P within the context of quantum gravitational theories is unclear, but interesting, and it raises many questions. Is it even appropriate to compare them since at T_c we have a singularity and $G \to \infty$, but the value of G in T_P is from the present epoch? What value of G *should* be used in the expression for the Planck temperature? If the maximum temperature that can be reached by any black body radiator is so much less than the Planck temperature with the value of G from the present epoch, can any thermal process by itself ever disrupt the structure of space-time?

At the lower energy end of the scale, Assis and Clemente (1993) have pointed out that Newton's inverse square law of gravitation must be modified to take into account heating of the interacting masses. They use the law of Dulong-Petit to calculate the fractional increase in the gravitational force between two test masses as they are heated, and find that for iron spheres, the fractional change in force would be $\delta F/F \approx 10^{-12}$ for a 100 K temperature change. This is close to the limit of what is possible to detect experimentally, Gillies (1997).

6. Other Aspects of G of Interest in Quantum Gravity

Much of the work on the development of quantum gravity has focused on black hole physics, since the general relativistic effects are large in such objects. Since most of the tests of general relativity result from studies of phenomena in the weak field limit there has historically been, by comparison, much less direct experimental support for strong field processes. However, a turning point in that situation came in last decade with the advent of observations on binary pulsar systems, and in particular with the work of Taylor and colleagues, see for instance Taylor et al. (1992). Among their many accomplishments, they used the timing observations from PSR 1913+16 and PSR 1534+12 to establish a strong field limit on the value of G. By evaluating the frequency of the arrival pulses, they were able to establish values of the Post-Keplarian Parameters β', c_1 and c_2 that appear in the expression $G_S = G\{1 + ½\beta'(c_1^2 + c_2^2)]$ where G_S is the strong-field value of G. After evaluating their data, the result was found to be $-0.11G \leq G_S \leq 0.14G$, thus constraining the value of G_S to within several percent of the laboratory value.

Others have attempted to explore strong field effects by carrying out experiments on test masses or quantum systems in large centrifugal fields. Beams (1970) looked for a variation in the inertial mass of platinum with respect to that of magnesium in centrifugal accelerations 150,000 times gravity, and found no differences to within 3 parts in 10^{10} per g. Of more relevance to gravitation in the quantum limit have been the searches for variations in nuclear decay rate as a function of (large) centrifugal field. These are particularly interesting in that the work seeks to reveal a connection between the nuclear forces and strong field gravity. The earliest such study was that of Rutherford and Compton (1919) who who showed that the rate of radon decay in centrifugal fields of 20,000g did not vary within 0.1% of the unaccelerated value. Freed, Jaffey and Schultz (1943) studied several isotopes in fields between 100,000 and 1,000,000 times g and found no such effect, but they did note a possible shift in Br^{80} but stated that it was likely within the total experimental error. By contrast, Whitaker (1982) did claim to see a non-negligible decay rate variation in ^{198}Au taken to accelerations of 196,000g, but the experiment does not seem to have been repeated. (It is difficult to see how such a large gravitational influence on nuclear decay rates could not affect, eg., nuclear processes in the Sun.)

Nieto and Goldman (1980) proposed an experiment in which the value of G/h could be measured directly using the AC Josephson effect. The essence of the proposal was to search for a gravity-driven change in the Josephson frequency produced by changing the orientation of a battery connected across the weakly coupled components of the junction. The authors discuss their work within the context of other experiments that have been aimed at seeing how the Earth's gravitational potential affects the behavior of a quantum mechanical system. We note that recent measurements of h, for example that of Williams et al. (1998), have uncertainties well below 100 ppb, and thus the determination of G/h can be an intrinsically high precision measurement, with possible implications for the reduction of the uncertainty in G.

A study entitled "Indirect Evidence for Quantum Gravity," was published by Page and Geilker (1981), in which a torsion balance was used in concert with a radioactive decay-based 'quantum decision process' that formed the basis for determining the mass configuration in the balance. The results of the experiment were claimed to be inconsistent with the simplest alternative to quantum gravity, and thus ultimately support gravitational field quantization. The actual meaning of the result was subsequently debated in the literature; for further discussion of it see Hawkins (1982), Ballentine (1982) and the associated responses for the authors. We do not find that there was any real quantum mechanical content in the experiment, since the radioactive decay process they used served only as a random number generator. Thus, it does not seem appropriate to draw any exclusive conclusions on quantum gravity based on the results that were obtained.

7. Conclusions and Future Possibilities

The weakness of the gravitational force makes any type of experimentation on it quite difficult, and especially so when it is being studied in concert with other physical phenomena. Even so, there have been many useful attempts to explore the interplay between general relativity and quantum mechanics within the context of measurements of the Newtonian gravitational constant. Moreover, several very interesting experiments have been proposed in this area, but are yet to be carried out. Table 1 summarizes the present state of the findings in quantum gravity that have involved determinations of G or other aspects of its role in modern physics.

The evolving technologies that will likely be significant in generating new studies include advanced superconducting systems, improved beam-balance and torsion balance systems, space-based probes of gravitational effects, and the interesting possibilities that may become available via nanotechnologies. For a general review of space-based determinations of G, see the monograph of Sanders and Gillies (1996), and for background on and a detailed discussion of torsion pendulum devices, see Gillies and Ritter (1993).

Acknowledgements

The hospitality of the E. Majorana Center for Scientific Culture is gratefully acknowledged, during the XVIIth Course of the School of Cosmology and Gravitation. Prof. Venzo de Sabbata and the staff of the Center have done much to make these Schools a significant venue for the presentation and discussion of fundamental topics.

References

Assis, A. K. T. and Clemente, R. A., *Il Nuov. Cim.* **108B** 713 (1993).

Ballentine, L. E., *Phys. Rev. Lett.* **48** 522 (1982).

Beams, J. W., *Phys. Rev. Lett.* **24** 840 (1970).

Cavendish, H., *Phil. Trans. R. Soc. London* **88** 469 (1798).

Cohen, E. R., and Taylor, B. N., *Rev. Mod. Phys.* **59** 1121 (1987).

Chiba, T., *Phys. Rev. D* **60** 083508 (1999).

Damour, T., *Meas. Sci. Technol.* **10** 467 (1999).

Dedov, V. P., Filimonov, B. P., Ryzhkin, Yu. V., and Sheikin, A. R., *Meas. Tech.* **42** 930 (1999).

Freed, S., Jaffey, A. H., and Schultz, M. L., *Phys. Rev.* **63** 12 (1943).

Gillies, G. T., *Metrologia* **24** (Suppl.) 1 (1987).

Gillies, G. T. and Ritter, R. C., *Rev. Sci. Instrum.* **64** 283 (1993).

Gillies, G. T., *Rep. Prog. Phys.* **60** 151 (1997).

Gundlach, J. H. and Merkowitz, S. M., *Phys. Rev. Lett.* **85** 2869 (2000).

Hawkins, B., *Phys. Rev. Lett.* **48** 520 (1982).

Jacobson, T., *Abstracts of GRG-10*, Vol. 2, ed. Bertotti, B., de Felice, F., and Pascolini, A., 990-992 (1983).

Institute of Physics (GB), Meas. Sci. Technol. **10**(6) (1999).

Kazakov, K. A., *Private communication* (2001).

Linde, A., *Phys. Lett.* **93B** 394 (1980).

Massa, C., *Helv. Phys. Acta* **62** 420 (1989).

Mohr, P. J. and Taylor, B. N., *Rev. Mod. Phys.* **72** 351 (2000).

Nieto, M. M. and Goldman, T., *Phys. Lett.* **79A** 449 (1980).

Page, D. N. and Geilker, C. D., *Phys. Rev. Lett.* **47** 979 (1981).

Pontikis, C., *C. R. Acad. Sci., Paris* **274** 437 (1972).

Quinn, T. J., *Nature* **408** 919 (2000).

Quinn, T. J., Speake, C. C., Richman, S. J., Davis, R. S., and Picard, A., *Phys. Rev. Lett.* (2001, submitted for publication).

Rooprai, N. and Lohiya, D., preprint: astro-ph/0101280 (2001).

Rutherford, E. and Compton, A. H., *Nature* **104** 412 (1919).

Sanders, A. J. and Gillies, G. T., *Riv. Nuov. Cim.* **19** 1 (1996).

Schurr, J., Nolting, F., and Kundig, W., *Phys. Lett. A* **248** 295 (1998a).

Schurr, J., Nolting, F., and Kundig, W., *Phys. Rev. Lett.* **80** 1142 (1998b).

Taylor, J. H., Wolszczan, A., Damour, T., and Weisberg, J. M., *Nature* **355** 132 (1992).

Unnikrishnan, C. S. and Gillies, G. T., *Phys. Rev. D* **61** 101101 (2000).

Unnikrishnan, C. S., Mohapatra, A. K., and Gillies, G. T., *Phys. Rev. D* **63** 062002 (2001).

Unnikrishnan, C. S. and Gillies, G. T., *Phys. Rev. Lett.* (2001a, submitted for publication).

Unnikrishnan, C. S. and Gillies, G. T., *Abstracts of GRG-16* (2001b, submitted for publication).

Unnikrishnan, C. S. and Gillies, G. T., *Gravity Research Foundation Essay* (2001c, submitted for publication).

Whitaker, D. H., Los Alamos National Laboratory Techical Report LA-9388-MS (June 1982).

Williams, E. R., Steiner, R. L., Newell, D. B., and Olsen, P. T., *Phys. Rev. Lett.* **81** 2404 (1998).

Table 1. Experimental studies and proposed experiments in quantum gravity involving the Newtonian gravitational constant.

Date	Authors	Particles/Materials	Observations	Results
1919	Rutherford and Compton	radon	nuclear decay rate shift vs. centrifugal field	< 0.1% up to 20,000g
1943	Freed, Jaffey and Schultz	various isotopes	nuclear decay rate shift vs. centrifugal field	no changes up to $\approx 10^{6}$g (possible effect for Br^{80})
1980	Nieto and Goldman	ionic species in a superconductor	observe gravity effects on Josephson frequency	determination of G/h (proposed)
1981	Page and Geilker	γ rays from Co^{60} plus torsion balance	quantum decision-based measurement of G	inconsistency with semi-classical gravity
1982	Whitaker	^{198}Au	nuclear decay rate shift vs. centrifugal field	decay probability exceeded control by 7.48 s.d.
1989	Massa	blackbody radiators	temperature dependent gravity leads to critical temperature T_c	$T_c = (\alpha G_0)^{-1/2} \approx 10^{12}$ K
1992	Taylor et al.	binary pulsars	strong field value for G from pulsar timing experiments	$-0.11G \leq G_S \leq 0.14G$
1999	Chiba	quintessence field	evaluation of $\dot{G}/G$ measurements as quintessence/gravity coupling	coupling strength limit is $\lvert\xi\rvert \leq 10^{-2}$
2001a-c	Unnikrishnan and Gillies	leptons	analysis of Univ. Zürich G result as test of violation of EAPM	no violation at ≈ 30 ppm

RECENT ADVANCES IN STOCHASTIC GRAVITY: THEORY AND ISSUES

B. L. HU
Department of Physics, University of Maryland
College Park, Maryland 20742-4111, U.S.A.
E-mail: hub@physics.umd.edu

AND

E. VERDAGUER
Departament de Física Fonamental, Universitat de Barcelona,
Av. Diagonal 647, 08028 Barcelona, Spain
E-mail: verdague@ffn.ub.es

1. Introduction

Whereas semiclassical gravity is based on the semiclassical Einstein equation with sources given by the expectation value of the stress-energy tensor of quantum fields, stochastic semiclassical gravity is based on the Einstein-Langevin equation, which has in addition sources due to the noise kernel. The noise kernel is the vacuum expectation value of the (operator-valued) stress-energy bi-tensor which describes the fluctuations of quantum matter fields in curved spacetimes. We show how a consistent stochastic semiclassical theory can be formulated as a perturbative generalization of semiclassical gravity which describes the back reaction of the lowest order stress-energy fluctuations. The original approach [1] used in the early investigations leading to the establishment of this field [2] (for a review emphasizing ideas, see [3]) was based on quantum open system concepts (where the metric field acts as the "system" of interest and the matter fields as part of its "environment") and the influence functional method. Here, following Refs. [4, 5] we first give an axiomatic derivation of the Einstein-Langevin equations and then show how they can also be derived by the original method based on the influence functional. As a first application we solve these equations following Ref. [6], and compute the two-point correlation functions for the linearized Einstein tensor and for the metric perturbations

P.G. Bergmann and V. de Sabbata (eds.),
Advances in the Interplay Between Quantum and Gravity Physics, 133–218.

in a Minkowski background. We then turn to the important issue of the validity of semiclassical gravity by examining the criteria based on the ratio of the variance of fluctuations of the stress-energy tensor of a quantum field to its mean. We show a calculation of these quantities performed in Ref. [7, 8] for a massless scalar field in the Minkowski and the Casimir vacua as a function of an intrinsic scale defined by introducing a smeared field or by point separation. Contrary to prior claims, the ratio of variance to mean-squared being of the order unity does not necessarily imply the failure of semiclassical gravity. Expressions for the variance to mean-squared ratio as a function of the ratio of the intrinsic to extrinsic scale (defined by the separation of the plates or the periodicity of space for the Casimir topology) identifying the spatial extent where negative energy density prevails are useful for studying quantum field effects in worm holes, baby universe and the design feasibility of 'time-machines'. This study also raises new questions into the very meaning of regularization of the stress-energy tensor for quantum fields in curved or ordinary flat spacetimes. It corroborates the view held by one of us [3] that at high energies it is more useful to consider spacetime as possesing an extended structure, with the stress-energy bitensor of quantum matter fields as the source. To compare with the more familiar point-defined quantum field theory operative at low energies, one needs to address the issue of regularization of the noise kernel. This we discuss following the work of Ref. [9] which uses the point-separation method to derive a general expression for a regularized noise-kernel for quantum fields in general curved spacetimes. From these expressions describing the behavior of fluctuations of quantum fields one could investigate a host of important problems in early universe and black hole physics. This includes addressing the viability of a vacuum-dominated phase which inflationary cosmology is predicated upon, examining how structures are seeded, how the horizon behavior of a black hole is altered and how the late time dynamics of black holes is affected with backreaction of Hawking radiation. We intend to discuss these applications in a later report.

1.1. OVERVIEW

Stochastic semiclassical gravity (STG) is a theory developed in the Nineties using semiclassical gravity (SCG, quantum fields in classical spacetimes, solved self-consistently) as the starting point and aiming at a theory of quantum gravity (QG) as the goal. While semiclassical gravity is based on the semiclassical Einstein equation with the source given by the expectation value of the stress-energy tensor of quantum fields, stochastic gravity (we will often use the shortened term *stochastic gravity* as there is no confusion in the source of stochasticity in gravity being due to quantum fields here, and not of a classical origin) includes also its fluctuations in a new stochastic semiclassical or the Einstein-Langevin equation. If the centerpiece in semiclassical gravity theory is the vacuum expectation value of the stress-energy tensor of a quantum field, and the central issues being how well the vacuum is defined and how the divergences can be controlled by regularization and renormalization, the centerpiece in stochastic semiclassical gravity theory is the stress-energy bi-tensor and its expectation value known as the noise kernel. The mathematical properties of this quantity and its physical content in relation to the behavior of fluctuations of quantum fields in curved spacetimes are the central issues of this new theory. How they induce metric fluctuations and seed the structures of the universe, how they affect the black hole horizons and the backreaction of Hawking radiance in black hole dynamics, including implications on trans-Planckian physics, are new horizons to explore. On the theoretical issues, stochastic gravity is the necessary foundation to investigate the validity of semiclassical gravity and the viability of inflationary cosmology based on the appearance and sustenance of a vacuum energy-dominated phase. It is also a useful beachhead supported by well-established low energy (sub-Planckian) physics to explore the connection with high energy (Planckian) physics in the realm of quantum gravity.

In these lectures we want to summarize our work on the theory aspects since 1998. (A review of ideas leading to stochastic gravity and further developments originating from it can be found in Refs. [1, 3]; a comprehensive formal description is given in Refs. [4, 5]). It is in the nature of a progress report rather than a review. In fact we will have room only to discuss two or three theoretical topics of basic importance. We will try to mention all related work so the reader can at least trace out the parallel and sequential developments. The references at the end of each topic below are representative work where one can seek out further treatments.

Stochastic gravity theory is built on three pillars: general relativity (GR), quantum fields (QF) and nonequilibrium (NEq) statistical mechanics. The first two uphold semiclassical gravity, the last two span statistical field theory, or NEqQF. Strictly speaking one can understand a great deal without appealing to statistical mechanics, and we will try to do so here. But concepts such as quantum open systems [10] and techniques such as the influence functional [11] (which is related to the closed-time-path effective action [12, 13]) were a great help in our understanding of the physical meaning of issues involved towards the construction of this new theory, foremost because quantum fluctuations and correlation have become the focus. Quantum statistical field theory and the statistical mechanics of quantum field theory [14, 15, 16, 17] also aided us in searching for the connection with quantum gravity through the retrieval of correlations and coherence.

We show the scope of stochastic gravity as follows:

I. Ingredients

A. From General Relativity [18, 19] to Semiclassical Gravity.
B. Quantum Field Theory in Curved Spacetimes [20, 21, 22, 23]:
  Stress-energy tensor: Regularization and renormalization.
  Self-consistent solution: Backreaction problems [24].
  Effective action: Closed time path, initial value formulation [12, 13].
  Equation of motion: Real and causal.
C. Nonequilibrium Statistical Mechanics:
  Open quantum systems [10].
  Influence Functional: Stochastic equations [11].
  Noise and Decoherence: Quantum to classical transition [25, 26].
D. Decoherence in Quantum Cosmology and Emergence of Classical Spacetimes [27]

II. Theory

A. Dissipation from Particle Creation [28, 29].
  Backreaction as Fluctuation-Dissipation Relation (FDR) [30].
B. Noise from Fluctuations of Quantum Fields [1, 32, 31].
C*. Einstein-Langevin Equations [2, 4, 5].
D*. Metric Fluctuations in Minkowski spacetime [6].

III. Issues

A*. Validity of Semiclassical Gravity [7, 8].
B. Viability of Vacuum Dominance and Inflationary Cosmology.
C*. Stress-Energy Bitensor and Noise Kernel: Regularization Reassessed [9].

IV. Applications: Early Universe and Black Holes

  [Test field treatment:]
A. Wave Propagation in Stochastic Geometry [33].
B. Black Hole Horizon Fluctuations: Spontaneous/Active versus Induced/Passive [34, 35, 36, 37, 38].
  [Including backreaction:]
C. Noise induced inflation [39].
D. Structure Formation: [40], Trace anomaly-driven inflation [41, 42, 43, 44].

E. Black Hole Backreaction as FDR [45, 46, 47, 48].

V. Related Topics

A. Metic Fluctuations and Trans-Planckian Problem [37, 49, 50].
B. Spacetime Foam [51].
C. Universal 'Metric Conductance' Fluctuations [52].

VI. Ideas

A. General Relativity as Geometro-Hydrodynamics [53].
B. Semiclassical Gravity as Mesoscopic Physics [54].
C. From Stochastic to Quantum Gravity:
via Correlation hierarchy of interacting quantum fields [3, 15, 17].
Possible relation to string theory and matrix theory [55].

We list only the latest work in the respective topics above describing ongoing research. The reader should consult the references therein for earlier work and the background material. We do not seek a complete coverage here, but will discuss only the selected topics in theory and issues (Parts II and III above), i.e., those topics marked with *s above, in Sections 2-5 below. In a later report we plan to discuss recent advances in the applications (Parts IV-V above). We use the $(+,+,+)$ sign conventions of Refs. [18, 19], and units in which $c = \hbar = 1$.

2. Stochastic Semiclassical Gravity

In this section we introduce stochastic semiclassical gravity. We will start with semiclasical gravity which assumes that the stress-energy fluctuations of the matter fields are negligible. When these fluctuations cannot be neglected we may perturbatively correct the semiclassical theory to take them into account. This will lead to the semiclassical Einstein-Langevin equations [2] as the dynamical equations which describe the back reaction of quantum stress-energy fluctuations on the gravitational field. We will introduce these equations by two independent methods. The first, based on Ref. [4], is axiomatic and motivated by the search of consistent equations to describe the back reaction of the quantum stress-energy fluctuations on the gravitational field. The second, based on Ref. [5], is by functional methods. It is motivated by the observation [1] that in some open quantum systems classicalization and decoherence [25] on the system may be brought about by interaction with an environment. The environment being in this case the matter fields and some degrees of freedom of the quantum gravitational field [27].

2.1. SEMICLASSICAL GRAVITY

Semiclassical gravity describes the interaccion of the gravitational field assumed to be a classical field with quantum matter fields. This theory cannot be derived as a limit in a quantum theory of gravity, but can be formaly derived in several ways. One of them is the leading $1/N$ approximation of quatum gravity interacting with N independent and identical free quantum fields [56]. When one assumes that the fields interact with gravity only and keep the value of NG finite, where G is Newton's gravitational constant, one arrives at a theory in which formaly the gravitational field can be treated as a c-number (i.e. quatized at tree level) but matter fields are fully quatized.

In the semiclassical theory theory the expectation value of the stress-energy tensor in a given quantum state is the source of the gravitational field. This theory may be summarized as follows. Let $(\mathcal{M}, g_{\mu\nu})$ be a globaly hyperbolic four-dimensional spacetime with metric $g_{\mu\nu}$ and consider a real scalar quantum field ϕ of mass m propagating on it (we assume a scalar field for simplicity but this is not essential).

The classical action S_m for this matter field is given by the functional

$$S_m[g,\phi] = -\frac{1}{2}\int d^4x\sqrt{-g}\left[g^{\mu\nu}\nabla_\mu\phi\nabla_\nu\phi + \left(m^2+\xi R\right)\phi^2\right], \tag{1}$$

where ∇_μ is the covariant derivative associated of the metric $g_{\mu\nu}$, ξ is a couplig parameter coupling the field to the scalar curvature R, and g under the square root stands for the determinat of $g_{\mu\nu}$.

The field may be quantized in the manifold using the standard canonical formalism [20, 21, 22]. The field operator in the Heisenberg reperesentation $\hat{\phi}$ is an operator valued distribution solution of the Klein-Gordon equation, the field equation of Eq. (1),

$$(\Box - m^2 - \xi R)\hat{\phi} = 0. \tag{2}$$

We will write the field operator as $\hat{\phi}[g](x)$ to indicate that it is a functional of the metric.

The classical stress-energy tensor is obtained by functional derivation of this action in the usual way $T^{\mu\nu}(x) = (2/\sqrt{-g})\delta S_m/\delta g_{\mu\nu}$, leading to

$$\begin{aligned} T_{\mu\nu}[g,\phi] &= \nabla_\mu\phi\nabla_\nu\phi - \frac{1}{2}g_{\mu\nu}\left(\nabla^\rho\phi\nabla_\rho\phi + m^2\phi^2\right) \\ &\quad +\xi\left(g_{\mu\nu}\Box - \nabla_\mu\nabla_\nu + G_{\mu\nu}\right)\phi^2, \end{aligned} \tag{3}$$

where $\Box = \nabla_\mu \nabla^\mu$ and $G_{\mu\nu}$ is the Einstein tensor. With the notation $T_{\mu\nu}[g, \phi]$ we explicity indicate that the stress-energy tensor is a functional of the metric $g_{\mu\nu}$ and the field ϕ.

The next step is to define a stress-energy tensor operator $\hat{T}_{\mu\nu}[g](x)$. Naively one would replace the classical field ϕ in the above functional by the quantum operator $\hat{\phi}[g]$, but this procedure involves taking the product of two distributions at the same spacetime point. This is ill-defined and we need a regularization procedure. There are several regularization methods which one may use, one is the point-splitting or point-separation regularizaton method in which one introduces a point y in a neighborhood of the point x and then uses as a regulator the vector tangent at the point x of the geodesic joining x and y; this method is discussed in detail in section 5. Another well known method is dimesional regularization in which one works in arbitray n dimensions, where n is not necessarily an integer, and then uses as the regulator the parameter $\epsilon = n - 4$; this method is used in this section and in section 3. The regularized stress-energy operator using the Weyl ordering prescription, *i.e.* symmetrical ordering, can be written as

$$\hat{T}^{\mu\nu}[g] = \frac{1}{2}\{\nabla^\mu \hat{\phi}[g] \, , \, \nabla^\nu \hat{\phi}[g]\} + \mathcal{D}^{\mu\nu}[g] \, \hat{\phi}^2[g], \tag{4}$$

where $\mathcal{D}^{\mu\nu}[g]$ is the differential operator

$$\mathcal{D}^{\mu\nu}_x \equiv \left(\xi - \frac{1}{4}\right) g^{\mu\nu}(x)\Box_x + \xi \left(R^{\mu\nu}(x) - \nabla^\mu_x \nabla^\nu_x\right). \tag{5}$$

Note that if dimensional regularization is used, the field operator $\hat{\phi}$ propagates in a n-dimensional spacetime. Once the regularization prescription has been introduced a renormalized and regularized stress-energy operator may be defined as

$$\hat{T}^R_{\mu\nu}[g](x) = \hat{T}_{\mu\nu}[g](x) + F^C_{\mu\nu}[g](x)\hat{I}, \tag{6}$$

where $\hat{I}$ is the identity operator and $F^C_{\mu\nu}[g]$ are some symmetric tensor counterterms which depend on the regulator and are local functionals of the metric. Here we are assuming that all these terms depend on the regulator, see Ref. [5] for details. These states can be chosen in such a way that for any pair of physically acceptable states, i.e. Hadamard states in the sense of Ref. [22], $|\psi\rangle$ and $|\varphi\rangle$ the matrix element $\langle\psi|T^R_{\mu\nu}|\varphi\rangle$ defined as the limit of the previous expression when the regulator takes the physical value, is finite and satisfies Wald's axioms [21, 57]. These counterterms can be extracted from the singular part of a Schwinger-DeWitt series [21, 58, 59]. The choice of these counterterms is not unique but this ambiguity can be absorbed into the renormalized coupling constants which appear in the equations of motion for the gravitational field.

The *semiclassical Einstein equations* for the metric $g_{\mu\nu}$ can then be written as

$$\frac{1}{8\pi G}\left(G_{\mu\nu}[g]+\Lambda g_{\mu\nu}\right)-2(\alpha A_{\mu\nu}+\beta B_{\mu\nu})[g]=\langle \hat{T}^R_{\mu\nu}\rangle[g], \tag{7}$$

where $\langle \hat{T}^R_{\mu\nu}\rangle[g]$ is the expectation value of the operator (6) after the regulator takes the physical value in some physically acceptable state of the field on $(\mathcal{M},g_{\mu\nu})$. Note that both the stress tensor and the quantum state are functionals of the metric, hence the notation. The parameters G, Λ, α and β are the renormalized coupling constants, respectively, the gravitational constant, the cosmological constant and two dimensionless coupling constants which are zero in the classical Einstein equations. These constants must be understoood as the result of "dressing" the bare constants which appear in the classical action before renormalization. The values of these constants must be determined by experiment. The left hand side of Eq. (7) may be derived from the gravitational action

$$S_g[g]=\int d^4x\sqrt{-g}\left[\frac{1}{16\pi G}(R-2\Lambda)+\alpha C_{\mu\nu\rho\sigma}C^{\mu\nu\rho\sigma}+\beta R^2\right], \tag{8}$$

where $C_{\mu\nu\rho\sigma}$ is the Weyl tensor. The tensors $A_{\mu\nu}$ and $B_{\mu\nu}$ come from the functional derivatives with respect to the metric of the terms quadratic in the curvature in Eq. (8), they are explicitly given by

$$\begin{aligned} A_{\mu\nu} &= \frac{1}{2}g_{\mu\nu}C_{\tau\epsilon\rho\sigma}C^{\tau\epsilon\rho\sigma}-2R_{\mu\rho\sigma\tau}R_{\nu}^{\ \rho\sigma\tau}+4R_{\mu\rho}R^{\rho}_{\ \nu}-\frac{2}{3}RR_{\mu\nu} \\ &\quad -2\Box R_{\mu\nu}+\frac{2}{3}\nabla_\mu\nabla_\nu R+\frac{1}{3}g_{\mu\nu}\Box R, \qquad (9)\\ B_{\mu\nu} &= \frac{1}{2}g_{\mu\nu}R^2-2RR_{\mu\nu}+2\nabla_\mu\nabla_\nu R-2g_{\mu\nu}\Box R, \qquad (10) \end{aligned}$$

where $R_{\mu\nu\rho\sigma}$ and $R_{\mu\nu}$ are the Riemann and Ricci tensors, respectively. These two tensors are, like the Einstein and metric tensors, symmetric and divergenceless: $\nabla^\mu A_{\mu\nu}=0=\nabla^\mu B_{\mu\nu}$. Note that a classical stress-energy can also be added to the right hand side of Eq. (7), but for simplicity we omit such a term.

A solution of semiclassical gravity consists of a spacetime $(\mathcal{M},g_{\mu\nu})$, a quantum field operator $\hat{\phi}[g]$ which satisfies (1), and a physically acceptable state $|\psi\rangle[g]$ for this field, such that Eq. (7) is satisfied when the expectation value of the renormalized stress-energy operator is evaluated in this state.

For a free quantum field this theory is robust in the sense that it is consistent and fairly well understood. As long as the gravitational field is assumed to be described by a classical metric, the above semiclassical Einstein equations seems to be the only plausible dynamical equation for this metric: the metric couples to mater fields via the stress-energy tensor and for a given quantum state the only physically observable c-number stress energy-tensor that one can construct is the above renormalized expectation value. However, lacking a full quantum gravity theory the scope and limits of the theory are not so well understood. It is assumed that the semiclassical theory should break down at Planck scales, which is when simple order of magnitude estimates suggest that the quantum effects of gravity should not be ignored because the energy of a quantum fluctuation in a Planck size region, as determined by the Heisenberg uncertainty principle, is comparable to the gravitational energy of that fluctuation.

The theory should also break down when the fluctuations of the stress-energy tensor are large. This has been emphasized by Ford and colaborators [60, 61]. It is less clear, however, how to quantify what a large fluctuation is, and different criteria have been proposed [61, 62, 7, 8]. In Sec. 4 we will discuss at length the issue of the validity of the semiclassical theory. One may illustrate the problem by the following example inspired in Ref. [60].

Let us assume a quantum state formed by an isolated sytem which consists of a superposition with equal amplitude of one configuration with mass M_1 and another with mass M_2. The semiclassical theory as described by Eq. (7) predicts that the gravitational field of the system is produced by the averaged mass $(M_1+M_2)/2$. However, one would expect that if we send a succession of test particles to probe the gravitational field of the above system half of the time they would react to the field of a mass M_1 and half of the time to the field of a mass M_2. If the two masses differ substantially the two predictions are clearly different, note that the fluctuation in mass of the quantum state is of order of $(M_1-M_2)^2$. Altough this example is suggestive a word of caution should be said in order not to take it too literaly. In fact, if the previous masses are macroscopic the quantum system decoheres very quickly [63] and instead of being described by a pure quantum state it is described by a density matrix which diagonalizes in a certain pointer basis. For observables associated to such a pointer basis the matrix density description is equivalnet to that provided by a statistical ensemble. In any case, though, the results will differ from the semiclassical prediction.

2.2. AXIOMATIC ROUTE TO STOCHASTIC SEMICLASSICAL GRAVITY

2.2.1. *The noise kernel*

The purpose of stochastical semiclassical gravity is to go beyond the semiclassical theory and account for the fluctuations of the stress-energy operator. But first, we have to give a physical observable that describes these fluctuations. To lowest order, these fluctuations are obviously described by the following bi-tensor, constructed with the two-point correlation of the stress-energy operator,

$$4N_{\mu\nu\rho\sigma}(x,y) = \frac{1}{2}\langle\{\hat{t}_{\mu\nu}(x), \hat{t}_{\rho\sigma}(y)\}\rangle[g], \tag{11}$$

where the curly brackets mean anticommutator, and where

$$\hat{t}_{\mu\nu}(x) \equiv \hat{T}_{\mu\nu}(x) - \langle\hat{T}_{\mu\nu}(x)\rangle. \tag{12}$$

This bi-tensor will be called *noise kernel* from now on. Note that we have defined it in terms of the unrenormalized stress-tensor operator $\hat{T}_{\mu\nu}[g](x)$ on a given background metric $g_{\mu\nu}$, thus a regulator is implicitly assumed on the r.h.s. of Eq. (11). However, for a linear quantum field the above kernel is free of ultraviolet divergencies because the ultraviolet behaviour of $\langle\hat{T}_{\mu\nu}(x)\hat{T}_{\rho\sigma}(y)\rangle$ is the same as that of $\langle\hat{T}_{\mu\nu}(x)\rangle\langle\hat{T}_{\rho\sigma}(y)\rangle$, thus following Eq. (6) one can replace $\hat{T}_{\mu\nu}$ by the renormalized operator $\hat{T}^R_{\mu\nu}$ in Eq. (11); an alternative proof of this result is given in section 5. The noise kernel should be thought of as a distribution function, the limit of coincidence points has meaning only in the sense of distributions. An analysis of the noise kernel based on the point-separation method is given in section 5.

The bi-tensor $N_{\mu\nu\rho\sigma}(x,y)$ is real and positive-semidefinite, as a consequence of $\hat{T}^R_{\mu\nu}$ being self-adjoint. A simple proof can be given as follows. Let $|\psi\rangle$ be a given quantum state and let $\hat{Q}$ be a selfadjoint operator, $\hat{Q}^\dagger = \hat{Q}$, then one can write $\langle\psi|\hat{Q}\hat{Q}|\psi\rangle = \langle\psi|\hat{Q}^\dagger Q|\psi\rangle = |\hat{Q}|\psi\rangle|^2 \geq 0$. Now let $\hat{t}(x)$ be a selfadjoint operator representing the operator $\hat{t}_{\mu\nu}$ of Eq. (11) but where the coordinate x now carries also the tensorial indices of Eq. (11), then if we define $\hat{Q} = \int dx f(x)\hat{t}(x)$ for an arbitrary well behaved "function" $f(x)$ (which carries also tensorial indices), the previous unequality can be written as $\int dx dy f(x)\langle\psi|\hat{t}(x)\hat{t}(y)|\psi\rangle f(y) \geq 0$, which is the condition for the noise kernel to be posite semi-definite. Note that when we will consider, later on, the inverse kernel $N^{-1}_{\mu\nu\rho\sigma}(x,y)$ we will implicitly assume that we are working in the subspace obtained from the eigenvectors which have strictly positive eigenvalues when we diagonalize the noise kernel.

2.2.2. *The Einstein-Langevin equation*

Our purpose now is to perturbatively modify the semiclassical theory. Thus we will assume that the background spacetime metric $g_{\mu\nu}$ is a solution of the semiclassical Einstein Eqs. (7). The stress-energy tensor will generally have quantum fluctuations on that background spacetime. One would expect that these fluctuations will have some effect on the gravitational field and that this may now be described by $g_{\mu\nu} + h_{\mu\nu}$, where we will assume that $h_{\mu\nu}$ is a linear perturbation to the background solution. The renormalized stress-energy operator and the state of the quatum field may now be denoted by $\hat{T}^R_{\mu\nu}[g+h]$ and $|\psi\rangle[g+h]$, respectively, and $\langle\hat{T}^R_{\mu\nu}\rangle[g+h]$ will be the corresponding expectation value.

Let us now introduce a Gaussian stochastic tensor field $\xi_{\mu\nu}$ defined by the following correlators:

$$\langle\xi_{\mu\nu}(x)\rangle_s = 0, \quad \langle\xi_{\mu\nu}(x)\xi_{\rho\sigma}(y)\rangle_s = N_{\mu\nu\rho\sigma}(x,y), \tag{13}$$

where $\langle\ldots\rangle_s$ means statistical average. The symmetry and positive semi-definite property of the noise kernel guarantees that the stochastic field tensor $\xi_{\mu\nu}$ just introduced is well defined. Note that this stochastic tensor does not capture the whole quantum nature of the fluctuations of the stress-energy operator since it assumes that cumulants of higher order are zero, but it does so at the Gaussian level.

An important property of this stochastic tensor is that it is covariantly conserved in the background spacetime $\nabla^\mu\xi_{\mu\nu} = 0$. In fact, as a consequence of the conservation of $\hat{T}^R_{\mu\nu}[g]$ one can seen that $\nabla^\mu_x N_{\mu\nu\rho\sigma}(x,y) = 0$. Taking the divergence in Eq. (13) one can then show that $\langle\nabla^\mu\xi_{\mu\nu}\rangle_s = 0$ and $\langle\nabla^\mu_x\xi_{\mu\nu}(x)\nabla^\rho_y\xi_{\rho\sigma}(y)\rangle_s = 0$ so that $\nabla^\mu\xi_{\mu\nu}$ is detreministic and represents with certainty the zero vector field in $\mathcal{M}$.

One can also see that for a conformal field, i.e. a field whose classical action is conformally invariant, $\xi_{\mu\nu}$ is traceless, *i.e.* $g^{\mu\nu}\xi_{\mu\nu} = 0$, so that for a conformal matter field the stochastic source gives no correction to the trace anomaly. In fact, from the trace anomaly result which states that $g^{\mu\nu}\hat{T}^R_{\mu\nu}[g]$ is in this case a local c-number functional of $g_{\mu\nu}$ times the identity operator, we have that $g^{\mu\nu}(x)N_{\mu\nu\rho\sigma}(x,y) = 0$. It then follows from Eq. (13) that $\langle g^{\mu\nu}\xi_{\mu\nu}\rangle_s = 0$ and $\langle g^{\mu\nu}(x)\xi_{\mu\nu}(x)g^{\rho\sigma}(y)\xi_{\rho\sigma(y)}\rangle_s = 0$; an alternative proof based on the point-separation method is given in section 5.

All these properties make it quite natural to incorporate into the Einstein equations the stress-energy fluctuations by using the stochastic tensor $\xi_{\mu\nu}$ as the source of the metric perturbations. Thus we will write the following equation.

$$\frac{1}{8\pi G}\left(G_{\mu\nu}[g+h] + \Lambda(g_{\mu\nu}+h_{\mu\nu})\right) - 2(\alpha A_{\mu\nu} + \beta B_{\mu\nu})[g+h] = \langle\hat{T}^R_{\mu\nu}\rangle[g+h] + 2\xi_{\mu\nu}. \tag{14}$$

This equation known as the *semiclassical Einstein-Langevin equation*, is a dynamical equation for the metric perturbation $h_{\mu\nu}$ to linear order. It describes the back-reaction of the metric to the quantum fluctuations of the stress-energy tensor of matter fields, and gives a first order correction to semiclassical gravity as described by the semiclassical Einstein equation (7). Note that the stochastic source $\xi_{\mu\nu}$ is not dynamical, it is independent of $h_{\mu\nu}$ since it decribes the fluctuations of the stress tensor on the semiclassical background $g_{\mu\nu}$.

An important property of the Einstein-Langevin equation is that it is gauge invariant. In the sense that if we change $h_{\mu\nu} \to h'_{\mu\nu} + \nabla_\mu \eta_\nu + \nabla_\nu \eta_\mu$ where η^μ is a stochastic vector field on the manifold $\mathcal{M}$. All the tensor which appear in the Eq. (14) transform as $R_{\mu\nu}[g + h'] = R_{\mu\nu}[g + h] + \mathcal{L}_\eta R_{\mu\nu}[g]$ to linear order in the perturbations, where $\mathcal{L}_\eta$ is the Lie derivative with respect to η^μ. If we substitute h by h' in Eq. (14), we get Eq. (14) plus the Lie derivative of the combination of the tensor which appear in Eq. (7). This last combination vanishes when Eq. (7) is satisfied, as we have assumed. It is thus necessary that the background metric $g_{\mu\nu}$ be a solution of semiclassical gravity.

A solution of Eq. (14) can be formally written as $h_{\mu\nu}[\xi]$. This solution is characterized by the whole family of its correlation functions. From the statistical average of this equation we have that $g_{\mu\nu} + \langle h_{\mu\nu} \rangle_s$ must be a solution of the semiclassical Einstein equation linearized around the background $g_{\mu\nu}$. The fluctuation of the metric around this average are described by the moments of the stochastic field $h^s_{\mu\nu}[\xi] = h_{\mu\nu}[\xi] - \langle h_{\mu\nu} \rangle_s$. Thus the solutions of the Einstein-Langevin equation will provide the two point metric correlators $\langle h^s_{\mu\nu}(x) h^s_{\rho\sigma}(y) \rangle_s$.

The stochastic theory may be understood as an intermediate step between the semiclassical theory and the full quantum theory. In the sense that whereas the semiclassical theory depends on the point-like value of the sress-energy operator, the stochastic theory carries information also on the two-point correlation of the stress-energy operator.

We should also emphasize that, even if the metric fluctuations are classical (stochastic), their origin is quantum not only because they are induced by the fluctuations of quantum matter, but also because they are supposed to describe some remnants of the quantum gravity fluctuations after some mechanism for decoherence and classicalization of the metric field [64, 65, 66, 67, 68]. From the formal assumption that such a mechanism is the Gell-Mann and Hartle mechanism of environment-induced decoherence of suitably coarse-grained system variables [64, 65], one may, in fact, derive the stochastic semiclassical theory [69]. Nevertheless, that derivation is of course formal, given that, due to the lack of the full quantum theory of gravity, the classicalization mechanism for the gravitational field is not understood.

2.2.3. *A toy model*

To illustrate the relation between the quantum and stochastic descriptions it is useful to introduce the following toy model. Let us assume that the gravitational equations are described by a linear field h coupled to a scalar source $T[\phi^2]$ independent of h, with the classical equations $\Box h = T$ in flat spacetime. We should emphasize that this model would describe a linear theory of gravity analogous to electromagnetism and does not mimic the linearized theory of gravity in which T is also linear in h. However it captures some of the key features of linearized gravity.

In the Heisenberg representation the quantum field $\hat{h}$ satisfies

$$\Box \hat{h} = \hat{T}. \tag{15}$$

The solutions of this equation, i.e. the field operator at the point x, $\hat{h}_x$, may be written in terms of the retarded propagator G_{xy} as,

$$\hat{h}_x = \hat{h}_x^0 + \int dx' G_{xx'} \hat{T}_{x'}, \tag{16}$$

where h^0 is the free field. From this solution we may compute, for instance, the following two point quantum correlation function (the anticommutator)

$$\frac{1}{2}\langle\{\hat{h}_x, \hat{h}_y\}\rangle = \frac{1}{2}\langle\{\hat{h}_x^0, \hat{h}_y^0\}\rangle + \frac{1}{2}\int\int dx' dy' G_{xx'} G_{yy'} \langle\{\hat{T}_{x'}, \hat{T}_{y'}\}\rangle, \tag{17}$$

where the expectation value is taken in the quantum state in which both fields ϕ and h have been quantized and we have used that for the free field $\langle \hat{h}^0 \rangle = 0$.

Let us now compare with the stochastic theory for this problem. As discussed previously we should write a classical equation for the field h where on the r.h.s. we substitute $\hat{T}$ by the expectation value $\langle \hat{T} \rangle$ plus a Gaussian stochastic field ξ defined by the quantum correlation function of T (the noise kernel). Note that in our model since T is independent of h we may simply renormalize the expectation value using time ordering, then for the vacuum state of the field $\hat{\phi}$, we would simply have $\langle \hat{T} \rangle_0 = 0$. Thus ξ is defined as in (13) by $\langle \xi \rangle_s = 0$ and $\langle \xi_x \xi_y \rangle_s = (1/2)\langle\{\hat{T}_x, \hat{T}_y\}\rangle - \langle \hat{T}_x \rangle \langle \hat{T}_y \rangle$. The semiclassical stochastic equation is thus

$$\Box h = \langle T \rangle + \xi. \tag{18}$$

The solution of this equation may be written in terms of the retarded propagator as,

$$h_x = h_x^0 + \int dx' G_{xx'} \left(\langle \hat{T}_{x'} \rangle + \xi_{x'} \right), \tag{19}$$

from where the two point correlation function for the classical field h, after using the definition of ξ and that $\langle h^0 \rangle_s = 0$, is given by

$$\langle h_x h_y \rangle_s = \langle h_x^0 h_y^0 \rangle_s + \frac{1}{2}\int\int dx' dy' G_{xx'} G_{yy'} \langle\{\hat{T}_{x'}, \hat{T}_{y'}\}\rangle. \tag{20}$$

Comparing (17) with (20) we see that the respective second terms on the right hand side are identical provided the expectation values are computed in the same quantum state for the field $\hat{\phi}$, note that we have assumed that T does not depend on h. The fact that the field h is also quantized in (17) does not change the previous statement. In the real theory of gravity T, in fact, depends also on h and then the previous statement is only true approximately, i.e perturbatively in h. The nature of the first terms on the right hand sides of equations (17) and (20) is different: in the first case it is the two point expectation value of the free quatum field $\hat{h}_0$ whereas in the second case it is the average of the two point classical average of the homogeneous field h_0, which depends on the initial conditions. Now we can still make these terms to be equal to each other if we assume for the homogeneous field h a distribution of initial conditions such that $\langle h_x^0 h_y^0 \rangle_s = (1/2)\langle\{\hat{h}_x^0, \hat{h}_y^0\}\rangle$. Thus, under this assumption on initial conditions for the field h the two point correlation function of (20) equal the quantum expectation value of (17) exactly.

Thus in a linear theory as in the model just described one may just use the statistical description given by (18) to compute the quantum two point function of equation (16). This does not mean that we can recover all quantum correlation functions with the stochastic description, see Ref. [70] for a general discussion about this point. Note that, for instance, the commutator of the classical stochastic field h is obviously zero, but the commutator of the quantum field $\hat{h}$ is not zero for timelike separated points; this is the prize we pay for the introduction of the classical field ξ to describe the quantum fluctuations. Furthermore, the statistical description is not able to account for the graviton-graviton effects which go beyond the linear approximation in $\hat{h}$.

2.3. FUNCTIONAL APPROACH

The Einstein-Langevin equation (14) may also be derived by a method based on functional techniques [5]. Functional techniques have a long history in semiclassical gravity. It started when effective action methods, which are so familiar in quantum field theory, were used to study the back-reaction of quantum fields in cosmological models [71]. These methods were of great help in the study of cosmological anisotropies since they allowed the introduction of familiar perturbative treatments into the subject. The most common method, the in-out effective action method, however, led to equations of motion which were not real because they were taylored to compute transition elements of quantum operators rather than expectation values. Fortunately the appropriate technique had already been developed by Schwinger and Keldysh [12] in the so called Closed Time Path (CTP) or in-in effective action. These techniques were soon adapted to the gravitational context [13] and were applied to different problems in cosmology. As a result one was now able to deduce the semiclassical Einstein equations by the CTP functional method: starting with an action for the interaction of gravity with matter fields, treating the matter fields as quantum fields and the gravitational field at tree level only.

Furthermore, in this case the CTP functional formalism turns out to be related [31, 72, 73, 74, 75, 5] to the influence functional formalism of Feynman and Vernon [76] since the full quantum system may be understood as consisting of a distinguished subsystem (the "system" of interest) interacting with an environment (the remaining degrees of freedom). The integration of the environment variables in a CTP path integral yields the influence functional, from which one can define an effective action for the dynamics of the system [77, 31, 30, 78, 79, 72].

In our case, we consider the metric field $g_{\mu\nu}(x)$ as the "system" degrees of freedom, and the scalar field $\phi(x)$ and also some "high-momentum" gravitational modes [68] as the "environment" variables. Unfortunately, since the form of a complete quantum theory of gravity interacting with matter is unknown, we do not know what these "high-momentum" gravitational modes are. Such a fundamental quantum theory might not even be a field theory, in which case the metric and scalar fields would not be fundamental objects [3]. Thus, in this case, we cannot attempt to evaluate the influence action of Feynman and Vernon starting from the fundamental quantum theory and performing the path integrations in the environment variables. Instead, we introduce the influence action for an effective quantum field theory of gravity and matter [80, 30], in which such "high-momentum" gravitational modes are assumed to have been already "integrated out." Adopting the usual procedure of effective field theories [81, 80], one has to take the effective action for the metric and the scalar field of the most general local form compatible with general covariance:

$$S[g,\phi] \equiv S_g[g] + S_m[g,\phi] + \cdots,$$

where $S_g[g]$ and $S_m[g,\phi]$ are given by Eqs. (8) and (1), respectively, and the dots stand for terms of order higher than two in the curvature and in the number of derivatives of the scalar field. Here, we shall neglect the higher order terms as well as self-interaction terms for the scalar field. The second order terms are necessary to renormalize one-loop ultraviolet divergencies of the scalar field stress-energy tensor, as we have already seen. Since $\mathcal{M}$ is a globally hyperbolic manifold, we can foliate it by a family of $t =$ constant Cauchy hypersurfaces Σ_t. We denote by $\mathbf{x}$ the coordinates on each of these hypersurfaces, and by t_i and t_f some initial and final times, respectively. The integration domain for the action terms must be understood as a compact region $\mathcal{U}$ of the manifold $\mathcal{M}$, bounded by the hypersurfaces Σ_{t_i} and Σ_{t_f}.

Assuming the form (1) for the effective action which couples the scalar and the metric fields, we can now introduce the corresponding influence functional. This is a functional of two copies of the metric field that we denote by $g^+_{\mu\nu}$ and $g^-_{\mu\nu}$. Let us assume that, in the quantum effective theory, the state of the full system (the scalar and the metric fields) in the Schrödinger picture at the initial time $t=t_i$ can be described by a factorizable density operator, *i.e.*, a density operator which can be written as the tensor product of two operators on the Hilbert spaces of the metric and of the scalar field. Let $\hat{\rho}^{\rm S}(t_i)$ be the density operator describing the initial state of the scalar field. If we consider the theory of a scalar field quantized in the classical background spacetime $(\mathcal{M}, g_{\mu\nu})$ through the action (1), a state in the Heisenberg representation described by a density operator $\hat{\rho}[g]$ corresponds to this state. Let $\{|\varphi(\mathbf{x})\rangle^{\rm S}\}$ be the basis of eigenstates of the scalar field operator $\hat{\phi}^{\rm S}(\mathbf{x})$ in the Schrödinger representation: $\hat{\phi}^{\rm S}(\mathbf{x})\,|\varphi\rangle^{\rm S} = \varphi(\mathbf{x})\,|\varphi\rangle^{\rm S}$. The matrix elements of $\hat{\rho}^{\rm S}(t_i)$ in this basis will be written as $\rho_i[\varphi,\tilde{\varphi}] \equiv {}^{\rm S}\langle\varphi|\,\hat{\rho}^{\rm S}(t_i)\,|\tilde{\varphi}\rangle^{\rm S}$. We can now introduce the *influence functional* as the following path integral over two copies of the scalar field:

$$\begin{aligned}\mathcal{F}_{\rm IF}[g^+,g^-] \;\equiv\; \int & \mathcal{D}[\phi_+]\,\mathcal{D}[\phi_-]\rho_i[\phi_+(t_i),\phi_-(t_i)]\;\delta[\phi_+(t_f)-\phi_-(t_f)] \\ & e^{i\left(S_m[g^+,\phi_+]-S_m[g^-,\phi_-]\right)}. \end{aligned} \tag{21}$$

The above double path integral can be rewritten as a closed time path (CTP) integral, namely, as an integral over a single copy of field paths with two different time branches, one going forward in time from t_i to t_f, and the other going backward in time from t_f to t_i. From this influence functional, the *influence action* of Feynman and Vernon, $S_{\rm IF}[g^+,g^-]$, is defined by

$$\mathcal{F}_{\rm IF}[g^+,g^-] \equiv e^{iS_{\rm IF}[g^+,g^-]}, \tag{22}$$

this action has all the relevant information on the matter fields. Then we can define the *effective action* for the gavitational field, $S_{\rm eff}[g^+,g^-]$, as

$$S_{\rm eff}[g^+,g^-] \equiv S_g[g^+] - S_g[g^-] + S_{\rm IF}[g^+,g^-]. \tag{23}$$

This is the action for the classical gravitational field in the CTP formalism. However, since the gravitational field is only treated at tree level, this is also the effective "classical" action from where the classical equations can be derived.

Expression (21) is ill-defined, it must be regularized to get a meaningful influence functional. We shall assume that we can use dimensional regularization, that is, that we can give sense to Eq. (21) by dimensional continuation of all the quantities that appear in this expression. For this we substitute the action S_m in (21) by some generalization to n spacetime dimensions, and similarly for $S_g[g]$. In this last case the parameters G, Λ, α and β of Eq. (8) are the bare parameters G_B, Λ_B, α_B and β_B, and instead of the square of the Weyl tensor one must use $(2/3)(R_{\mu\nu\rho\sigma}R^{\mu\nu\rho\sigma} - R_{\mu\nu}R^{\mu\nu})$ which by the Gauss-Bonnet theorem leads the same equations of motion as the action (8) when $n = 4$. The form of S_g in n dimensions is suggested by the Schwinger-DeWitt analysis of the ultraviolet divergencies in the matter stress-energy tensor using dimensional regularization. One can then write the effective action of Feynman and Vernon, $S_{\rm eff}[g^+, g^-]$, in Eq. (23) in dimensional regularization. Since both S_m and S_g contain second order derivatives of the metric, one should also add some boundary terms [19, 30]. The effect of these terms is to cancel out the boundary terms which appear when taking variations of $S_{\rm eff}[g^+, g^-]$ keeping the value of $g^+_{\mu\nu}$ and $g^-_{\mu\nu}$ fixed on the boundary of $\mathcal{U}$. Alternatively, in order to obtain the equations of motion for the metric in the semiclassical regime, we can work with the action terms without boundary terms and neglect all boundary terms when taking variations with respect to $g^{\pm}_{\mu\nu}$. From now on, all the functional derivatives with respect to the metric will be understood in this sense.

Now we can derive the semiclassical Einstein equations (7). Using the definition of the stress-energy tensor $T^{\mu\nu}(x) = (2/\sqrt{-g})\delta S_m/\delta g_{\mu\nu}$ and the definition of the influence functional, Eqs. (21) and (22), we see that

$$\langle \hat{T}^{\mu\nu}(x)\rangle[g] = \frac{2}{\sqrt{-g(x)}} \left. \frac{\delta S_{\rm IF}[g^+, g^-]}{\delta g^+_{\mu\nu}(x)} \right|_{g^+=g^-=g}, \tag{24}$$

where the expectation value is taken in the n-dimensional spacetime generalization of the state described by $\hat{\rho}[g]$. Therefore, differentiating $S_{\rm eff}[g^+, g^-]$ in Eq. (23) with respect to $g^+_{\mu\nu}$, and then setting $g^+_{\mu\nu} = g^-_{\mu\nu} = g_{\mu\nu}$, we get the semiclassical Einstein equation in n dimensions. This equation is then renormalized by absorbing the divergencies in the regularized $\langle \hat{T}^{\mu\nu}\rangle[g]$ into the bare parameters, and taking the limit $n \to \infty$ we get the physical semiclassical Einstein equations (7).

2.4. INFLUENCE FUNCTIONAL ROUTE TO STOCHASTIC SEMICLASSICAL GRAVITY

In this subsection we derive the semiclassical Einstein-Langevin equation (14) by means of the influence functional. We also work out the semiclassical Einstein-Langevin equations more explicitly, in a form more suitable for specific calculations.

In the spirit of the previous derivation of the Einstein-Langevin equations, we now seek a dynamical equation for a linear perturbation $h_{\mu\nu}$ to a semiclassical metric $g_{\mu\nu}$, solution of Eq. (7). Strictly speaking if we use dimensional regularization we must consider the n-dimensional version of that equation; see Ref. [5] for details. From the result of the previous subsection, if such equation were simply a linearized semiclassical Einstein equation, it could be obtained from an expansion of the effective action $S_{\rm eff}[g+h^+, g+h^-]$. In particular, since, from Eq. (24), we have that

$$\langle \hat{T}_n^{\mu\nu}(x)\rangle[g+h] = \frac{2}{\sqrt{-\det(g+h)(x)}} \left.\frac{\delta S_{\rm IF}[g+h^+, g+h^-]}{\delta h^+_{\mu\nu}(x)}\right|_{h^+=h^-=h}, \quad (25)$$

the expansion of $\langle \hat{T}_n^{\mu\nu}\rangle[g+h]$ to linear order in $h_{\mu\nu}$ can be obtained from an expansion of the influence action $S_{\rm IF}[g+h^+, g+h^-]$ up to second order in $h^\pm_{\mu\nu}$.

To perform the expansion of the influence action, we have to compute the first and second order functional derivatives of $S_{\rm IF}[g^+, g^-]$ and then set $g^+_{\mu\nu} = g^-_{\mu\nu} = g_{\mu\nu}$. If we do so using the path integral representation (21), we can interpret these derivatives as expectation values of operators. The relevant second order derivatives are

$$\begin{aligned}
\frac{1}{\sqrt{-g(x)}\sqrt{-g(y)}} \left.\frac{\delta^2 S_{\rm IF}[g^+, g^-]}{\delta g^+_{\mu\nu}(x)\delta g^+_{\rho\sigma}(y)}\right|_{g^+=g^-=g} &= -H_{\rm S}^{\mu\nu\rho\sigma}[g](x,y) \\
&\quad -K^{\mu\nu\rho\sigma}[g](x,y) \\
&\quad +iN^{\mu\nu\rho\sigma}[g](x,y), \\
\frac{1}{\sqrt{-g(x)}\sqrt{-g(y)}} \left.\frac{\delta^2 S_{\rm IF}[g^+, g^-]}{\delta g^+_{\mu\nu}(x)\delta g^-_{\rho\sigma}(y)}\right|_{g^+=g^-=g} &= -H_{\rm A}^{\mu\nu\rho\sigma}[g](x,y) \\
&\quad -iN^{\mu\nu\rho\sigma}[g](x,y), \quad (26)
\end{aligned}$$

where

$$\begin{aligned}
N^{\mu\nu\rho\sigma}[g](x,y) &\equiv \frac{1}{8}\left\langle \left\{\hat{t}^{\mu\nu}(x),\, \hat{t}^{\rho\sigma}(y)\right\}\right\rangle[g], \\
H_{\rm S}^{\mu\nu\rho\sigma}[g](x,y) &\equiv \frac{1}{4}\,{\rm Im}\left\langle {\rm T}^*\!\left(\hat{T}^{\mu\nu}(x)\hat{T}^{\rho\sigma}(y)\right)\right\rangle[g], \\
H_{\rm A}^{\mu\nu\rho\sigma}[g](x,y) &\equiv -\frac{i}{8}\left\langle \left[\hat{T}^{\mu\nu}(x),\, \hat{T}^{\rho\sigma}(y)\right]\right\rangle[g], \\
K^{\mu\nu\rho\sigma}[g](x,y) &\equiv \frac{-1}{\sqrt{-g(x)}\sqrt{-g(y)}}\left\langle \left.\frac{\delta^2 S_m[g,\phi]}{\delta g_{\mu\nu}(x)\delta g_{\rho\sigma}(y)}\right|_{\phi=\hat{\phi}}\right\rangle[g],
\end{aligned}$$

with $\hat{t}^{\mu\nu}$ defined in Eq. (12), and where we use a Weyl ordering prescription for the operators in the last of these expressions. Here, [,] means the commutator, and we use the symbol T* to denote that, first, we have to time order the field operators $\hat{\phi}$ and then apply the derivative operators which appear in each term of the product $T^{\mu\nu}(x)T^{\rho\sigma}(y)$, where $T^{\mu\nu}$ is the functional (3). This T* "time ordering" arises because we have path integrals containing products of derivatives of the field, which can be expressed as derivatives of the path integrals which do not contain such derivatives. Notice, from their definitions, that all the kernels which appear in expressions (26) are real and also that $H_{\rm A}^{\mu\nu\rho\sigma}$ is free of ultraviolet divergencies in the limit $n \to 4$.

From Eqs. (26) it is clear that the imaginary part of the influence action, which does not contribute to the semiclassical Einstein equation (7) because the expectation value of $\hat{T}^{\mu\nu}[g]$ is real, contains information on the fluctuations of this operator. From (24) and (26), taking into account that $S_{\rm IF}[g,g]=0$ and that $S_{\rm IF}[g^-,g^+]=-S_{\rm IF}^*[g^+,g^-]$, we can write the expansion for the influence action $S_{\rm IF}[g{+}h^+,g{+}h^-]$ around a background metric g_{ab} in terms of the previous kernels. Taking into account that these kernels satisfy the symmetry relations

$$\begin{aligned} H_{\rm S}^{\mu\nu\rho\sigma}(x,y) = H_{\rm S}^{\rho\sigma\mu\nu}(y,x), \qquad & H_{\rm A}^{\mu\nu\rho\sigma}(x,y) = -H_{\rm A}^{\rho\sigma\mu\nu}(y,x), \\ K^{\mu\nu\rho\sigma}(x,y) \;\; = \;\; & K^{\rho\sigma\mu\nu}(y,x); \end{aligned} \tag{27}$$

and introducing the new kernel

$$H^{\mu\nu\rho\sigma}(x,y) \equiv H_{\rm S}^{\mu\nu\rho\sigma}(x,y) + H_{\rm A}^{\mu\nu\rho\sigma}(x,y), \tag{28}$$

the expansion of $S_{\rm IF}$ can be finally written as

$$\begin{aligned} S_{\rm IF}[g{+}h^\pm] = & \frac{1}{2}\int d^4x\,\sqrt{-g(x)}\;\langle \hat{T}^{\mu\nu}(x)\rangle[g]\;[h_{\mu\nu}(x)] \\ & +\frac{i}{2}\int d^4x\,d^4y\,\sqrt{-g(x)}\sqrt{-g(y)}\;[h_{\mu\nu}(x)]\,N^{\mu\nu\rho\sigma}[g](x,y)\,[h_{\rho\sigma}(y)] \\ & -\frac{1}{2}\int d^4x\,d^4y\,\sqrt{-g(x)}\sqrt{-g(y)}\;[h_{\mu\nu}(x)]\,(H^{\mu\nu\rho\sigma}[g](x,y) \\ & \qquad\qquad +K^{\mu\nu\rho\sigma}[g](x,y))\,\{h_{\rho\sigma}(y)\} + 0(h^3), \end{aligned} \tag{29}$$

where we have used the notation

$$[h_{\mu\nu}] \equiv h^+_{\mu\nu} - h^-_{\mu\nu}, \qquad \{h_{\mu\nu}\} \equiv h^+_{\mu\nu} + h^-_{\mu\nu}. \tag{30}$$

We are now in the position to carry out the formal derivation of the semiclassical Einstein-Langevin equation. The procedure is well known [31, 30, 73, 82], it consists of deriving a new "stochastic" effective action using the the following identity:

$$\begin{aligned} & e^{-\frac{1}{2}\int d^4x\,d^4y\,\sqrt{-g(x)}\sqrt{-g(y)}\,[h_{\mu\nu}(x)]\,N^{\mu\nu\rho\sigma}(x,y)\,[h_{\rho\sigma}(y)]} \\ & \qquad\qquad = \int \mathcal{D}[\xi]\,\mathcal{P}[\xi]\,e^{i\int d^4x\,\sqrt{-g(x)}\,\xi^{\mu\nu}(x)\,[h_{\mu\nu}(x)]}, \end{aligned} \tag{31}$$

where $\mathcal{P}[\xi]$ is the probability distribution functional of a Gaussian stochastic tensor $\xi^{\mu\nu}$ characterized by the correlators (13) with N^{abcd} given by Eq. (11), and where the path integration measure is assumed to be a scalar under diffeomorphisms of $(\mathcal{M}, g_{\mu\nu})$. The above identity follows from the identification of the right hand side of (31) with the characteristic functional for the stochastic field $\xi^{\mu\nu}$. In fact, by differentiation of this expression with respect to $[h_{\mu\nu}]$, it can be checked that this is the characteristic functional of a stochastic field characterized by the correlators (13). When $N^{\mu\nu\rho\sigma}(x,y)$ is strictly positive definite, the probability distribution functional for $\xi^{\mu\nu}$ is explicitly given by

$$\mathcal{P}[\xi] = \frac{e^{-\frac{1}{2}\int d^4x\, d^4y\, \sqrt{-g(x)}\sqrt{-g(y)}\, \xi^{\mu\nu}(x)\, N^{-1}_{\mu\nu\rho\sigma}(x,y)\, \xi^{\rho\sigma}(y)}}{\int \mathcal{D}[\bar{\xi}]\, e^{-\frac{1}{2}\int d^4z\, d^4w\, \sqrt{-g(z)}\sqrt{-g(w)}\, \bar{\xi}^{\tau\omega}(z)\, N^{-1}_{\tau\omega\epsilon\kappa}(z,w)\, \bar{\xi}^{\epsilon\kappa}(w)}}, \tag{32}$$

where $N^{-1}_{\mu\nu\rho\sigma}[g](x,y)$ is the inverse of $N^{\mu\nu\rho\sigma}[g](x,y)$ defined by

$$\int d^4z\, \sqrt{-g(z)}\, N^{\mu\nu\tau\omega}(x,z) N^{-1}_{\tau\omega\rho\sigma}(z,y) = \frac{1}{2}\left(\delta^{\mu}_{\rho}\delta^{\nu}_{\sigma} + \delta^{\mu}_{\sigma}\delta^{\nu}_{\rho}\right) \frac{\delta^4(x-y)}{\sqrt{-g(x)}}. \tag{33}$$

We may now introduce the *stochastic effective action* as

$$S^{s}_{\rm eff}[h^+, h^-; g; \xi] \equiv S_g[g+h^+] - S_g[g+h^-] + S^{s}_{\rm IF}[h^+, h^-; g; \xi], \tag{34}$$

where the "stochastic" influence action is defined as

$$S^{s}_{\rm IF}[h^{\pm}; g; \xi_n] \equiv {\rm Re}\, S_{\rm IF}[g+h^{\pm}] + \int d^4x\, \sqrt{-g(x)}\, \xi^{\mu\nu}(x)\, [h_{\mu\nu}(x)] + O(h^3). \tag{35}$$

Note that the influence functional as defined from the influence action (29) can be written as a statistical average over $\xi^{\mu\nu}$:

$$\mathcal{F}_{\rm IF}[g+h^+, g+h^-] = \left\langle e^{iS^{s}_{\rm IF}[h^+,h^-;g;\xi_n]} \right\rangle_s. \tag{36}$$

Thus, the effect of the imaginary part of the influence action (29) on the corresponding influence functional is equivalent to the averaged effect of the stochastic source $\xi^{\mu\nu}$ coupled linearly to the perturbations $h^{\pm}_{\mu\nu}$

The stochastic equations of motion for $h_{\mu\nu}$ can now be derived as

$$\frac{1}{\sqrt{-\det(g+h)(x)}} \left.\frac{\delta S^{s}_{\rm eff}[h^+, h^-; g; \xi_n]}{\delta h^{+}_{ab}(x)}\right|_{h^+=h^-=h} = 0. \tag{37}$$

Then, from (25), taking into account that only the real part of the influence action contributes to the expectation value of the stress-energy tensor, we get, to linear order in $h_{\mu\nu}$ the stochastic semiclassical equations (14). To be precise we get first the regularized n-dimensional equations with the bare parameters, and where instead of the tensor $A^{\mu\nu}$ we get $(2/3)D^{\mu\nu}$ where

$$\begin{aligned} D^{\mu\nu} &\equiv \frac{1}{\sqrt{-g}}\frac{\delta}{\delta g_{\mu\nu}}\int d^n x\,\sqrt{-g}\,(R_{\rho\sigma\tau\omega}R^{\rho\sigma\tau\omega}-R_{\rho\sigma}R^{\rho\sigma}) \\ &= \frac{1}{2}\,g^{\mu\nu}(R_{\rho\sigma\tau\omega}R^{\rho\sigma\tau\omega}-R_{\rho\sigma}R^{\rho\sigma}+\Box R)-2R^{\mu\rho\sigma\tau}R^{\nu}{}_{\rho\sigma\tau} \\ &\quad -2R^{\mu\rho\nu\sigma}R_{\rho\sigma}+4R^{\mu\rho}R_{\rho}{}^{\nu}-3\Box R^{\mu\nu}+\nabla^{\mu}\nabla^{\nu}R. \end{aligned} \tag{38}$$

Of course, when $n=4$ these tensors are related, $A^{\mu\nu}=(2/3)D^{\mu\nu}$. After that we renormalize and take the limit $n\to 4$ to obtain the Einstein-Langevin equations in the physical spacetime.

2.4.1. *Explicit linear form of the Eintein-Langevin equation*

We can write the Einstein-Langevin equation in a more explicit form by working out the expansion of $\langle\hat{T}^{\mu\nu}\rangle[g{+}h]$ up to linear order in the perturbation $h_{\mu\nu}$. From Eq. (25), we see that this expansion can be easily obtained from (29). The result is

$$\begin{aligned} \langle\hat{T}_n^{\mu\nu}(x)\rangle[g{+}h] &= \langle\hat{T}_n^{\mu\nu}(x)\rangle[g]+\langle\hat{T}_n^{(1)\mu\nu}[g;h](x)\rangle[g] \\ &\quad -2\int d^n y\,\sqrt{-g(y)}\,H_n^{\mu\nu\rho\sigma}[g](x,y)h_{\rho\sigma}(y)+0(h^2), \end{aligned} \tag{39}$$

where the operator $\hat{T}_n^{(1)\mu\nu}$ is defined from the term of first order in the expansion of $T^{ab}[g+h]$ as

$$T^{\mu\nu}[g{+}h]=T^{\mu\nu}[g]+T^{(1)\mu\nu}[g;h]+0(h^2), \tag{40}$$

using, as always, a Weyl ordering prescription for the operators in the last definition. Here we use a subscript n on a given tensor to indicate that we are explicitly working in n-dimensions, as we use dimensional regularization, and we also use the superindex (1) to generally indicate that the tensor is the first order correction, linear in $h_{\mu\nu}$, in a perturbative expansion around the background $g_{\mu\nu}$.

Using the Klein-Gordon equation (2), and expressions (3) for the stress-energy tensor and the corresponding operator operator, we can write

$$\hat{T}_n^{(1)\mu\nu}[g;h]=\left(\frac{1}{2}\,g^{\mu\nu}h_{\rho\sigma}-\delta^{\mu}_{\rho}h^{\nu}_{\sigma}-\delta^{\nu}_{\rho}h^{\mu}_{\sigma}\right)\hat{T}_n^{\rho\sigma}[g]+\mathcal{F}^{\mu\nu}[g;h]\,\hat{\phi}_n^2[g], \tag{41}$$

where $\mathcal{F}^{\mu\nu}[g;h]$ is the differential operator

$$\begin{aligned} \mathcal{F}^{\mu\nu} \equiv \left(\xi-\frac{1}{4}\right)\left(h^{\mu\nu}-\frac{1}{2}\,g^{\mu\nu}h^{\rho}_{\rho}\right)\Box+\frac{\xi}{2}\Big[\nabla^{\rho}\nabla^{\mu}h^{\nu}_{\rho}+\nabla^{\rho}\nabla^{\nu}h^{\mu}_{\rho}-\Box h^{\mu\nu} \\ -\nabla^{\mu}\nabla^{\nu}h^{\rho}_{\rho}-g^{\mu\nu}\nabla^{\rho}\nabla^{\sigma}h_{\rho\sigma}+g^{\mu\nu}\Box h^{\rho}_{\rho}+\Big(\nabla^{\mu}h^{\nu}_{\rho}+\nabla^{\nu}h^{\mu}_{\rho} \\ -\nabla_{\rho}h^{\mu\nu}-2g^{\mu\nu}\nabla^{\sigma}h_{\rho\sigma}+g^{\mu\nu}\nabla_{\rho}h^{\sigma}_{\sigma}\Big)\nabla^{\rho}-g^{\mu\nu}h_{\rho\sigma}\nabla^{\rho}\nabla^{\sigma}\Big]. \end{aligned} \tag{42}$$

It is understood that indices are raised with the background inverse metric $g^{\mu\nu}$ and that all the covariant derivatives are associated to the metric $g_{\mu\nu}$.

Substituting (39) into the n-dimensional version of the Einstein-Langevin Eq. (15), taking into account that $g_{\mu\nu}$ satisfies the semiclassical Einstein equation (7), and substituting expression (41) we can write the Einstein-Langevin equation in dimensional regularization as

$$
\begin{aligned}
&\frac{1}{8\pi G_B}\left[G^{(1)\mu\nu}-\frac{1}{2}\,g^{\mu\nu}G^{\rho\sigma}h_{\rho\sigma}+G^{\mu\rho}h^{\nu}_{\rho}+G^{\nu\rho}h^{\mu}_{\rho}+\Lambda_B\left(h^{\mu\nu}-\frac{1}{2}\,g^{\mu\nu}h^{\rho}_{\rho}\right)\right]\\
&\quad-\frac{4}{3}\,\alpha_B\left(D^{(1)\mu\nu}-\frac{1}{2}\,g^{\mu\nu}D^{\rho\sigma}h_{\rho\sigma}+D^{\mu\rho}h^{\nu}_{\rho}+D^{\nu\rho}h^{\mu}_{\rho}\right)(x)-2\beta_B\left(B^{(1)\mu\nu}\right.\\
&\quad\left.-\frac{1}{2}g^{\mu\nu}B^{\rho\sigma}h_{\rho\sigma}+B^{\mu\rho}h^{\nu}_{\rho}+B^{\nu\rho}h^{\mu}_{\rho}\right)(x)-\,\mu^{-(n-4)}\,\mathcal{F}^{\mu\nu}_{x}\langle\hat{\phi}^2_n(x)\rangle[g]\\
&\quad+2\int d^n y\,\sqrt{-g(y)}\,\mu^{-(n-4)}H^{\mu\nu\rho\sigma}_{n}[g](x,y)\,h_{\rho\sigma}(y)=2\mu^{-(n-4)}\xi^{\mu\nu}_{n}(x), \qquad (43)
\end{aligned}
$$

where the tensors $G^{\mu\nu}$, $D^{\mu\nu}$ and $B^{\mu\nu}$ are computed from the semiclassical metric $g_{\mu\nu}$, and where we have omitted the functional dependence on $g_{\mu\nu}$ and $h_{\mu\nu}$ in $G^{(1)\mu\nu}$, $D^{(1)\mu\nu}$, $B^{(1)\mu\nu}$ and $\mathcal{F}^{\mu\nu}$ to simplify the notation. The parameter μ is a mass scale which relates the dimensions of the physical field ϕ with the dimensions of the corresponding field in n-dimensions, $\phi_n=\mu^{(n-4)/2}\phi$. Notice that, in Eq. (43), all the ultraviolet divergencies in the limit $n\to 4$, which must be removed by renormalization of the coupling constants, are in $\langle\hat{\phi}^2_n(x)\rangle$ and the symmetric part $H^{\mu\nu\rho\sigma}_{\mathrm{S}n}(x,y)$ of the kernel $H^{\mu\nu\rho\sigma}_{n}(x,y)$, whereas the kernels $N^{\mu\nu\rho\sigma}_{n}(x,y)$ and $H^{\mu\nu\rho\sigma}_{\mathrm{A}n}(x,y)$ are free of ultraviolet divergencies. If we introduce the bi-tensor $F^{\mu\nu\rho\sigma}_{n}[g](x,y)$ defined by

$$F^{\mu\nu\rho\sigma}_{n}[g](x,y)\equiv\left\langle\hat{t}^{\mu\nu}_{n}(x)\,\hat{t}^{\rho\sigma}_{n}(y)\right\rangle[g] \qquad (44)$$

where $\hat{t}^{\mu\nu}$ is defined by Eq. (12), then the kernels N and H_A can be written as

$$
\begin{aligned}
N^{\mu\nu\rho\sigma}_{n}[g](x,y) &= \frac{1}{4}\,\mathrm{Re}\,F^{\mu\nu\rho\sigma}_{n}[g](x,y),\\
H^{\mu\nu\rho\sigma}_{\mathrm{A}n}[g](x,y) &= \frac{1}{4}\,\mathrm{Im}\,F^{\mu\nu\rho\sigma}_{n}[g](x,y), \qquad (45)
\end{aligned}
$$

where we have used that

$$2\left\langle\hat{t}^{\mu\nu}(x)\,\hat{t}^{\rho\sigma}(y)\right\rangle=\left\langle\left\{\hat{t}^{\mu\nu}(x),\,\hat{t}^{\rho\sigma}(y)\right\}\right\rangle+\left\langle\left[\hat{t}^{\mu\nu}(x),\,\hat{t}^{\rho\sigma}(y)\right]\right\rangle,$$

and the fact that the first term on the right hand side of this identity is real, whereas the second one is pure imaginary. Once we perform the renormalization procedure in Eq. (43), setting $n = 4$ will yield the physical semiclassical Einstein-Langevin equation. Due to the presence of the kernel $H_n^{\mu\nu\rho\sigma}(x,y)$, this equation will be usually non-local in the metric perturbation.

2.4.2. *The kernels for the vacuum state*

When the expectation values in the Einstein-Langevin equation are taken in a vacuum state $|0\rangle$, such as, for instance, an "in" vacuum, we can go further. Since then we can write these expectation values in terms of the Wightman and Feynman functions, defined as

$$G_n^+(x,y) \equiv \langle 0|\, \hat{\phi}_n(x)\hat{\phi}_n(y)\, |0\rangle[g], \;\; iG_{Fn}(x,y) \equiv \langle 0|\, \mathrm{T}\Big(\hat{\phi}_n(x)\hat{\phi}_n(y)\Big)\, |0\rangle[g]. \tag{46}$$

These expressions for the kernels in the Einstein-Langevin equation will be very useful for explicit calculations. To simplify the notation, we omit the functional dependence on the semiclassical metric $g_{\mu\nu}$, which will be understood in all the expressions below.

From Eqs. (45), we see that the kernels $N_n^{\mu\nu\rho\sigma}(x,y)$ and $H_{\mathrm{A}n}^{\mu\nu\rho\sigma}(x,y)$ are the real and imaginary parts, respectively, of the bi-tensor $F_n^{\mu\nu\rho\sigma}(x,y)$. From the expression (4) we see that the stress-energy operator $\hat{T}_n^{\mu\nu}$ can be written as a sum of terms of the form $\Big\{\mathcal{A}_x\hat{\phi}_n(x),\, \mathcal{B}_x\hat{\phi}_n(x)\Big\}$, where $\mathcal{A}_x$ and $\mathcal{B}_x$ are some differential operators. It then follows that we can express the bi-tensor $F_n^{\mu\nu\rho\sigma}(x,y)$ in terms of the Wightman function as

$$\begin{aligned}
F_n^{\mu\nu\rho\sigma}(x,y) &= \nabla_x^\mu\nabla_y^\rho G_n^+(x,y)\nabla_x^\nu\nabla_y^\sigma G_n^+(x,y) \\
&+\nabla_x^\mu\nabla_y^\sigma G_n^+(x,y)\nabla_x^\nu\nabla_y^\rho G_n^+(x,y) + 2\mathcal{D}_x^{\mu\nu}\big(\nabla_y^\rho G_n^+(x,y)\nabla_y^\sigma G_n^+(x,y)\big) \\
&+2\,\mathcal{D}_y^{\rho\sigma}\big(\nabla_x^\mu G_n^+(x,y)\nabla_x^\nu G_n^+(x,y)\big) + 2\,\mathcal{D}_x^{\mu\nu}\mathcal{D}_y^{\rho\sigma}\big(G_n^{+2}(x,y)\big),
\end{aligned} \tag{47}$$

where $\mathcal{D}_x^{\mu\nu}$ is the differential operator (5). From this expression and the relations (45), we get expressions for the kernels N_n and $H_{\mathrm{A}n}$ in terms of the Wightman function $G_n^+(x,y)$.

Similarly the kernel $H_{\mathrm{S}n}^{\mu\nu\rho\sigma}(x,y)$, can be written in terms of the Feynman function as

$$\begin{aligned}
H_{\mathrm{S}n}^{\mu\nu\rho\sigma}(x,y) &= -\frac{1}{4}\,\mathrm{Im}\Big[\nabla_x^\mu\nabla_y^\rho G_{Fn}(x,y)\nabla_x^\nu\nabla_y^\sigma G_{Fn}(x,y) \\
&+\nabla_x^\mu\nabla_y^\sigma G_{Fn}(x,y)\nabla_x^\nu\nabla_y^\rho G_{Fn}(x,y) - g^{\mu\nu}(x)\nabla_x^\tau\nabla_y^\rho G_{Fn}(x,y)\nabla_\tau^x\nabla_y^\sigma G_{Fn}(x,y) \\
&-g^{\rho\sigma}(y)\nabla_x^\mu\nabla_y^\tau G_{Fn}(x,y)\nabla_x^\nu\nabla_\tau^y G_{Fn}(x,y) + \mathcal{K}_x^{\mu\nu}\big(2\nabla_y^\rho G_{Fn}(x,y)\nabla_y^\sigma G_{Fn}(x,y) \\
&+\frac{1}{2}\,g^{\mu\nu}(x)g^{\rho\sigma}(y)\nabla_x^\tau\nabla_y^\kappa G_{Fn}(x,y)\nabla_\tau^x\nabla_\kappa^y G_{Fn}(x,y) \\
&-g^{\rho\sigma}(y)\nabla_y^\tau G_{Fn}(x,y)\nabla_\tau^y G_{Fn}(x,y)\big) + \mathcal{K}_y^{\rho\sigma}\big(2\nabla_x^\mu G_{Fn}(x,y)\nabla_x^\nu G_{Fn}(x,y)
\end{aligned}$$

$$-g^{\mu\nu}(x)\nabla^{\tau}_{x}G_{Fn}(x,y)\nabla^{x}_{\tau}G_{Fn}(x,y)) + 2\mathcal{K}^{\mu\nu}_{x}\mathcal{K}^{\rho\sigma}_{y}(G^{2}_{Fn}(x,y))\Big], \tag{48}$$

where $\mathcal{K}^{\mu\nu}_{x}$ is the differential operator

$$\mathcal{K}^{\mu\nu}_{x} \equiv \xi\left(g^{\mu\nu}(x)\Box_{x} - \nabla^{\mu}_{x}\nabla^{\nu}_{x} + G^{\mu\nu}(x)\right) - \frac{1}{2}m^{2}g^{\mu\nu}(x). \tag{49}$$

Note that, in the vacuum state $|0\rangle$, the term $\langle\hat{\phi}^{2}_{n}(x)\rangle$ in equation (43) can also be written as $\langle\hat{\phi}^{2}_{n}(x)\rangle = iG_{Fn}(x,x) = G^{+}_{n}(x,x)$.

Finally, the causality of the Einstein-Langevin equation (43) can be explicitly seen as follows. The non-local terms in that equation are due to the kernel $H(x,y)$ which is defined in Eq. (28) as the sum of $H_{S}(x,y)$ and $H_{A}(x,y)$. Now, when the points x and y are spacelike separated, $\hat{\phi}_{n}(x)$ and $\hat{\phi}_{n}(y)$ commute and, thus, $G^{+}_{n}(x,y) = iG_{Fn}(x,y) = (1/2)\langle 0|\{\hat{\phi}_{n}(x),\hat{\phi}_{n}(y)\}|0\rangle$, which is real. Hence, from the above expressions, we have that $H^{\mu\nu\rho\sigma}_{\mathrm{A}n}(x,y) = H^{\mu\nu\rho\sigma}_{\mathrm{S}n}(x,y) = 0$, and thus $H^{\mu\nu\rho\sigma}_{n}(x,y) = 0$ This fact is not surprising since, from the causality of the expectation value of the stress-energy operator, we know that the non-local dependence on the metric perturbation in the Einstein-Langevin equation, see Eq. (14), must be causal.

2.4.3. *Discussion*

In this section, based on Refs. [4, 5] we have shown how a consistent stochastic semiclassical theory of gravity can be formulated. This theory is a perturbative generalization of semiclassical gravity which describes the back reaction of the lowest order stress-energy fluctuations of quantum matter fields on the gravitational field through the semiclassical Einstein-Langevin equation. We have shown that this equation can be formally derived with a method based on the influence functional of Feynman and Vernon, where one considers the metric field as the "system" of interest and the matter fields as part of its "environment". An explicit linear form of the Einstein-Langevin equations has been given in terms of some kernels which depend on the Wightman and Feynman functions when a vacuum state is considered.

In Ref. [5] the fluctuation-dissipation relations and particle creation in the stochastic gravity context was also discussed. When the background solution of semiclassical gravity consists of a stationary spacetime and a scalar field in a thermal equilibrium state, it is possible to identify a dissipation kernel in the Einstein-Langevin equation which is related to the noise kernel by a fluctuation-dissipation relation [83, 84]. The same relation was previously derived by Mottola [46] using a linear response theory approach, and it is in agreement with previous findings on a Minkowski background [74]; see, however, comments in [47]. This result was also generalized to the case of a conformal scalar field in a conformally stationary background solution of semiclassical gravity.

Particle creation by stochastic metric perturbations in stationary and conformally stationary background solutions of semiclassical gravity was also considered. It is possible to express the total probability of particle creation and the number of created particles (the expectation value of the number operator for "out" particles in the "in" vacuum) in terms of the vacuum noise kernel. Remarkably the averaged value of those quantities is enhanced, over the semiclassical result, by the presence of stochastic metric fluctuations. In the particular cases of a Minkowski background and a conformal field in a spatially flat RW background, the energy of the created particles can be expressed in terms of the vacuum dissipation kernels; see also Refs. [31, 85].

3. Metric Fluctuations in Minkowski Spacetime

In this section we describe the first application of the full stochastic semiclassical theory of gravity, where we evaluate the stochastic gravitational fluctuations in a Minkowski background. In order to do so, we first use the method developed in section 2 to derive the semiclassical Einstein-Langevin equation around a class of solutions of semiclassical gravity consisting of Minkowski spacetime and a linear real scalar field in its vacuum state, which may be considered the ground state of semiclassical gravity. Although the Minkowski vacuum is an eigenstate of the total four-momentum operator of a field in Minkowski spacetime, it is not an eigenstate of the stress-energy operator. Hence, even for these solutions of semiclassical gravity, for which the expectation value of the stress-energy operator can always be chosen to be zero, the fluctuations of this operator are non-vanishing. This fact leads to consider the stochastic corrections to these solutions described by the semiclassical Einstein-Langevin equation.

We then solve the Einstein-Langevin equation for the linearized Einstein tensor and compute the associated two-point correlation functions. Even though, in this case, we expect to have negligibly small values for these correlation functions for points separated by lengths larger than the Planck length, there are several reasons why it is worth carrying out this calculation.

On the one hand, these are the first back-reaction solutions of the full semiclassical Einstein-Langevin equation. There are analogous solutions to a "reduced" version of this equation inspired in a "mini-superspace" model [86], and there is also a previous attempt to obtain a solution to the Einstein-Langevin equation in Ref.[73], but, there, the non-local terms in the Einstein-Langevin equation were neglected.

On the other hand, the results of this calculation, which confirm our expectations that gravitational fluctuations are negligible at length scales larger than the Planck length, but also predict that the fluctuations are strongly suppressed on small scales, can be considered a first test of stochastic semiclassical gravity. In addition, we can extract conclusions on the possible qualitative behavior of the solutions to the Einstein-Langevin equation. Thus, it is interesting to note that the correlation functions at short scales are characterized by correlation lengths of the order of the Planck length; furthermore, such correlation lengths enter in a non-analytic way in the correlation functions. This kind of non-analytic behavior is actually quite common in the solutions to Langevin-type equations with dissipative terms and hints at the possibility that correlation functions for other solutions to the Einstein-Langevin equation are also non-analytic in their characteristic correlation lengths.

3.1. PERTURBATIONS AROUND MINKOWSKI SPACETIME

The Minkowski metric $\eta_{\mu\nu}$ in a manifold $\mathcal{M}$ which is topologically $\mathbb{R}^4$ and the usual Minkowski vacuum, denoted as $|0\rangle$, are the class of simplest solutions to the semiclassical Einstein equation (7), the so called trivial solutions of semiclassical gravity [87]. Note that each possible value of the parameters (m^2, ξ) leads to a different solution. In fact, we can always choose a renormalization scheme in which the renormalized expectation value $\langle 0|\, \hat{T}_R^{\mu\nu}\, |0\rangle[\eta] = 0$. Thus, Minkowski spacetime $(\mathbb{R}^4, \eta_{\mu\nu})$ and the vacuum state $|0\rangle$ are a solution to the semiclassical Einstein equation with renormalized cosmological constant $\Lambda = 0$. The fact that the vacuum expectation value of the renormalized stress-energy operator in Minkowski spacetime should vanish was originally proposed by Wald [57] and it may be understood as a renormalization convention [21, 23]. There are other possible renormalization prescriptions in which such vacuum expectation value is proportional to $\eta^{\mu\nu}$, and this would determine the value of the cosmological constant Λ in the semiclassical equation. Of course, all these renormalization schemes give physically equivalent results: the total effective cosmological constant, *i.e.*, the constant of proportionality in the sum of all the terms proportional to the metric in the semiclassical Einstein and Einstein-Langevin equations, has to be zero.

As we have already mentioned the vacuum $|0\rangle$ is an eigenstate of the total four-momentum operator in Minkowski spacetime, but not an eigenstate of $\hat{T}^R_{\mu\nu}[\eta]$. Hence, even in the Minkowski background, there are quantum fluctuations in the stress-energy tensor and, as a result, the noise kernel does not vanish. This fact leads to consider the stochastic corrections to this class of trivial solutions of semiclassical gravity. Since, in this case, the Wightman and Feynman functions (46), their values in the two-point coincidence limit, and the products of derivatives of two of such functions appearing in expressions (47) and (48) are known in dimensional regularization, we can compute the semiclassical Einstein-Langevin equation using the method outlined in section 2.

In Minkowski spacetime, the components of the classical stress-energy tensor (3) reduce to

$$T^{\mu\nu}[\eta,\phi] = \partial^\mu\phi\partial^\nu\phi - \frac{1}{2}\eta^{\mu\nu}\partial^\rho\phi\partial_\rho\phi - \frac{1}{2}\eta^{\mu\nu}m^2\phi^2 + \xi\left(\eta^{\mu\nu}\Box - \partial^\mu\partial^\nu\right)\phi^2, \tag{50}$$

where $\Box \equiv \partial_\mu\partial^\mu$, and the formal expression for the components of the corresponding "operator" in dimensional regularization, see Eq. (4), is

$$\hat{T}_n^{\mu\nu}[\eta] = \frac{1}{2}\{\partial^\mu\hat{\phi}_n, \partial^\nu\hat{\phi}_n\} + \mathcal{D}^{\mu\nu}\hat{\phi}_n^2, \tag{51}$$

where $\mathcal{D}^{\mu\nu}$ is the differential operator (5), with $g_{\mu\nu} = \eta_{\mu\nu}$, $R_{\mu\nu} = 0$, and $\nabla_\mu = \partial_\mu$. The field $\hat{\phi}_n(x)$ is the field operator in the Heisenberg representation in a n-dimensional Minkowski spacetime, which satisfies the Klein-Gordon equation (2). We use here a stress-energy tensor which differs from the canonical one, which corresponds to $\xi = 0$, both tensors, however, define the same total momentum.

The Wightman and Feynman functions (46) when $g_{\mu\nu} = \eta_{\mu\nu}$, are well known:

$$G_n^+(x,y) = i\Delta_n^+(x-y), \qquad G_{Fn}(x,y) = \Delta_{Fn}(x-y), \tag{52}$$

with

$$\begin{aligned} \Delta_n^+(x) &= -2\pi i \int \frac{d^nk}{(2\pi)^n}\, e^{ikx}\,\delta(k^2+m^2)\,\theta(k^0), \\ \Delta_{Fn}(x) &= -\int \frac{d^nk}{(2\pi)^n}\,\frac{e^{ikx}}{k^2+m^2-i\epsilon}, \qquad \epsilon\to 0^+, \end{aligned} \tag{53}$$

where $k^2 \equiv \eta_{\mu\nu}k^\mu k^\nu$ and $kx \equiv \eta_{\mu\nu}k^\mu x^\nu$. Note that the derivatives of these functions satisfy $\partial_\mu^x\Delta_n^+(x-y) = \partial_\mu\Delta_n^+(x-y)$ and $\partial_\mu^y\Delta_n^+(x-y) = -\partial_\mu\Delta_n^+(x-y)$, and similarly for the Feynman propagator $\Delta_{Fn}(x-y)$.

To write down the semiclassical Einstein equation (7) in n-dimensions for this case, we need to compute the vacuum expectation value of the stress-energy operator components (51). Since, from (52), we have that $\langle 0|\hat{\phi}_n^2(x)|0\rangle = i\Delta_{Fn}(0) = i\Delta_n^+(0)$, which is a constant (independent of x), we have simply

$$\begin{aligned} \langle 0|\,\hat{T}_n^{\mu\nu}\,|0\rangle[\eta] &= -i\left(\partial^\mu\partial^\nu\Delta_{Fn}\right)(0) \\ &= -i\int \frac{d^nk}{(2\pi)^n}\,\frac{k^\mu k^\nu}{k^2+m^2-i\epsilon} = \frac{\eta^{\mu\nu}}{2}\left(\frac{m^2}{4\pi}\right)^{n/2}\Gamma\!\left(-\frac{n}{2}\right), \end{aligned} \tag{54}$$

where the integrals in dimensional regularization have been computed in the standard way (see Ref. [6]) and where $\Gamma(z)$ is the Euler's gamma function. The semiclassical Einstein equation (7) in n-dimensions before renormalization reduces now to

$$\frac{\Lambda_B}{8\pi G_B}\,\eta^{\mu\nu} = \mu^{-(n-4)}\langle 0|\hat{T}_n^{\mu\nu}|0\rangle[\eta]. \tag{55}$$

This equation, thus, simply sets the value of the bare coupling constant Λ_B/G_B. Note, from (54), that in order to have $\langle 0|\,\hat{T}_R^{\mu\nu}\,|0\rangle[\eta]=0$, the renormalized and regularized stress-energy tensor "operator" for a scalar field in Minkowski spacetime, see Eq. (6), has to be defined as

$$\hat{T}_R^{\mu\nu}[\eta] = \mu^{-(n-4)}\,\hat{T}_n^{\mu\nu}[\eta] - \frac{\eta^{\mu\nu}}{2}\,\frac{m^4}{(4\pi)^2}\left(\frac{m^2}{4\pi\mu^2}\right)^{\frac{n-4}{2}}\Gamma\!\left(-\frac{n}{2}\right), \tag{56}$$

which corresponds to a renormalization of the cosmological constant

$$\frac{\Lambda_B}{G_B} = \frac{\Lambda}{G} - \frac{2}{\pi}\,\frac{m^4}{n(n-2)}\,\kappa_n + O(n-4), \tag{57}$$

where

$$\kappa_n \equiv \frac{1}{(n-4)}\left(\frac{e^{\gamma}m^2}{4\pi\mu^2}\right)^{\frac{n-4}{2}} = \frac{1}{n-4} + \frac{1}{2}\ln\!\left(\frac{e^{\gamma}m^2}{4\pi\mu^2}\right) + O(n-4), \tag{58}$$

being γ the Euler's constant. In the case of a massless scalar field, $m^2=0$, one simply has $\Lambda_B/G_B = \Lambda/G$. Introducing this renormalized coupling constant into Eq. (55), we can take the limit $n\to 4$. We find again that, for $(\mathbb{R}^4, \eta_{ab}, |0\rangle)$ to satisfy the semiclassical Einstein equation, we must take $\Lambda=0$.

We can now write down the Einstein-Langevin equations for the components $h_{\mu\nu}$ of the stochastic metric perturbation in dimensional regularization. In our case, using $\langle 0|\hat{\phi}_n^2(x)|0\rangle = i\Delta_{Fn}(0)$ and the explicit expression of Eq. (43) we obtain

$$\begin{aligned}\frac{1}{8\pi G_B}\left[G^{(1)\mu\nu} + \Lambda_B\left(h^{\mu\nu} - \frac{1}{2}\eta^{\mu\nu}h\right)\right](x) &- \frac{4}{3}\,\alpha_B D^{(1)\mu\nu}(x)\\ -2\beta_B B^{(1)\mu\nu}(x) &- \xi\, G^{(1)\mu\nu}(x)\mu^{-(n-4)}\,i\Delta_{Fn}(0)\\ +2\int d^n y\,\mu^{-(n-4)}\,H_n^{\mu\nu\alpha\beta}(x,y)\,h_{\alpha\beta}(y) &= 2\xi^{\mu\nu}(x).\end{aligned} \tag{59}$$

The indices in $h_{\mu\nu}$ are raised with the Minkowski metric and $h \equiv h^{\rho}_{\rho}$, and here a superindex (1) denotes the components of a tensor linearized around the flat metric.

Note that in n-dimensions the correlator for the field $\xi^{\mu\nu}$ is written as

$$\langle \xi^{\mu\nu}(x)\xi^{\alpha\beta}(y)\rangle_s = \mu^{-2(n-4)} N_n^{\mu\nu\alpha\beta}(x,y), \tag{60}$$

Explicit expressions for $D^{(1)\mu\nu}$ and $B^{(1)\mu\nu}$ are given by

$$D^{(1)\mu\nu}(x) = \frac{1}{2}\left(3\mathcal{F}_x^{\mu\alpha}\mathcal{F}_x^{\nu\beta} - \mathcal{F}_x^{\mu\nu}\mathcal{F}_x^{\alpha\beta}\right) h_{\alpha\beta}(x), \;\; B^{(1)\mu\nu}(x) = 2\mathcal{F}_x^{\mu\nu}\mathcal{F}_x^{\alpha\beta} h_{\alpha\beta}(x), \tag{61}$$

where $\mathcal{F}_x^{\mu\nu}$ is the differential operator $\mathcal{F}_x^{\mu\nu} \equiv \eta^{\mu\nu}\Box_x - \partial_x^\mu \partial_x^\nu$.

3.2. THE KERNELS IN THE MINKOWSKI BACKGROUND

3.2.1. *The noise and dissipation kernels*

Since the two kernels (45) are free of ultraviolet divergencies in the limit $n \to 4$, we can deal directly with the $F^{\mu\nu\alpha\beta}(x-y) \equiv \lim_{n\to 4} \mu^{-2(n-4)} F_n^{\mu\nu\alpha\beta}$ in Eq. (44). The kernels $4N^{\mu\nu\alpha\beta}(x,y) = \mathrm{Re}\, F^{\mu\nu\alpha\beta}(x-y)$ and $4H_{\mathrm{A}}^{\mu\nu\alpha\beta}(x,y) = \mathrm{Im}\, F^{\mu\nu\alpha\beta}(x-y)$ are actually the components of the "physical" noise and dissipation kernels that will appear in the Einstein-Langevin equations once the renormalization procedure has been carried out. The bi-tensor $F^{\mu\nu\alpha\beta}$ can be expressed in terms of the Wightman function in four spacetime dimensions, in the following way:

$$\begin{aligned} F^{\mu\nu\alpha\beta}(x) &= -2\Big[\partial^\mu\partial^{(\alpha}\Delta^+(x)\,\partial^{\beta)}\partial^\nu\Delta^+(x) + \mathcal{D}^{\mu\nu}\Big(\partial^\alpha\Delta^+(x)\,\partial^\beta\Delta^+(x)\Big) \\ &\quad + \mathcal{D}^{\alpha\beta}\left(\partial^\mu\Delta^+(x)\,\partial^\nu\Delta^+(x)\right) + \mathcal{D}^{\mu\nu}\mathcal{D}^{\alpha\beta}\Big(\Delta^{+2}(x)\Big)\Big]. \end{aligned} \tag{62}$$

The different terms in Eq. (62) can be easily computed using the integrals

$$I(p) \equiv \int \frac{d^4k}{(2\pi)^4}\, \delta(k^2+m^2)\,\theta(-k^0)\,\delta[(k-p)^2+m^2]\,\theta(k^0-p^0), \tag{63}$$

and $I^{\mu_1\cdots\mu_r}(p)$ which are defined as the previous one by inserting the momenta $k^{\mu_1}\dots k^{\mu_r}$ with $r=1,2,3,4$ in the integrand. All these integral can be expressed in terms of $I(p)$; see Ref. [6] for the explicit expressions. It is convenient to separate $I(p)$ in its even and odd parts with respect to the variables p^μ as

$$I(p) = I_{\mathrm{S}}(p) + I_{\mathrm{A}}(p), \tag{64}$$

where $I_{\mathrm{S}}(-p) = I_{\mathrm{S}}(p)$ and $I_{\mathrm{A}}(-p) = -I_{\mathrm{A}}(p)$. These two functions are explicitly given by

$$\begin{aligned} I_{\mathrm{S}}(p) &= \frac{1}{8\,(2\pi)^3}\,\theta(-p^2-4m^2)\sqrt{1+4\,\frac{m^2}{p^2}}, \\ I_{\mathrm{A}}(p) &= \frac{-1}{8\,(2\pi)^3}\,\mathrm{sign}\,p^0\,\theta(-p^2-4m^2)\sqrt{1+4\,\frac{m^2}{p^2}}. \end{aligned} \tag{65}$$

After some manipulations, we find

$$F^{\mu\nu\alpha\beta}(x) = \frac{\pi^2}{45}(3\mathcal{F}_x^{\mu(\alpha}\mathcal{F}_x^{\beta)\nu} - \mathcal{F}_x^{\mu\nu}\mathcal{F}_x^{\alpha\beta})\int\frac{d^4p}{(2\pi)^4}\,e^{-ipx}\left(1+4\,\frac{m^2}{p^2}\right)^2 I(p)$$
$$+\frac{8\pi^2}{9}\,\mathcal{F}_x^{\mu\nu}\mathcal{F}_x^{\alpha\beta}\int\frac{d^4p}{(2\pi)^4}\,e^{-ipx}\left(3\,\Delta\xi+\frac{m^2}{p^2}\right)^2 I(p), \qquad (66)$$

where $\Delta\xi \equiv \xi - 1/6$. The real and imaginary parts of the last expression, which yield the noise and dissipation kernels, are easily recognized as the terms containing $I_{\rm S}(p)$ and $I_{\rm A}(p)$, respectively. To write them explicitly, it is useful to introduce the new kernels

$$N_{\rm A}(x) \equiv \frac{1}{1920\pi}\int\frac{d^4p}{(2\pi)^4}e^{ipx}\theta(-p^2-4m^2)\sqrt{1+4\frac{m^2}{p^2}}\left(1+4\frac{m^2}{p^2}\right)^2,$$

$$N_{\rm B}(x) \equiv \frac{1}{288\pi}\int\frac{d^4p}{(2\pi)^4}e^{ipx}\theta(-p^2-4m^2)\sqrt{1+4\frac{m^2}{p^2}}\left(3\Delta\xi+\frac{m^2}{p^2}\right)^2,$$

$$D_{\rm A}(x) \equiv \frac{-i}{1920\pi}\int\frac{d^4p}{(2\pi)^4}e^{ipx}{\rm sign}p^0\theta(-p^2-4m^2)$$
$$\sqrt{1+4\,\frac{m^2}{p^2}}\left(1+4\,\frac{m^2}{p^2}\right)^2,$$

$$D_{\rm B}(x) \equiv \frac{-i}{288\pi}\int\frac{d^4p}{(2\pi)^4}\,e^{ipx}\,{\rm sign}\,p^0\theta(-p^2-4m^2)$$
$$\sqrt{1+4\,\frac{m^2}{p^2}}\left(3\Delta\xi+\frac{m^2}{p^2}\right)^2, \qquad (67)$$

and we finally get

$$N^{\mu\nu\alpha\beta}(x,y) = \frac{1}{6}\,(3\mathcal{F}_x^{\mu(\alpha}\mathcal{F}_x^{\beta)\nu} - \mathcal{F}_x^{\mu\nu}\mathcal{F}_x^{\alpha\beta})N_{\rm A}(z) + \mathcal{F}_x^{\mu\nu}\mathcal{F}_x^{\alpha\beta}N_{\rm B}(z),$$
$$H_{\rm A}^{\mu\nu\alpha\beta}(x,y) = \frac{1}{6}(3\mathcal{F}_x^{\mu(\alpha}\mathcal{F}_x^{\beta)\nu} - \mathcal{F}_x^{\mu\nu}\mathcal{F}_x^{\alpha\beta})D_{\rm A}(z) + \mathcal{F}_x^{\mu\nu}\mathcal{F}_x^{\alpha\beta}D_{\rm B}(z), \qquad (68)$$

where $z = x - y$. Notice that the noise and dissipation kernels defined in (67) are actually real because, for the noise kernels, only the $\cos px$ terms of the exponentials e^{ipx} contribute to the integrals, and, for the dissipation kernels, the only contribution of such exponentials comes from the $i\sin px$ terms.

3.2.2. *The kernel* $H_S^{\mu\nu\alpha\beta}(x,y)$

The evaluation of the kernel components $H_{Sn}^{\mu\nu\alpha\beta}(x,y)$ is a much more cumbersome task. Since these quantities contain divergencies in the limit $n\to 4$, we shall compute them using dimensional regularization. Using Eq. (48), these components can be written in terms of the Feynman propagator (53) as

$$\mu^{-(n-4)} H_{Sn}^{\mu\nu\alpha\beta}(x,y) = \frac{1}{4}\,\mathrm{Im}\,K^{\mu\nu\alpha\beta}(x-y), \tag{69}$$

where

$$\begin{aligned} K^{\mu\nu\alpha\beta}(x) \equiv -\mu^{-(n-4)}\Big\{ & 2\partial^\mu\partial^{(\alpha}\Delta_{Fn}(x)\,\partial^{\beta)}\partial^\nu\Delta_{Fn}(x) \\ & +2\mathcal{D}^{\mu\nu}\Big(\partial^\alpha\Delta_{Fn}(x)\partial^\beta\Delta_{Fn}(x)\Big) + 2\mathcal{D}^{\alpha\beta}\Big(\partial^\mu\Delta_{Fn}(x)\,\partial^\nu\Delta_{Fn}(x)\Big) \\ & +2\mathcal{D}^{\mu\nu}\mathcal{D}^{\alpha\beta}\Big(\Delta_{Fn}^2(x)\Big) + \Big[\eta^{\mu\nu}\partial^{(\alpha}\Delta_{Fn}(x)\partial^{\beta)} + \eta^{\alpha\beta}\partial^{(\mu}\Delta_{Fn}(x)\partial^{\nu)} \\ & +\Delta_{Fn}(0)\left(\eta^{\mu\nu}\mathcal{D}^{\alpha\beta} + \eta^{\alpha\beta}\mathcal{D}^{\mu\nu}\right) \\ & +\frac{1}{4}\,\eta^{\mu\nu}\eta^{\alpha\beta}\left(\Delta_{Fn}(x)\Box - m^2\Delta_{Fn}(0)\right)\Big]\,\delta^n(x)\Big\}. \end{aligned} \tag{70}$$

Let us define the integrals

$$J_n(p) \equiv \mu^{-(n-4)}\int\frac{d^nk}{(2\pi)^n}\,\frac{1}{(k^2+m^2-i\epsilon)\,[(k-p)^2+m^2-i\epsilon]}, \tag{71}$$

and $J_n^{\mu_1\ldots\mu_r}(p)$ obtained by inserting the momenta $k^{\mu_1}\ldots k^{\mu_r}$ into the previous integral, together with

$$I_{0n} \equiv \mu^{-(n-4)}\int\frac{d^nk}{(2\pi)^n}\,\frac{1}{(k^2+m^2-i\epsilon)}, \tag{72}$$

and $I_{0n}^{\mu_1\ldots\mu_r}$ which are also obtained by inserting momenta in the integrand. Then, the different terms in Eq. (70) can be computed. These integrals are explicitly given in Ref. [6]. It is found that $I_{0n}^{\mu}=0$ and the remaining integrals can be written in terms of I_{0n} and $J_n(p)$. It is useful to introduce the projector $P^{\mu\nu}$ orthogonal to p^μ as

$$p^2P^{\mu\nu} \equiv \eta^{\mu\nu}p^2 - p^\mu p^\nu \tag{73}$$

then the action of the operator $\mathcal{F}_x^{\mu\nu}$ is simply written as $\mathcal{F}_x^{\mu\nu}\int d^np\,e^{ipx}\,f(p) = -\int d^np\,e^{ipx}\,f(p)\,p^2P^{\mu\nu}$ where $f(p)$ is an arbitrary function of p^μ.

After a rather long but straightforward calculation, we get, expanding around $n=4$,

$$K^{\mu\nu\alpha\beta}(x) = \frac{i}{(4\pi)^2}\Big\{\kappa_n\left[\frac{1}{90}\,(3\mathcal{F}_x^{\mu(\alpha}\mathcal{F}_x^{\beta)\nu} - \mathcal{F}_x^{\mu\nu}\mathcal{F}_x^{\alpha\beta})\delta^n(x)\right.$$

$$
\begin{aligned}
&+4\,\Delta\xi^2\,\mathcal{F}_x^{\mu\nu}\mathcal{F}_x^{\alpha\beta}\delta^n(x)+\frac{2}{3}\frac{m^2}{(n-2)}\left(\eta^{\mu\nu}\eta^{\alpha\beta}\Box_x-\eta^{\mu(\alpha}\eta^{\beta)\nu}\Box_x\right.\\
&\left.+\eta^{\mu(\alpha}\partial_x^{\beta)}\partial_x^{\nu}+\eta^{\nu(\alpha}\partial_x^{\beta)}\partial_x^{\mu}-\eta^{\mu\nu}\partial_x^{\alpha}\partial_x^{\beta}-\eta^{\alpha\beta}\partial_x^{\mu}\partial_x^{\nu}\right)\delta^n(x)\\
&\left.+\frac{4m^4}{n(n-2)}\left(2\eta^{\mu(\alpha}\eta^{\beta)\nu}-\eta^{\mu\nu}\eta^{\alpha\beta}\right)\delta^n(x)\right]\\
&+\frac{1}{180}\left(3\mathcal{F}_x^{\mu(\alpha}\mathcal{F}_x^{\beta)\nu}-\mathcal{F}_x^{\mu\nu}\mathcal{F}_x^{\alpha\beta}\right)\int\frac{d^np}{(2\pi)^n}\,e^{ipx}\left(1+4\,\frac{m^2}{p^2}\right)^2\bar{\phi}(p^2)\\
&+\frac{2}{9}\mathcal{F}_x^{\mu\nu}\mathcal{F}_x^{\alpha\beta}\int\frac{d^np}{(2\pi)^n}\,e^{ipx}\left(3\,\Delta\xi+\frac{m^2}{p^2}\right)^2\bar{\phi}(p^2)\\
&-\left[\frac{4}{675}\left(3\mathcal{F}_x^{\mu(\alpha}\mathcal{F}_x^{\beta)\nu}-\mathcal{F}_x^{\mu\nu}\mathcal{F}_x^{\alpha\beta}\right)+\frac{1}{270}\left(60\xi-11\right)\mathcal{F}_x^{\mu\nu}\mathcal{F}_x^{\alpha\beta}\right]\delta^n(x)\\
&\left.-m^2\left[\frac{2}{135}\left(3\mathcal{F}_x^{\mu(\alpha}\mathcal{F}_x^{\beta)\nu}-\mathcal{F}_x^{\mu\nu}\mathcal{F}_x^{\alpha\beta}\right)+\frac{1}{27}\,\mathcal{F}_x^{\mu\nu}\mathcal{F}_x^{\alpha\beta}\right]\Delta_n(x)\right\}+\ldots,
\end{aligned}
\tag{74}
$$

where the dots stand for $O(n-4)$, κ_n has been defined in (58), and $\bar{\phi}(p^2)$ and $\Delta_n(x)$ are given by

$$
\begin{aligned}
\bar{\phi}(p^2) &\equiv \int_0^1 d\alpha\;\ln\!\left(1+\frac{p^2}{m^2}\,\alpha(1-\alpha)-i\epsilon\right)\\
&= -i\pi\,\theta(-p^2-4m^2)\sqrt{1+4\,\frac{m^2}{p^2}}+\varphi(p^2),
\end{aligned}
\tag{75}
$$

$$
\Delta_n(x) \equiv \int\frac{d^np}{(2\pi)^n}\,e^{ipx}\,\frac{1}{p^2},
\tag{76}
$$

where $\varphi(p^2)\equiv\int_0^1 d\alpha\;\ln|1+\frac{p^2}{m^2}\,\alpha(1-\alpha)|$.

The imaginary part of (74) [which, using (69), gives the kernel components $\mu^{-(n-4)}H_{\mathrm{S}n}^{\mu\nu\alpha\beta}(x,y)$] can be easily obtained multiplying this expression by $-i$ and retaining only the real part, $\varphi(p^2)$, of the function $\bar{\phi}(p^2)$. Making use of this result, it is easy to compute the contribution of these kernel components to the Einstein-Langevin equation.

3.3. THE EINSTEIN-LANGEVIN EQUATIONS

With the results of the previous subsections we can now write the n-dimensional Einstein-Langevin equation (59), previous to the renormalization. Taking into account that, from Eqs. (54) and (55),

$$
\frac{\Lambda_B}{8\pi G_B}=-\frac{1}{4\pi^2}\,\frac{m^4}{n(n-2)}\,\kappa_n+O(n-4),
\tag{77}
$$

we may finally write:

$$
\begin{aligned}
&\frac{1}{8\pi G_B}\, G^{(1)\mu\nu}(x) - \frac{4}{3}\,\alpha_B D^{(1)\mu\nu}(x) - 2\beta_B B^{(1)\mu\nu}(x) \\
&+\frac{\kappa_n}{(4\pi)^2}\left[-4\Delta\xi\,\frac{m^2}{(n-2)}\, G^{(1)\mu\nu} + \frac{1}{90}\, D^{(1)\mu\nu} + \Delta\xi^2 B^{(1)\mu\nu}\right](x) \\
&+\frac{1}{2880\pi^2}\left\{-\frac{16}{15}\, D^{(1)\mu\nu}(x) + \left(\frac{1}{6} - 10\,\Delta\xi\right) B^{(1)\mu\nu}(x)\right. \\
&+\int d^n y \int \frac{d^n p}{(2\pi)^n}\, e^{ip(x-y)}\, \varphi(p^2)\left[\left(1 + 4\,\frac{m^2}{p^2}\right)^2 D^{(1)\mu\nu}(y)\right. \\
&\left. +10\left(3\Delta\xi + \frac{m^2}{p^2}\right)^2 B^{(1)\mu\nu}(y)\right] \\
&\left.-\frac{m^2}{3}\int d^n y\, \Delta_n(x-y)\left(8D^{(1)\mu\nu} + 5B^{(1)\mu\nu}\right)(y)\right\} \\
&+2\int d^n y\, \mu^{-(n-4)} H^{\mu\nu\alpha\beta}_{\mathrm{A}n}(x,y)\, h_{\alpha\beta}(y) + O(n-4) = 2\xi^{\mu\nu}(x).
\end{aligned} \tag{78}
$$

Notice that the terms containing the bare cosmological constant have canceled. These equations can now be renormalized, that is, we can now write the bare coupling constants as renormalized coupling constants plus some suitably chosen counterterms and take the limit $n \to 4$. In order to carry out such a procedure, it is convenient to distinguish between massive and massless scalar fields. We shall evaluate these two cases in different subsections.

3.3.1. *Massive field*

In the case of a scalar field with mass $m \neq 0$, we can use, as we have done in Eq. (57) for the cosmological constant, the renormalized coupling constants $1/G$, α and β as

$$
\begin{aligned}
\frac{1}{G_B} &= \frac{1}{G} + \frac{2}{\pi}\,\Delta\xi\,\frac{m^2}{(n-2)}\,\kappa_n + O(n-4), \\
\alpha_B &= \alpha + \frac{1}{(4\pi)^2}\,\frac{1}{120}\,\kappa_n + O(n-4), \\
\beta_B &= \beta + \frac{\Delta\xi^2}{32\pi^2}\,\kappa_n + O(n-4).
\end{aligned} \tag{79}
$$

Note that for conformal coupling, $\Delta\xi = 0$, one has $1/G_B = 1/G$ and $\beta_B = \beta$, that is, only the coupling constant α and the cosmological constant need renormalization.

Let us introduce the two new kernels

$$H_{\rm A}(x;m^2) \equiv \frac{1}{1920\pi^2}\int \frac{d^4p}{(2\pi)^4}\, e^{ipx} \left\{ \left(1+4\,\frac{m^2}{p^2}\right)^2 \right.$$
$$\left. \times\left[-i\pi \,{\rm sign}\, p^0\; \theta(-p^2-4m^2)\sqrt{1+4\,\frac{m^2}{p^2}} + \varphi(p^2)\right] - \frac{8}{3}\,\frac{m^2}{p^2}\right\},$$
$$H_{\rm B}(x;m^2,\Delta\xi) \equiv \frac{1}{288\pi^2}\int \frac{d^4p}{(2\pi)^4}\, e^{ipx} \left\{ \left(3\,\Delta\xi+\frac{m^2}{p^2}\right)^2 \right.$$
$$\left. \times\left[-i\pi \,{\rm sign}\, p^0\; \theta(-p^2-4m^2)\sqrt{1+4\,\frac{m^2}{p^2}} + \varphi(p^2)\right] - \frac{1}{6}\,\frac{m^2}{p^2}\right\}, \quad (80)$$

where $\varphi(p^2)$ is given by the restriction to $n = 4$ of expression (75). Substituting expressions (79) into Eq. (78), we can now take the limit $n \to 4$, using the fact that, for $n = 4$, $D^{(1)\mu\nu}(x) = (3/2)\, A^{(1)\mu\nu}(x)$, we obtain the semiclassical Einstein-Langevin equations for the physical stochastic perturbations $h_{\mu\nu}$ in the four-dimensional manifold $\mathcal{M} \equiv {\rm I\!R}^4$:

$$\frac{1}{8\pi G}\, G^{(1)\mu\nu}(x) - 2\left(\alpha A^{(1)\mu\nu}(x) + \beta B^{(1)\mu\nu}(x)\right)$$
$$+ \frac{1}{2880\pi^2}\left[-\frac{8}{5}\, A^{(1)\mu\nu}(x) + \left(\frac{1}{6} - 10\,\Delta\xi\right) B^{(1)\mu\nu}(x)\right]$$
$$+ \int d^4y \left[H_{\rm A}(x-y;m^2) A^{(1)\mu\nu}(y) + H_{\rm B}(x-y;m^2,\Delta\xi) B^{(1)\mu\nu}(y)\right]$$
$$= 2\xi^{\mu\nu}(x), \quad (81)$$

where $\xi^{\mu\nu}$ are the components of a Gaussian stochastic tensor of vanishing mean value and two-point correlation function $\langle \xi^{\mu\nu}(x)\xi^{\alpha\beta}(y)\rangle_s = N^{\mu\nu\alpha\beta}(x,y)$, given by one of the Eqs. (68). Note that the two kernels defined in (80) are real and can be split into an even part and an odd part with respect to the variables x^μ, with the odd terms being the dissipation kernels $D_{\rm A}(x;m^2)$ and $D_{\rm B}(x;m^2,\Delta\xi)$ defined in (67). In spite of appearances, one can show that the Fourier transforms of the even parts of these kernels are finite in the limit $p^2 \to 0$ and, hence, the kernels $H_{\rm A}$ and $H_{\rm B}$ are well defined distributions.

We should mention that in Ref. [88], the same Einstein-Langevin equations were calculated using rather different methods. The way in which the result is written makes difficult a direct comparison with our equations (81). For instance, it is not obvious that there is some analog of the dissipation kernels.

3.3.2. *Massless field*

In this subsection, we consider the limit $m \to 0$ of equations (78). The renormalization scheme used in the previous subsection becomes singular in the massless limit because the expressions (79) for α_B and β_B diverge when $m \to 0$. Therefore, a different renormalization scheme is needed in this case. First, note that we may separate κ_n in (58) as $\kappa_n = \tilde{\kappa}_n + \frac{1}{2}\ln(m^2/\mu^2) + O(n-4)$, where

$$\tilde{\kappa}_n \equiv \frac{1}{(n-4)}\left(\frac{e^\gamma}{4\pi}\right)^{\frac{n-4}{2}} = \frac{1}{n-4} + \frac{1}{2}\ln\left(\frac{e^\gamma}{4\pi}\right) + O(n-4), \tag{82}$$

and that, from Eq. (75), we have

$$\lim_{m^2\to 0}\left[\varphi(p^2) + \ln(m^2/\mu^2)\right] = -2 + \ln\left|\frac{p^2}{\mu^2}\right|. \tag{83}$$

Hence, in the massless limit, equations (78) reduce to

$$\begin{aligned}
&\frac{1}{8\pi G_B}G^{(1)\mu\nu}(x) - \frac{4}{3}\alpha_B D^{(1)\mu\nu}(x) - 2\beta_B B^{(1)\mu\nu}(x)\\
&+\frac{1}{(4\pi)^2}(\tilde{\kappa}_n - 1)\left[\frac{1}{90}D^{(1)\mu\nu} + \Delta\xi^2 B^{(1)\mu\nu}\right](x)\\
&+\frac{1}{2880\pi^2}\left\{-\frac{16}{15}D^{(1)\mu\nu}(x) + \left(\frac{1}{6} - 10\Delta\xi\right)B^{(1)\mu\nu}(x)\right.\\
&\left.+\int d^n y\int\frac{d^n p}{(2\pi)^n}e^{ip(x-y)}\ln\left|\frac{p^2}{\mu^2}\right|\left[D^{(1)\mu\nu}(y) + 90\Delta\xi^2 B^{(1)\mu\nu}(y)\right]\right\}\\
&+\lim_{m^2\to 0} 2\int d^n y\,\mu^{-(n-4)}H_{\mathrm{A}n}^{\mu\nu\alpha\beta}(x,y)\,h_{\alpha\beta}(y) + O(n-4) = 2\xi^{\mu\nu}(x).
\end{aligned}\tag{84}$$

These equations can be renormalized by introducing the renormalized coupling constants $1/G$, α and β as

$$\frac{1}{G_B} = \frac{1}{G}, \qquad \alpha_B = \alpha + \frac{1}{120(4\pi)^2}(\tilde{\kappa}_n - 1) + O(n-4),$$
$$\beta_B = \beta + \frac{\Delta\xi^2}{32\pi^2}(\tilde{\kappa}_n - 1) + O(n-4). \tag{85}$$

Thus, in the massless limit, the Newtonian gravitational constant is not renormalized and, in the conformal coupling case, $\Delta\xi = 0$, we have again that $\beta_B = \beta$. Introducing the last expressions into Eq. (84), we can take the limit $n \to 4$. Note that, by making $m = 0$ in (67), the noise and dissipation kernels can be written as

$$\begin{aligned}
N_{\mathrm{A}}(x; m^2 = 0) &= N(x), \qquad N_{\mathrm{B}}(x; m^2 = 0, \Delta\xi) = 60\Delta\xi^2 N(x),\\
D_{\mathrm{A}}(x; m^2 = 0) &= D(x), \qquad D_{\mathrm{B}}(x; m^2 = 0, \Delta\xi) = 60\Delta\xi^2 D(x),
\end{aligned}\tag{86}$$

where

$$N(x) \equiv \frac{1}{1920\pi} \int \frac{d^4p}{(2\pi)^4}\, e^{ipx}\, \theta(-p^2),$$
$$D(x) \equiv \frac{-i}{1920\pi} \int \frac{d^4p}{(2\pi)^4}\, e^{ipx}\, \mathrm{sign}\, p^0\, \theta(-p^2). \qquad (87)$$

It is now convenient to introduce the new kernel

$$\begin{aligned} H(x;\mu^2) \;&\equiv\; \frac{1}{1920\pi^2} \int \frac{d^4p}{(2\pi)^4}\, e^{ipx} \left[\ln\left|\frac{p^2}{\mu^2}\right| - i\pi\, \mathrm{sign}\, p^0\, \theta(-p^2) \right] \\ &=\; \frac{1}{1920\pi^2} \lim_{\epsilon\to 0+} \int \frac{d^4p}{(2\pi)^4}\, e^{ipx}\, \ln\left(\frac{-(p^0+i\epsilon)^2 + p^i p_i}{\mu^2} \right). \end{aligned} \qquad (88)$$

Again, this kernel is real and can be written as the sum of an even part and an odd part in the variables x^μ, where the odd part is the dissipation kernel $D(x)$. The Fourier transforms (87) and (88) can actually be computed and, thus, in this case, we have explicit expressions for the kernels in position space. For $N(x)$ and $D(x)$, we get (see, for instance, Ref. [89])

$$N(x) = \frac{1}{1920\pi} \left[\frac{1}{\pi^3}\, \mathcal{P}f\left(\frac{1}{(x^2)^2} \right) + \delta^4(x) \right],$$
$$D(x) = \frac{1}{1920\pi^3}\, \mathrm{sign}\, x^0\, \frac{d}{d(x^2)}\, \delta(x^2), \qquad (89)$$

where $\mathcal{P}f$ denotes a distribution generated by the Hadamard finite part of a divergent integral, see Refs. [90] for the definition of these distributions. The expression for the kernel $H(x;\mu^2)$ can be found in Refs. [91, 92] and it is given by

$$\begin{aligned} H(x;\mu^2) &= \frac{1}{960\pi^2} \left\{ \mathcal{P}f\left(\frac{1}{\pi}\, \theta(x^0)\, \frac{d}{d(x^2)}\, \delta(x^2) \right) + (1-\gamma-\ln\mu)\, \delta^4(x) \right\} \\ &= \frac{1}{960\pi^2} \lim_{\lambda\to 0+} \left\{ \frac{1}{\pi} \theta(x^0)\theta(|\mathbf{x}|-\lambda) \frac{d}{d(x^2)} \delta(x^2) + [1-\gamma-\ln(\mu\lambda)]\, \delta^4(x) \right\} \end{aligned} \qquad (90)$$

See Ref. [91] for the details on how this last distribution acts on a test function.

Finally, the semiclassical Einstein-Langevin equations for the physical stochastic perturbations $h_{\mu\nu}$ in the massless case are

$$\begin{aligned} &\frac{1}{8\pi G}\, G^{(1)\mu\nu}(x) - 2\left(\alpha A^{(1)\mu\nu}(x) + \beta B^{(1)\mu\nu}(x)\right) + \\ &\frac{1}{2880\pi^2} \left[-\frac{8}{5}\, A^{(1)\mu\nu}(x) + \left(\frac{1}{6} - 10\Delta\xi \right) B^{(1)\mu\nu}(x) \right] \\ &+ \int d^4y\, H(x-y;\mu^2) \left[A^{(1)\mu\nu}(y) + 60\Delta\xi^2 B^{(1)\mu\nu}(y) \right] = 2\xi^{\mu\nu}(x), \end{aligned} \qquad (91)$$

where the Gaussian stochastic source components $\xi^{\mu\nu}$ have zero mean value and

$$\langle \xi^{\mu\nu}(x)\xi^{\alpha\beta}(y)\rangle_s = \lim_{m\to 0} N^{\mu\nu\alpha\beta}(x,y)$$
$$= \left[\frac{1}{6}\left(3\mathcal{F}_x^{\mu(\alpha}\mathcal{F}_x^{\beta)\nu} - \mathcal{F}_x^{\mu\nu}\mathcal{F}_x^{\alpha\beta}\right) + 60\,\Delta\xi^2 \mathcal{F}_x^{\mu\nu}\mathcal{F}_x^{\alpha\beta}\right] N(x-y). \quad (92)$$

It is interesting to consider the conformally coupled scalar field, *i.e.*, the case $\Delta\xi = 0$, of particular interest because of its similarities with the electromagnetic field, and also because of its interest in cosmology: massive fields become conformally invariant when their masses are negligible compared to the spacetime curvature. We have already mentioned that for a conformally coupled, field, the stochastic source tensor must be "traceless" (up to first order in perturbation theory around semiclassical gravity), in the sense that the stochastic variable $\xi_\mu^\mu \equiv \eta_{\mu\nu}\xi^{\mu\nu}$ behaves deterministically as a vanishing scalar field. This can be directly checked by noticing, from Eq. (92), that, when $\Delta\xi = 0$, one has $\langle \xi_\mu^\mu(x)\xi^{\alpha\beta}(y)\rangle_s = 0$, since $\mathcal{F}_\mu^\mu = 3\Box$ and $\mathcal{F}^{\mu\alpha}\mathcal{F}_\mu^\beta = \Box\mathcal{F}^{\alpha\beta}$. The semiclassical Einstein-Langevin equations for this particular case (and generalized to a spatially flat Robertson-Walker background) were first obtained in Ref. [73], where the coupling constant β was fixed to be zero.

Note that the expectation value of the renormalized stress-energy tensor for a scalar field can be obtained by identification of Eqs. (81) and (91) with the components of the physical Einstein-Langevin equation (14). The explicit expressions are given in Ref. [6] The results agree with the general form found by Horowitz [92, 93] using an axiomatic approach and coincides with that given in Ref. [87]. The particular cases of conformal coupling, $\Delta\xi = 0$, and minimal coupling, $\Delta\xi = -1/6$, are also in agreement with the results for this cases given in Refs. [92, 93, 94, 29, 95] (modulo local terms proportional to $A^{(1)\mu\nu}$ and $B^{(1)\mu\nu}$ due to different choices of the renormalization scheme). For the case of a massive minimally coupled scalar field, $\Delta\xi = -1/6$, our result is equivalent to that of Ref. [96].

3.4. CORRELATION FUNCTIONS FOR GRAVITATIONAL PERTURBATIONS

In this section, we solve the semiclassical Einstein-Langevin equations (81) and (91) for the components $G^{(1)\mu\nu}$ of the linearized Einstein tensor. In the first subsection we use these solutions to compute the corresponding two-point correlation functions, which give a measure of the gravitational fluctuations predicted by the stochastic semiclassical theory of gravity in the present case. Since the linearized Einstein tensor is invariant under gauge transformations of the metric perturbations, these two-point correlation functions are also gauge invariant. Once we have computed the two-point correlation functions for the linearized Einstein tensor, we find the solutions for the metric perturbations in the next subsection and we show how the associated two-point correlation functions can be computed. This procedure to solve the Einstein-Langevin equations is similar to the one used by Horowitz [92], see also Ref. [87], to analyze the stability of Minkowski spacetime in semiclassical gravity.

We first note that the tensors $A^{(1)\mu\nu}$ and $B^{(1)\mu\nu}$ can be written in terms of $G^{(1)\mu\nu}$ as

$$A^{(1)\mu\nu} = \frac{2}{3}\,(\mathcal{F}^{\mu\nu} G^{(1)\alpha}{}_{\alpha} - \mathcal{F}^{\alpha}_{\alpha} G^{(1)\mu\nu}), \qquad B^{(1)\mu\nu} = 2\mathcal{F}^{\mu\nu} G^{(1)\alpha}{}_{\alpha}, \tag{93}$$

where we have used that $3\Box = \mathcal{F}^{\alpha}_{\alpha}$. Therefore, the Einstein-Langevin equations (81) and (91) can be seen as linear integro-differential stochastic equations for the components $G^{(1)\mu\nu}$. These Einstein-Langevin equations can be written in a unified form, in both cases for $m\neq 0$ and for $m=0$, as

$$\frac{1}{8\pi G}\, G^{(1)\mu\nu}(x) - 2\,(\bar{\alpha} A^{(1)\mu\nu}(x) + \bar{\beta} B^{(1)\mu\nu}(x))$$
$$+\int d^4y\, \left[H_{\rm A}(x-y) A^{(1)\mu\nu}(y) + H_{\rm B}(x-y) B^{(1)\mu\nu}(y)\right] = 2\xi^{\mu\nu}(x), \tag{94}$$

where the new constants $\bar{\alpha}$ and $\bar{\beta}$, and the kernels $H_{\rm A}(x)$ and $H_{\rm B}(x)$ can be identified in each case by comparison of this last equation with Eqs. (81) and (91). For instance, when $m=0$, we have $H_{\rm A}(x) = H(x;\mu^2)$ and $H_{\rm B}(x) = 60\Delta\xi^2 H(x;\mu^2)$. In this case, we can use the arbitrariness of the mass scale μ to eliminate one of the parameters $\bar{\alpha}$ or $\bar{\beta}$.

In order to find solutions to Eq. (94), it is convenient to Fourier transform them. Introducing Fourier transforms with the following convention $\tilde{f}(p) = \int d^4x e^{-ipx} f(x))$ for a given field $f(x)$, one finds, from (93),

$$\tilde{A}^{(1)\mu\nu}(p) = 2p^2 \tilde{G}^{(1)\mu\nu}(p) - \frac{2}{3}\, p^2 P^{\mu\nu} \tilde{G}^{(1)\alpha}{}_{\alpha}(p),$$
$$\tilde{B}^{(1)\mu\nu}(p) = -2p^2 P^{\mu\nu} \tilde{G}^{(1)\alpha}{}_{\alpha}(p). \tag{95}$$

Using these relations, the Fourier transform of the Einstein-Langevin Eq. (94) reads

$$F^{\mu\nu}{}_{\alpha\beta}(p)\, \tilde{G}^{(1)\alpha\beta}(p) = 16\pi G\, \tilde{\xi}^{\mu\nu}(p), \tag{96}$$

where

$$F^{\mu\nu}{}_{\alpha\beta}(p) \equiv F_1(p)\, \delta^{\mu}_{(\alpha}\delta^{\nu}_{\beta)} + F_2(p)\, p^2 P^{\mu\nu} \eta_{\alpha\beta}, \tag{97}$$

with

$$F_1(p) \equiv 1 + 16\pi G\, p^2 \left[\tilde{H}_{\rm A}(p) - 2\bar{\alpha}\right],$$
$$F_2(p) \equiv -\frac{16}{3}\,\pi G \left[\tilde{H}_{\rm A}(p) + 3\tilde{H}_{\rm B}(p) - 2\bar{\alpha} - 6\bar{\beta}\right]. \tag{98}$$

In the Fourier transformed Einstein-Langevin Eq. (96), $\tilde{\xi}^{\mu\nu}(p)$, the Fourier transform of $\xi^{\mu\nu}(x)$, is a Gaussian stochastic source of zero average and

$$\langle\tilde{\xi}^{\mu\nu}(p)\tilde{\xi}^{\alpha\beta}(p')\rangle_s = (2\pi)^4\, \delta^4(p+p')\, \tilde{N}^{\mu\nu\alpha\beta}(p), \tag{99}$$

where we have introduced the Fourier transform of the noise kernel. The explicit expression for $\tilde{N}^{\mu\nu\alpha\beta}(p)$ is found from (68) and (67) to be

$$\tilde{N}^{\mu\nu\alpha\beta}(p) = \frac{1}{2880\pi}\, \theta(-p^2-4m^2)\sqrt{1+4\,\frac{m^2}{p^2}}\, \left[\frac{1}{4}\left(1+4\,\frac{m^2}{p^2}\right)^2 (p^2)^2 \right.$$
$$\left. \times\left(3P^{\mu(\alpha}P^{\beta)\nu} - P^{\mu\nu}P^{\alpha\beta}\right) + 10\left(3\Delta\xi + \frac{m^2}{p^2}\right)^2 (p^2)^2 P^{\mu\nu}P^{\alpha\beta}\right], \tag{100}$$

which in the massless case reduces to

$$\lim_{m\to 0}\tilde{N}^{\mu\nu\alpha\beta}(p) = \frac{1}{1920\pi}\, \theta(-p^2)\left[\frac{1}{6}\,(p^2)^2\,(3P^{\mu(\alpha}P^{\beta)\nu} - P^{\mu\nu}P^{\alpha\beta}) \right.$$
$$\left. +60\,\Delta\xi^2(p^2)^2 P^{\mu\nu}P^{\alpha\beta}\right]. \tag{101}$$

3.4.1. *Correlation functions for the linearized Einstein tensor*
In general, we can write $G^{(1)\mu\nu} = \langle G^{(1)\mu\nu}\rangle_s + G_{\rm f}^{(1)\mu\nu}$, where $G_{\rm f}^{(1)\mu\nu}$ is a solution to Eqs. (94) (or, in the Fourier transformed version, (96)) with zero average. The averages $\langle G^{(1)\mu\nu}\rangle_s$ must be a solution of the linearized semiclassical Einstein equations obtained by averaging Eqs. (94) [or (96)]. Solutions to these equations (specially in the massless case, $m=0$) have been studied by several authors [92, 93, 97, 98, 99, 95, 87], particularly in connection with the problem of the stability of the ground state of semiclassical gravity. The two-point correlation functions for the linearized Einstein tensor are defined by

$$\begin{aligned} \mathcal{G}^{\mu\nu\alpha\beta}(x,x') &\equiv \langle G^{(1)\mu\nu}(x)G^{(1)\alpha\beta}(x')\rangle_s - \langle G^{(1)\mu\nu}(x)\rangle_s\langle G^{(1)\alpha\beta}(x')\rangle_s \\ &= \langle G_{\rm f}^{(1)\mu\nu}(x)G_{\rm f}^{(1)\alpha\beta}(x')\rangle_s. \end{aligned} \tag{102}$$

Now we shall seek the family of solutions to the Einstein-Langevin equations which can be written as a linear functional of the stochastic source and whose Fourier transform, $\tilde{G}^{(1)\mu\nu}(p)$, depends locally on $\tilde{\xi}^{\alpha\beta}(p)$. Each of such solutions is a Gaussian stochastic field and, thus, it can be completely characterized by the averages $\langle G^{(1)\mu\nu}\rangle_s$ and the two-point correlation functions (102). For such a family of solutions, $\tilde{G}_{\rm f}^{(1)\mu\nu}(p)$ is the most general solution to Eq. (96) which is linear, homogeneous and local in $\tilde{\xi}^{\alpha\beta}(p)$. It can be written as

$$\tilde{G}_{\rm f}^{(1)\mu\nu}(p) = 16\pi G\, D^{\mu\nu}{}_{\alpha\beta}(p)\, \tilde{\xi}^{\alpha\beta}(p), \tag{103}$$

where $D^{\mu\nu}{}_{\alpha\beta}(p)$ are the components of a Lorentz invariant tensor field distribution in Minkowski spacetime (by "Lorentz invariant" we mean invariant under the transformations of the orthochronous Lorentz subgroup; see Ref. [92] for more details on the definition and properties of these tensor distributions), symmetric under the interchanges $\alpha \leftrightarrow \beta$ and $\mu \leftrightarrow \nu$, which is the most general solution of

$$F^{\mu\nu}{}_{\rho\sigma}(p)\, D^{\rho\sigma}{}_{\alpha\beta}(p) = \delta^{\mu}_{(\alpha}\delta^{\nu}_{\beta)}. \tag{104}$$

In addition, we must impose the conservation condition to the solutions: $p_\nu \tilde{G}_{\rm f}^{(1)\mu\nu}(p) = 0$, where this zero must be understood as a stochastic variable which behaves deterministically as a zero vector field. We can write $D^{\mu\nu}{}_{\alpha\beta}(p) = D_p^{\mu\nu}{}_{\alpha\beta}(p) + D_h^{\mu\nu}{}_{\alpha\beta}(p)$, where $D_p^{\mu\nu}{}_{\alpha\beta}(p)$ is a particular solution to Eq. (104) and $D_h^{\mu\nu}{}_{\alpha\beta}(p)$ is the most general solution to the homogeneous equation. Consequently, see Eq. (103), we can write $\tilde{G}_{\rm f}^{(1)\mu\nu}(p) = \tilde{G}_p^{(1)\mu\nu}(p) + \tilde{G}_h^{(1)\mu\nu}(p)$. To find the particular solution, we try an ansatz of the form

$$D_p^{\mu\nu}{}_{\alpha\beta}(p) = d_1(p)\, \delta^{\mu}_{(\alpha}\delta^{\nu}_{\beta)} + d_2(p)\, p^2 P^{\mu\nu} \eta_{\alpha\beta}. \tag{105}$$

Substituting this ansatz into Eqs. (104), it is easy to see that it solves these equations if we take

$$d_1(p) = \left[\frac{1}{F_1(p)}\right]_r, \qquad d_2(p) = -\left[\frac{F_2(p)}{F_1(p)F_3(p)}\right]_r, \tag{106}$$

with

$$F_3(p) \equiv F_1(p) + 3p^2 F_2(p) = 1 - 48\pi G\, p^2 \left[\tilde{H}_{\rm B}(p) - 2\bar{\beta}\right], \tag{107}$$

and where the notation $[\ \ ]_r$ means that the zeros of the denominators are regulated with appropriate prescriptions in such a way that $d_1(p)$ and $d_2(p)$ are well defined Lorentz invariant scalar distributions. This yields a particular solution to the Einstein-Langevin equations:

$$\tilde{G}_p^{(1)\mu\nu}(p) = 16\pi G\, D_p^{\mu\nu}{}_{\alpha\beta}(p)\, \tilde{\xi}^{\alpha\beta}(p), \tag{108}$$

which, since the stochastic source is conserved, satisfies the conservation condition. Note that, in the case of a massless scalar field, $m=0$, the above solution has a functional form analogous to that of the solutions of linearized semiclassical gravity found in the Appendix of Ref. [87]. Notice also that, for a massless conformally coupled field, $m=0$ and $\Delta\xi=0$, the second term in the right hand side of Eq. (105) will not contribute in the correlation functions (102), since in this case the stochastic source is traceless.

Next, we can work out the general form for $D_h^{\mu\nu}{}_{\alpha\beta}(p)$, which is a linear combination of terms consisting of a Lorentz invariant scalar distribution times one of the products $\delta^{\mu}_{(\alpha}\delta^{\nu}_{\beta)}$, $p^2P^{\mu\nu}\eta_{\alpha\beta}$, $\eta^{\mu\nu}\eta_{\alpha\beta}$, $\eta^{\mu\nu}p^2P_{\alpha\beta}$, $\delta^{(\mu}_{(\alpha}p^2P^{\nu)}_{\beta)}$ and $p^2P^{\mu\nu}p^2P_{\alpha\beta}$. However, taking into account that the stochastic source is conserved, we can omit some terms in $D_h^{\mu\nu}{}_{\alpha\beta}(p)$ and simply write

$$\tilde{G}_h^{(1)\mu\nu}(p) = 16\pi G\, D_h^{\mu\nu}{}_{\alpha\beta}(p)\, \tilde{\xi}^{\alpha\beta}(p), \tag{109}$$

where

$$D_h^{\mu\nu}{}_{\alpha\beta}(p) = h_1(p)\, \delta^{\mu}_{(\alpha}\delta^{\nu}_{\beta)} + h_2(p)\, p^2P^{\mu\nu}\eta_{\alpha\beta} + h_3(p)\, \eta^{\mu\nu}\eta_{\alpha\beta}, \tag{110}$$

with $h_1(p)$, $h_2(p)$ and $h_3(p)$ being Lorentz invariant scalar distributions. From the fact that $D_h^{\mu\nu}{}_{\alpha\beta}(p)$ must satisfy the homogeneous equation corresponding to Eq. (104), we find that $h_1(p)$ and $h_3(p)$ have support on the set of points $\{p^\mu\}$ for which $F_1(p)=0$, and that $h_2(p)$ has support on the set of points $\{p^\mu\}$ for which $F_1(p)=0$ or $F_3(p)=0$. Moreover, the conservation condition for $\tilde{G}_h^{(1)\mu\nu}(p)$ implies that the term with $h_3(p)$ is only allowed in the case of a massless conformally coupled field, $m=0$ and $\Delta\xi=0$. From (99), we get

$$\langle \tilde{G}_h^{(1)\mu\nu}(p)\, \tilde{\xi}^{\alpha\beta}(p')\rangle_s = (2\pi)^4\, 16\pi G\, \delta^4(p+p')\, D_h^{\mu\nu}{}_{\rho\sigma}(p)\, \tilde{N}^{\rho\sigma\alpha\beta}(p). \tag{111}$$

Note, from expressions (100) and (101), that the support of $\tilde{N}^{\mu\nu\alpha\beta}(p)$ is on the set of points $\{p^\mu\}$ for which $-p^2 \geq 0$ when $m=0$, and for which $-p^2-4m^2>0$ when $m\neq 0$. At such points, using expressions (98), (107), (88) and (80), it is easy to see that $F_1(p)$ and $F_3(p)$ are always different from zero, except for some particular values of $\Delta\xi$ and $\bar{\beta}$: (a) when $m=0$, $\Delta\xi=0$ and $\bar{\beta}>0$; and (b) when $m\neq 0$, $0<\Delta\xi<(1/12)$ and $\bar{\beta}=(\Delta\xi/32\pi^2)\,[\pi/(Gm^2)+1/36]$.

In the case (a), $F_3(p)=0$ for the set of points $\{p^\mu\}$ satisfying $-p^2=1/(96\pi G\bar{\beta})$; in the case (b), $F_3(p)=0$ for $\{p^\mu\}$ such that $-p^2=m^2/(3\Delta\xi)$. Hence, except for the above cases (a) and (b), the intersection of the supports of $\tilde{N}^{\mu\nu\alpha\beta}(p)$ and $D_h^{\rho\sigma}{}_{\lambda\gamma}(p)$ is an empty set and, thus, the correlation function (111) is zero. In the cases (a) and (b), we can have a contribution to (111) coming from the term with $h_2(p)$ in (110) of the form $D_h^{\mu\nu}{}_{\rho\sigma}(p)\,\tilde{N}^{\rho\sigma\alpha\beta}(p) = H_3(p;\{C\})\,p^2P^{\mu\nu}\,\tilde{N}^{\alpha\beta\rho}{}_{\rho}(p)$, where $H_3(p;\{C\})$ is the most general Lorentz invariant distribution satisfying $F_3(p)\,H_3(p;\{C\})=0$, which depends on a set of arbitrary parameters represented as $\{C\}$. However, from (100), we see that $\tilde{N}^{\alpha\beta\rho}{}_{\rho}(p)$ is proportional to $\theta(-p^2-4m^2)\,(1+4m^2/p^2)^{(1/2)}\,(3\Delta\xi+m^2/p^2)^2$. Thus, in the case (a), we have $\tilde{N}^{\alpha\beta\rho}{}_{\rho}(p)=0$ and, in the case (b), the intersection of the supports of $\tilde{N}^{\alpha\beta\rho}{}_{\rho}(p)$ and of $H_3(p;\{C\})$ is an empty set. Therefore, from the above analysis, we conclude that $\tilde{G}_h^{(1)\mu\nu}(p)$ gives no contribution to the correlation functions (102), since $\langle \tilde{G}_h^{(1)\mu\nu}(p)\, \xi^{\alpha\beta}(p')\rangle_s = 0$, and we have simply $\mathcal{G}^{\mu\nu\alpha\beta}(x,x') = \langle G_p^{(1)\mu\nu}(x) G_p^{(1)\alpha\beta}(x')\rangle_s$, where $G_p^{(1)\mu\nu}(x)$ is the inverse Fourier transform of (108).

Therefore the correlation functions (102) can then be computed from

$$\langle \tilde{G}_p^{(1)\mu\nu}(p)\, \tilde{G}_p^{(1)\alpha\beta}(p')\rangle_s = 64\,(2\pi)^6\, G^2\, \delta^4(p+p')\, D_p^{\mu\nu}{}_{\rho\sigma}(p)\, D_p^{\alpha\beta}{}_{\lambda\gamma}(-p)\, \tilde{N}^{\rho\sigma\lambda\gamma}(p). \quad (112)$$

It is easy to see from the above analysis that the prescriptions $[\ \ ]_r$ in the factors D_p are irrelevant in the last expression and, thus, they can be suppressed. Taking into account that $F_l(-p) = F_l^*(p)$, with $l = 1,2,3$, we get from Eqs. (105) and (106)

$$\begin{aligned}\langle \tilde{G}_p^{(1)\mu\nu}(p)\, \tilde{G}_p^{(1)\alpha\beta}(p')\rangle_s &= 64(2\pi)^6\, G^2\, \frac{\delta^4(p+p')}{|F_1(p)|^2}\left[\tilde{N}^{\mu\nu\alpha\beta}(p) - \frac{F_2(p)}{F_3(p)}\, p^2 P^{\mu\nu}\, \tilde{N}^{\alpha\beta\rho}{}_{\rho}(p)\right.\\ &\left. - \frac{F_2^*(p)}{F_3^*(p)}\, p^2 P^{\alpha\beta}\, \tilde{N}^{\mu\nu\rho}{}_{\rho}(p) + \frac{|F_2(p)|^2}{|F_3(p)|^2}\, p^2 P^{\mu\nu}\, p^2 P^{\alpha\beta}\, \tilde{N}^{\rho}{}_{\rho}{}^{\sigma}{}_{\sigma}(p)\right]. \quad (113)\end{aligned}$$

This last expression is well defined as a bi-distribution and can be easily evaluated using Eq. (100). The final explicit result for the Fourier transformed correlation function for the Einstein tensor is thus

$$\begin{aligned}\langle \tilde{G}_p^{(1)\mu\nu}(p)\, \tilde{G}_p^{(1)\alpha\beta}(p')\rangle_s &= \frac{2}{45}(2\pi)^5\, G^2\, \frac{\delta^4(p+p')}{|F_1(p)|^2}\, \theta(-p^2-4m^2)\sqrt{1+4\,\frac{m^2}{p^2}}\\ &\times\left[\frac{1}{4}\left(1+4\,\frac{m^2}{p^2}\right)^2 (p^2)^2\,(3P^{\mu(\alpha}P^{\beta)\nu} - P^{\mu\nu}P^{\alpha\beta})\right.\\ &\left. + 10\left(3\Delta\xi + \frac{m^2}{p^2}\right)^2 (p^2)^2 P^{\mu\nu}P^{\alpha\beta}\left|1 - 3p^2\,\frac{F_2(p)}{F_3(p)}\right|^2\right]. \quad (114)\end{aligned}$$

To obtain the correlation functions in coordinate space, Eq. (102), we have to take the inverse Fourier transform of the above result, the final result is:

$$\mathcal{G}^{\mu\nu\alpha\beta}(x,x') = \frac{\pi}{45}\, G^2\, \mathcal{F}_x^{\mu\nu\alpha\beta}\, \mathcal{G}_{\rm A}(x-x') + \frac{8\pi}{9}\, G^2\, \mathcal{F}_x^{\mu\nu}\mathcal{F}_x^{\alpha\beta}\, \mathcal{G}_{\rm B}(x-x'), \quad (115)$$

with

$$\tilde{\mathcal{G}}_{\rm A}(p) \equiv \theta(-p^2-4m^2)\sqrt{1+4\,\frac{m^2}{p^2}}\left(1+4\,\frac{m^2}{p^2}\right)^2\frac{1}{|F_1(p)|^2},$$

$$\tilde{\mathcal{G}}_{\rm B}(p) \equiv \theta(-p^2-4m^2)\sqrt{1+4\,\frac{m^2}{p^2}}\left(3\Delta\xi+\frac{m^2}{p^2}\right)^2 \times\frac{1}{|F_1(p)|^2}\left|1-3p^2\,\frac{F_2(p)}{F_3(p)}\right|^2, \qquad (116)$$

and $\mathcal{F}_x^{\mu\nu\alpha\beta} \equiv 3\mathcal{F}_x^{\mu(\alpha}\mathcal{F}_x^{\beta)\nu} - \mathcal{F}_x^{\mu\nu}\mathcal{F}_x^{\alpha\beta}$, and where $F_l(p)$, $l=1,2,3$, are given in (98) and (107). Notice that, for a massless field ($m=0$), we have

$$F_1(p) = 1+16\pi G p^2\,\tilde{H}(p;\bar{\mu}^2),$$
$$F_2(p) = -\frac{16}{3}\,\pi G\left[(1+180\Delta\xi^2)\,\tilde{H}(p;\bar{\mu}^2)-6\Upsilon\right],$$
$$F_3(p) = 1-48\pi G p^2\left[60\Delta\xi^2\,\tilde{H}(p;\bar{\mu}^2)-2\Upsilon\right], \qquad (117)$$

with $\bar{\mu} \equiv \mu\,\exp(1920\pi^2\bar{\alpha})$ and $\Upsilon \equiv \bar{\beta}-60\Delta\xi^2\,\bar{\alpha}$, and where $\tilde{H}(p;\mu^2)$ is the Fourier transform of $H(x;\mu^2)$ given in (88).

3.4.2. *Correlation functions for the metric perturbations*

Starting from the solutions found for the linearized Einstein tensor, which are characterized by the two-point correlation functions (115) [or, in terms of Fourier transforms, (114)], we can now solve the equations for the metric perturbations. Working in the harmonic gauge, $\partial_\nu \bar{h}^{\mu\nu}=0$ (this zero must be understood in a statistical sense) where $\bar{h}_{\mu\nu} \equiv h_{\mu\nu}-(1/2)\eta_{\mu\nu}h^\alpha_\alpha$, the equations for the metric perturbations in terms of the Einstein tensor are

$$\Box\bar{h}^{\mu\nu}(x) = -2G^{(1)\mu\nu}(x), \qquad (118)$$

or, in terms of Fourier transforms, $p^2\tilde{\bar{h}}^{\mu\nu}(p) = 2\tilde{G}^{(1)\mu\nu}(p)$. Similarly to the analysis of the equation for the Einstein tensor, we can write $\bar{h}^{\mu\nu} = \langle\bar{h}^{\mu\nu}\rangle_s + \bar{h}^{\mu\nu}_{\rm f}$, where $\bar{h}^{\mu\nu}_{\rm f}$ is a solution to these equations with zero average, and the two-point correlation functions are defined by

$$\begin{aligned}\mathcal{H}^{\mu\nu\alpha\beta}(x,x') &\equiv \langle\bar{h}^{\mu\nu}(x)\bar{h}^{\alpha\beta}(x')\rangle_s - \langle\bar{h}^{\mu\nu}(x)\rangle_s\langle\bar{h}^{\alpha\beta}(x')\rangle_s \\ &= \langle\bar{h}^{\mu\nu}_{\rm f}(x)\bar{h}^{\alpha\beta}_{\rm f}(x')\rangle_s. \end{aligned} \qquad (119)$$

We can now seek solutions of the Fourier transform of Eq. (118) of the form $\tilde{\bar{h}}^{\mu\nu}_{\rm f}(p) = 2D(p)\tilde{G}^{(1)\mu\nu}_{\rm f}(p)$, where $D(p)$ is a Lorentz invariant scalar distribution in Minkowski spacetime, which is the most general solution of

$p^2 D(p)=1$. Note that, since the linearized Einstein tensor is conserved, solutions of this form automatically satisfy the harmonic gauge condition. As in the previous subsection, we can write $D(p)=[1/p^2]_r+D_h(p)$, where $D_h(p)$ is the most general solution to the associated homogeneous equation and, correspondingly, we have $\tilde{\bar{h}}^{\mu\nu}_{\rm f}(p)=\tilde{\bar{h}}^{\mu\nu}_{p}(p)+\tilde{\bar{h}}^{\mu\nu}_{h}(p)$. However, since $D_h(p)$ has support on the set of points for which $p^2=0$, it is easy to see from Eq. (114) (from the factor $\theta(-p^2-4m^2)$) that $\langle\tilde{\bar{h}}^{\mu\nu}_{h}(p)\tilde{G}^{(1)\alpha\beta}_{\rm f}(p')\rangle_s=0$ and, thus, the two-point correlation functions (119) can be computed from $\langle\tilde{\bar{h}}^{\mu\nu}_{\rm f}(p)\tilde{\bar{h}}^{\alpha\beta}_{\rm f}(p')\rangle_s=\langle\tilde{\bar{h}}^{\mu\nu}_{p}(p)\tilde{\bar{h}}^{\alpha\beta}_{p}(p')\rangle_s$. From Eq. (114) and due to the factor $\theta(-p^2-4m^2)$, it is also easy to see that the prescription $[\ ]_r$ is irrelevant in this correlation function and we obtain

$$\langle\tilde{\bar{h}}^{\mu\nu}_{p}(p)\tilde{\bar{h}}^{\alpha\beta}_{p}(p')\rangle_s=\frac{4}{(p^2)^2}\,\langle\tilde{G}^{(1)\mu\nu}_{p}(p)\,\tilde{G}^{(1)\alpha\beta}_{p}(p')\rangle_s, \tag{120}$$

where $\langle\tilde{G}^{(1)\mu\nu}_{p}(p)\,\tilde{G}^{(1)\alpha\beta}_{p}(p')\rangle_s$ is given by Eq. (114). The right hand side of this equation is a well defined bi-distribution, at least for $m\neq 0$ (the θ function provides the suitable cutoff). In the massless field case, since the noise kernel is obtained as the limit $m\to 0$ of the noise kernel for a massive field, it seems that the natural prescription to avoid the divergencies on the lightcone $p^2=0$ is a Hadamard finite part, see Refs. [90] for its definition. Taking this prescription, we also get a well defined bi-distribution for the massless limit of the last expression.

The final result for the two-point correlation function for the field $\bar{h}^{\mu\nu}$ is:

$$\begin{aligned}\mathcal{H}^{\mu\nu\alpha\beta}(x,x') &= \frac{4\pi}{45}\,G^2\,\mathcal{F}^{\mu\nu\alpha\beta}_x\,\mathcal{H}_{\rm A}(x-x')\\ &\quad+\frac{32\pi}{9}\,G^2\,\mathcal{F}^{\mu\nu}_x\mathcal{F}^{\alpha\beta}_x\,\mathcal{H}_{\rm B}(x-x'),\end{aligned} \tag{121}$$

where $\tilde{\mathcal{H}}_{\rm A}(p)\equiv[1/(p^2)^2]\,\tilde{\mathcal{G}}_{\rm A}(p)$ and $\tilde{\mathcal{H}}_{\rm B}(p)\equiv[1/(p^2)^2]\,\tilde{\mathcal{G}}_{\rm B}(p)$, with $\tilde{\mathcal{G}}_{\rm A}(p)$ and $\tilde{\mathcal{G}}_{\rm B}(p)$ given by (116). The two-point correlation functions for the metric perturbations can be easily obtained using $h_{\mu\nu}=\bar{h}_{\mu\nu}-(1/2)\,\eta_{\mu\nu}\,\bar{h}^{\alpha}_{\alpha}$.

3.4.3. *Conformally coupled field*

For a conformally coupled field, *i.e.*, when $m=0$ and $\Delta\xi=0$, the previous correlation functions are greatly simplified and can be approximated explicitly in terms of analytic functions. The detailed results are given in Ref. [6], here we outline the main features.

When $m=0$ and $\Delta\xi=0$ we have that $\mathcal{G}_{\rm B}(x)=0$ and $\tilde{\mathcal{G}}_{\rm A}(p)=\theta(-p^2)\,|F_1(p)|^{-2}$. Thus the two-point correlations functions for the Einstein tensor is

$$\mathcal{G}^{\mu\nu\alpha\beta}(x,x') = \frac{\pi}{45} G^2 \mathcal{F}_x^{\mu\nu\alpha\beta} \int \frac{d^4p}{(2\pi)^4} \frac{e^{ip(x-x')}\,\theta(-p^2)}{|1+16\pi G p^2 \tilde{H}(p;\bar{\mu}^2)|^2}, \tag{122}$$

where $\tilde{H}(p,\mu^2) = (1920\pi^2)^{-1} \ln[-((p^0+i\epsilon)^2 + p^i p_i)/\mu^2]$, see Eq. (88).

To estime this integral let us consider spacelike separated points $(x - x')^\mu = (0,\mathbf{x}-\mathbf{x}')$, and define $\mathbf{y} = \mathbf{x} - \mathbf{x}'$. We may now formaly change the momentum variable p^μ by the dimensionless vector s^μ: $p^\mu = s^\mu/|\mathbf{y}|$, then the previous integral denominator is $|1 + 16\pi(L_P/|\mathbf{y}|)^2 s^2 \tilde{H}(s)|^2$, where we have introduced the Planck length $L_P = \sqrt{G}$. It is clear that we can consider two regimes: (a) when $L_P \ll |\mathbf{y}|$, and (b) when $|\mathbf{y}| \sim L_P$. In case (a) the correlation funtion, for the 0000 component, say, will be of the order

$$\mathcal{G}^{0000}(\mathbf{y}) \sim \frac{L_P^4}{|\mathbf{y}|^8}.$$

In case (b) when the denominator has zeros a detailed calculation carried out in Ref. [6] shows that:

$$\mathcal{G}^{0000}(\mathbf{y}) \sim e^{-|\mathbf{y}|/L_P} \left(\frac{L_P}{|\mathbf{y}|^5} + \ldots + \frac{1}{L_P^2 |\mathbf{y}|^2} \right),$$

which indicates an exponiential decay at distances around the Planck scale. Thus short scale fluctuations are strongly suppressed.

For the two-point metric correlation the results are similar. In this case we have

$$\mathcal{H}^{\mu\nu\alpha\beta}(x,x') = \frac{4\pi}{45} G^2 \mathcal{F}_x^{\mu\nu\alpha\beta} \int \frac{d^4p}{(2\pi)^4} \frac{e^{ip(x-x')}\theta(-p^2)}{(p^2)^2|1+16\pi G p^2 \tilde{H}(p;\bar{\mu}^2)|^2}. \tag{123}$$

The integrand has the same behavior of the correlation function of Eq. (122) thus matter fields tends to supress the short scale metric perturbations. In this case we find, as for the correlation of the Einstein tensor, that for case (a) above we have,

$$\mathcal{H}^{0000}(\mathbf{y}) \sim \frac{L_P^4}{|\mathbf{y}|^4},$$

and for case (b) we have

$$\mathcal{H}^{0000}(\mathbf{y}) \sim e^{-|\mathbf{y}|/L_P} \left(\frac{L_P}{|\mathbf{y}|} + \ldots \right).$$

It is interesting to write expression (123) in an alternative way. If we introduce the dimensionless tensor $P^{\mu\nu\alpha\beta} \equiv 3P^{\mu(\alpha}P^{\beta)\nu} - P^{\mu\nu}P^{\alpha\beta}$, where

$P^{\mu\nu}$ is the projector defined in Eq. (73), to account for the effect of the operator $\mathcal{F}_x^{\mu\nu\alpha\beta}$, we can write

$$\mathcal{H}^{\mu\nu\alpha\beta}(x,x') = \frac{4\pi}{45} G^2 \int \frac{d^4p}{(2\pi)^4} \frac{e^{ip(x-x')}\, P^{\mu\nu\alpha\beta}\, \theta(-p^2)}{|1+16\pi G p^2 \tilde{H}(p;\bar{\mu}^2)|^2}. \tag{124}$$

This expression allows a direct comparison with the graviton propagator for linearized quantum gravity in the $1/N$ approximation found by Tomboulis [100]. One can see that the imaginary part of the graviton propagator leads, in fact, to Eq. (124).

3.4.4. *Discussion*

The main results of this section are the correlation functions (115) and (121). In the case of a conformal field, the correlation functions of the linearized Einstein tensor have been explicitly estimated. From the exponential factors $e^{-|\mathbf{y}/L_P}$ in these results for scales near the Planck length, we see that the correlation functions of the linearized Einstein tensor have the Planck length as the correlation length. A similar behavior is found for the correlation functions of the metric perturbations. Since these fluctuations are induced by the matter fluctuations we infer that the effect of the matter fields is to suppress the fluctuations of the metric at very small scales. On the other hand, at scales much larger than the Planck length the induced metric fluctuations are small compared with the free graviton propagator which goes like $L_P^2/|\mathbf{y}|^2$, since the action for the free graviton goes like $S_h \sim \int d^4x\, L_P^{-2} h \Box h$.

It is interesting to note, however, that these results for correlation functions are non-analytic in their characteristic correlation lengths. This kind of non-analytic behavior is actually quite typical of the solutions of Langevin-type equations with dissipative terms. An example in the context of a reduced version of the semiclassical Einstein-Langevin equation is given in Ref. [86].

For background solutions of semiclassical gravity with other scales present apart from the Planck scales (for instance, for matter fields in a thermal state), stress-energy fluctuations may be important at larger scales. For such backgrounds, stochastic semiclassical gravity might predict correlation functions with characteristic correlation lengths larger than the Planck scales. It seems quite plausible, nevertheless, that these correlation functions would remain non-analytic in their characteristic correlation lengths. This would imply that these correlation functions could not be obtained from a calculation involving a perturbative expansion in the characteristic correlation lengths. In particular, if these correlation lengths are proportional to the Planck constant $\hbar$, the gravitational correlation functions could not be obtained from an expansion in $\hbar$. Hence, stochastic semiclassical gravity might predict a behavior for gravitational correlation functions different from that of the analogous functions in perturbative quantum gravity [80]. This is not necessarily inconsistent with having neglected action terms of higher order in $\hbar$ when considering semiclassical gravity as an effective theory [87].

4. Fluctuations of Energy Density and Validity of Semiclassical Gravity

We now turn our attention to some basic issues involving vacuum energy density fluctuations invoking only the simplest spacetimes, Minkowski and Casimir. Recent years saw the beginning of serious studies of the fluctuations of the energy momentum tensor (EMT) $\hat{T}_{\mu\nu}$ of quantum fields in spacetimes with boundaries [20, 21, 22] (such as Casimir effect [101]) [102, 61], nontrivial topology (such as imaginary time thermal field theory) or nonzero curvature (such as the Einstein universe) [62]. A natural measure of the strength of fluctuations is χ [7], the ratio of the variance $\Delta\rho^2$ of fluctuations in the energy density (expectation value of the $\hat{\rho}^2$ operator minus the square of the mean $\hat{\rho}$ taken with respect to some quantum state) to its mean-squared (square of the expectation value of $\hat{\rho}$):

$$\chi \equiv \frac{\langle\hat{\rho}^2\rangle - \langle\hat{\rho}\rangle^2}{\langle\hat{\rho}\rangle^2} \equiv \frac{\Delta\rho^2}{\bar{\rho}^2} \tag{125}$$

Alternatively, we can use the quantity introduced by Kuo and Ford [61]

$$\Delta \equiv \frac{\langle\hat{\rho}^2\rangle - \langle\hat{\rho}\rangle^2}{\langle\hat{\rho}^2\rangle} = \frac{\chi}{\chi+1} \tag{126}$$

Assuming a positive definite variance $\Delta\rho^2 \geq 0$, then $0 \leq \chi \leq \infty$ and $0 \leq \Delta \leq 1$ always, with $\Delta \ll 1$ falling in the classical domain. Kuo and Ford (KF) displayed a number of quantum states (vacuum plus 2 particle state, squeezed vacuum and Casimir vacuum) with respect to which the expectation value of the energy momentum tensor (00 component) gives rise to negative local energy density. For these states Δ is of order unity. From this result they drew the implications, amongst other interesting inferences, that semiclassical gravity (SCG) [24] could become invalid under these conditions The validity of semiclassical gravity in the face of fluctuations of quantum fields as source is an important issue which has caught the attention of many authors. Amongst others Phillips and Hu (PH) [8] hold a different viewpoint on this issue from KF. This section is a summary of their investigations on this issue.

To begin with it may not be so surprising that states which are more quantum (e.g., squeezed states) in nature than classical (e.g., coherent states) [103] may lead to large fluctuations in energy density comparable to the mean[1]. Such a condition exists peacefully with the underlying spacetime at least at the low energy of today's universe. PH calculated the variance of fluctuations to mean-squared ratio of a quantum field for the simplest case of Minkowski spacetime i.e., for ordinary quantum field theory to be $\Delta = 2/5$. This is a simple counter-example to the claim of KF, since $\Delta = O(1)$ holds also for Minkowski space, where SCG is known to be valid at large scales. PH do not see sufficient ground to question the validity of SCG at energy below the Planck energy when the spacetime is depictable by a manifold structure, approximated locally by the Minkowski space. To them the fluctuations to mean being of the order unity arises from the quantum nature of the vacuum state and says little about the compatibility of the field source with the spacetime the quantum field lives in[2].

PH pointed out that one should refer to a scale (of interaction or for probing accuracy) when measuring the validity of SCG. The conventional belief is that when reaching the Planck scale from below, QFTCST will break down because, amongst other things happening, graviton production at that energy will become significant so as to render the classical background spacetime unstable, and the mean value of quantum field taken as a source for the Einstein equation becomes inadequate. To address this issue as well as the issue of the spatial extent where negative energy density can exist, PH view it necessary to introduce a scale in the spacetime regions where quantum fields are defined to monitor how the mean value and the fluctuations of the energy momentum tensor change.

In conventional field theories the stress tensor built from the product of a pair of field operators evaluated at a single point in the spacetime manifold is, strictly speaking, ill-defined. Point separation is a well-established method which suits the present concern very well, and we will discuss this method in section 5. For here we will use a simpler method to introduce a scale in the quantum field theory, i.e., by introducing a (spatial) smearing function $f(\mathbf{x})$ to define smeared field operators $\hat{\phi}_t(f_{\mathbf{x}})$. Using a Gaussian smearing function (with variance σ^2) PH derive expressions for the EM

[1]This can be seen even in the ratio of expectation values of moments of the displacement operators in simple quantum harmonic oscillators

[2]One should draw a distinction between quantum fields in curved spacetime QFCST and semiclassical gravity: the former is a test field situation with quantum fields propagating in a fixed background space while in the latter both the field and the spacetime are determined self-consistently by solving the semiclassical Einstein equation. The cases studied in Kuo and Ford [61] as well as many others [62] are of a test-field nature, where backreaction is not considered. So KF's criterion pertains more to QFTCST than to SCG, where in the former the central issue is compatibility, which is a weaker condition than consistency in the latter.

bi-tensor operator, its mean and its fluctuations as functions of σ, for a massless scalar field in both the Minkowski and the Casimir spacetimes. The interesting result PH find is that while both the vacuum expectation value and the fluctuations of energy density grow as $\sigma \to 0$, the ratio of the variance of the fluctuations to its mean-squared remains a constant χ_d (d is the spatial dimension of spacetime) which is independent of σ. The measure Δ_d ($= \chi_d/(\chi_d + 1)$) depends on the dimension of space and is of the order unity. It varies only slightly for spacetimes with boundary or nontrivial topology. For example Δ for Minkowski is $2/5$, while for Casimir is $6/7$ (cf, from [62]). Add to this our prior result for the Einstein Universe, $\Delta = 111/112$, independent of curvature, and that for hot flat space [104], we see that invariably the fluctuations to mean ratio is of the order unity.

These results allow us to address three interrelated issues in quantum field theory in curved spacetime in the light of fluctuations of quantum stress energy : 1) Fluctuation to mean ratio of vacuum energy density and the validity of semiclassical gravity. 2) The spatial extent where negative energy density can exist and its implications for quantum effects of worm holes, baby universes and time travel. 3) Dependence of fluctuations on intrinsic (defined by smearing or point-separation) and the extrinsic scale (such as the Casimir or finite temperature periodicity). 4) The circumstances when and how divergences appear and the meaning of regularization in point-defined field theories versus theories defined at separated points and/or smeard fields. This includes also the issue of the cross term.

We begin by defining the smeared field operators and their products and construct from them the smeared energy density and its fluctuations. We then calculate the ratio of the fluctuations to the mean for a flat space (Minkowski geometry) followed by a Casimir geometry of one periodic spatial dimension. Finally we discuss the meaning of our finding in relation to the issues raised above.

4.1. SMEARED FIELD OPERATORS

Since the field operator in conventional point-defined quantum field theory is an operator-valued distribution, products of field operators at a point become problematic. This parallels the problem with defining the square of a delta function $\delta^2(x)$. Distributions are defined via their integral against a test function: they live in the space dual to the test function space. By going from the field operator $\hat{\phi}(x)$ to its integral against a test function, $\hat{\phi}(f) = \int \hat{\phi}\, f$, we can now readily consider products.

When we take the test functions to be spatial Gaussians, we are smearing the field operator over a finite spatial region. Physically we see smearing as representing the necessarily finite extent of an observer's probe, or the intrinsic limit of resolution in carrying out a measurement at a low energy (compared to Planck scale). In contrast to the ordinary point-defined quantum field theory, where ultraviolet divergences occur in the energy momentum tensor, smeared fields give no ultraviolet divergence. This is because smearing is equivalent to a regularization scheme which imparts an exponential suppression to the high momentum modes and restricts the contribution of the high frequency modes in the mode sum.

With this in mind, we start by defining the spatially smeared field operator

$$\hat{\phi}_t(f_{\mathbf{x}}) = \int \hat{\phi}(t,\mathbf{x}')f_{\mathbf{x}}(\mathbf{x}')d\mathbf{x}' \tag{127}$$

where $f_{\mathbf{x}}(\mathbf{x}')$ is a suitably smooth function. With this, the two point operator becomes

$$\left(\hat{\phi}_t(f_{\mathbf{x}})\right)^2 = \int\int \hat{\phi}(t,\mathbf{x}')\hat{\phi}(t,\mathbf{x}'')f_{\mathbf{x}}(\mathbf{x}')f_{\mathbf{x}}(\mathbf{x}'')d\mathbf{x}\,d\mathbf{x}' \tag{128}$$

which is now finite. In terms of the vacuum $|0\rangle$ ($\hat{a}_{\mathbf{k}}\,|0\rangle = 0$, for all $\mathbf{k}$) we have the usual mode expansion

$$\hat{\phi}\,(t_1,\mathbf{x}_1) = \int d\mu\,(\mathbf{k}_1)\left(\hat{a}_{\mathbf{k}_1}\,u_{\mathbf{k}_1}(t_1,\mathbf{x}_1) + \hat{a}^\dagger_{\mathbf{k}_1}\,u^*_{\mathbf{k}_1}(t_1,\mathbf{x}_1)\right) \tag{129}$$

with

$$u_{\mathbf{k}_1}\,(t_1,\mathbf{x}_1) = N_{k_1}e^{i\,(\mathbf{k}_1\cdot\mathbf{x}_1 - t_1\,\omega_1)}, \quad \omega_1 = |\mathbf{k}_1|\,, \tag{130}$$

where the integration measure $\int d\mu\,(\mathbf{k}_1)$ and the normalization constants $N_{\mathbf{k}_1}$ are given for a Minkowski and Casimir spaces by (140) and (147) respectively.

Consider a Gaussian smearing function

$$f_{\mathbf{x}_0}(\mathbf{x}) = \left(\frac{1}{4\,\pi\,\sigma^2}\right)^{\frac{d}{2}} e^{-\left(\frac{\mathbf{x}_0-\mathbf{x}}{2\sigma}\right)^2}, \tag{131}$$

which has the following the properties $\int f_{\mathbf{x}_0}(\mathbf{x}')\,d\mathbf{x}' = 1$, $\int \mathbf{x}'\,f_{\mathbf{x}_0}(\mathbf{x}')\,d\mathbf{x}' = \mathbf{x}_0$ and $\int |\mathbf{x}'|^2\,f_{\mathbf{x}_0}(\mathbf{x}')\,d\mathbf{x}' = 2d\sigma^2 + |\mathbf{x}_0|^2$. Using

$$\begin{aligned}\int u_{\mathbf{k}_1}(t,\mathbf{x})\,f_{\mathbf{x}_1}(\mathbf{x})d\mathbf{x} \;&=\; N_{k_1}e^{-it\omega_1}\prod_{i=1}^{d}\left(\frac{1}{2\,\sqrt{\pi}\,\sigma}\int e^{+ik_{1i}x_i-\left(\frac{x_{1i}-x_i}{2\sigma}\right)^2}dx_i\right)\\ &=\; N_{k_1}e^{-i\,t\,w+i\,\mathbf{k}_1\cdot\mathbf{x}_1-\sigma^2\,k_1{}^2}\end{aligned} \tag{132}$$

we get the smeared field operator

$$\begin{aligned}\hat{\phi}_{t_1}\,(f_{\mathbf{x}_1}) = \int d\mu\,(\mathbf{k}_1)\,N_{k_1}e^{-i\,\mathbf{k}_1\cdot\mathbf{x}_1-\sigma^2\,k_1{}^2-i\,t_1\,\omega_1}&\\ \times\left(e^{2\,i\,\mathbf{k}_1\cdot\mathbf{x}_1}\,\hat{a}_{\mathbf{k}_1} + e^{2\,i\,t_1\,\omega_1}\,\hat{a}^\dagger_{\mathbf{k}_1}\right)&\end{aligned} \tag{133}$$

and their derivatives

$$\left(\partial_{t_1}\hat{\phi}_{t_1}\right)(f_{\mathbf{x}_1}) = \int \left(\partial_{t_1}\hat{\phi}\left(t_1,\mathbf{x}'\right)\right) f_{\mathbf{x}_1}\left(\mathbf{x}'\right) d\mathbf{x}'$$
$$= i\int d\mu(\mathbf{k}_1) N_{k_1}\,\omega_1 e^{-i\,\mathbf{k}_1\cdot\mathbf{x}_1-\sigma^2\,{k_1}^2-i\,t_1\,\omega_1}$$
$$\times\left(e^{2\,i\,t_1\,\omega_1}\,\hat{a}^\dagger_{\mathbf{k}_1}-e^{2\,i\,\mathbf{k}_1\cdot\mathbf{x}_1}\,\hat{a}_{\mathbf{k}_1}\right) \quad (134)$$

$$\left(\vec{\nabla}_{\mathbf{x}_1}\hat{\phi}_{t_1}\right)(f_{\mathbf{x}_1}) = \int \left(\vec{\nabla}_{\mathbf{x}'}\hat{\phi}\left(t_1,\mathbf{x}'\right)\right) f_{\mathbf{x}_1}\left(\mathbf{x}'\right) d\mathbf{x}'$$
$$= -i\int d\mu(\mathbf{k}_1)\mathbf{k}_1\, N_{k_1} e^{-i\,\mathbf{k}_1\cdot\mathbf{x}_1-\sigma^2\,{k_1}^2-i\,t_1\,\omega_1}$$
$$\times\left(e^{2\,i\,t_1\,\omega_1}\,\hat{a}^\dagger_{\mathbf{k}_1}-e^{2\,i\,\mathbf{k}_1\cdot\mathbf{x}_1}\,\hat{a}_{\mathbf{k}_1}\right) \quad (135)$$

From this we can calculate the two point function of the field which make up the energy density and their correlation function. Letting $x \equiv (t,\mathbf{x}) = (t_2,\mathbf{x}_2) - (t_1,\mathbf{x}_1)$, they are given by

$$\rho\,(t,\mathbf{x};\sigma) = \int d\mu\,(\mathbf{k})\,N_k^2\,\omega^2\,e^{-2\,\sigma^2\,k^2}\,\cos(\mathbf{x}\cdot\mathbf{k}-t\,\omega) \quad (136)$$

$$\Delta\rho^2\,(t,\mathbf{x};\sigma) = \frac{1}{2}\int d\mu\,(\mathbf{k}_1,\mathbf{k}_2)\,N_{k_1}^2\,N_{k_2}^2\,(\mathbf{k}_1\cdot\mathbf{k}_2+\omega_1\,\omega_2)^2$$
$$\times e^{-2\,\sigma^2\,(k_1^2+k_2^2)-i\,\mathbf{x}\cdot(\mathbf{k}_1+\mathbf{k}_2)+i\,t\,(\omega_1+\omega_2)}. \quad (137)$$

Setting $x = 0$ we obtain the smeared vacuum energy density at one point

$$\rho\,(\sigma) = \int d\mu\,(\mathbf{k}_1)\,N_{k_1}^2\,\omega_1^2\,e^{-2\,\sigma^2\,k_1^2} \quad (138)$$

4.1.1. *Smeared-Field Energy Density and Fluctuations in Minkowski Space*

We consider a Minkowski space $R^1 \times R^d$ with d-spatial dimensions. For this space the mode density is

$$\int d\mu(\mathbf{k}) = \int_0^\infty k^{d-1}\,dk \int_{S^{d-1}} d\Omega_{d-1} \quad \text{with} \quad \int_{S^{d-1}} d\Omega_{d-1} = \frac{2\,\pi^{\frac{d}{2}}}{\Gamma\left(\frac{d}{2}\right)} \quad (139)$$

and the mode function normalization constant is

$$N_{k_1} = 1/\sqrt{2^{d+1}\,\pi^d\,\omega_1}. \quad (140)$$

We introduce the angle between two momenta in phase space, γ, via

$$\mathbf{k}_1 \cdot \mathbf{k}_2 = k_1 k_2 \cos(\gamma) = \omega_1 \omega_2 \cos(\gamma). \tag{141}$$

The averages of the cosine and cosine squared of this angle over a pair of unit spheres are

$$\int_{S^{d-1}} d\Omega_1 \int_{S^{d-1}} d\Omega_2 \cos(\gamma) \;=\; 0 \tag{142}$$

$$\int_{S^{d-1}} d\Omega_1 \int_{S^{d-1}} d\Omega_2 \cos^2(\gamma) \;=\; \frac{4\,\pi^d}{d\,\Gamma\left(\frac{d}{2}\right)^2}. \tag{143}$$

The smeared energy density (138) becomes

$$\begin{aligned}\rho(\sigma) &= \frac{1}{2^d\,\pi^{\frac{d}{2}}\,\Gamma\left(\frac{d}{2}\right)} \int_0^\infty dk_1 \frac{{k_1}^d}{e^{2\,\sigma^2\,{k_1}^2}} \\ &= \frac{\Gamma\left(\frac{d+1}{2}\right)}{2^{\frac{3\,(d+1)}{2}}\,\pi^{\frac{d}{2}}\,\sigma^{d+1}\Gamma\left(\frac{d}{2}\right)}\end{aligned} \tag{144}$$

For the fluctuations of the smeared energy density operator, we evaluate (137) for this space and find

$$\begin{aligned}\Delta\rho^2(\sigma) &= \frac{1}{2^{(2d+3)}\pi^{2\,d}} \int_0^\infty \int_0^\infty \int_{S^{d-1}} \int_{S^{d-1}} d\Omega_1\, d\Omega_2\, dk_1\, dk_2 \\ &\qquad \times \frac{\left(1+\cos(\gamma)\right)^2 k_1^d\, k_2^d}{e^{2\,\sigma^2\,(k_1^2+k_2^2)}} \\ &= \frac{(d+1)\;\Gamma\left(\frac{d+1}{2}\right)^2}{2^{(3d+4)}\,d\,\pi^d\,\sigma^{2\,(d+1)}\,\Gamma\left(\frac{d}{2}\right)^2}\end{aligned} \tag{145}$$

Putting these together we obtain for the Minkowski space

$$\Delta_{\text{Minkowski}}(d) = \frac{1+d}{1+3\,d} \tag{146}$$

which has the particular values:

$$\frac{1}{2} \text{ for } d=1, \quad \frac{2}{5} \text{ for } d=3, \quad \frac{3}{8} \text{ for } d=5, \quad \frac{1}{3} \text{ for } d=\infty.$$

4.1.2. *Smeared-Field in Casimir Topology*

The Casimir topology is obtained from a flat space (with d spatial dimensions, i.e., $R^1 \times R^d$) by imposing periodicity L in one of its spatial dimensions, say, z, thus endowing it with a $R^1 \times R^{d-1} \times S^1$ topology. We decompose $\mathbf{k}$ into a component along the periodic dimension and call the remaining components $\mathbf{k}_\perp$:

$$\begin{aligned} \mathbf{k} &= \left(\mathbf{k}_\perp, \frac{2\pi n}{L}\right) = (\mathbf{k}_\perp, ln)\,, l \equiv 2\pi/L \\ \omega_1 &= \sqrt{k_1^2 + l^2\, n_1{}^2}, \end{aligned} \tag{147}$$

The normalization and momentum measure are

$$\int d\mu(\mathbf{k}) = \int_0^\infty k^{d-2}\, dk \int_{S^{d-2}} d\Omega_{d-2} \sum_{n=-\infty}^{\infty} \tag{148}$$

$$N_{k_1} = \frac{1}{\sqrt{2^d\, L\, \pi^{d-1}\, \omega_1}} \tag{149}$$

With this, the energy density (138) becomes

$$\rho_L(\sigma) = \frac{l}{2^d \pi^{\frac{d+1}{2}} \Gamma\left(\frac{d-1}{2}\right)} \sum_{n_1=-\infty}^{\infty} \int_0^\infty dk_1 k_1^{d-2} \left(k_1^2 + l^2 n_1^2\right)^{\frac{1}{2}} \times e^{-2\sigma^2\left(k_1^2 + l^2 n_1^2\right)}, \tag{150}$$

we can write this as the sum of the two smeared Green function derivatives

$$\begin{aligned} \rho_L(\sigma) &= \left\langle 0_L \left| \left((\nabla_\perp \phi_t)\,(f_{\mathbf{x}}) \right)^2 \right| 0_L \right\rangle + \left\langle 0_L \left| \left((\partial_z \phi_t)\,(f_{\mathbf{x}}) \right)^2 \right| 0_L \right\rangle \\ &= G_L(\sigma)_{,x_\perp x_\perp} + G_L(\sigma)_{,zz} \end{aligned} \tag{151}$$

where $|0_L\rangle$ is the Casimir vacuum.

Since $G_L(\sigma)_{,i} = G_{L,i}^{\text{div}} + G_{L,i}^{\text{fin}}$ ($i = x_\perp x_\perp$ or zz) we see how to split the smeared energy density into a $\sigma \to 0$ divergent term and the finite contribution:

$$\rho_L(\sigma) = \rho_L^{\text{div}} + \rho_L^{\text{fin}} \tag{152}$$

where

$$\begin{aligned} \rho_L^{\text{div}} &= G_{L,x_\perp x_\perp}^{\text{div}} + G_{L,zz}^{\text{div}} \\ &= \frac{\Gamma\left(\frac{d+1}{2}\right)}{2^{\frac{3(d+1)}{2}} \pi^{\frac{d}{2}} \sigma^{d+1} \Gamma\left(\frac{d}{2}\right)} \\ &= \rho(\sigma) \end{aligned} \tag{153}$$

and

$$\begin{aligned}\rho_L^{\text{fin}} &= G_{L,x_\perp x_\perp}^{\text{fin}} + G_{L,zz}^{\text{fin}} \\ &= -\frac{d\Gamma\left(-\frac{d}{2}\right)\Gamma\left(\frac{d}{2}\right)}{(4\pi)^{(d+3)/2} l^{d+1}} \sum_{p=1}^{\infty} (-1)^p (2\,l)^{2p} p(2p-1)^2 \sigma^{2(p-1)} \\ &\quad \times \frac{B_{2p+d-1}}{2p+d-1} \frac{(2p-3)!!}{(2p)!} \frac{\Gamma\left(p-\frac{1}{2}\right)}{\Gamma\left(p+\frac{d}{2}\right)}. \end{aligned} \tag{154}$$

With this we define the regularized energy density

$$\begin{aligned}\rho_{L,\text{reg}} &\equiv \lim_{\sigma\to 0} \left(\rho_L(\sigma) - \rho(\sigma)\right) \\ &= \frac{d\,\pi^{\frac{d}{2}}\, B_{d+1}\, \Gamma\left(-\frac{d}{2}\right)\, \Gamma\left(\frac{d}{2}\right)}{2\,(d+1)\, L^{d+1}\, \Gamma\left(\frac{d}{2}+1\right)} \end{aligned} \tag{155}$$

and get the usual results:

$$-\frac{\pi}{6\,L^2} \text{ for } d=1, \quad -\frac{\pi^2}{90\,L^4} \text{ for } d=3, \quad -\frac{2\,\pi^3}{945\,L^6} \text{ for } d=5.$$

For the d-dimensional Casimir geometry, the fluctuations are

$$\begin{aligned}\Delta\rho_L^2(\sigma) = \frac{l^2}{2^{2d+3}\,\pi^{2\,d}} \sum_{n_1=-\infty}^{\infty} \sum_{n_2=-\infty}^{\infty} \int_0^\infty k_1^{d-2}\, dk_1 \int_0^\infty k_2^{d-2}\, dk_2 \int_{S^{d-2}} d\Omega_1 \int_{S^{d-2}} d\Omega_2 \\ \times \frac{e^{-2\,\sigma^2\left(\omega_1^2+\omega_2^2\right)}}{\omega_1\,\omega_2} \left(\cos(\gamma)\, k_1\, k_2 + l^2\, n_1\, n_2 + \omega_1\,\omega_2\right)^2 \end{aligned} \tag{156}$$

This expression can again be written in terms of products of the Green functions derivatives used above:

$$\Delta\rho_L^2(\sigma) = \frac{d\left(G_L(\sigma)_{,x_\perp x_\perp}\right)^2}{2\,(d-1)} + G_L(\sigma)_{,zz}\left(G_L(\sigma)_{,x_\perp x_\perp} + G_L(\sigma)_{,zz}\right) \tag{157}$$

and split into three general terms

$$\Delta\rho_L^2(\sigma) = \Delta\rho_L^{2,\text{div}} + \Delta\rho_L^{2,\text{cross}} + \Delta\rho_L^{2,\text{fin}} \tag{158}$$

The full expression for each of these terms are given in the Appendix of [8] The first term contains only the divergent parts of the Green functions while the last term contains only the finite parts. This is similar to the split we used for the smeared energy density above. What is new here is the middle term $\Delta\rho_L^{2,\text{cross}}$. This comes about from the products of the divergent part of one Green function and the finite part of the other. That this term arises for computations of the energy density fluctuations is a generic feature. We will discuss in greater detail the meaning of this term later.

The results of Appendix A in [8] give

$$\Delta\rho_L^{2,\text{div}} = \chi_d \left(\rho_L^{\text{div}}\right)^2 = \chi_d \left(\rho\left(\sigma\right)\right)^2, \tag{159}$$

$$\Delta\rho_L^{2,\text{cross}} = 2\chi_d \, \rho_L^{\text{div}} \, \rho_L^{\text{fin}}, \tag{160}$$

$$\Delta\rho_L^{2,\text{fin}} = \frac{d\,(d+1)}{2} \left(\rho_{L,\text{reg}}\right)^2. \tag{161}$$

From this we see the divergent and cross terms can be related to the smeared energy density via

$$\Delta\rho_L^{2,\text{div}} + \Delta\rho_L^{2,\text{cross}} = \chi_d \left\{ \left(\rho_L^{\text{div}}\right)^2 + 2\rho_L^{\text{div}} \rho_L^{\text{fin}} \right\} \tag{162}$$

where χ_d is the function that relates the fluctuations of the energy density to the mean energy density when the boundaries are not present, i.e., Minkowski space. This leads us to interpret these terms as due to the vacuum fluctuations that are always present. With this in mind, we define the regularized fluctuations of the energy density

$$\begin{aligned} \Delta\rho_{L,\text{reg}}^2 &= \lim_{\sigma\to 0} \left(\Delta\rho_L^2\,(f) - \chi_d \left\{ \left(\rho_L^{\text{div}}\right)^2 + 2\rho_L^{\text{div}} \rho_L^{\text{fin}} \right\} \right) \\ &= \chi_{d,L} \left(\rho_{L,\text{reg}}\right)^2 \end{aligned} \tag{163}$$

where

$$\chi_{d,L} \equiv \frac{d\,(d+1)}{2}. \tag{164}$$

We also define a regularized version of the dimensionless measure Δ:

$$\Delta_{L,\text{Reg}} \equiv \frac{\Delta\rho_{L,\text{Reg}}^2}{\Delta\rho_{L,\text{Reg}}^2 + \left(\rho_{L,\text{Reg}}\right)^2} = \frac{d\,(d+1)}{2+d+d^2} \tag{165}$$

and note the values:

$$\frac{1}{2} \text{ for } d=1, \quad \frac{6}{7} \text{ for } d=3, \quad \frac{15}{16} \text{ for } d=5, \quad 1 \text{ for } d=\infty.$$

Following the procedures described in Appendix B of [8], Phillips and Hu have made two plots, Fig. 1 of $\Delta(\sigma, L)$ and $\Delta_{L,Reg}$ versus σ/L, (which we call σ' here for short); and Fig. 2 of $\rho_{L,Reg}$ and $\sqrt{\Delta\rho_{L,Reg}^2}$ versus σ'. The range of σ' is limited to ≤ 0.4 because going any further would make the meaning of a local energy density ill-defined, as the smearing of the field extends to the Casimir boundary in space. (The infrared limit also carry important physical meaning in reference to the structure of spacetime.)

Let us ponder on the meaning they convey. In Fig. 1, we first note that both curves are of the order unity. But the behavior of Δ (recall that the energy density fluctuations thus defined include the cross term along with the finite part and the state independent divergent part) is relatively insensitive to the smearing width, whereas $\Delta_{L,Reg}$, which measures only the finite part of the energy density fluctuations to the mean, has more structure. In particular, it saturates its upper bound of 1 around $\sigma' = 0.24$. Note that if one adheres to the KF criterion [61] one would say that semiclassical gravity fails, but all that is happening here is that $\rho_{L,\text{Reg}} = 0$ while $\Delta\rho^2_{L,\text{Reg}}$ shows no special feature. The real difference between these two functions is the cross term, which is responsible for their markedly different structure and behavior.

In Fig. 2, the main feature to notice is that the regularized energy density crosses from negative to positive values at around $\sigma' = 0.24$. The negative Casimir energy density calculated in a point-wise field theory which corresponds to small ranges of σ' is expected, and is usually taken to signify the quantum nature of the Casimir state. As σ' increases we are averaging the field operator over a larger region, and thus sampling the field theory from the ultraviolet all the way to the infrared region. At large σ' finite size effect begins to set in. The difference and relation of these two effects are explained in [105]: Casimir effect arises from summing up the quantum fluctuations of ALL modes (as altered by the boundary), with no insignificant short wavelength contributions, whereas finite size effect has dominant contributions from the LONGEST wavelength modes, and thus reflect the large scale behavior. As the smearing moves from a small scale to the far boundary of space, the behavior of the system is expected to shift from a Casimir-dominated to a finite size-dominated effect. This could be the underlying reason in the crossover behavior of $\rho_{L,Reg}$.

4.2. FLUCTUATION TO MEAN RATIO AND SPATIAL EXTENT OF NEGATIVE ENERGY DENSITY

Now that we have the results we can return to the issues raised earlier. We discuss the first two here, i.e., 1) Fluctuations of the energy density and validity of semiclassical gravity, 2) The spatial extent where negative energy density can exist. We will discuss the regularization of energy density fluctuations and the issue of the cross term in the next subsection.

4.2.1. *Fluctuation to Mean ratio and Validity of SCG*

From these results we see that i) the fluctuations of the energy density as well as its mean both increase with decreasing distance (or probing scale), while ii) the ratio of the variance of the fluctuations in EMT to its mean-squared is of the order unity. We view the first but not the second feature as linked to the question of the validity of SCG. The second feature represents something quite different, pertaining more to the quantum nature of the vacuum state than to the validity of SCG.

If we adopt the criterion of Kuo and Ford [61] that the variance of the fluctuation relative to the mean-squared (vev taken with respect to the ordinary Minkowskian vacuum) being of the order unity be an indicator of the failure of SCG, then all spacetimes studied above would indiscriminately fall into that category, and SCG fails wholesale, regardless of the scale these physical quantities are probed. This contradicts with the common expectation that SCG is valid at scales below Planck energy. We believe the criterion for the validity or failure of a theory should depend on the range or the energy probed. The findings of Phillips and Hu [8] related here seem to confirm this: Both the mean (the vev of EMT with respect to the Minkowski vacuum) AND the fluctuations of EMT increase as the scale deceases. As one probes into an increasingly finer scale or higher energy the expectation value of EMT grows in value and the induced metric fluctuations become important, leading to the failure of SCG. A generic scale for this to 'happen is the Planck length. At such energy densities and above, particle creation from the quantum field vacuum would become copious and their backreaction on the background spacetime would become important [24]. Fluctuations in the quantum field EMT entails these quantum processes. The induced metric fluctuations [3] render the smooth manifold structure of spacetime inadequate, spacetime foams [106] including topological transitions [107] begin to appear and SCG no longer can provide an adequate description of these dominant processes. This picture first conjured by Wheeler is consistent with the common notion adopted in SCG, and we believe it is a valid one.

4.2.2. *Extent of Negative Energy Density*

It is well known that negative energy density exists in Casimir geometry, moving mirrors, black holes and worm holes. Proposals have also been conjured to use the negative energy density for the design of time machines [108]. Our results (Figures 1, 2) provide an explicit scale dependence of the regularized vacuum energy density $\rho_{L,reg}$ and its fluctuations $\Delta_{L,reg}$, specifically σ/L, the ratio of the smearing length (field scale) to that of the Casimir length (geometry scale). For example, Fig. 2 shows that only for $\sigma/L < 0.24$ is $\rho_{L,reg} < 0$. Recall σ gives the spatial extent the field is probed or smeared. Ordinary pointwise quantum field theory which probes the field only at a point does not carry information about the spatial extent where negative energy density sustains. These results have direct implications on wormhole physics (and time machines, if one gets really serious about these fictions [108]). If L is the scale characterizing the size ('throat') of the wormhole where one thinks negative energy density might prevail, and designers of 'time machines' wish to exploit for 'time-travel' , the results of Phillips and Hu can provide a limit on the size of the probe (spaceship in the case of time-travel) in ratio to L where such conditions may exist. It could also provide a quantum field-theoretical bound on the probability of spontaneous creation of baby universes from quantum field energy fluctuations.

4.3. DEPENDENCE OF FLUCTUATIONS ON INTRINSIC AND EXTRINSIC SCALES

One may ask why the fluctuations of the energy density to its mean for the many cases calculated by PH and KF should be the fractional numbers as they are. Is there any simple reason behind the following features observed in Phillips and Hu's calculations? .

a) $\Delta = O(1)$

b) The specific numeric values of Δ for the different cases.

c) For the Minkowski vacuum the ratio of the variance to the mean-squared, calculated from the coincidence limit, is identical to the value of the Casimir case at the same limit for spatial point separation while identical to the value of a hot flat space result with a temporal point-separation.

Point a) has also been shown by earlier calculations [61, 62], and our understanding is that this is true only for states of quantum nature, including the vacuum and certain squeezed states, but probably not true for states of a more classical nature like the coherent state. We also emphasized that this result should not be used as a criterion for the validity of semiclassical gravity.

For point b), we can trace back the calculation of the fluctuations (second moment) of the energy momentum tensor in ratio to its mean (first moment) to the integral of the term containing the inner product of two momenta $\mathbf{k}_1 \cdot \mathbf{k}_2$ summed over all participating modes. The modes contributing to this are different for different geometries, e.g., Minkowski versus Casimir boundary –for the Einstein universe this enters as 3j symbols – and could account for the difference in the numerical values of Δ for the different cases.

For point c), to begin with, it is well-known that the regularization by taking the coincidence limit of a spatial versus a temporal point separation will give different results. The case of temporal split involves integration of three spatial dimensions while the case of spatial split involve integration of two remaining spatial and one temporal dimension, which would give different results. The calculation of fluctuations involves the second moment of the field and is in this regard similar to what enters into the calculation of moments of inertia for rotating objects. We suspect that the difference between the temporal and the spatial results is similar, to the extent this analogy holds, to the difference in the moment of inertia of the same object but taken with respect to different axes of rotation.

It may be surprising that in a Minkowski calculation the result of Casimir geometry or thermal field should appear, as both cases involve a scale – the former in the spatial dimension and the latter in the (imaginary) temporal dimension. (Both cases have the same topology $R^3 \times S^1$, with the S^1 in the (imaginary) time for the former and in the space for the latter.) But if we note that the results for Casimir geometry or thermal field are obtained at the coincidence (ultraviolet) limit, when the scale (infrared) of the problem does not intercede in any major way, then the main components of the calculations for these two cases would be similar to the two cases of taking the coincidence limit in the spatial and temporal directions in Minkowski space.

All of these cases are effectively devoid of scale as far as the pointwise field theory is concerned. As soon as we depart from this limit the effect of the presence of a scale shows up. The point-separated or field-smeared results for the Casimir calculation shows clearly that the boundary scale enters in a major way and the result for the fluctuations and the ratio are different from those obtained at the coincident limit. For other cases where a scale enters intrinsically in the problem such as that of a massive or non-conformally coupled field it would show a similar effect in these regards as the present cases (of Casimir and thermal field) where a periodicity condition exists (in the spatial and temporal directions respectively). We expect a similar strong disparity between point-coincident and point-separated cases. The field theory changes its nature in a fundamental and physical way when this limit is taken. This brings us to an even more fundamental issue made clear in this investigation, i.e., the appearance of divergences and the meaning of regularization in the light of a point-separated versus a point-defined quantum field theory.

4.4. REGULARIZATION IN THE FLUCTUATIONS OF EMT AND THE ISSUE OF THE CROSS TERM

An equally weighty issue brought to light in the study of Phillips and Hu is the meaning of regularization in the face of EMT fluctuations. Since the point-separated or smeared field expressions of the EMT and its fluctuations become available one can study how they change as a function of separation or smearing scale in addition to how divergences arise at the coincidence limit. Whether certain cross terms containing divergences have physical meaning is a question raised by the recent studies of Wu and Ford [35]. We can use these calculations to examine these issues and ask the broader question of what exactly regularization means and entails, where divergences arise and why they need to be, and not just how they ought to be treated.

Recall the smeared energy density fluctuations for the Casimir topology has the form

$$\Delta\rho_L^2(\sigma) = \Delta\rho_L^{\rm div} + \Delta\rho_L^{\rm cross} + \Delta\rho_L^{\rm fin} \tag{166}$$

with

$$\Delta\rho_L^{\rm div} = \chi_d \left(\rho_L^{\rm div}\right)^2 = \chi_d(\rho(\sigma))^2 \tag{167}$$

$$\Delta\rho_L^{\rm cross} = 2\chi_d \rho_L^{\rm div}\rho_L^{\rm fin} \tag{168}$$

$$\Delta\rho_L^{\rm fin} = 2\chi_{d,L}\left(\rho_L^{\rm fin}\right)^2 + \text{terms that vanish as } \sigma \to 0 \tag{169}$$

where χ_d is the ratio between the fluctuations for Minkowski space and the square of the corresponding energy density: $\Delta\rho^2 = \chi_d(\rho(\sigma))^2$. Our results show that $\Delta\rho_L^2(\sigma)$ diverges as the width σ of the smearing function shrinks to zero with contributions from the truly divergent and the cross terms. We also note that the divergent term $\Delta\rho^{\rm div}$ is state independent, in the sense that it is independent of L, while the cross term $\Delta\rho^{\rm cross}$ is state dependent, as is the finite term $\Delta\rho^{\rm fin}$.

If we want to ask about the strength of fluctuations of the energy density, the relevant quantity to study is the energy density correlation function $H(x,y) = \langle\hat\rho(x)\hat\rho(y)\rangle - \langle\hat\rho(x)\rangle\langle\hat\rho(y)\rangle$. It is finite at $x \neq y$ for a linear quantum theory (this happens since the divergences for $\langle\hat\rho(x)\hat\rho(y)\rangle$ are exactly the same as the product $\langle\hat\rho(x)\rangle\langle\hat\rho(y)\rangle$), but diverges as $y \to x$, corresponding to the coincident or unsmeared limit $\sigma \to 0$.

To define a procedure for rendering $\Delta\rho_L^2(\sigma)$ finite, one can see that there exists choices – which means ambiguities in the regularization scheme. Three possibilities present themselves: The first is to just drop the state independent $\Delta\rho^{\rm div}$. This is easily seen to fail since we are left with the divergences from the cross term. The second is to neglect all terms that diverge as $\sigma \to 0$. This is too rash a move since $\Delta\rho^{\rm cross}$ has, along with its divergent parts, ones that are finite in the $\sigma \to 0$ limit. This comes about since it is of the form $\rho_L^{\rm div}\rho_L^{\rm fin}$ and the negative powers of σ present in $\rho_L^{\rm div}$ will cancel out against the positive powers in $\rho_L^{\rm fin}$. Besides, they yield results in disagreement with earlier results using well-tested methods such as normal ordering in flat space [61] and zeta-function regularization in curved space [62].

The third choice is the one adopted by PH [8]. For the energy density, we can think of regularization as computing the contribution "above and beyond" the Minkowski vacuum contribution, same for regularizing the fluctuations. So we need to first determine for Minkowski space vacuum how the fluctuations of the energy density are related to the vacuum energy density $\Delta\rho^2 = \chi\ (\rho)^2$. This we obtained for finite smearing. For Casimir topology the sum of the divergent and cross terms take the form

$$\Delta\rho_L^{\text{div}} + \Delta\rho_L^{\text{cross}} = \chi\left\{\left(\rho_L^{\text{div}}\right)^2 + 2\rho_L^{\text{div}}\rho_L^{\text{fin}}\right\} = \chi\left\{(\rho_L)^2 - \left(\rho_L^{\text{fin}}\right)^2\right\} \quad (170)$$

where χ is the ratio derived for Minkowski vacuum. We take this to represent the (state dependent) vacuum contribution. What we find interesting is that to regularize the smeared energy density fluctuations, a state dependent subtraction must be used. With this, just the $\sigma \to 0$ limit of the finite part $\Delta\rho_L^{\text{fin}}$ is identified as the regularized fluctuations $\Delta\rho_{L,\text{Reg}}^2$. The ratio χ_L thus obtained gives exactly the same result as derived by Kuo and Ford for $d = 3$ via normal ordering [61] and by PH for arbitrary d via the ζ-function [62].

That this procedure is the one to follow can be seen by considering the problem from the point separation method, see section 5. For this method, the energy density expectation value is defined as the $x' \to x$ limit of

$$\rho(x, x') = \mathcal{D}_{x,x'} G(x, x') \quad (171)$$

for the suitable Green function $G(x, x')$ and $\mathcal{D}_{x,x'}$ is a second order differential operator. In the limit $x' \to x$, $G(x, x')$ is divergent. The Green function is regularized by subtracting from it a Hadamard form $G^L(x, x')$: $G_{\text{Reg}}(x, x') = G(x, x') - G^L(x, x')$ [109]. With this, the regularized energy density can be obtained

$$\rho_{\text{Reg}}(x) = \lim_{x' \to x} \left(\mathcal{D}_{x,x'} G_{\text{Reg}}(x, x')\right) \quad (172)$$

Or, upon re-arranging terms, define the divergent and finite pieces as

$$G^{\text{div}}(x, x') = G^L(x, x'),\ G^{\text{fin}}(x, x') = G_{\text{Reg}}(x, x') = G(x, x') - G^L(x, x') \quad (173)$$

and

$$\rho(x, x') = \rho^{\text{div}}(x, x') + \rho^{\text{fin}}(x, x') \quad (174)$$

$$\rho^{\text{div}}(x, x') = \mathcal{D}_{x,x'} G^{\text{div}}(x, x') \quad \text{and} \quad \rho^{\text{fin}}(x, x') = \mathcal{D}_{x,x'} G^{\text{fin}}(x, x')$$

so that $\rho_{\text{Reg}}(x) = \lim_{x' \to x} \rho^{\text{fin}}(x, x')$, which corresponds to the $\sigma \to 0$ limit in PH's computation of the Casimir energy density.

Now turning to the fluctuations, we have the point-separated expression for the correlation function

$$H(x,y) = \lim_{x' \to x} \lim_{y' \to y} \mathcal{D}_{x,x'} \mathcal{D}_{y,y'} G(x,x',y,y') \tag{175}$$

where $G(x,x',y,y')$ is the suitable four point function. For linear theories we can use Wick's Theorem to express this in terms of products of Green functions $G(x,x',y,y') = G(x,y)G(x',y')$+permutations of$(x,x',y,y')$. Excluded from the permutations is $G(x,x')G(y,y')$. (Details are in [104, 9], which includes correct identifications of needed permutations and Green functions.) The general form is

$$H(x,y) = \lim_{x' \to x} \lim_{y' \to y} \mathcal{D}_{x,x'} \mathcal{D}_{y,y'} G(x,y)G(x',y') \ + \ \text{permutations} \tag{176}$$

The $(x',y') \to (x,y)$ limits are only retained to keep track of which derivatives act on which Green functions, but we can see there are no divergences for $y \neq x$. However, to get the point-wise fluctuations of the energy density, the divergences from $\lim_{y \to x} G(x,y)$ will present a problem. Splitting the Green function into its finite and divergent pieces, we can recognize terms leading to those we found for $\Delta\rho_L^2(\sigma)$:

$$H(x,y) = H^{\text{div}}(x,y) + H^{\text{cross}}(x,y) + H^{\text{fin}}(x,y) \tag{177}$$

where

$$H^{\text{div}}(x,y) = \lim_{x' \to x} \lim_{y' \to y} \mathcal{D}_{x,x'} \mathcal{D}_{y,y'} G^{\text{div}}(x,y) G^{\text{div}}(x',y') \tag{178}$$

$$H^{\text{cross}}(x,y) = 2 \lim_{x' \to x} \lim_{y' \to y} \mathcal{D}_{x,x'} \mathcal{D}_{y,y'} G^{\text{div}}(x,y) G^{\text{fin}}(x',y') \tag{179}$$

$$H^{\text{fin}}(x,y) = \lim_{x' \to x} \lim_{y' \to y} \mathcal{D}_{x,x'} \mathcal{D}_{y,y'} G^{\text{fin}}(x,y) G^{\text{fin}}(x',y'), \tag{180}$$

plus permutations. Thus we see the origin of both the divergent and cross terms. When the un-regularized Green function is used, we must get a cross term, along with the expected divergent term. If the fluctuations of the energy density is regularized via point separation, i.e. $G(x,x')$ is replaced by $G_{\text{Reg}}(x,x') = G^{\text{fin}}(x,y)$, then we should do the same replacement for the fluctuations. When this is done, it is only the finite part above that will be left and we can define the point-wise fluctuations as

$$\Delta\rho^2_{\text{Reg}} = \lim_{y \to x} H^{\text{fin}}(x,y) \tag{181}$$

The parallel with the smeared-field derivation presented here can be seen when the analysis of $G_L(\sigma)_{,zz}$ and $G_L(\sigma)_{,x_\perp x_\perp}$ (given in the Appendix of [8]) is considered. There it is shown they are derivatives of Green functions and can be separated into state-independent divergent part and state-dependent finite contribution: $G_L(\sigma)_{,i} = G^{\rm div}_{L,i} + G^{\rm fin}_{L,i}$, same as the split hereby shown for the Green function.

When analyzing the energy density fluctuations, discarding the divergent piece is the same as subtracting from the Green function its divergent part. If this is done, we also no longer have the cross term, just as viewing the problem from the point separation method outlined above. We feel this makes it problematic to analyze the cross term without also including the divergent term. At the same time, regularization of the fluctuations involving the subtraction of state dependent terms as realized in this calculation raises new issues on regularization which merits further investigations. In a recent work Wu and Ford [110] showed a connection between the cross term and radiation pressure.

5. Noise Kernel, Stress-Energy Bi-Tensor and Point Separation

As pointed out by one of us before [3] the stress energy bi-tensor could be the starting point for a new quantum field theory constructed on spacetimes with extended structures. But for comparison with ordinary phenomena at low energy we need to find out if it behaves normally in the limit of ordinary (point-defined) quantum field theory. The method of point-separation is best suited for this purpose. The task is to seek a regularized noise-kernel for quantum fields in general curved spacetimes upon taking the coincidence limit. This was carried out by Phillips and Hu [9]. The following is a summary of their work.

PH began with a discussion of the procedures for dealing with the quantum stress tensor bi-operator at two separated points and the noise kernel and end with a general expression for the noise kernel in terms of the quantum field's Green function and its covariant derivatives up to the fourth order . (The stress tensor involves up to two covariant derivatives.) This result holds for $x \neq y$ without recourse to renormalization of the Green function, showing that $N_{abc'd'}(x, y)$ is always finite for $x \neq y$ (and off the light cone for massless theories). In particular for a massless conformally coupled free scalar field on a four dimensional manifold they computed the trace of the noise kernel at both points and found this double trace vanishes identitically. This implies that there is no stochastic correction to the trace anomaly for massless conformal fields, in agreement with results arrived at in Refs. [31, 73, 4].

Now to obtain the point-defined quantities, one needs to deal with the divergences in the coincidence limit. For this PH adopted the "modified" point separation scheme [109, 111, 112] to get a regularized Green function.

In this procedure, the naive Green function is rendered finite by assuming the divergences present for $y \to x$ are state independent and can be removed by subtraction of a Hadamard form. They showed that the noise kernel in the $y \to x$ limit is meaningful for an arbitrary curved spacetime by explicitly deriving a general expression for the noise kernel and its coincident form.

After following these expositions of PH, we will end our discussion with a reflection on the issues related to regularization we started addressing in the previous section, reiterating the important role the point-separated quantities can play in a new approach to a quantum theory of gravity. There is a fundamental shift of viewpoint in the nature and meaning of the point separation scheme: in the 70's it was used as a technique (many practitioners may still view it as a trick, even a clumpsy one) for the purpose of identifying the ultraviolet divergences. Now in the new approach (as advocated by one of us [3, 55]) we want to use the point-separated expressions to construct a quantum field theory for extended spacetimes.

5.1. POINT SEPARATION

The point separation scheme introduced in the 60's by DeWitt [113] was brought to more popular use in the 70's in the context of quantum field theory in curved spacetimes [114, 58] as a means for obtaining a finite quantum stress tensor[3]. Since the stress-energy tensor is built from the product of a pair of field operators evaluated at a single point, it is not well-defined. In this scheme, one introduces an artificial separation of the single point x to a pair of closely separated points x and x'. The problematic terms involving field products such as $\hat{\phi}(x)^2$ becomes $\hat{\phi}(x)\hat{\phi}(x')$, whose expectation value is well defined. If one is interested in the low energy behavior captured by the point-defined quantum field theory – as the effort in the 70's was directed – one takes the coincidence limit. Once the divergences present are identified, they may be removed (regularization) or moved (by renormalizing the coupling constants), to produce a well-defined, finite stress tensor at a single point.

Thus the first order of business is the construction of the stress tensor and then derive the symmetric stress-energy tensor two point function, the noise kernel, in terms of the Wightman Green function. In this section we

[3]We know there are several regularization methods developed for the removal of ultraviolet divergences in the stress energy tensor of quantum fields in curved spacetime [20, 21, 22]. Their mutual relations are known, and discrepancies explained. This formal structure of regularization schemes for quantum fields in curved spacetime should remain intact as we apply them to the regularization of the noise kernel in general curved spacetimes. Specific considerations will of course enter for each method. But for the methods employed so far, such as zeta-function, point separation, dimensional, smeared-field [62, 9] applied to simple cases (Casimir, Einstein, thermal fields) there is no new inconsistency or discrepancy.

will use the traditional notation for index tensors in the point-separation context.

5.1.1. *n-tensors and end-point expansions*

An object like the Green function $G(x, y)$ is an example of a *bi-scalar*: it transforms as scalar at both points x and y. We can also define a *bi-tensor* $T_{a_1 \cdots a_n\, b'_1 \cdots b'_m}(x, y)$: upon a coordinate transformation, this transforms as a rank n tensor at x and a rank m tensor at y. We will extend this up to a *quad-tensor* $T_{a_1 \cdots a_{n_1}\, b'_1 \cdots b'_{n_2}\, c''_1 \cdots c''_{n_3}\, d'''_1 \cdots d'''_{n_4}}$ which has support at four points x, y, x', y', transforming as rank n_1, n_2, n_3, n_4 tensors at each of the four points. This also sets the notation we will use: unprimed indices referring to the tangent space constructed above x, single primed indices to y, double primed to x' and triple primed to y'. For each point, there is the covariant derivative ∇_a at that point. Covariant derivatives at different points commute and the covariant derivative at, say, point x', does not act on a bi-tensor defined at, say, x and y:

$$T_{ab';c;d'} = T_{ab';d';c} \quad \text{and} \quad T_{ab';c''} = 0. \tag{182}$$

To simplify notation, henceforth we will eliminate the semicolons after the first one for multiple covariant derivatives at multiple points.

Having objects defined at different points, the coincident limit is defined as evaluation "on the diagonal", in the sense of the spacetime support of the function or tensor, and the usual shorthand $[G(x, y)] \equiv G(x, x)$ is used. This extends to n-tensors as

$$\left[T_{a_1 \cdots a_{n_1}\, b'_1 \cdots b'_{n_2}\, c''_1 \cdots c''_{n_3}\, d'''_1 \cdots d'''_{n_4}}\right] = T_{a_1 \cdots a_{n_1}\, b_1 \cdots b_{n_2}\, c_1 \cdots c_{n_3}\, d_1 \cdots d_{n_4}}, \tag{183}$$

i.e., this becomes a rank $(n_1 + n_2 + n_3 + n_4)$ tensor at x. The multi-variable chain rule relates covariant derivatives acting at different points, when we are interested in the coincident limit:

$$\left[T_{a_1 \cdots a_m\, b'_1 \cdots b'_n}\right]_{;c} = \left[T_{a_1 \cdots a_m\, b'_1 \cdots b'_n ;c}\right] + \left[T_{a_1 \cdots a_m\, b'_1 \cdots b'_n ;c'}\right]. \tag{184}$$

This result is referred to as *Synge's theorem* in this context; we follow Fulling's [21] discussion.

The bi-tensor of *parallel transport* ${g_a}^{b'}$ is defined such that when it acts on a vector $v_{b'}$ at y, it parallel transports the vector along the geodesics connecting x and y. This allows us to add vectors and tensors defined at different points. We cannot directly add a vector v_a at x and vector $w_{a'}$ at y. But by using ${g_a}^{b'}$, we can construct the sum $v^a + {g_a}^{b'} w_{b'}$. We will also need the obvious property $\left[{g_a}^{b'}\right] = {g_a}^{b}$.

The main bi-scalar we need is the *world function* $\sigma(x, y)$. This is defined as a half of the square of the geodesic distance between the points x and y. It satisfies the equation

$$\sigma = \frac{1}{2}\sigma^{;p}\sigma_{;p} \tag{185}$$

Often in the literature, a covariant derivative is implied when the world function appears with indices: $\sigma^a \equiv \sigma^{;a}$, *i.e.*taking the covariant derivative at x, while $\sigma^{a'}$ means the covariant derivative at y. This is done since the vector $-\sigma^a$ is the tangent vector to the geodesic with length equal the distance between x and y. As σ^a records information about distance and direction for the two points this makes it ideal for constructing a series expansion of a bi-scalar. The *end point* expansion of a bi-scalar $S(x, y)$ is of the form

$$S(x,y) = A^{(0)} + \sigma^p A^{(1)}_p + \sigma^p\sigma^q A^{(2)}_{pq} + \sigma^p\sigma^q\sigma^r A^{(3)}_{pqr} + \sigma^p\sigma^q\sigma^r\sigma^s A^{(4)}_{pqrs} + \cdots \tag{186}$$

where, following our convention, the expansion tensors $A^{(n)}_{a_1\cdots a_n}$ with unprimed indices have support at x and hence the name end point expansion. Only the symmetric part of these tensors contribute to the expansion. For the purposes of multiplying series expansions it is convenient to separate the distance dependence from the direction dependence. This is done by introducing the unit vector $p^a = \sigma^a/\sqrt{2\sigma}$. Then the series expansion can be written

$$S(x,y) = A^{(0)} + \sigma^{\frac{1}{2}} A^{(1)} + \sigma A^{(2)} + \sigma^{\frac{3}{2}} A^{(3)} + \sigma^2 A^{(4)} + \cdots \tag{187}$$

The expansion scalars are related to the expansion tensors via $A^{(n)} = 2^{n/2} A^{(n)}_{p_1\cdots p_n} p^{p_1} \cdots p^{p_n}$.

The last object we need is the *Van Vleck-Morette* determinant $D(x, y)$, defined as $D(x,y) \equiv -\det\left(-\sigma_{;ab'}\right)$. The related bi-scalar

$$\Delta^{1/2} = \left(\frac{D(x,y)}{\sqrt{g(x)g(y)}}\right)^{\frac{1}{2}} \tag{188}$$

satisfies the equation

$$\Delta^{1/2}\left(4 - \sigma_{;p}{}^{p}\right) - 2\Delta^{1/2}{}_{;p}\sigma^{;p} = 0 \tag{189}$$

with the boundary condition $\left[\Delta^{1/2}\right] = 1$.

Further details on these objects and discussions of the definitions and properties are contained in [58] and [115]. There it is shown how the defining equations for σ and $\Delta^{1/2}$ are used to determine the coincident limit expression for the various covariant derivatives of the world function ($[\sigma_{;a}]$, $[\sigma_{;ab}]$, *etc.*) and how the defining differential equation for $\Delta^{1/2}$ can be used to determine the series expansion of $\Delta^{1/2}$. We show how the expansion tensors $A^{(n)}_{a_1 \cdots a_n}$ are determined in terms of the coincident limits of covariant derivatives of the bi-scalar $S(x, y)$. (Ref. [115] details how point separation can be implemented on the computer to provide easy access to a wider range of applications involving higher derivatives of the curvature tensors.)

5.2. STRESS ENERGY BI-TENSOR OPERATOR AND NOISE KERNEL

Even though we believe that the point-separated results are more basic in the sense that it reflects a deeper structure of the quantum theory of spacetime, we will nevertheless start with quantities defined at one point because they are more familiar in conventional quantum field theory. We will use point separation to introduce the biquantities. The key issue here is thus the distinction between point-defined (*pt*) and point-separated (*bi*) quantities.

For a free classical scalar field ϕ with the action $S_m[g, \phi]$ defined in Eq. (1), the classical stress-energy tensor is

$$\begin{aligned} T_{ab} &= (1-2\xi)\, \phi_{;a}\, \phi_{;b} + \left(2\xi - \frac{1}{2}\right) \phi_{;p}\, \phi^{;p}\, g_{ab} + 2\xi\, \phi\, \left(\phi_{;p}{}^{p} - \phi_{;ab}\, g_{ab}\right) \\ & + \phi^2\, \xi \left(R_{ab} - \frac{1}{2} R\, g_{ab}\right) - \frac{1}{2} m^2\, \phi^2\, g_{ab}, \end{aligned} \qquad (190)$$

which is equivalent to the tensor of Eq. (3) but written in a slightly different form for convenience. When we make the transition to quantum field theory, we promote the field $\phi(x)$ to a field operator $\hat{\phi}(x)$. The fundamental problem of defining a quantum operator for the stress tensor is immediately visible: the field operator appears quadratically. Since $\hat{\phi}(x)$ is an operator-valued distribution, products at a single point are not well-defined. But if the product is point separated ($\hat{\phi}^2(x) \rightarrow \hat{\phi}(x)\hat{\phi}(x')$), they are finite and well-defined.

Let us first seek a point-separated extension of these classical quantities and then consider the quantum field operators. Point separation is symmetrically extended to products of covariant derivatives of the field according to

$$(\phi_{;a})\,(\phi_{;b}) \rightarrow \frac{1}{2}\left(g_a{}^{p'} \nabla_{p'} \nabla_b + g_b{}^{p'} \nabla_a \nabla_{p'}\right) \phi(x)\phi(x'), \qquad (191)$$

$$\phi\,(\phi_{;ab}) \rightarrow \frac{1}{2}\left(\nabla_a \nabla_b + g_a{}^{p'} g_b{}^{q'} \nabla_{p'} \nabla_{q'}\right) \phi(x)\phi(x'). \qquad (192)$$

The bi-vector of parallel displacement $g_a{}^{a'}(x,x')$ is included so that we may have objects that are rank 2 tensors at x and scalars at x'.

To carry out point separation on (190), we first define the differential operator

$$\begin{aligned}\mathcal{T}_{ab} &= \frac{1}{2}(1-2\xi)\left(g_a{}^{a'}\nabla_{a'}\nabla_b + g_b{}^{b'}\nabla_a\nabla_{b'}\right) + \left(2\xi - \frac{1}{2}\right) g_{ab}g^{cd'}\nabla_c\nabla_{d'} \\ &\quad -\xi\left(\nabla_a\nabla_b + g_a{}^{a'}g_b{}^{b'}\nabla_{a'}\nabla_{b'}\right) + \xi g_{ab}\left(\nabla_c\nabla^c + \nabla_{c'}\nabla^{c'}\right) \\ &\quad +\xi\left(R_{ab} - \frac{1}{2}g_{ab}R\right) - \frac{1}{2}m^2 g_{ab} \end{aligned} \tag{193}$$

from which we obtain the classical stress tensor as

$$T_{ab}(x) = \lim_{x'\to x} \mathcal{T}_{ab}\phi(x)\phi(x'). \tag{194}$$

That the classical tensor field no longer appears as a product of scalar fields at a single point allows a smooth transition to the quantum tensor field. From the viewpoint of the stress tensor, the separation of points is an artificial construct so when promoting the classical field to a quantum one, neither point should be favored. The product of field configurations is taken to be the symmetrized operator product, denoted by curly brackets:

$$\phi(x)\phi(y) \to \frac{1}{2}\left\{\hat{\phi}(x),\hat{\phi}(y)\right\} = \frac{1}{2}\left(\hat{\phi}(x)\hat{\phi}(y) + \hat{\phi}(y)\hat{\phi}(x)\right) \tag{195}$$

With this, the point separated stress energy tensor operator is defined as

$$\hat{T}_{ab}(x,x') \equiv \frac{1}{2}\mathcal{T}_{ab}\left\{\hat{\phi}(x),\hat{\phi}(x')\right\}. \tag{196}$$

While the classical stress tensor was defined at the coincidence limit $x' \to x$, we cannot attach any physical meaning to the quantum stress tensor at one point until the issue of regularization is dealt with, which will happen in the next section. For now, we will maintain point separation so as to have a mathematically meaningful operator.

The expectation value of the point-separated stress tensor can now be taken. This amounts to replacing the field operators by their expectation value, which is given by the Hadamard (or Schwinger) function

$$G^{(1)}(x,x') = \langle\left\{\hat{\phi}(x),\hat{\phi}(x')\right\}\rangle. \tag{197}$$

and the point-separated stress tensor is defined as

$$\langle\hat{T}_{ab}(x,x')\rangle = \frac{1}{2}\mathcal{T}_{ab}G^{(1)}(x,x') \tag{198}$$

where, since $\mathcal{T}_{ab}$ is a differential operator, it can be taken "outside" the expectation value. The expectation value of the point-separated quantum stress tensor for a free, massless ($m = 0$) conformally coupled ($\xi = 1/6$) scalar field on a four dimension spacetime with scalar curvature R is

$$\begin{aligned}\langle \hat{T}_{ab}(x,x')\rangle &= \frac{1}{6}\left(g^{p'}{}_b\, G^{(1)}{}_{;p'a} + g^{p'}{}_a\, G^{(1)}{}_{;p'b}\right) - \frac{1}{12} g^{p'}{}_q\, G^{(1)}{}_{;p'}{}^q\, g_{ab} \\ &-\frac{1}{12}\left(g^{p'}{}_a\, g^{q'}{}_b\, G^{(1)}{}_{;p'q'} + G^{(1)}{}_{;ab}\right) + \frac{1}{12}\left(\left(G^{(1)}{}_{;p'}{}^{p'} + G^{(1)}{}_{;p}{}^{p}\right) g_{ab}\right) \\ &+\frac{1}{12} G^{(1)}\left(R_{ab} - \frac{1}{2} R\, g_{ab}\right) \end{aligned} \tag{199}$$

5.2.1. *Finiteness of Noise Kernel*

We now turn our attention to the noise kernel introduced in Eq. (11), which is the symmetrized product of the (mean subtracted) stress tensor operator:

$$\begin{aligned} 8N_{ab,c'd'}(x,y) &= \langle\left\{\hat{T}_{ab}(x) - \langle\hat{T}_{ab}(x)\rangle, \hat{T}_{c'd'}(y) - \langle\hat{T}_{c'd'}(y)\rangle\right\}\rangle \\ &= \langle\left\{\hat{T}_{ab}(x), \hat{T}_{c'd'}(y)\right\}\rangle - 2\langle\hat{T}_{ab}(x)\rangle\langle\hat{T}_{c'd'}(y)\rangle \end{aligned} \tag{200}$$

Since $\hat{T}_{ab}(x)$ defined at one point can be ill-behaved as it is generally divergent, one can question the soundness of these quantities. But as will be shown later, the noise kernel is finite for $y \neq x$. All field operator products present in the first expectation value that could be divergent are canceled by similar products in the second term. We will replace each of the stress tensor operators in the above expression for the noise kernel by their point separated versions, effectively separating the two points (x, y) into the four points (x, x', y, y'). This will allow us to express the noise kernel in terms of a pair of differential operators acting on a combination of four and two point functions. Wick's theorem will allow the four point functions to be re-expressed in terms of two point functions. From this we see that all possible divergences for $y \neq x$ will cancel. When the coincidence limit is taken divergences do occur. The above procedure will allow us to isolate the divergences and obtain a finite result.

Taking the point-separated quantities as more basic, one should replace each of the stress tensor operators in the above with the corresponding point separated version (196), with $\mathcal{T}_{ab}$ acting at x and x' and $\mathcal{T}_{c'd'}$ acting at y and y'. In this framework the noise kernel is defined as

$$8N_{ab,c'd'}(x,y) = \lim_{x'\to x}\lim_{y'\to y} \mathcal{T}_{ab}\mathcal{T}_{c'd'}\, G(x,x',y,y') \tag{201}$$

where the four point function is

$$G(x,x',y,y') \;=\; \frac{1}{4}\left[\langle\left\{\left\{\hat{\phi}(x), \hat{\phi}(x')\right\}, \left\{\hat{\phi}(y), \hat{\phi}(y')\right\}\right\}\rangle\right.$$

$$-2\,\langle\left\{\hat{\phi}(x),\hat{\phi}(x')\right\}\rangle\langle\left\{\hat{\phi}(y),\hat{\phi}(y')\right\}\rangle\Big]\,. \quad (202)$$

We assume the pairs (x, x') and (y, y') are each within their respective Riemann normal coordinate neighborhoods so as to avoid problems that possible geodesic caustics might be present. When we later turn our attention to computing the limit $y \to x$, after issues of regularization are addressed, we will want to assume all four points are within the same Riemann normal coordinate neighborhood.

Wick's theorem, for the case of free fields which we are considering, gives the simple product four point function in terms of a sum of products of Wightman functions (we use the shorthand notation $G_{xy} \equiv G_+(x, y) = \langle\hat{\phi}(x)\,\hat{\phi}(y)\rangle$):

$$\langle\hat{\phi}(x)\,\hat{\phi}(y)\,\hat{\phi}(x')\,\hat{\phi}(y')\rangle = G_{xy'}\,G_{yx'} + G_{xx'}\,G_{yy'} + G_{xy}\,G_{x'y'} \quad (203)$$

Expanding out the anti-commutators in (202) and applying Wick's theorem, the four point function becomes

$$G(x, x', y, y') = G_{xy'}\,G_{x'y} + G_{xy}\,G_{x'y'} + G_{yx'}\,G_{y'x} + G_{yx}\,G_{y'x'} \quad (204)$$

We can now easily see that the noise kernel defined via this function is indeed well defined for the limit $(x', y') \to (x, y)$:

$$G(x, x, y, y) = 2\left(G_{xy}^2 + G_{yx}^2\right). \quad (205)$$

From this we can see that the noise kernel is also well defined for $y \neq x$; any divergence present in the first expectation value of (202) have been cancelled by those present in the pair of Green functions in the second term, in agreement with the results of section 2.

5.2.2. *Explicit Form of the Noise Kernel*

We will let the points separated for a while so we can keep track of which covariant derivative acts on which arguments of which Wightman function. As an example (the complete calculation is quite long), consider the result of the first set of covariant derivative operators in the differential operator (193), from both $\mathcal{T}_{ab}$ and $\mathcal{T}_{c'd'}$, acting on $G(x, x', y, y')$:

$$\frac{1}{4}(1-2\xi)^2\left(g_a{}^{p''}\nabla_{p''}\nabla_b + g_b{}^{p''}\nabla_{p''}\nabla_a\right)$$
$$\times\left(g_{c'}{}^{q'''}\nabla_{q'''}\nabla_{d'} + g_{d'}{}^{q'''}\nabla_{q'''}\nabla_{c'}\right)G(x, x', y, y') \quad (206)$$

(Our notation is that ∇_a acts at x, $\nabla_{c'}$ at y, $\nabla_{b''}$ at x' and $\nabla_{d'''}$ at y'). Expanding out the differential operator above, we can determine which derivatives act on which Wightman function:

$$\frac{(1-2\,\xi)^2}{4} \times \Big[g_{c'}{}^{p'''}\, g^{q''}{}_{a}\,\big(G_{xy';bp'''}\, G_{x'y;q''d'} + G_{xy;bd'}\, G_{x'y';q''p'''} + G_{yx';q''d'}\, G_{y'x;bp'''} + G_{yx;bd'}\, G_{y'x';q''p'''}\big)$$
$$+g_{d'}{}^{p'''}\, g^{q''}{}_{a}\,\big(G_{xy';bp'''}\, G_{x'y;q''c'} + G_{xy;bc'}\, G_{x'y';q''p'''} + G_{yx';q''c'}\, G_{y'x;bp'''} + G_{yx;bc'}\, G_{y'x';q''p'''}\big)$$
$$+g_{c'}{}^{p'''}\, g^{q''}{}_{b}\,\big(G_{xy';ap'''}\, G_{x'y;q''d'} + G_{xy;ad'}\, G_{x'y';q''p'''} + G_{yx';q''d'}\, G_{y'x;ap'''} + G_{yx;ad'}\, G_{y'x';q''p'''}\big)$$
$$+g_{d'}{}^{p'''}\, g^{q''}{}_{b}\,\big(G_{xy';ap'''}\, G_{x'y;q''c'} + G_{xy;ac'}\, G_{x'y';q''p'''} + G_{yx';q''c'}\, G_{y'x;ap'''} + G_{yx;ac'}\, G_{y'x';q''p'''}\big)\Big]\,. \tag{207}$$

If we now let $x' \to x$ and $y' \to y$ the contribution to the noise kernel is (including the factor of $\frac{1}{8}$ present in the definition of the noise kernel):

$$\frac{1}{8}\Big\{(1-2\,\xi)^2\, \left(G_{xy;ad'}\, G_{xy;bc'} + G_{xy;ac'}\, G_{xy;bd'}\right) +(1-2\,\xi)^2\, \left(G_{yx;ad'}\, G_{yx;bc'} + G_{yx;ac'}\, G_{yx;bd'}\right)\Big\} \tag{208}$$

That this term can be written as the sum of a part involving G_{xy} and one involving G_{yx} is a general property of the entire noise kernel. It thus takes the form

$$N_{abc'd'}(x,y) = N_{abc'd'}\left[G_{+}(x,y)\right] + N_{abc'd'}\left[G_{+}(y,x)\right]. \tag{209}$$

We will present the form of the functional $N_{abc'd'}\,[G]$ shortly. First we note, for x and y time-like separated, the above split of the noise kernel allows us to express it in terms of the Feynman (time ordered) Green function $G_F(x,y)$ and the Dyson (anti-time ordered) Green function $G_D(x,y)$:

$$N_{abc'd'}(x,y) = N_{abc'd'}\left[G_F(x,y)\right] + N_{abc'd'}\left[G_D(x,y)\right] \tag{210}$$

[4] The complete form of the functional $N_{abc'd'}\,[G]$ is

$$N_{abc'd'}\,[G] = \tilde{N}_{abc'd'}\,[G] + g_{ab}\tilde{N}_{c'd'}\,[G] + g_{c'd'}\tilde{N}'_{ab}\,[G] + g_{ab}g_{c'd'}\tilde{N}\,[G] \tag{212}$$

[4]This can be connected with the zeta function approach to this problem [62] as follows: Recall when the quantum stress tensor fluctuations determined in the Euclidean section is analytically continued back to Lorentzian signature ($\tau \to it$), the time ordered product results. On the other hand, if the continuation is $\tau \to -it$, the anti-time ordered product results. With this in mind, the noise kernel is seen to be related to the quantum stress tensor fluctuations derived via the effective action as

$$16 N_{abc'd'} = \Delta T^2_{abc'd'}\Big|_{t=-i\tau,t'=-i\tau'} + \Delta T^2_{abc'd'}\Big|_{t=i\tau,t'=i\tau'}\,. \tag{211}$$

with

$$
\begin{aligned}
8\tilde{N}_{abc'd'}[G] \;=\;& (1-2\xi)^2 \left(G_{;c'b}\,G_{;d'a} + G_{;c'a}\,G_{;d'b}\right) \\
&+4\xi^2 \left(G_{;c'd'}\,G_{;ab} + G\,G_{;abc'd'}\right) \\
&-2\xi\,(1-2\xi)\,(G_{;b}\,G_{;c'ad'} + G_{;a}\,G_{;c'bd'} \\
&+G_{;d'}\,G_{;abc'} + G_{;c'}\,G_{;abd'}) \\
&+2\xi\,(1-2\xi)\left(G_{;a}\,G_{;b}\,R_{c'd'} + G_{;c'}\,G_{;d'}\,R_{ab}\right) \\
&-4\xi^2 \left(G_{;ab}\,R_{c'd'} + G_{;c'd'}\,R_{ab}\right) G \\
&+2\xi^2\,R_{c'd'}\,R_{ab} G^2
\end{aligned}
\tag{213}
$$

$$
\begin{aligned}
8\tilde{N}'_{ab}[G] \;=\;& 2\,(1-2\xi)\Bigg(\left(2\xi-\frac{1}{2}\right) G_{;p'b}\,G_{;}{}^{p'}{}_{a} \\
&+\xi\left(G_{;b}\,G_{;p'a}{}^{p'} + G_{;a}\,G_{;p'b}{}^{p'}\right)\Bigg) \\
&-4\xi\Bigg(\left(2\xi-\frac{1}{2}\right) G_{;}{}^{p'}\,G_{;abp'} \\
&+\xi\left(G_{;p'}{}^{p'}\,G_{;ab} + G\,G_{;abp'}{}^{p'}\right)\Bigg) \\
&-\left(m^2+\xi R'\right)\left((1-2\xi)\,G_{;a}\,G_{;b} - 2G\,\xi\,G_{;ab}\right) \\
&+2\xi\left(\left(2\xi-\frac{1}{2}\right) G_{;p'}\,G_{;}{}^{p'} + 2G\,\xi\,G_{;p'}{}^{p'}\right) R_{ab} \\
&-\left(m^2+\xi R'\right)\xi\,R_{ab}G^2
\end{aligned}
\tag{214}
$$

$$
\begin{aligned}
8\tilde{N}[G] \;=\;& 2\left(2\xi-\frac{1}{2}\right)^2 G_{;p'q}\,G_{;}{}^{p'q} + 4\xi^2\left(G_{;p'}{}^{p'}\,G_{;q}{}^{q} + G\,G_{;p}{}^{p}{}_{q'}{}^{q'}\right) \\
&+4\xi\left(2\xi-\frac{1}{2}\right)\left(G_{;p}\,G_{;q'}{}^{pq'} + G_{;}{}^{p'}\,G_{;q}{}^{q}{}_{p'}\right) \\
&-\left(2\xi-\frac{1}{2}\right)\left(\left(m^2+\xi R\right) G_{;p'}\,G_{;}{}^{p'} + \left(m^2+\xi R'\right) G_{;p}\,G^{;p}\right) \\
&-2\xi\left(\left(m^2+\xi R\right) G_{;p'}{}^{p'} + \left(m^2+\xi R'\right) G_{;p}{}^{p}\right) G \\
&\frac{1}{2}\left(m^2+\xi R\right)\left(m^2+\xi R'\right) G^2
\end{aligned}
\tag{215}
$$

5.2.3. *Trace of the Noise Kernel*

One of the most interesting and surprising results to come out of the investigations undertaken in the 1970's of the quantum stress tensor was the discovery of the trace anomaly[116]. When the trace of the stress tensor $T = g^{ab}T_{ab}$ is evaluated for a field configuration that satisties the field equation (2) the trace is seen to vanish for massless conformally coupled fields. When this analysis is carried over to the renormalized expectation value of the quantum stress tensor, the trace no longer vanishes. Wald [112] showed this was due to the failure of the renormalized Hadamard function $G_{\rm ren}(x,x')$ to be symmetric in x and x', implying it does not necessarily satisfy the field equation (2) in the variable x'. The definition of $G_{\rm ren}(x,x')$ in the context of point separation will come next.)

With this in mind, we can now determine the noise associated with the trace. Taking the trace at both points x and y of the noise kernel functional (210):

$$\begin{aligned} N\,[G] &= g^{ab}\,g^{c'd'}\,N_{abc'd'}\,[G] \\ &= -3\,G\,\xi\left\{\left(m^2+\frac{1}{2}\xi R\right)\,G_{;p'}{}^{p'}+\left(m^2+\frac{1}{2}\xi R'\right)\,G_{;p}{}^{p}\right\} \\ &\quad +\frac{9\,\xi^2}{2}\left\{G_{;p'}{}^{p'}\,G_{;p}{}^{p}+G\,G_{;p}{}^{p}{}_{p'}{}^{p'}\right\} \\ &\quad +\left(m^2+\frac{1}{2}\xi R\right)\left(m^2+\frac{1}{2}\xi R'\right)G^2 \\ &\quad +3\left(\frac{1}{6}-\xi\right)\left\{+3\left(\frac{1}{6}-\xi\right)G_{;p'p}\,G_{;}{}^{p'p}\right. \\ &\quad -3\xi\left(G_{;p}\,G_{;p'}{}^{pp'}+G_{;p'}\,G_{;p}{}^{pp'}\right) \\ &\quad \left.\left(m^2+\frac{1}{2}\xi R\right)G_{;p'}\,G_{;}{}^{p'}+\left(m^2+\frac{1}{2}\xi R'\right)G_{;p}\,G^{;p}\right\} \end{aligned} \tag{216}$$

For the massless conformal case, this reduces to

$$\begin{aligned} N\,[G] &= \frac{1}{144}\left\{RR'G^2-6G\,(R\Box'+R'\Box)\,G\right. \\ &\qquad \left.+18\,((\Box G)\,(\Box' G)+\Box'\Box G)\right\} \end{aligned} \tag{217}$$

which holds for any function $G(x,y)$. For G being the Green function, it satisfies the field equation (2):

$$\Box G=(m^2+\xi R)G \tag{218}$$

We will only assume the Green function satisfies the field equation in its first variable. Using the fact $\Box' R=0$ (because the covariant derivatives act at a different point than at which R is supported), it follows that

$$\Box'\Box G=(m^2+\xi R)\Box' G. \tag{219}$$

With these results, the noise kernel trace becomes

$$\begin{aligned} N\,[G] &= \frac{1}{2}\left(m^2\,(1-3\,\xi)+3\,R\,\left(\frac{1}{6}-\xi\right)\,\xi\right) \\ &\quad \times\left\{G^2\,\left(2\,m^2+R'\,\xi\right)+(1-6\,\xi)\,G_{;p'}\,G_{;}{}^{p'}-6\,G\,\xi\,G_{;p'}{}^{p'}\right\} \\ &\quad +\frac{1}{2}\left(\frac{1}{6}-\xi\right)\left\{3\,\left(2\,m^2+R'\,\xi\right)\,G_{;p}\,G^{;p}-18\,\xi\,G_{;p}\,G_{;p'}{}^{pp'}\right. \end{aligned}$$

$$+18\left(\frac{1}{6}-\xi\right)G_{;p'p}\,G_{;}{}^{p'p}\Big\}\qquad(220)$$

which vanishes for the massless conformal case. We have thus shown, based solely on the definition of the point separated noise kernel, there is no noise associated with the trace anomaly. This result obtained in Ref. [9] is completely general since it is assumed that the Green function is only satisfying the field equations in its first variable; an alternative proof of this result was given in Ref. [4]. This condition holds not just for the classical field case, but also for the regularized quantum case, where one does not expect the Green function to satisfy the field equation in both variables. One can see this result from the simple observation used in section 2: since the trace anomaly is known to be locally determined and quantum state independent, whereas the noise present in the quantum field is non-local, it is hard to find a noise associated with it. This general result is in agreement with previous findings [31, 30, 73], derived from the Feynman-Vernon influence functional formalism [76, 77] for some particular cases.

5.3. REGULARIZATION OF THE NOISE KERNEL

We pointed out earlier that field quantities defined at two separated points possess important information which could be the starting point for probes into possible extended structures of spacetime. Moving in the other (homeward) direction, it is of interest to see how fluctuations of energy momentum (loosely, noise) would enter in the ordinary (point-wise) quantum field theory in helping us to address a new set of issues such as a) whether the fluctuations to mean ratio can act as a criterion for the validity of semiclassical gravity. b) Whether the fluctuations in the vacuum energy density which drives inflationary cosmology violates the positive energy condition, c) How can we derive structure formation from quantum fluctuations, or d) General relativity as a low energy effective theory in the geometro- hydrodynamic limit [53, 3].

For these inquires we need to construct regularization procedures to remove the ultraviolet divergences in the coincidence limit. The goal is to obtain a finite expression for the noise kernel in this limit.

We can see from the point separated form of the stress tensor (198) what we need to regularize is the Green function $G^{(1)}(x,x')$. Once the Green function has been regularized such that it is smooth and has a well defined $x'\to x$ limit, the stress tensor will be well defined. In Minkowski space, this issue is easily resolved by a "normal ordering" prescription, which hinges on the existence of a unique vacuum. Unfortunately, for a general curved spacetime, there is no unique vacuum, and hence, no unique mode expansion on which to build a normal ordering prescription. But we can still ask if there is a way to determine a contribution we can subtract to yield a unique quantum stress tensor. Here we follow the prescription of Wald [109], and Adler *et. al.* [111] (with corrections [112]) summarized in [22]. We give a short synopsis below as it will be referred to in subsequent discussions.

The idea builds on the fact that for $G(x,x')_\omega = \langle\omega|\hat{\phi}(x)\hat{\phi}(x')|\omega\rangle$, the function

$$F(x,x') = G(x,x')_{\omega_1} - G(x,x')_{\omega_2} \tag{221}$$

is a smooth function of x and x', where ω_1 and ω_2 denote two different states. This means the difference between the stress tensor for two states is well defined for the point separation scheme, *i.e.*,

$$\mathcal{F}_{ab} = \frac{1}{2}\lim_{x'\to x} \mathcal{T}_{ab}\left(F(x,x') + F(x',x)\right) \tag{222}$$

is well defined. So a bi-distribution $G^L(x,x')$ might be useful for the vacuum subtraction. At first, it would seem unlikely we could find such a unique bi-distribution. Wald found that with the introduction of four axioms for the regularized stress tensor

$$\langle\hat{T}_{ab}(x)\rangle_{\text{ren}} = \lim_{x'\to x}\frac{1}{2}\mathcal{T}_{ab}\left(G^{(1)}(x,x') - G^L(x,x')\right) \tag{223}$$

$G^L(x,x')$ is uniquely determined, up to a local conserved curvature term. The Wald axioms are [112, 22]:

1. The difference between the stress tensor for two states should agree with (222);
2. The stress tensor should be local with respect to the state of the field;
3. For all states, the stress tensor is conserved: $\nabla^a\langle T_{ab}\rangle = 0$;
4. In Minkowski space, the result $\langle 0|T_{ab}|0\rangle = 0$ is recovered.

We are still left with the problem of determining the form of such a bi-distribution. Hadamard [117] showed that the Green function for a large class of states takes the form (in four spacetime dimensions)

$$G^L(x,x') = \frac{1}{8\pi^2}\left(\frac{2U(x,x')}{\sigma} + V(x,x')\log\sigma + W(x,x')\right) \tag{224}$$

with $U(x,x')$, $V(x,x')$ and $W(x,x')$ being smooth functions[5]. We refer to Eqn (224) as the "Hadamard ansatz".

[5]When working in the Lorentz sector of a field theory, *i.e.*, when the metric signature is $(-,+,+,+)$, as opposed to the Euclidean sector with the signature $(+,+,+,+)$, we must modify the above function to account for null geodesics, since $\sigma(x,x') = 0$ for null separated x and x'. This problem can be overcome by using $\sigma \to \sigma + 2i\epsilon(t-t') + \epsilon^2$. Here, we will work only with geometries that possess Euclidean sectors and carry out our analysis with Riemannian geometries and only at the end continue back to the Lorentzian geometry. Nonetheless, this presents no difficulty. At any point in the analysis the above replacement for σ can be performed.

Since the functions $V(x,x')$ and $W(x,x')$ are smooth functions, they can be expanded as

$$V(x,x') = \sum_{n=0}^{\infty} v_n(x,x')\sigma^n, \tag{225}$$

$$W(x,x') = \sum_{n=0}^{\infty} w_n(x,x')\sigma^n, \tag{226}$$

with the v_n's and w_n's themselves smooth functions. These functions and $U(x,x')$ are determined by substituting $G^L(x,x')$ in the wave equation $KG^L(x,x') = 0$ and equating to zero the coefficients of the explicitly appearing powers of σ^n and $\sigma^n \log\sigma$. Doing so, we get the infinite set of equations

$$\begin{aligned} U(x,x') &= \Delta^{1/2}; && (227)\\ 2H_0 v_0 + K\Delta^{1/2} &= 0; && (228)\\ 2nH_n v_n + Kv_{n-1} &= 0, \quad n \geq 1; && (229)\\ 2H_{2n} v_n + 2nH_n w_n + Kw_{n-1} &= 0, \quad n \geq 1 && (230) \end{aligned}$$

with

$$H_n = \sigma^{;p}\nabla_p + \left(n - 1 + \frac{1}{2}(\Box\sigma)\right). \tag{231}$$

From Eqs (230), the functions v_n are completely determined. In fact, they are symmetric functions, and hence $V(x,x')$ is a symmetric function of x and x'. On the other hand, the field equations only determine w_n, $n \geq 1$, leaving $w_0(x,x')$ completely arbitrary. This reflects the state dependence of the Hadamard form above. Moreover, even if $w_0(x,x')$ is chosen to be symmetric, this does not guarantee that $W(x,x')$ will be. By using axiom (4) $w_0(x,x') \equiv 0$. With this choice, the Minkowski spacetime limit is

$$G^L = \frac{1}{(2\pi)^2}\frac{1}{\sigma}, \tag{232}$$

where now $2\sigma = (t-t')^2 - (\mathbf{x} - \mathbf{x}')^2$ and this corresponds to the correct vacuum contribution that needs to be subtracted.

With this choice though, we are left with a $G^L(x,x')$ which is not symmetric and hence does not satisfy the field equation at x', for fixed x. Wald [112] showed this in turn implies axiom (3) is not satisfied. He resolved this problem by adding to the regularized stress tensor a term which cancels that which breaks the conservation of the old stress tensor:

$$\langle T_{ab}^{\text{new}}\rangle = \langle T_{ab}^{\text{old}}\rangle + \frac{1}{2(4\pi)^2} g_{ab}\,[v_1], \tag{233}$$

where $[v_1] = v_1(x,x)$ is the coincident limit of the $n = 1$ solution of Eq (230). This yields a stress tensor which satisfies all four axioms and produces the well known trace anomaly $\langle T_a{}^a \rangle = [v_1]/8\pi^2$. We can view this redefinition as taking place at the level of the stress tensor operator via

$$\hat{T}_{ab} \to \hat{T}_{ab} + \frac{\hat{1}}{2(4\pi)^2} g_{ab} [v_1] \tag{234}$$

Since this amounts to a constant shift of the stress tensor operator, it will have no effect on the noise kernel or fluctuations, as they are the variance about the mean. This is further supported by the fact that there is no noise associated with the trace. Since this result was derived by only assuming that the Green function satisfies the field equation in *one* of its variables, it is independent of the issue of the lack of symmetry in the Hadamard ansatz (224).

Using the above formalism we now derive the coincident limit expression for the noise kernel (5.2.2). To get a meaningful result, we must work with the regularization of the Wightman function, obtained by following the same procedure outlined above for the Hadamard function:

$$G_{\rm ren}(x,y) \equiv G_{\rm ren,+}(x,y) = G_+(x,y) - G^L(x,y) \tag{235}$$

In doing this, we assume the singular structure of the Wightman function is the same as that for the Hadamard function. In all applications, this is indeed the case. Moreover, when we compute the coincident limit of $N_{abc'd'}$, we will be working in the Euclidean section where there is no issue of operator ordering. For now we only consider spacetimes with no time dependence present in the final coincident limit result, so there is also no issue of Wick rotation back to a Minkowski signature. If this was the case, then care must be taken as to whether we are considering $[N_{abc'd'}[G_{\rm ren,+}(x,y)]]$ or $[N_{abc'd'}[G_{\rm ren,+}(y,x)]]$.

We now have all the information we need to compute the coincident limit of the noise kernel (5.2.2). Since the point separated noise kernel $N_{abc'd'}(x,y)$ involves covariant derivatives at the two points at which it has support, when we take the coincident limit we can use Synge's theorem (184) to move the derivatives acting at y to ones acting at x. Due to the long length of the noise kernel expression, we will only give an example by examining a single term.

Consider a typical term from the noise kernel functional (5.2.2):

$$G_{{\rm ren};c'b}\, G_{{\rm ren};d'a} + G_{{\rm ren};c'a}\, G_{{\rm ren};d'b} \tag{236}$$

Recall the noise kernel itself is related to the noise kernel functional via

$$N_{abc'd'} = N_{abc'd'}[G_{ren}(x,y)] + N_{abc'd'}[G_{ren}(y,x)]\,. \tag{237}$$

This is implemented on our typical term by adding to it the same term, but now with the roles of x and y reversed, so we have to consider

$$G_{\mathrm{ren};c'b}\,G_{\mathrm{ren};d'a}+G_{\mathrm{ren};a'd}\,G_{\mathrm{ren};b'c}+G_{\mathrm{ren};c'a}\,G_{\mathrm{ren};d'b}+G_{\mathrm{ren};a'c}\,G_{\mathrm{ren};b'd} \tag{238}$$

It is to this form that we can take the coincident limit:

$$\begin{aligned}&[G_{\mathrm{ren};c'b}]\,[G_{\mathrm{ren};d'a}]+[G_{\mathrm{ren};a'd}]\,[G_{\mathrm{ren};b'c}]\\&+[G_{\mathrm{ren};c'a}]\,[G_{\mathrm{ren};d'b}]+[G_{\mathrm{ren};a'c}]\,[G_{\mathrm{ren};b'd}]\end{aligned} \tag{239}$$

We can now apply Synge's theorem:

$$\begin{aligned}&([G_{\mathrm{ren};a}]_{;d}-[G_{\mathrm{ren};ad}])\,([G_{\mathrm{ren};b}]_{;c}-[G_{\mathrm{ren};bc}])\\&+([G_{\mathrm{ren};d}]_{;a}-[G_{\mathrm{ren};ad}])\,([G_{\mathrm{ren};c}]_{;b}-[G_{\mathrm{ren};bc}])\\&+([G_{\mathrm{ren};a}]_{;c}-[G_{\mathrm{ren};ac}])\,([G_{\mathrm{ren};b}]_{;d}-[G_{\mathrm{ren};bd}])\\&+([G_{\mathrm{ren};c}]_{;a}-[G_{\mathrm{ren};ac}])\,([G_{\mathrm{ren};d}]_{;b}-[G_{\mathrm{ren};bd}])\,.\end{aligned} \tag{240}$$

This is the desired form for once we have an end point expansion of G_{ren}, it will be straightforward to compute the above expression. The details of such an evaluation in the context of symbolic computations can be found in [115].

The final result for the coincident limit of the noise kernel is broken down into a rank four and rank two tensor and a scalar according to

$$[N_{abc'd'}]=N'_{abcd}+g_{ab}N''_{cd}+g_{cd}N''_{ab}+g_{ab}g_{cd}N'''. \tag{241}$$

The complete expression is given in Ref. [9].

5.4. SUMMARY STATEMENTS

5.4.1. *Further Developments*

In this section we showed how to obtain a general expression for the noise kernel, or the vacuum expectation value of the stress energy bi-tensor for a quantum scalar field in a general curved space time using the point separation method. The general form is expressed as products of covariant derivatives of the quantum field's Green function. It is finite when the noise kernel is evaluated for distinct pairs of points (and non-null points for a massless field). We also have shown the trace of the noise kernel vanishes for massless conformal fields, confirming there is no noise associated with the trace anomaly. This holds regardless of issues of regularization of the noise kernel.

The noise kernel as a two point function of the stress energy tensor diverges as the pair of points are brought together, representing the "standard" ultraviolet divergence present in the (point-defined) quantum field theory. By using the modified point separation regularization method we render the field's Green function finite in the coincident limit. This in turn permits the derivation of the formal expression for the regularized coincident limit of the noise kernel.

The general results obtained here are now applied by Phillips and Hu to compute the regularized noise kernel for three different groups of spacetimes: 1) ultrastatic metrics including the Einstein universe [62], hot flat space and optical Schwarzschild spacetimes. 2) Robertson-Walker universe and Schwarzschild black holes, from which structure formation from quantum fluctuations [40] and backreaction of Hawking radiation on the black hole spacetime [47] can be studied. 3) de Sitter and anti-de Sitter spacetimes: The former is necessary for scrutinizing primordial fluctuations in the cosmic background radiation while the latter is related to black hole phase transition and AdS/CFT issues in string theory.

When the Green function is available in closed analytic form, as is the case for optical metrics including hot flat space and Einstein Universe, conformally static spaces such as Robertson-Walker Unvierses, and maximally symmetric spaces such as de Sitter and Anti de Sitter , one can carry out an end point expansion according to (186), displaying the ultraviolet divergence. Subtraction of the Hadamard ansatz (224), expressed as a series expansion (5.3), will render this Green function finite in the coincident limit. With this, one can calculate the noise kernel for a variety of spacetimes.

5.4.2. *The Meaning of Regularization Revisited*

Thus far we have focussed on isolating and removing the divergences in the stress energy bitensor and the noise kernel in the coincident limit. In closing, we would like to redirect the reader's attention to the intrinsic values of the point-separated geometric quantities like the bitensor. As stressed earlier [3], in the stochastic gravity program, point separated expression of stress energy bi-tensor have fundamental physical meaning as it contains information on the fluctuations and correlation of quantum fields, and by consistency with the gravity sector, can provide a probe into the coherent properties of quantum spacetimes. Taking this view, we may also gain a new perspective on ordinary quantum field theory defined on single points: The coincidence limit depicts the low energy limit of the full quantum theory of matter and spacetimes. Ordinary (pointwise) quantum field theory, classical general relativity and semiclassical gravity are the lowest levels of approximations and should be viewed not as fundamental, but only as effective theories. As such, even the way how the conventional point-defined field theory emerges from the full theory when the two points (e.g., x and y in the noise kernel) are brought together is interesting. For example, one can ask if there will also be a quantum to classical transition in spacetime accompanying the coincident limit? Certain aspects like decoherence has been investigated before (see, e.g., [27]), but here the non-local structure of spacetime and their impact on quantum field theory become the central issue. (This may also be a relevant issue in noncommutative geometry). The point-wise limit of field theory of course has ultraviolet divergences and requires regularization. A new viewpoint towards regularization evolved from this perspective of treating conventional pointwise field theory as an effective theory in the coincident limit of the point-separated theory of extended spacetime.

To end this discussion, we venture one philosophical point we find resounding in these investigations reported here. It has to do with the meaning of a point-defined versus a point-separated field theory, the former we take as an effective theory coarse-grained from the latter, the point-separated theory reflecting a finer level of spacetime structure. It bears on the meaning of regularization, not just at the level of technical procedures, but related to finding an effective description and matching with physics observed at a coarser scale or lower energy.

In particular, we feel that finding a finite energy momentum tensor (and its fluctuations as we do here) which occupied the center of attention in the research of quantum field theory in curved spacetime in the 70's is only a small part of a much larger and richer structure of theories of fields and spacetimes . We come to understand that whatever regularization method one uses to get these finite parts in a point-wise field theory should not be viewed as universally imparting meaning beyond its specified function, i.e., to identify the divergent pieces and provide a prescription for their removal. We believe the extended structure of spacetime (e.g., via point-separation or smearing) and the field theory defined therein has its own much fuller meaning beyond just reproducing the well-recognized result in ordinary quantum field theory as we take the point-wise or coincident limit. In this way of thinking, the divergence- causing terms are only 'bad' when they are forced to a point-wise limit, because of our present inability to observe or resolve otherwise . If we accord them with the full right of existence beyond this limit, and acknowledge that their misbehavior is really due to our own inability to cope, we will be rewarded with the discovery of new physical phenomena and ideas of a more intricate world. (Maybe this is just another way to appreciate the already well-heeded paths of string theory.)

Acknowledgements

BLH wishes to thank the organizers of this course for the invitation to lecture and their warm hospitality at Erice. The materials presented here originated from several research papers, three of BLH with Nicholas Phillips and three of EV with Rosario Martín. We thank Drs. Martin and Phillips as well as Antonio Campos, Andrew Matacz, Sukanya Sinha, Tom Shiokawa, Albert Roura and Yuhong Zhang for fruitful collaboration and their cordial friendship since their Ph. D. days. We enjoy lively discussions with our long-time collaborator and dear friend Esteban Calzetta, who contributed greatly to the establishment of this field. We acknowledge useful discussions with Paul Anderson, Larry Ford, Ted Jacobson, Renaud Parentani and Raphael Sorkin. This work is supported in part by NSF grant PHY98-00967 and the CICYT Reserach Project No. AEN98-0431. EV also acknowledges support from the Spanish Ministery of Education under the FPU grant PR2000-0181 and the University of Maryland for hospitality.

References

1. B. L. Hu, Physica A **158**, 399 (1989).
2. E. Calzetta and B.L. Hu, Phys. Rev. D **49**, 6636 (1994); B. L. Hu and A. Matacz, Phys. Rev. D **51**, 1577 (1995); B. L. Hu and S. Sinha, Phys. Rev. D **51**, 1587 (1995); A. Campos and E. Verdaguer, Phys. Rev. D **53**, 1927 (1996); F. C. Lombardo and F. D. Mazzitelli, Phys. Rev. D **55**, 3889 (1997).
3. B. L. Hu, Int. J. Theor. Phys. **38**, 2987 (1999); gr-qc/9902064.
4. R. Martín and E. Verdaguer, Phys. Lett. B **465**, 113 (1999).
5. R. Martín and E. Verdaguer, Phys. Rev. D **60**, 084008 (1999).
6. R. Martín and E. Verdaguer, Phys. Rev. D **61**, 124024 (2000).
7. B. L. Hu and N. G. Phillips, Int. J. Theor. Phys. **39**, 1817 (2000); gr-qc/0004006.
8. N. G. Phillips and B. L. Hu, Phys. Rev. D **62**, 084017 (2000).
9. N. G. Phillips and B. L. Hu, Phys. Rev. D **63**, 104001 (2001).
10. See, e.g., E. B. Davies, *The Quantum Theory of Open Systems* (Academic Press, London, 1976); K. Lindenberg and B. J. West, *The Nonequilibrium Statistical Mechanics of Open and Closed Systems* (VCH Press, New York, 1990); U. Weiss, *Quantum Dissipative Systems* (World Scientific, Singapore, 1993)
11. R. P. Feynman and F. L. Vernon, Ann. Phys. (NY) **24**, 118 (1963); R. P. Feynman and A. R. Hibbs, *Quantum Mechanics and Path Integrals*, (McGraw-Hill, New York, 1965); A. O. Caldeira and A. J. Leggett, Physica **121A**, 587 (1983); Ann. Phys. (NY) **149**, 374 (1983); H. Grabert, P. Schramm and G. L. Ingold, Phys. Rep. **168**, 115 (1988); B. L. Hu, J. P. Paz and Y. Zhang, Phys. Rev. D **45**, 2843 (1992); *ibid.* **47**, 1576 (1993).
12. J. Schwinger, J. Math. Phys. **2** (1961) 407; P. M. Bakshi and K. T. Mahanthappa, J. Math. Phys. **4**, 1 (1963); **4**, 12 (1963); L. V. Keldysh, Zh. Eksp. Teor. Fiz. **47** , 1515 (1964) [Engl. trans. Sov. Phys. JEPT **20**, 1018 (1965)]; K. Chou, Z. Su, B. Hao and L. Yu, Phys. Rep. **118**, 1 (1985); Z. Su, L. Y. Chen, X. Yu and K. Chou, Phys. Rev. B **37**, 9810 (1988). E. Calzetta and B. L. Hu, Phys. Rev. D **40**, 656 (1989).
13. B. S. DeWitt, in *Quantum Concepts in Space and Time* ed. R. Penrose and C. J. Isham (Claredon Press, Oxford, 1986). R. D. Jordan, Phys. Rev. D **33**, 444 (1986); E. Calzetta and B. L. Hu, Phys. Rev. D **35**, 495 (1987); R. D. Jordan, Phys. Rev. D **36**, 3593 (1987); J. P. Paz, Phys. Rev. D **41**, 1054 (1990); A. Campos and E. Verdaguer, Phys. Rev. D **49**, 1861 (1994).
14. E. Calzetta and B. L. Hu, Phys. Rev. D **37**, 2878 (1988).
15. E. Calzetta and B. L. Hu, "Decoherence of Correlation Histories" in *Directions in General Relativity, Vol II: Brill Festschrift*, eds B. L. Hu and T. A. Jacobson (Cambridge University Press, Cambridge, 1993).

16. E. Calzetta and B. L. Hu, "Correlations, Decoherence, Disspation and Noise in Quantum Field Theory", in *Heat Kernel Techniques and Quantum Gravity*, ed. S. Fulling (Texas A& M Press, College Station 1995); hep-th/9501040.
17. E. Calzetta and B. L. Hu, Phys. Rev. D **61**, 025012 (2000).
18. C. W. Misner, K. S. Thorne and J. A. Wheeler, *Gravitation* (Freeman, San Francisco, 1973).
19. R. M. Wald, *General Relativity* (The University of Chicago Press, Chicago, 1984).
20. N. D. Birrell and P. C. W. Davies, *Quantum fields in curved space* (Cambridge University Press, Cambridge, England, 1982).
21. S. A. Fulling, *Aspects of quantum field theory in curved spacetime* (Cambridge University Press, Cambridge, England, 1989).
22. R. M. Wald, *Quantum field theory in curved spacetime and black hole thermodynamics* (Cambridge University Press, Cambridge, England, 1994).
23. A. A. Grib, S. G. Mamayev and V. M. Mostepanenko, *Vacuum Quantum Effects in Strong Fields* (Friedmann Laboratory Publishing, St. Petersburg, 1994).
24. Ya. Zel'dovich and A. Starobinsky, Zh. Eksp. Teor. Fiz **61**, 2161 (1971) [Sov. Phys.-JETP **34**, 1159 (1971)]; L. Grishchuk, Ann. N. Y. Acad. Sci. 302, 439 (1976); B. L. Hu and L. Parker, Phys. Lett. A **63**, 217 (1977); B. L. Hu and L. Parker, Phys. Rev. D **17**, 933 (1978); F. V. Fischetti, J. B. Hartle and B. L. Hu, Phys. Rev. D **20**, 1757 (1979); J. B. Hartle and B. L. Hu, Phys. Rev. D **20**, 1772 (1979); *ibid.* **21**, 2756 (1980); J. B. Hartle, Phys. Rev. D **23**, 2121 (1981); P. A. Anderson, Phys. Rev. D **28**, 271 (1983); *ibid.* **29**, 615 (1984).
25. W. H. Zurek, Phys. Rev. D **24**, 1516 (1981); D26, 1862 (1982); in *Frontiers of Nonequilibrium Statistical Physics*, ed. G. T. Moore and M. O. Scully (Plenum, N. Y., 1986); Physics Today **44**, 36 (1991); E. Joos and H. D. Zeh, Z. Phys. B **59**, 223 (1985); A. O. Caldeira and A. J. Leggett, Phys. Rev. A **31**, 1059 (1985); W. G. Unruh and W. H. Zurek, Phys. Rev. D **40**, 1071 (1989); B. L. Hu, J. P. Paz and Y. Zhang, Phys. Rev. D **45**, 2843 (1992); W. H. Zurek, Prog. Theor. Phys. **89**, 281 (1993); D. Giulini *et al*, *Decoherence and the Appearance of a Classical World in Quantum Theory* (Springer Verlag, Berlin, 1996); J. T. Whelan, Phys. Rev. D **57**, 768 (1998).
26. R. B. Griffiths, J. Stat. Phys. **36**, 219 (1984); R. Omnés, J. Stat Phys. **53**, 893 (1988); *ibid.* **53** 933 (1988); *ibid.* **53** 957 (1988); *ibid.* **54**, 357 (1988); Ann. Phys. (NY) **201**, 354 (1990); Rev. Mod. Phys. **64**, 339 (1992); *The Interpretation of Quantum Mechanics* (Princeton UP, Princeton, 1994); M. Gell-Mann and J. B. Hartle, in *Complexity, Entropy and the Physics of Information*, ed. by W. H. Zurek (Addison-Wesley, Reading, 1990); Phys. Rev. D **47**, 3345 (1993); J. B. Hartle, "Quantum Mechanics of Closed Systems" in *Directions in General Relativity* Vol. 1, eds B. L. Hu, M. P. Ryan and C. V. Vishveswara (Cambridge Univ., Cambridge, 1993); H. F. Dowker and J. J. Halliwell, Phys. Rev. D **46**, 1580 (1992); J. J. Halliwell, Phys. Rev. D **48**, 4785 (1993); *ibid.* **57**, 2337 (1998); T. Brun, Phys. Rev. D **47**, 3383 (1993); J. P. Paz and W. H. Zurek, Phys. Rev. D **48** 2728 (1993); J. Twamley, Phys. Rev. D **48**, 5730 (1993); C. J. Isham, J. Math. Phys. **35**, 2157 (1994); C. J. Isham and N. Linden, *ibid.* **35**, 5452 (1994); *ibid.* **36**, 5392 (1994); J. J. Halliwell, Ann. N.Y. Acad. Sc. **755**, 726 (1995); F. Dowker and A. Kent, Phys. Rev. Lett **75**, 3038 (1995); J. Stat. Phys. **82**, 1575 (1996); A. Kent, Phys. Rev. A **54**, 4670 (1996); Phys. Rev. Lett **78**, 2874 (1997); *ibid.* **81**, 1982 (1998); C. J. Isham, N. Linden, K. Savvidou and S. Schreckenberg, J. Math. Phys. **39**, 1818 (1998).
27. C. Kiefer, Clas. Quant. Grav. **4**, 1369 (1987); J. J. Halliwell, Phys. Rev. D **39**, 2912 (1989); T. Padmanabhan, *ibid.* 2924 (1989); B. L. Hu "Quantum and Statistical Effects in Superspace Cosmology" in *Quantum Mechanics in Curved Space time*, ed. J. Audretsch and V. de Sabbata (Plenum, London 1990); E. Calzetta, Class. Quan. Grav. **6**, L227 (1989); Phys. Rev. D **43**, 2498 (1991); J. P. Paz and S. Sinha, Phys. Rev. D **44**, 1038 (1991); *ibid* **45**, 2823 (1992); B. L. Hu, J. P.

Paz and S. Sinha, "Minisuperspace as a Quantum Open System" in *Directions in General Relativity* Vol. 1, (Misner Festschrift) eds B. L. Hu, M. P. Ryan and C.V. Vishveswara (Cambridge University Press, Cambridge, England, 1993).
28. E. Calzetta and B. L. Hu, Phys. Rev. D **35**, 495 (1987).
29. A. Campos and E. Verdaguer, Phys. Rev. D **49**, 1861 (1994).
30. B.L. Hu and S. Sinha, Phys. Rev. D **51**, 1587 (1995).
31. E. Calzetta and B.L. Hu, Phys. Rev. D **49**, 6636 (1994).
32. B.L. Hu, in *Proceedings of the Third International Workshop on Thermal Fields and its Applications*, CNRS Summer Institute, Banff, August 1993, edited by R. Kobes and G. Kunstatter (World Scientific, Singapore, 1994), gr-qc/9403061.
33. B. L. Hu and K. Shiokawa, Phys. Rev. D **57**, 3474 (1998).
34. L. H. Ford and N. F. Svaiter, Phys. Rev. D. **56**, 2226 (1997).
35. C.-H. Wu and L. H. Ford, Phys. Rev. D **60**, 104013 (1999).
36. R. Sorkin, "How Wrinkled is the Surface of a Black Hole?", in *Proceedings of the First Australasian Conference on General Relativity and Gravitation* February 1996, Adelaide, Australia, edited by David Wiltshire, pp. 163-174 (University of Adelaide, 1996); gr-qc/9701056; R. D. Sorkin and D. Sudarsky, Class. Quantum Grav. **16**, 3835 (1999).
37. C. Barrabès, V. Frolov and R. Parentani, Phys. Rev. D **62**, 044020 (2000).
38. S. Massar and R. Parentani, Nucl. Phys. B **575**, 333 (2000).
39. E. Calzetta and E. Verdaguer, Phys. Rev. D **59**, 083513 (1999); E. Calzetta, Int. J. Theor. Phys. **38**, 2755 (1999).
40. E. Calzetta and B. L. Hu, Phys. Rev. D **52**, 6770 (1995); A. Matacz, Phys. Rev. D **55**, 1860 (1997); E. Calzetta and S. Gonorazky, Phys. Rev. D **55**, 1812 (1997); A. Roura and E. Verdaguer, Int. J. Theor. Phys. **38**, 3123 (1999); *ibid.* **39**, 1831 (2000).
41. A. A. Starobinsky, Phys. Lett. B **91**, 99 (1980).
42. A. Vilenkin, Phys. Rev. D **32**, 2511 (1985).
43. S. W. Hawking, T. Hertog and H. S. Reall, Phys. Rev. D **63**, 083504 (2001).
44. B. L. Hu and E. Verdaguer, in preparation (2001).
45. P. Candelas and D. W. Sciama, Phys. Rev. Lett. **38**, 1372 (1977).
46. E. Mottola, Phys. Rev. D **33**, 2136 (1986).
47. B. L. Hu, A. Raval and S. Sinha, "Notes on Black Hole Fluctuations and Backreaction" in *Black Holes, Gravitational Radiation and the Universe: Essays in honor of C. V. Vishveshwara* eds. B. Iyer and B. Bhawal (Kluwer Academic Publishers, Dordrecht, 1998); gr-qc/9901010.
48. A. Campos and B. L. Hu, Phys. Rev. D **58** (1998) 125021; Int. J. Theor. Phys. 38 (1999) 1253.
49. R. Parentani, Phys. Rev. D **63**, 041503 (2001).
50. J. C. Niemeyer and R. Parentani, astro-ph/0101451.
51. S. Carlip, Phys. Rev. Lett. 79, 4071 (1997); Class. Quan. Grav. 15, 2629 (1998); L. J. Garay, Phys. Rev. Lett. **80**, 2508 (1998); Phys. Rev. D **58**, 124015 (1998); Int. J. Mod. Phys. A **14**, 4079 (1999).
52. K. Shiokawa, Phys. Rev. D **62**, 024002 (2000).
53. B. L. Hu, "General Relativity as Geometro-Hydrodynamics" Invited talk at the Second Sakharov Conference, Moscow, May, 1996; gr-qc/9607070.
54. B. L. Hu, "Semiclassical Gravity and Mesoscopic Physics" in *Quantum Classical Correspondence* eds. D. S. Feng and B. L. Hu (International Press, Boston, 1997); gr-qc/9511077.
55. B. L. Hu, Invited Talk at Peyresq 6 (June 2001), Int. J. Theor. Phys. (2002).
56. J.B. Hartle and G.T. Horowitz, Phys. Rev. D **24**, 257 (1981).
57. R. M. Wald, Commun. Math. Phys. **54**, 1 (1977).
58. S. M. Christensen, Phys. Rev. D **14**, 2490 (1976); *ibid.* **17**, 946 (1978).
59. T. S. Bunch, J. Phys. A **12**, 517 (1979).
60. L. H. Ford, Ann. Phys. (N.Y.) **144**, 238 (1982).

61. C.-I. Kuo and L. H. Ford, Phys. Rev. D **47**, 4510 (1993).
62. N. G. Phillips and B. L. Hu, Phys. Rev. D **55**, 6132 (1997).
63. W. H. Zurek, Physics Today **44**, 36 (1991); Prog. Theor. Phys. **81**, 281 (1993).
64. M. Gell-Mann and J. B. Hartle, Phys. Rev. D **47**, 3345 (1993).
65. J. B. Hartle, in *Gravitation and Quantizations*, proceedings of the 1992 Les Houches Summer School, edited by B. Julia and J. Zinn-Justin (North-Holland, Amsterdam, 1995); gr-qc/9304006, and references therein.
66. H. F. Dowker and J. J. Halliwell, Phys. Rev. D **46**, 1580 (1992).
67. J. J. Halliwell, Phys. Rev. D **48**, 4785 (1993); *ibid.* **57**, 2337 (1998).
68. J. T. Whelan, Phys. Rev. D **57**, 768 (1998).
69. R. Martín and E. Verdaguer, Int. J. Theor. Phys. **38**, 3049 (1999).
70. E. Calzetta, A. Roura and E. Verdaguer, quant-ph/0011097.
71. J. B. Hartle, Phys. Rev. Lett. **39**, 1373 (1977); M. Fischetti, J. B. Hartle and B. L. Hu, Phys. Rev. D **20**, 1757 (1979); J. B. Hartle and B. L. Hu, Phys. Rev. D **20**, 1772 (1979).
72. C. Greiner and B. Müller, Phys. Rev. D **55**, 1026 (1997).
73. A. Campos and E. Verdaguer, Phys. Rev. D **53**, 1927 (1996).
74. A. Campos and B. L. Hu, Phys. Rev. D **58**, 125021 (1998).
75. M. Morikawa, Phys. Rev. D **33**, 3607 (1986); D.-S. Lee and D. Boyanovsky, Nucl. Phys. B **406**, 631 (1993).
76. R. P. Feynman and F. L. Vernon, Ann. Phys. **24**, 118 (1963).
77. R. P. Feynman and A. R. Hibbs, *Quantum mechanics and path integrals* (McGraw-Hill, New York, 1965).
78. B. L. Hu, J. P. Paz and Y. Zhang, Phys. Rev. D **45**, 2843 (1992).
79. B. L. Hu and A. Matacz, Phys. Rev. D **49**, 6612 (1994).
80. J. F. Donoghue, Phys. Rev. Lett. **72**, 2996 (1994); Phys. Rev. D **50**, 3874 (1994); Helv. Phys.Acta **69**, 269 (1996); in *Advanced School on Effective Theories*, edited by F. Cornet and M. J. Herrero (World Scientific, Singapore, 1996), gr-qc/9512024; gr-qc/9712070.
81. S. Weinberg, *The Quantum Theory of Fields*, vols. I and II (Cambridge University Press, Cambridge, England, 1995 and 1996).
82. M. Gleiser and R. O. Ramos, Phys. Rev. D **50**, 2441 (1994); D. Boyanovsky, H. J. de Vega, R. Holman, D.S-Lee and A. Singh, Phys. Rev. D **51**, 4419 (1995); E. Calzetta and B. L. Hu, Phys. Rev. D **55**, 3536 (1997); M. Yamaguchi and J. Yokoyama, Phys. Rev. D **56**, 4544 (1997); S. A. Ramsey, B. L. Hu and A. M. Stylianopoulos, Phys. Rev. D **57**, 6003 (1998).
83. R. Kubo, J. Phys. Soc. (Japan) **12**, 570 (1957); in *Lectures in Theoretical Physics*, vol. I, proceedings of the 1958 Summer Institute for Theoretical Physics, University of Colorado, Boulder, edited by W. E. Brittin and L. G. Dunham (Interscience, New York, 1959); Rep. Prog. Phys. **29**, 255 (1966).
84. P. C. Martin and J. Schwinger, Phys. Rev. **115**, 1342 (1959); L. P. Kadanoff and P. C. Martin, Ann. Phys. **24**, 419 (1963); M. Plischke and B. Bergersen, *Equilibrium Statistical Physics*, 2nd ed. (World Scientific, Singapore, 1994).
85. A. Roura and E. Verdaguer, Phys. Rev. D **60**, 107503 (1999).
86. E. Calzetta, A. Campos and E. Verdaguer, Phys. Rev. D **56**, 2163 (1997); A. Campos and E. Verdaguer, Int. J. Theor. Phys. **36**, 2525 (1997).
87. É. É. Flanagan and R. M. Wald, Phys. Rev. D **54**, 6233 (1996).
88. F. C. Lombardo and F. D. Mazzitelli, Phys. Rev. D **55**, 3889 (1997).
89. D. S. Jones, *Generalised functions* (McGraw-Hill, New York, 1966).
90. L. Schwartz, *Théorie des distributions*, Tomes I et II (Hermann, Paris, 1957 and 1959); A. H. Zemanian, *Distribution Theory and Transform Analysis* (Dover, New York, 1987).
91. A. Campos, R. Martín and E. Verdaguer, Phys. Rev. D **52**, 4319 (1995).
92. G. T. Horowitz, Phys. Rev. D **21**, 1445 (1980).

93. G. T. Horowitz, in *Quantum Gravity 2: A Second Oxford Symposium*, edited by C. J. Isham, R. Penrose and D. W. Sciama (Clarendon Press, Oxford, 1981).
94. G. T. Horowitz and R. M. Wald, Phys. Rev. D **21**, 1462 (1980); *ibid.* **25**, 3408 (1982); A. A. Starobinsky, Pis'ma Zh. Eksp. Teor. Fiz. **34**, 460 (1981) [JEPT Lett. **34**, 438, (1981)].
95. R. D. Jordan, Phys. Rev. D **36**, 3593 (1987).
96. W. Tichy and É. É. Flanagan, Phys. Rev. D **58**, 124007 (1998).
97. G. T. Horowitz and R. M. Wald, Phys. Rev. D **17**, 414 (1978); S. Randjbar-Daemi, J. Phys. A **14**, L229 (1981); *ibid.* **15**, 2209 (1982); W.-M. Suen, Phys. Rev. Lett. **62**, 2217 (1989); Phys. Rev. D **40**, 315 (1989).
98. J. B. Hartle and G. T. Horowitz, Phys. Rev. D **24**, 257 (1981).
99. J. Z. Simon, Phys. Rev. D **43**, 3308 (1991); see also the discussion in W.-M. Suen, gr-qc/9210018.
100. E. Tomboulis, Phys. Lett. B **70**, 361 (1977).
101. H. B. Casimir, Proc. Kon. Ned. Akad. Wet. **51**, 793 (1948).
102. G. Barton, J. Phys. A. **24**, 991 (1991); *ibid.* **24** 5533 (1991).
103. See any textbook in quantum optics, e.g., D. F. Walls and G. J. Milburn, *Quantum Optics* (Springer-Verlag, Berlin, 1994); M. O. Scully and S. Zubairy, *Quantum Optics* (Cambridge University Press, Cambridge 1997); P. Meystre and M. Sargent III, *Elements of Quantum Optics* (Springer-Verlag, Berlin, 1990).
104. N. G. Phillips, Ph. D. Thesis, University of Maryland (1999)
105. B. L. Hu and D. J. O'Connor, Phys. Rev. D **36**, 1701 (1987).
106. J. A. Wheeler, Ann. Phys. (N. Y.) **2**, 604 (1957); *Geometrodynamics* (Academic Press, London, 1962); in Relativity, Groups and Topology, eds B. DeWitt and C. DeWitt (Gordon and Breach, New York, 1964).
107. S. W. Hawking, Nucl. Phys. B **144**, 349 (1978). S. W. Hawking, D. N. Page and C. N. Pope, Nucl. Phys. B **170** 283 (1980).
108. K. S. Thorne, *Black Holes and Time Warps* (Norton Books, 1994).
109. R. M. Wald, Commun. Math. Phys. **45**, 9 (1975).
110. C. H. Wu and L. H. Ford, gr-qc/0102063.
111. S. L. Adler, J. Lieberman and Y. J. Ng, Ann. Phys. (N.Y.) **106**, 279 (1977).
112. R. M. Wald, Phys. Rev. D **17**, 1477 (1977).
113. B. S. DeWitt, *Dynamical Theory of Groups and Fields* (Gordon and Breach, 1965).
114. B. S. DeWitt, Phys. Rep. **19C** 297 (1975).
115. N. G. Phillips, "Symbolic Computation of Higher Order Correlation Functions of Quantum Fields in Curved Spacetimes" (in preparation).
116. D. M. Capper and M. J. Duff, Nouvo Cimento **23A**, 173 (1974); M. J. Duff, *Quantum Gravity: An Oxford symposium*, ed. C. J. Isham, R. Penrose and D. W. Sciama (Oxford University Press, Oxford, 1975).
117. J. Hadamard, *Lectures on Cauchy's Problem in Linear Partial Differential Equations* (Yale University Press, New Haven, 1923).

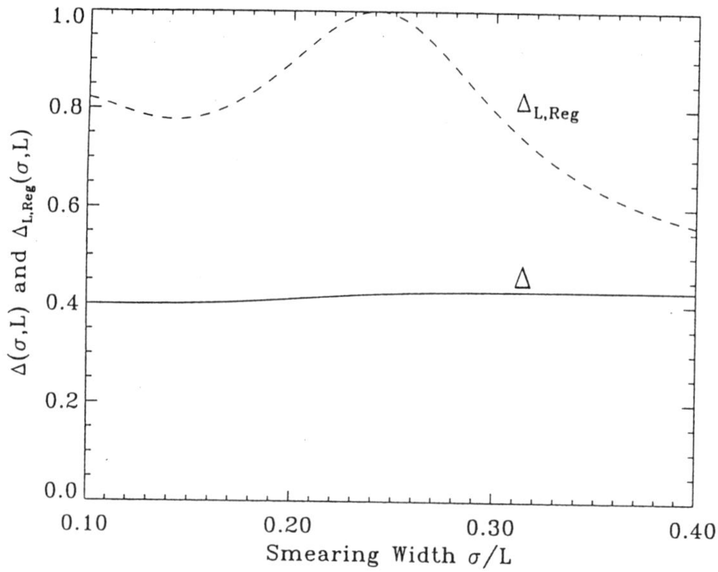

Figure 1. The dimensionless fluctuation measure $\Delta \equiv \left(\langle\hat{\rho}^2\rangle - \langle\hat{\rho}\rangle^2\right) / \langle\hat{\rho}^2\rangle$ for the Casimir topology, along with $\Delta_{L,\mathrm{Reg}}$. The topology is that of a flat three spatial dimension manifold with one periodic dimension of period $L = 1$. The smearing width σ represents the sampling width of the energy density operator $\hat{\rho}(\sigma)$. Δ is for the complete fluctuations, including divergent and cross terms, while $\Delta_{L,\mathrm{Reg}}$ is just for the finite parts of the mean energy density and fluctuations.

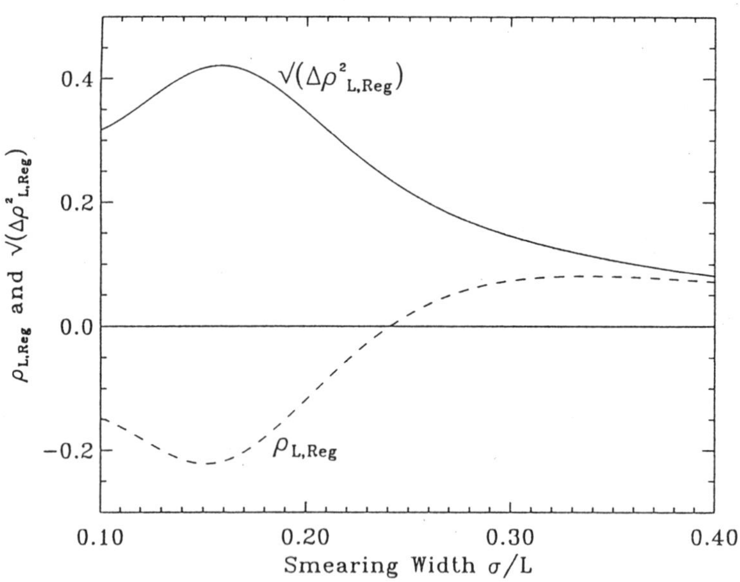

Figure 2. The finite parts of the mean energy density $\rho^{\mathrm{fin}}(\sigma, L)$ and the fluctuations $\Delta\rho^{\mathrm{fin}}(\sigma, L)$ for the Casimir topology, as a function of the smearing width.

QUANTUM COSMOLOGY: NEW RESULTS AND PROSPECTS

A.YU. KAMENSHCHIK
L.D. Landau Institute for Theoretical Physics of Russian Academy of Sciences
Kosygin str 2, 117334, Moscow, Russia
and
Landau Network - Centro Volta
Villa Olmo, via Cantoni 1, 22100, Como, Italy

Abstract. We consider some new results and problems of quantum cosmology. The lecture is concentrated about four topics: 1. How to describe the quantization of the geometry of spacetime and matter fields in a consistent way and what restrictions on the possible models of particle physics this quantization imposes. 2. How to define the wave function of the Universe and take into account the degrees of freedom responsible for the formation of the large scale structure of the Universe. 3. How to describe and estimate the chaos in isotropic cosmological models. 4. In what way the quantum birth of the Universe results in the modern classical world. What is the quantum decoherence and how arise classical properties of quantum objects.

Keywords: Quantum cosmology, quantization, wave function of the Universe, chaos, decoherence.

1. Introduction

Quantum cosmology occupies now the unique position in the modern theoretical physics being connected on one hand with observational cosmology and on the other hand with such theories as those of string and branes. To our mind one can separate at least three groups of motives explaining the importance of quantum cosmology.

1. Quantization of gravity is necessary for the construction of unified theory of elementrary particles and fundamental interactions. Quantum cosmology which treats the Universe as an unique quantum object

P.G. Bergmann and V. de Sabbata (eds.),
Advances in the Interplay Between Quantum and Gravity Physics, 219–242.

represents the most consistent and radical approach to the problem of quantization of gravity.

2. Quantum cosmology is connected with inflationary cosmology [1, 2] which has acquired observational status after the discovery of anisotropy of cosmic microwave background relic radiation $\frac{\delta T}{T}$ [3], predicted many years ago [4]. Here we would like to remind that the main idea of the inflationary cosmology consists in the suggestion that at the beginning of the cosmological evolution the Universe was expanding according to (quasi)- exponential law:

$$ds^2 = dt^2 - a^2(t)dl^2 \tag{1}$$

$$a(t) \sim \exp(Ht) \tag{2}$$

$$H = \frac{\dot{a}}{a}, \quad H^2 = \frac{\Lambda}{3}, \tag{3}$$

where dl^2 is a spatial distance, $a(t)$ is a cosmological radius, H is the Hubble constant, while Λ is an effective cosmological constant, generated by the scalar field, which is usually called inflaton. At the beginning of evolution inflaton is varying very slowly (slow-roll regime) and

$$\Lambda \sim \frac{1}{2}V(\Phi), \tag{4}$$

where $V(\Phi)$ is the potential of the inflaton Φ. Inflationary comology resolves successfully many traditional problems of the theory of the hot Universe, however, it is not able to predict the initial conditions for cosmological evolution. Prediction of an initial value of the field Φ, which in turn determines these initial conditions is one of the main problems of quantum cosmology.

3. Mathematical structure of quantum cosmology is keen to that of string and brane theories [5]. Indeed, all these theories (gravity, strings, branes) are theories with first-class constraints, possessing reparametrization invariance.

Let us now remind the basic notions of quantum cosmology. The Hilbert-Einstein action for the gravity looks like

$$S = \int d^{N+1}x\sqrt{-g}\left(\frac{1}{l_P^{N-1}}R + L_{matter}\right), \tag{5}$$

where N is the dimensionality of space, l_P is the Planck length.

Making the transition to the Hamiltonian formalism one has

$$S = \int dt(\pi^{ij}g_{ij} + p_a\dot{\varphi}^a - N^0H_\perp - \dot{N}^iH_i + boundary\ term, \tag{6}$$

where g_{ij} is a spatial metric, π^{ij} its conjugate momenta, φ and p are matter field and their conjugate momenta, N^0 and N^i are lapse and shift functions, while $H_\perp$ and H_i are super-Hamiltonian and supermomentum constraints correspondigly. We shall write down their explicit form:

$$H_\perp = l_P^{N-1} G_{ij,kl} \pi^{ij} \pi^{kl} - \frac{1}{l_P^{N-1}} \overset{(N)}{g} {}^{(N)}R$$
$$+H_{\perp matter}, \tag{7}$$

$$H_i = -2 g_{ij} \pi^{ij}_{,k} - (g_{ij,m} + g_{im,k} - g_{km,i}) \pi^{mk}$$
$$+H_{i\ matter}. \tag{8}$$

These constraints constitute the system of the first-class constraints, i.e. belong to the closed (pseudo) algebra in respect to Poisson brackets:

$$\{H_\perp(x), H_\perp(x')\} = g g^{ij} H_i(x) \frac{\partial}{\partial x^j} \delta(x, x')$$
$$-g g^{ij} H_i(x') \frac{\partial}{\partial x'^j} \delta(x, x'), \tag{9}$$

$$\{H_i(x), H_\perp(x')\} = H_\perp(x) \frac{\partial}{\partial x^j} \delta(x, x')$$
$$-H_\perp(x') \frac{\partial}{\partial x'^i} \delta(x, x'), \tag{10}$$

$$\{H_i(x), H_j(x')\} = H_j(x) \frac{\partial}{\partial x^i} \delta(x, x')$$
$$-H_i \frac{\partial}{\partial x'^j} \delta(x, x'). \tag{11}$$

The traditional approach to the construction quantum cosmology ascends to the Dirac quantization of systems with constraints [6] and consists in the requirement that the quantized constraints (7),(8) should eliminate the quantum state of the Universe $|\Psi\rangle$:

$$\hat{H}_\perp |\Psi\rangle = 0, \tag{12}$$

$$\hat{H}_i |\Psi\rangle = 0. \tag{13}$$

The first of these equations is called the Wheeler-DeWitt equation [7] and is the main equation of quantum cosmology. The second equation reflects the independence of the wave function of the Universe on the concrete choice of spatial coordinate system.

Now we are in a position to formulate the questions which seems to us to be very important for the development of quantum cosmology.

1. How to construct the wave function of the Universe?
2. Is the Dirac quantization of gravity and cosmology possible?
3. Is it possible to escape the cosmological singularity?
4. How can one explain the fact the Universe born as a quantum object has become the classical world which we observe today?

These questions motivate the choice of topics for this lecture which are the following ones:

1. Investigation of conditions of quantum consistency of the system of equations in quantum cosmology. The elaboration of the scheme of BRST quantization for closed Cosmological models. The critical relations between dimensionality of space and spectrum of matter fields for the simple cosmological model (see Ref. [8]).
2. Construction of cosmological wave function in one-loop approximation. Normalizability of wave function. Quantum scale of inflation and restrictions on the spectrum of matter fields (see Refs. [9, 10, 11, 12, 13, 14, 15, 16]).
3. Investigation of quantum and classical dynamics in cosmological models with a scalar field. Tunneling, quantum birth, nonsingular trajectories, chaos etc. (See Refs. [17, 18, 19, 20, 21, 22, 23]).
4. Decoherence and ultraviolet divergences in quantum cosmology. Classicalization of the quantum Universe (see Refs. [24, 25, 26]).

2. Hamiltonian BRST quantization of closed cosmological models

In this part of the talk we describe an attempt of the Hamiltonian BFV–BRST quantization of closed cosmological models by exploiting the keenship between the string and gravity gauge symmetries [8].

Apparently, a phase-space constrained description is very similar for string theory and gravity while they are being carried out locally with dynamical variables and constraints defined at a point of a space-like section of world-sheet or spacetime, respectively. However, there is an essential distinction in habitual approaches to quantization of strings and gravity.

Quantizing strings, one may work with a discrete set of modes, describing string excitations, and a discrete set of constraints, which form Virasoro algebra. In doing so, one observes that quantum commutators of the constraints acquire central extension, and the quantum consistency of the theory is provided for the critical values of dimensionality and intercept. One of the most efficient methods to observe this phenomenon is the BFV-BRST Hamiltonian quantization of constrained systems [27, 28, 29]. In the framework of this method one can show how the nilpotency condition for the BRST charge defines the values of critical parameters for strings [30], membranes and W_3 gravity. Actually, the use of discrete set of modes and constraints is merely a convenient tool for taking into account the topology of the compact world-sheet spacelike section, when the method is applied.

Canonical quantization of gravity and cosmology is usually treated by either perturbative consideration of path integral or operatorial quantization according to Dirac scheme with constraints imposed onto physical states. Basically, such a treatment implements continuous basis of dynamical modes and constraints, which can satisfactory describe the dynamics locally but are hardly well-defined globally on compact spacelike sections of the spacetime. Meanwhile, the consideration of cosmological perturbations in terms of globally defined harmonics was used for the investigation of different problems in classical and quantum cosmology. However, a globally defined discrete constraint basis for quantum cosmology of closed Universes has not been hitherto discussed. Consistent canonical quantization requires both the dynamical modes and constraints to be globally defined, since this is essential for the adjustment of the ordering of the modes to the quantum algebra of the constraints. For example, in the case of the Wick ordering of bosonic string modes, the quantum corrections to the constraint commutators give rise to the central extension of the Virasoro algebra representation in the Fock space. Then the quantum constraint algebra can be made consistent provided the proper basis of constraints is chosen (L and $\bar{L}$ are employed instead of habitual for cosmologists lapse and shift generators $H_\perp$, $H_{||}$ and moreover, only one half of L_n and $\bar{L}_m$ should annihilate physical states). At the mean time, the straightforward pursuing the Dirac scheme, with $H_\perp$ and $H_{||}$ being imposed on the physical subspace, leads to a contradiction.

Here, we suggest a new scheme for canonical quantization of the closed cosmological models. The distinguishing features of this treatment are as follows: the decomposition of both dynamical variables and constraints into harmonics which are the eigenfunctions of the Laplace operator for the maximally symmetric space of given topology; separation of subalgebra of area-preserving diffeomorphisms from a total set of gravity constraints whereas the rest of the constraints form Virasoro-like generators; quantization of this constrained theory is performed in the framework of the Hamiltonian BFV-BRST formalism with due regard to the ordering of gravity, matter and ghost harmonics; studying the quantum nilpotency condition for the BRST operator, we apply a certain perturbative expansion of the constraints and structure functions with a small parameter $l_P/V^{1/N}$, where l_P is a Planck length and V is a spatial volume of the Universe, N is a dimensionality of the space. Implementing this scheme, we focus the consideration mostly at the stationary Bianchi-I type Universe, although any other closed cosmological model could be treated in a similar way. The specificity of a particular topology is encoded in the constraint algebra structure functions which appear to be expressed via Clebsch–Gordan coefficients of the corresponding symmetry group representation. In the case of Bianchi-I cosmology, these coefficients are the simplest, as the group is $U(1)^N$. Although we take this particular type of a closed cosmological model for the sake of technical simplicity, it is, however, of a certain physical interest. On the other hand, N–torus allows to exploit a straightforward analogy with the closed bosonic string sigma–model when the nilpotency of the quantum BRST charge is examined. The most striking outcome of this quantization procedure, being applied to the Einstein gravity coupled to the matter fields, is that the number of the matter degrees of freedom appears to be correlated with the dimensionality of the space N. For example, when there are d massless scalar fields only, already the first quantum correction gives rise to the relation between d and N:

$$d = 30 + \frac{5}{2}(N+1)(N-2), \tag{14}$$

which is a necessary condition for nilpotency of the quantum BRST charge. Mention the curious fact: if one puts $N = 1$ (thereby torus reduces to a circle and the constraints form the Virasoro algebra) then relation (14) results in $d = 25$. This result can be naturally understood from the standpoint of the string sigma–model without Weyl invariance [31]. As a matter of fact, bosonic string theory should not be thought about as one-dimensional limit of the Einstein gravity coupled to a set of d scalar fields, because string possesses an extra gauge symmetry, namely Weyl invariance, besides the diffeomorphisms, whereas the mentioned σ-model does not have Weyl invariance. As is known [31] the critical dimensionality of the σ-model is 25, in contrast with 26 for strings. The discrepancy in these dimensionalities can be easily explained if one remembers that Weyl invariance gauges out conformal mode of 1+1-metric, while in the non-conformal σ-model case this mode contributes to the Virasoro generators on an equal footing with string excitations [31]. Thus, in some sense, the critical dimensionality for the both cases is 26, but for the σ-model this number consists of 25 string coordinates and 1 gravity conformal mode.

In its basic features the suggested scheme should, probably, be actual for other closed cosmological models, including those with non- stationary classical background. But the critical relations analogous to (1) may change their form for each particular model.

Now, we give some comments concerning the most peculiar in our opinion features of proposed approach.

First of all, we would like to describe the structure of vacuum for graviton modes and the construction of Fock space for them. While the choice of creation and annihilation operators for scalar field is trivial and is fixed by the natural requirement of diagonal form of Hamiltonian an analogous choice for gravitons is more complicated. Let us write down the corresponding expressions for metric g_{ij} and the conjugate momenta π^{ij} (see for detail [8]):

$$\begin{aligned} g_{ab}(\vec{x}) &= \sqrt{\frac{2l_P^{N-1}}{V}} g_{ab}(0) \\ &+ \sum_{\vec{k}} \sqrt{\frac{2l_P^{N-1}}{V}} \sqrt{\frac{1}{2\omega_{\vec{k}}}} (a_{ab}(\vec{k}) \exp(-i\vec{k}\vec{x}) \\ &+ a_{ab}^{+}(\vec{k}) \exp(i\vec{k}\vec{x}) \\ &+ \bar{a}_{ab}(\vec{k}) \exp(i\vec{k}\vec{x}) \\ &+ \bar{a}_{ab}^{+}(\vec{k}) \exp(-i\vec{k}\vec{x})), \end{aligned} \tag{15}$$

$$\begin{aligned} \pi^{cd}(\vec{x}) &= \sqrt{\frac{1}{2l_P^{N-1}V}} \pi^{cd}(0) \\ &+ \sum_{\vec{k}} iG^{ab,cd} \sqrt{\frac{1}{2l_P^{N-1}V}} \sqrt{\frac{\omega_{\vec{k}}}{2}} (-a_{ab}(\vec{k}) \exp(-i\vec{k}\vec{x}) \\ &+ a_{ab}^{+}(\vec{k}) \exp(i\vec{k}\vec{x}) - \bar{a}_{ab}(\vec{k}) \exp(i\vec{k}\vec{x}) \\ &+ \bar{a}_{ab}^{+}(\vec{k}) \exp(-i\vec{k}\vec{x})). \end{aligned} \tag{16}$$

It is the choice of creation and annihilation operators which provides the diagonal form for the part of Hamiltonian $H_{\perp}(\vec{0})$ describing physical degrees of freedom for gravitons, i.e. transverse-traceless modes. The presence of Planck length l_P provides us with a proper perturbative expansion of the super-Hamiltonian constraint in the powers of small dimensionless parameter equal to the ratio of this length to the linear size of the Universe. It is important to stress that while the choice of coefficients in Eqs. (2)–(3) is motivated by "multi-dimensional" considerations it guarantees the existence of the correct $N = 1$-dimensional limit $d = 25$ described above.

Implementation of formulae (15) and (16) gives us an opportunity to calculate the contribution into central extension of the commutator

$$[L(\vec{n}), L(-\vec{n})], \tag{17}$$

given by gravitons. This contribution has the form

$$-\frac{1}{12}\left(4 + \frac{5}{2}(N+1)(N-2)\right). \tag{18}$$

It is easy to see that the number

$$\frac{(N+1)(N-2)}{2} \tag{19}$$

is the number of physical degrees of freedom (transverse-traceless gravitons) in the spacetime of the corresponding dimensionality. The appearance of such structure looks quite natural and reminds the corresponding contribution for scalar field but differs from it by the coefficient (-5) which need the explanation. Besides the in (18) we have the constant term which is independent of dimensionality N and whose explanation is still absent as well.

Another important feature of the presented approach which should survive at the consideration of other cosmological models is connected with the choice of the discrete basis of constraints and the rules of quantization for the corresponding ghosts. Instead of consideration of traditional for gravity and cosmology super-Hamiltonian $H_\perp$ and supermomenta H_i constraints we use their linear combinations

$$L = \frac{1}{2}(H_\perp + H_{||}) \tag{20}$$

and

$$\bar{L} = \frac{1}{2}(H_\perp - H_{||}) \tag{21}$$

which due to their analogy with the constraints in the theory of closed bosonic string can be called "Virasoro-like" generators. In Eqs. (20)–(21) $H_{||}$ denotes longitudinal component of supermomentum. The rest of components of supermomentum constitute the subalgebra of area-preserving diffeomorphisms which are well-known in p-brane theories.

The next step consists in the choice of ordering for the ghosts corresponding to Virasoro-like generators and area-preserving diffeomorphisms. Pursuing the analogy with string one can require ghosts corresponding to Virasoro-like generators obey Wick quantization, while those corresponding to area-preserving diffeomorphisms obey Weyl quantization rule. Namely this choice of quantization allows to write down the quantum BRST-generator nilpotency condition in a treatable (though rather a cumbersome) form and comparing the contribution of ghosts into the central extension of commutator (17) with that of scalar and graviton fields and to come to formula (14).

Formula (1) can be generalized for the model including massless vector and spinor fields. The corresponding critical relation looks as follows:

$$\begin{aligned} & d + d_V(N-1) + d_F 2^{[(N-1)/2-1]} \\ & = 30 + \frac{5}{2}(N-2)(N+1). \end{aligned} \tag{22}$$

Thus, we may claim that this short relation eventually represents a necessary condition for a nilpotency of the BRST charge in the first quantum approximation. As is seen it imposes some correlations between number of scalars d, vectors d_V, spinors d_F and dimensionality of the space. Although, the restrictions imposed by Eq. (22) do not predict the unique spectrum and spatial dimensionality, one can observe some curious consequences, for example: stationary Bianchi-I closed cosmology is inconsistent without matter at any positive integer N. Notice, also, that in $N = 3$ case, the admissible matter spectra do not contain sufficient number of degrees of freedom to be compatible with the Standard Model or its generalizations. It is not too disappointing, because the stationary closed Bianchi-I model itself could hardly be expected to describe the observable cosmology. On the other hand, the very existence of such a relation allows to hope that analogous conditions, being deduced for more realistic models (with non-stationarity, other topology, massive particles and, perhaps, with an extended symmetry), can display a better correspondence with the realistic particle physics.

One more (a little bit speculative) way of fine-tuning the number of matter degrees of freedom to the dimensionality of space could be found in a generalized Kaluza-Klein ideology. In this way, Eq. (22) can be thought of as valid for the fundamental multidimensional theory. That provides much more opportunities for the matter spectrum in low-dimensional effective theory by means of a proper compactification.

While an opportunity of interplay between matter content of the Universe and its geometrical characteristics is of certain interest, we would like to think that the designed scheme of Hamiltonian treatment and quantization of closed cosmological models may open some other promising prospects for studying various aspects of quantum cosmology.

3. Wave function of the Universe in the one-loop approximation

The most common approach to the solution of the Wheeler-DeWitt equation (12) and the construction of the wave function of the Universe is based on considering the so-called mini-superspaces which only involve a finite number of cosmological degrees of freedom, the rest being "frozen". Another useful trick consists in the semiclassical expansion of the functional integral formally representing the solution of the Wheeler-DeWitt equation. This approach has been actively pursued since the early 1980s, when two ideologically close prescriptions for constructiting the wave function of the Universe were proposed [32, 33, 34, 35, 36, 37, 38, 39]. Both approaches draw an analogy between the birth of the Universe "from nothing" And the processes of tunneling in quantum mechanics and quantum field theory, treated in the semiclassical approximation. The probabilities of such processes can be evaluated by calculating the Euclidean action on the instantons - the solutions of the Euclidean equations of motion for the systems in question. The so-called "no-boundary" prescription for the wave function of the Universe [32, 33] can be formulated in terms of path integral

$$\Psi_{NB}({}^{(3)}g, \varphi) = \int DgD\varphi \exp(-I[{}^{(4)}g, \varphi], \tag{23}$$

where the path integral is taken on all the compact four-geometries and matter -field configurations matching the three geometry ${}^{(3)}g$ and the matter-field configuration φ on it. In other words one can describe the wave function (23) as an object representing the transition from the Euclidean "embrional" state of the Universe with the metric

$$ds^2 = d\tau^2 + a^2(\tau)d^2\Omega^{(3)},\ a(\tau) = \frac{1}{H}\sin H\tau \tag{24}$$

to the Lorentzian DeSitter Universe

$$ds^2 = -dt^2 + a^2(t)d^2\Omega^{(3)},\ a(t) = \frac{1}{H}\cosh Ht \tag{25}$$

by means of the analytical continuation of time

$$\tau = \frac{\pi}{2H} + it. \tag{26}$$

In the tree-level approximation the path integral (23) is reduced to

$$\Psi_{NB} \sim \exp(-I). \tag{27}$$

Notice that for gravity the classical Euclidean action I is negative and the obtained wave function of the Universe is contro - intuitive one. Indeed, it is not only non normalizable but also predicts that the birth of the Universe with a large initial radius is more probable than the birth of the Universe with a small one.

The tunneling wave function prescription formulated in the most consistent way in the papers [34, 35, 36] fixes the boundary conditions for the wave function of the Universe requiring the absence of ingoing wave after the underbarrier exit of the Universe. In the tree-level approximation it gives

$$\Psi_T \sim \exp(+I). \tag{28}$$

This function is less contro-intuitive predicting the small values of initial cosmological radius but still non normalizable.

In the series of papers [9, 10, 11, 12, 13, 14, 15, 16] the wave function of the Universe was considered in one-loop approximation and with inclusion of inhomogeneous degrees of freedom. Consideration of inhomogeneous degrees of freedom in classical cosmology has rather a long history (see [40]) and was applied also to quantum cosmology [41, 42]. We have used unitarity approach to quantum cosmology, which allows to construct the corresponding Hilbert space for the quantum state of the Universe [43]. In this case the wave function of the Universe can be written in the form

$$\Psi_T^{NB}(a, \Phi, f_i) = \exp\left(\mp I_0 - W_{1-loop} - \frac{1}{2} f_i \frac{\dot{v}_i}{v_i} f_i\right), \tag{29}$$

where I_0 is the classical Euclidean action for homogeneous modes (i.e. for the cosmological radius a and the inflaton scalar field Φ), f_i are the amplitudes of inhomogeneous modes, v_i - solutions of linearized equations of motion for inhomogeneous modes, and W_{1-loop} is a one-loop correction to the effective action. One can consider now the transition to Lorentzian time and to show that probability distribution for an initial value of inflaton scalar field Φ, which in turn determines an initial radius of the Universe is given by the expression

$$P_T^{NB} \sim \frac{\exp \mp 2I_0 - \Gamma_{1-loop}}{H^2(\Phi)}. \tag{30}$$

Here Γ_{1-loop} is a one-loop effective action on a closed (quasi) DeSitter instanton. For the limit of small cosmological raiduses (or large values of the inflaton scalar field Φ) one can write

$$\Gamma_{1-loop}|_{\Phi\to\infty} \sim Z \ln \frac{H}{\mu}, \tag{31}$$

where Z is an anomalous scaling of the theory under cosnideration depending on its field content, and μ is a massive constant reflecting the renormalization ambiguity. The most convenient way of calculation of the anomalous scaling is the implementation of the Schwinger-DeWitt expansion [44, 45] and the well.known fact that the anomalosous scaling is equal to thr second coefficient of this expansion

$$Z = A_2. \tag{32}$$

Analysing the expressions (30),(31) it is easy to come to the criterium of the normalizability of the wave function of the Universe in one-loop approximation [9] which reads as

$$Z > -1. \tag{33}$$

This criterium provides a tool for the selection between different particle physics models from the point of view of the possibility to have a normalizable wave function of the Universe. The simple analysis [10, 15] shows that the models possessing a large number of scalar fields are preferable from this point of view.

It was shown also [11, 12, 13, 14, 16] that in the model with a strong nonminimal coupling between inflaton and gravity [46, 47, 48]

$$\frac{1}{2}|\xi|R\Phi^2 \tag{34}$$

there is a probablity peak for the tunneling wave function of the Universe, while for the no-boundary wave function of the Universe such a peak does not exists. It is interesting that the scale at which this probability peak arises is 10^5 less than the Planck scale and thus coincides with the supposed Grand Unification scale.

Let us add also that in the model with a strong nonminimal coupling arise rather rigid restrictions on the constants of interaction between the inflaton scalar field and other matter fiels.

4. Classical and quantum dynamics of cosmological models with a scalar field

In the preceding section we have shown that the consideration of the wave fucntion of the Universe in one-loop approximation allows one to construct the normalizable wave function and to predict the most probable initial conditions for the inflation. It is interesting that using a little bit more complicated models with a scalar field one can get the probability peaks for initial conditions for cosmological evolution estimating the wave function of the Universe already in tree-level approximation. One of such models is the model with a complex scalar field. The action of this models can be written as follows:

$$S = \int d^4x\sqrt{-g}\left(\frac{m_P^2}{16\pi}(R-2\Lambda) + \frac{1}{2}g^{\mu\nu}\Phi^*_{,\nu}\Phi_{,\mu}\right.$$
$$\left. +\frac{1}{2}\xi R\Phi^*\Phi - \frac{1}{2}m^2\Phi^*\Phi - \frac{1}{4!}\lambda(\Phi^*\Phi)^2\right). \tag{35}$$

If one introduces the notation

$$\Phi = x\exp(i\theta) \tag{36}$$

one can see that exists a conserved charge

$$Q = a^3x^2\dot{\theta}, \tag{37}$$

which in turn implies the appearance the centrifugal term in the potential:

$$\frac{Q^2}{2a^3x^2}. \tag{38}$$

The presence of the centrifugal term (38) implies the compactification of the Euclidean or classically forbidden region for the model under consideration which in turn garanties the presence of instanton solutions [49]. In papers [50, 51, 52] the cosmological model with a minimally coupled complex scalar field was constructed and it was shown that the Euclidean region in the minisuperspace parametrized by two parameters a and x has the closed banana-like form. It was shown that the instanton solutions correspond to existence of the peaks in the probability conditions for inflation in the no-boundary prescription for the wave function of the Universe.

In the papers [17, 18, 22, 23] the general cosmological model with a complex scalar field non-minimally coupled to gravity was considered and it was shown that manipulating by parameters of the model one can create different types of Euclidean regions. They can be connected and disconnected, can have stocking-like or bag-like form and can have different number of instantons (0,1 or 2). It is interesting that some of these instantons describe peaks in the probability distributions generated by a no-boundary wave fucntion of the Universe while other correspond to peaks following from the tunneling wave fucntion.

Speaking about the cosmological models with a scalar field, it is worth noticing that already the simplest isotropic cosmological models with a massive minimally coupled real scalar field has a very interesting classical dynamics [53]. Especially interesting is this dynamics for a closed model because in this case there is possibility of bouncing and the Universe, contracting can escape singularity [53, 54]. The studying of dynamics of closed isotropic cosmological model with a scalar field has rather a long story. First, it was noticed that in such a model with a minimally coupled massive scalar field there is opportunity to escape singularity at contraction [55, 56]. Then the periodical trajectories escaping singularity were studied [57]. In paper [54] it was shown that the set of infinitely bouncing aperiodical trajectories has a fractal nature. Later this result in other terms was reproduced in our papers [19, 20, 22, 23].

Here we would like to describe briefly the approach presented in [19]. The main idea consists in the fact that in the closed isotropic model with a minimally coupled massive scalar field all the trajectories have the point of maximal expansion. The localization of points of maximal expansion on the configuration plane (a, ϕ), where a is a cosmological radius, while ϕ is a scalar field could be found analytically. Then the trajectories could be classified according to localization of their points of maximal expansion, which is equivalent to a particular choice of a Poincare section for our dynamical system. The area of points of maximal expansion is located inside the Euclidean region. Numerical investigation show that this area has a quasiperiodical structure, zones corresponding to the falling to singularity are intermingled with zones in which are placed points of maximal expansion of trajectories having the so called "bounce" or a point of minimal contraction. Then studying the substracture of these zones from the point of view of possibility to have two bounces one can see that this substructure repeats on the qualitative level the structure of the whole region of the possible points of maximal expansion. Continuing this procedure *ad infinitum* one can observe that the field of the localization of the points of maximal expansion corresponding to trajectories escaping singularity can be found as a result of an infinite procedure at each stage of which we encounter self-similar structures. Such a self-similarity of structures appearing at different scales points out on the fractal nature of the set obtained as the result of infinite procedure [58]. Thus, while the set of trajectories escaping singularity and infinitely oscillating between points of minimal contraction and maximal expansion has vanishing measure in the set of all possible trajectories it can at the same time have a non-trivial fractal dimensionality. Unfortunately, it is difficult to calculate it explicitly and it is necessary to use other methods for quantitative characterization of chaos. One of them is connected with the so called topological entropy.

The Poincare section described above gives us an opportunity to see that there is also the set of periodical trajectories. All these periodical trajectories contain bounces intermingled with series of oscillations of the value of scalar field ϕ. It is important that there is no restrictions on the lengths of series of oscillations in this case. In paper by Cornish and Shellard [59] the topological entropy was calculated for this case and it was shown that it is positive, that signifies that the dynamics is chaotic.

The topological entropy measures the growth of the number of periodic trajectories as their period increases. One can do it using the so called symbolic dynamics. In the case of isotropic cosmological models in the minisuperspace this symbolic dynamics has a very simple form. To characterize any periodic trajectory it is enough to introduce two symbols (letters): one, corresponding to a bounce and another corresponding to changing the sign of the Scalar field or, in other words, to crossing the line $\phi = 0$. When the finite sequence of these two letters corresponds to a certain periodical trajectory and vice versa [59]. The topological entropy is defined as

$$H_T = \lim_{k \to \infty} \frac{1}{k} \ln N(k), \tag{39}$$

where $N(k)$ is a number of periodic orbits of length k. For the simplest cosmological model this quantity was calculated in the paper [59] and it was shown that was equal to

$$H_{T_0} = \ln\left(\frac{1+\sqrt{5}}{2}\right). \tag{40}$$

At the calculation of expression (40) was used the fact that two bounces cannot follow each other without oscillations of the scalar field between them (a prohibition rule).

In our paper [21] the technique of the calculation of the topological entropy was generalized for the models containing together with a scalar field also hydrodynamical matter. In this case, the form of the Euclidean region is changed and the rules of intermingling between bounces and oscillations of the scalar field can be rather complicated. Nevertheless, it is possible for any concrete set of rules of construction of periodic trajectories to write a system of recurrence relations, which gives an opportunity to get an algebraic equation, the logarithm of the biggest root of which gives the value of the topological entropy.

Here, we shall give as an example one particular algebraic equation, which corresponds to the model where the mean number of oscillations between two bounces is equal to ten and the maximal number of oscillations between two bounces is 19. The equation for the calculation of the topological entropy is the following one:

$$\begin{aligned}
&x^{209} - x^{207} - x^{206} - x^{205} - x^{204} - x^{203} - x^{202} - x^{201} - x^{200}\\
&-x^{199} - x^{198} - x^{195} - 2x^{194} - 3x^{193} - 4x^{192} - 5x^{191} - 6x^{190}\\
&-7x^{189} - 8x^{188} - 9x^{187} + 9x^{184} + 16x^{183} + 21x^{182} + 24x^{181}\\
&+25x^{180} + 24x^{179} + 21x^{178} + 16x^{177} + 9x^{176} + 8x^{172} + 21x^{171}\\
&+36x^{170} + 50x^{169} + 60x^{168} + 63x^{167} + 56x^{166} + 36x^{165} - 36x^{161}\\
&-84x^{160} - 126x^{159} - 150x^{158} - 150x^{157} - 126x^{156} - 84x^{155}\\
&-36x^{154} - 28x^{149} - 84x^{148} - 150x^{147} - 200x^{146} - 210x^{145}\\
&-168x^{144} - 84x^{143} + 84x^{142} + 224x^{137} + 350x^{136} + 400x^{135}\\
&+350x^{134} + 224x^{133} + 84x^{132} + 56x^{126} + 175x^{125} + 300x^{124}\\
&+350x^{123} + 280x^{122} + 126x^{121} - 126x^{115} - 350x^{114} - 525x^{113}\\
&-525x^{112} - 350x^{111} - 126x^{110} - 70x^{103} - 210x^{102} - 315x^{101}\\
&-280x^{100} - 126x^{99} + 126x^{92} + 336x^{91} + 441x^{90} + 336x^{89}\\
&+126x^{88} + 56x^{80} + 147x^{79} + 168x^{78} + 84x^{77} - 84x^{69}\\
&-196x^{68} - 196x^{67} - 84x^{66} - 28x^{57} - 56x^{56} - 36x^{55}\\
&+36x^{46} + 64x^{45} + 36x^{44} + 8x^{34} + 9x^{33} - 9x^{23}\\
&-9x^{22} - x^{11} + 1 = 0.
\end{aligned} \tag{41}$$

Resolving numerically Eq. (41) one can find the biggest root which is equal to

$$\lambda \approx 1.61771. \tag{42}$$

Correspondingly the topological entropy is given by logarithm of the biggest root.

One can show that similar calculations can be done for every value of n_C. Here we can give the list of numbers λ corresponding to different values of mean number of oscillations n_C:

$$\begin{aligned}
&n_C = 2, \quad \lambda \approx 1.37747,\\
&n_C = 3, \quad \lambda \approx 1.51714,\\
&n_C = 4, \quad \lambda \approx 1.57388,\\
&n_C = 5, \quad \lambda \approx 1.59837,\\
&n_C = 10, \quad \lambda \approx 1.67771,\\
&n_C = \infty, \quad \lambda = (1+\sqrt{5})/2 \approx 1.61803.
\end{aligned} \tag{43}$$

Thus, the topological entropy tends to the golden mean rather rapidly.

The scheme described above can be applied to many different physical models, obeying the different sets of prohibition rules governing the structure of periodical trajectories.

Concluding this section, we would like to stress that the study of chaos in classical cosmology is important since the quantum to classical transition is qualitatively different for chaotic and non-chaotic systems.

5. Decoherence in quantum cosmology and the problem of ultraviolet divergences

All activity connected with the studying of the decoherence in quantum cosmology is in essence an attempt to answer the question: "How the Universe born as a quantum object becomes the classical world which we observe today?" This question is closely connected with a related question: "How a certain result of a quantum measurement is chosen?" In other terms, how the reduction of the wave packet takes place.

The accepted approach to this question is the decoherence approach [60], which we shall illustrate using the simplest quantum-mechanical example - Stern-Gerlach experiment. At the beginning an atom is in a state, which is a linear combination of two eigenstates with spin directed up and spin directed down. The quantum state of the system, including an atom and measuring device is

$$|\Psi\rangle = |\Phi\rangle(c_1|\psi_\uparrow\rangle + c_2|\psi_\downarrow\rangle). \tag{44}$$

As a result of measurement this state transforms as follows

$$|\Psi\rangle_{final} = c_1|\Phi_\uparrow\rangle|\psi_\uparrow\rangle + c_2|\Phi_\downarrow\rangle|\psi_\downarrow\rangle, \tag{45}$$

where $|\Phi\rangle$ is an initial quantum state of measuring device and $|\Phi_\uparrow\rangle$ and $|\Phi_\downarrow\rangle$ the states of device correlated with the corresponding states of an atom.

But do we see only one outcome of this quantum experiment? To answer this question one writes down the density matrix of the state $|\Psi\rangle_{final}$:

$$\begin{aligned}\hat{\rho} = |\Psi\rangle_{final}\langle\Psi|_{final} = & |c_1|^2||\Phi_\uparrow\rangle\langle\Phi_\uparrow||\Psi_\uparrow\rangle\langle\Psi_\uparrow| \\ & +|c_2|^2||\Phi_\downarrow\rangle\langle\Phi_\downarrow||\Psi_\downarrow\rangle\langle\Psi_\downarrow| \\ & +c_1c_2^*|\Phi_\uparrow\rangle\langle\Phi_\downarrow||\Psi_\uparrow\rangle\langle\Psi_\downarrow| \\ & +c_2c_1^*|\Phi_\downarrow\rangle\langle\Phi_\uparrow||\Psi_\downarrow\rangle\langle\Psi_\uparrow|. \end{aligned} \tag{46}$$

The last two terms in the expression (46) describe the quantum interference and prevent us from interpretating the result of quantum measurement as an appearance of statistical mixture.

However, we should remember that besides a quantum object and measuring device always exists an environment - the rest of the Universe which also interacts with the measuring device. In fact, one has

$$|\Psi\rangle = |\chi\rangle|\Phi\rangle(c_1|\psi_\uparrow\rangle + c_2|\psi_\downarrow\rangle), \tag{47}$$

which after measurement becomes

$$|\Psi\rangle_{final} = c_1|\chi_\uparrow\rangle|\Phi_\uparrow\rangle|\psi_\uparrow\rangle + c_2|\chi_\downarrow\rangle|\Phi_\downarrow\rangle|\psi_\downarrow\rangle. \tag{48}$$

Here, we should remember that we do not observe an environment. Thus, to describe the observed reality one can calculate a "reduced density matrix" tracing out the environment.

$$\hat{\rho}_{reduced} = Tr_{\{\chi\}}|\Psi\rangle\langle\Psi|, \tag{49}$$

Due to huge number of degrees of freedom the states of environment, corresponding to different outcomes of experiment are mutually orthogonal:

$$\langle\chi_\uparrow||\chi_\downarrow\rangle = 0, \langle\chi_\uparrow||\chi_\uparrow\rangle = 1, \langle\chi_\downarrow||\chi_\downarrow\rangle = 1. \tag{50}$$

Using these relation one has the following reduced density matrix, which describes already classical statistical mixture:

$$\begin{aligned}\hat{\rho} = |\Psi\rangle_{final}\langle\Psi|_{final} = & |c_1|^2||\Phi_\uparrow\rangle\langle\Phi_\uparrow||\Psi_\uparrow\rangle\langle\Psi_\uparrow| \\ & +|c_2|^2||\Phi_\downarrow\rangle\langle\Phi_\downarrow||\Psi_\downarrow\rangle\langle\Psi_\downarrow|. \end{aligned} \tag{51}$$

The decoherence approach has proved its efficiency in quantum mechanics on both, theoretical and experimental level [61] and it is natural to try to apply it to cosmology. First attempts of such an application were carried out in [62].

The first question which arises when one tries to explain the classicalization of quantum world using decoherence approach, is connected with the definition of an environment. Indeed, in contrast to usual description of quantum - mechanical experiment in laboratory, there is no external environment, because the object of quantum cosmology is the whole Universe. Thus, we should treat some part of degrees of freedom as essential and observables, while others could be treated as an environment with subsequent tracing them out of reduced density matrix.

It is natural to believe that inhomogeneous degrees of freedom play the role of environment while macroscopic variables such as a cosmological radius or an initial value of inflaton scalar field are treated as observables. It is easy to guess that there are an infinite number of environmental degrees of freedom and hence, calculating a reduced density matrix one can encounter the problem of ultraviolet divergences.

To tackle this problem we should consider the wave function of the universe in one-loop approximation describing simultaneously with homogeneous also inhomogeneous degrees of freedom (see the second section of the present paper).

It is convenient to write down this function in the following form:

$$\Psi(t|\varphi, f) = \frac{1}{\sqrt{v_\varphi^*(t)}} \exp(\mp I(\varphi)/2 + iS(t,\varphi)) \times \prod_n \psi_n(t,\varphi|f_n), \tag{52}$$

$$\psi_n(t,\varphi|f_n) = \frac{1}{\sqrt{v_n^*(t)}} \exp\left(-\frac{1}{2}\Omega_n(t) f_n^2\right), \tag{53}$$

$$\Omega_n(t) = -ia^k(t)\frac{\dot{v}_n^*(t)}{v_n^*(t)}. \tag{54}$$

Here, the sign minus or plus before an Euclidean action $I(\varphi)$ in the exponent of (52) corresponds to the no-boundary and to the tunneling wave functions of the Universe, respectively, f_n describe amplitudes of inhomogeneous modes, while v_n are corresponding to these modes solutions of linearized second-order differential equations. The exponent k in the expression for the frequency function Ω_n depends on the spin s of the field under consideration and on its parametrization. For the "standard" parametrization $k = 3 - 2s$.

An information about decoherence behaviour of the system is contained in the off-diagonal elements of the density matrix.In our case

$$\rho(t|\varphi,\varphi') = C\left(\frac{\Delta_\varphi \Delta_{\varphi'}}{v_\varphi v^*_{\varphi'}}\right)^{1/4}$$

$$\times \exp\left(-\frac{1}{2}\Gamma - \frac{1}{2}\Gamma' + i(S - S')\right) D(t|\varphi,\varphi). \tag{55}$$

Here $D(t|\varphi,\varphi)$ is the so called decoherence factor defined by the formula

$$D(t|\varphi,\varphi) = \prod_n \left(\frac{4Re\Omega_n Re\Omega'^*_n}{(\Omega_n + \Omega'^*_n)^2}\right)^{1/4} \left(\frac{v_n v'^*_n}{v^*_n v'_n}\right)^{1/4} \tag{56}$$

or

$$|D(t|\varphi,\varphi)| = \exp\left(\frac{1}{4}\sum_n \ln\frac{4Re\Omega_n Re\Omega'^*_n}{(\Omega_n + \Omega'^*_n)^2}\right). \tag{57}$$

But how to cope with ultraviolet divergences appearing in the sum of this type ? This question was already discussed in [63, 64, 65]. In our papers [24, 25, 26] we have tried to use a dimensional regularization [66]. As usual, the main effect of a dimensional regularization consists in the changing of the number of degrees of freedom involved in summation. For example, for the scalar field, the degeneracy number of harmonics in the spacetime of dimensionality d from the well-known value [40]

$$dim(n,4) = n^2 \tag{58}$$

becomes

$$\dim(n,d) = \frac{(2n+d-4)\Gamma(n+d-3)}{\Gamma(n)\Gamma(d-1)}. \tag{59}$$

Making analytical continuation and discarding the poles $\frac{1}{d-4}$ one can have finite values for $|D(t|\varphi,\varphi')|$, however, for scalar, photon and graviton fields one get the pathalogical behaviour:

$$|D(t|\varphi,\varphi')| \to \infty, \; when \; |\varphi - \varphi'| \to \infty. \tag{60}$$

For example, for massive scalar field

$$\ln|D(t|\varphi,\varphi')| \approx \frac{7}{64} m^3 \bar{a}(a - a')^2, \tag{61}$$

where

$$a = \frac{1}{H(\varphi)} \cosh H(\varphi)t, \tag{62}$$

$$a' = \frac{1}{H(\varphi')} \cosh H(\varphi')t, \tag{63}$$

$$\bar{a} = \frac{a + a'}{2}. \tag{64}$$

Such a form of a decoherence factor not only does not correspond to the decoherence, but also render a density matrix ill-defined, breaking the condition $Tr\hat{\rho}^2 \leq 1$.

However, there is way out of this situation: using the reparametrization of a bosonic scalar field

$$f \to \tilde{f} = a^{\mu} f;\ v_n \to \tilde{v}_n = a^{\mu} v_n \tag{65}$$

one can get the new form of frequancy function

$$\Omega_n(t) = -i a^{3-2\mu}(t) \frac{\dot{\tilde{v}}_n^*(t)}{\tilde{v}_n^*(t)}. \tag{66}$$

In such a way one can suppress ultraviolet divergences. For the so called conformal parametrization, $\mu = 1$, for the massive scalar field one has

$$\ln |\tilde{D}(t|\varphi, \varphi')| = -\frac{m^3 \pi \bar{a}(a - a')^2}{64}, \tag{67}$$

which manifests already a good decoherent behaviour. For a massles, conformally coupled scalar field and for an electromagnetic field, one has

$$|\tilde{D}| = 1. \tag{68}$$

Unfortunately, for the case of fermions these tricks do not work [?]. One can check that the dimensional regularization gives contradictory results just like in the case of boson degrees of freedom. Moreover, one cannot use the conformal reparametrization in this case because standard fermion variables are already presented in the conformal parametrization.

However, there is a way out also in this case. One can perform a non-local Bogoliubov transformation mixing Grassmann variables x and y. This transformation modifies an effective mass of fermions. Choosing it in a certain way one can suppress ultraviolet divergences. The reasonable idea is to fix this transformation requiring the absence of decoherence in the stationary spacetime. Then:

$$|D(a, \varphi|a', \varphi')|_{ren} = \exp\left(-\frac{\pi^2 m^2 (a - a')^2}{192}\right), \tag{69}$$

and is finite.

The main conclusion which it is possible to draw out from our considerations consists in the fact that the requirement of consistency of a reduced density matrix determines the very definition of the environment in quantum cosmology.

6. Conclusion

Concluding the paper we would like to make some reference to the String Conference, 2000 (see the paper [67]), where outstanding experts in the field of modern theoretical physics have tried to select the ten most important problems in fundamental physics. It is interesting that between these ten problems, there are five topics which are directly connected with quantum cosmology. Here we give the list of these 5 problems.

1. Are all the (measurable) dimensionless parameters that characterize the physical Universe calculable in principle?
2. How can quantum gravity help to explain the origin of the Universe?
3. Why does the Universe appear to have one time and three space dimensions?
4. Why does the cosmological constant have the value that it has, is it zero and is it really constant?
5. What physics explains the enormous disparity between the gravitational scale and the typical mass scale of the elementary particles?

Acknowledgement

This work was partially supported by RFBR via grants No 99-02-18409 and No 99-02-16224. I am grateful to Cariplo Science Foundation for the financial support.

References

1. Starobinsky A.A. (1986) Stochastic DeSitter (inflationary) stage in the early universe, in H.J. De Vega and N. Sanchez (eds.), *Current trends in field theory, quantum gravity and strings, Lecture Notes in Physics* **246**, Springer Verlag, Heidelberg, pp. 107-126.
2. Linde A.D. (1990) *Particle Physics and Inflationary Cosmology*, Harwood Academic, New York.
3. Jaffe A.H. et al (2001) Cosmology from MAXIMA-1, BOOMERANG and COBE/DMR CMB observations, *Physical Review Letters* **86**, 3475-3479.
4. Starobinsky A.A. (1982) Dynamics of phase transition in the new inflationary Universe, scenario and generation of perturbations, *Physics Letters* **B 117**, 175-178.
5. Brink L. and Henneaux M. (1988) *Principles of String Theory*, Plenum Press, New York.
6. Dirac P.A.M. (1964) *Lectures on Quantum Mechanics*, Belfer Graduate School of Science, Yeshiva University, New York.

7. DeWitt B.S. (1967) Qauntum Theory of Gravity. I. The canonical theory, *Physical Review 160*, 1113-1148.
8. Kamenshchik A.Yu. and Lyakhovich S.L. (1997) Hamiltonian BFV-BRST theory of of closed quantum cosmological models, *Nuclear Physics* **B 495**, 309-328.
9. Barvinsky A.O. and Kamenshchik A.Yu. (1990) One-loop quantum cosmology: the normalizability of the Hartle-Hawking wave function and the probability of inflation, *Classical and Quantum Gravity* **7**, L181-L186.
10. Kamenshchik A.Yu. (1993) Normalizability of the wave function of the Universe, particle physics and supersymmetry, *Physics Letters* **B 316**, 45-50.
11. Barvinsky A.O. and Kamenshchik A.Yu. (1994) Quantum scale of inflation, *Physics Letters* **B 332**, 270-276.
12. Barvinsky A.O. and Kamenshchik A.Yu. (1994) Tunneling geometries: analiticity, unitarity and instantons in quantum cosmology, *Physical Review* **D 50**, 5093-5114.
13. Barvinsky A.O. and Kamenshchik A.Yu. (1996) Quantum origin of the energy scale of inflation: no-boundary vs tunneling wavefunctions, *International Journal of Modern Physics* **D 5**, 825-843.
14. Barvinsky A.O., Kamenshchik A.Yu. and Mishakov I.V. (1997) Quantum origin of the early inflationary universe, *Nuclear Physics* **B 491**, 387-426.
15. Esposito G., Kamenshchik A.Yu. and Miele G. (1997) Unified gauge models and one-loop quantum cosmology, *Physical Review* **D 56**, 1328-1331.
16. Barvinsky A.O. and Kamenshchik A.Yu. (1998) Effective equations of motion and initial conditions for inflation in quantum cosmology, *Nuclear Physics* **B 532**, 339-360.
17. Kamenshchik A.Yu., Khalatnikov I.M. and Toporensky A.V. (1995) Non-minimally coupled complex scalar field in classical and quantum cosmology, *Physics Letters* **B 357**, 36-42.
18. Kamenshchik A.Yu., Khalatnikov I.M. and Toporensky A.V. (1997) Complex inflaton field in quantum cosmology, *International Journal of Modern Physics* **D 6**, 649-671.
19. Kamenshchik A.Yu., Khalatnikov I.M. and Toporensky A.V. (1997) Simplest cosmological model with the scalar field, *International Journal of Modern Physics* **D 6**, 673-691.
20. Kamenshchik A.Yu., Khalatnikov I.M. and Toporensky A.V. (1998) Simplest cosmological model with the scalar field II. Influence of cosmological constant, *International Journal of Modern Physics* **D 7**, 129-138.
21. Kamenshchik A.Yu., Khalatnikov I.M., Savchenko S.V. and Toporensky A.V. (1999) Topological entropy in some isotropic cosmological models, *Physical Review* **D 59**, 123516.
22. Khalatnikov I.M. and Kamenshchik A.Yu. (1997) Singularity, initial conditions, quantum tunneling and scalar field in modern cosmology, *Physics Reports* **288**, 513-543.
23. Khalatnikov I.M. and Kamenshchik A.Yu. (1998) Singularity, initial conditions and quantum tunneling in modern cosmology, *Physics-Uspekhi* **41**, 525-542.
24. Barvinsky A.O., Kamenshchik A.Yu., Kiefer C. and Mishakov I.V. (1999) Decoherence in quantum cosmology at the onset of inflation, *Nuclear Physics* **B 551**, 374-396.
25. Barvinsky A.O., Kamenshchik A.Yu. and Kiefer C. (1999) Effective action and decoeherence by fermions in quantum cosmology, *Nuclear Physics* **B 552**, 420-444.
26. Barvinsky A.O., Kamenshchik A.Yu. and Kiefer C. (1999) Origin of the inflationary Universe, *Modern Physics Letters* **A 14**, 1083-1088.
27. Fradkin E.S. and Vilkovisky G.A. (1975) Quantization of relativistic systems with constraints, *Physics Letters* **B 55**, 224-226.
28. Batalin I.A. and Vilkovisky G.A. (1977) Operator quantization of relativistic dynamical systems subject to first class constraints, *Physics Letters* **B 69**, 309-312.
29. Batalin I.A. and Fradkin E.S. (1988) Operatorial quantization of dynamical systems

subject to constraints. A further study of the construction, *Annales Institute Henri Poincare* **49**, 145-214.
30. Hwang S. (1983) Covariant quantization of the string in dimensions $D \leq 26$ using a BRS formulation, *Physical Review* **D 28**, 2614-2620.
31. Buchbinder I.L, Fradkin E.S., Lyakhovich S.L. and Pershin V.D. (1991) Generalized canonical quantization of bosonic strings in background fields, *International Journal of Modern Physics* **A 6**, 1211-1231.
32. Hartle J.B. and Hawking S.W. (1983) Wave function of the Universe, *Physical Review* **D 28**, 2960-2975.
33. Hawking S.W. (1984) The quantum state of the Universe, *Nuclear Physics* **B 239**, 257-276.
34. Vilenkin A. (1982) Creation of the universes from nothing, *Physics Letters* **B 117**, 25-28.
35. Vilenkin A. (1984) Quantum creation of universes, *Physical Review* **D 30**, 509-511.
36. Vilenkin A. (1988) Quantum cosmology and the initial state of the Universe, *Physical Review* **D 37**, 888-897.
37. Linde A.D. (1984) Quantum creation of the inflationary Universe, *Letters in Nuovo Cimento* **39**, 401-405.
38. Zeldovich Ya.B. and Starobinsky A.A. (1984) A Universe with a nontrivial topology and the possibility of its quantum birth, *Soviet Astronomical Letters* **10**, 135-137.
39. Rubakov V.A. (1984) Quantum mechanics in the tunneling universe, *Physics Letters* **B 148**, 280-286.
40. Lifshitz E.M. and Khalatnikov I.M. (1963) Investigations in relativistic cosmology, *Advances in Physics* **12**, 185-249.
41. Halliwell J.J and Hawking S.W. (1985) The origin of the structure in the Universe, *Physical Review* **D 31**, 1777-1791.
42. Laflamme R.(1987) The Euclidean vacuum, justification from quantum cosmology, *Physics Letters* **B 198**, 156-160.
43. Barvinsky A.O. (1993) Unitarity approach to quantum cosmology, *Physics Report* **230**, 237-367.
44. DeWitt B.S. (1965) *Dynamical Theory of Groups and Fields*, Gordon and Breach, New York.
45. Barvinsky A.O. and Vilkovisky G.A. (1985) The generalized Schwinger-DeWitt technique in gauge theories and quantum gravity, *Physics Reports* **119**, 1-74.
46. Spokoiny B.L. (1984) Inflation and generation of perturbations in broken- symmetric theory of gravity, *Physics Letters* **B 147**, 39-43.
47. Fakir R. and Unruh W.G. (1990) Improvement of cosmological chaotic inflation through nonminimal coupling, *Physical Review* **D 41**, 1783-1791.
48. Salopek D.S., Bond J.R. and Bardeen J.M. (1989) Designing density spectra fluctuations, *Physical Review* **D 40**, 1753-1788.
49. Lavrelashvili G.V., Rubakov V.A., Serebryakov M.S. and Tinyakov P.G. (1990) Negative Euclidean action: instantons and pair creation in strong background fields, *Nuclear Physics* **B**, 98-116.
50. Khalatnikov I.M. and Mezhlumian A. (1992) The classical and quantum cosmology with a complex scalar field, *Physics Letters* **A 169**, 308-312.
51. Khalatnikov I.M. and Schiller P. (1993) From instanton to inflationary Universe, *Physics Letters* **B 302**, 176-182.
52. Amendola L.,Khalatnikov I.M., Litterio M. and Occhionero F. (1994) Quantum cosmology with a complex field, *Physical Review* **D 49**, 1881-1885.
53. Belinsky V.A., Grishchuk L.P., Zeldovich Ya.B. and Khalatnikov I.M. (1985) Inflationary stages in cosmological models with a scalar field, *Soviet Physics - JETP* **62**, 195-209.
54. Page D.N. (1984) A fractal set of infinitely bouncing universes ?, it Classical and Quantum Gravity **1**, 417-428.
55. Parker L. and Fulling S.A. (1973) Quantized matter fields and the avoidance of

singularities in general relativity, *Physical Review* **D 7**, 2357-2374.
56. Starobinsky A.A. (1978) About one nonlinear isotropic cosmological model, *Soviet Astronomical Letters* **4**, 82-86.
57. Hawking S.W. (1984) Quantum cosmology, in B.S. DeWitt and R. Stora (eds.) *Relativity, Groups and Topology II*, North-Holland Publishing, Amsterdam, pp. 333-380.
58. Mandelbrot B.B. (1992) *Fractal Geometry of Nature*, W.H. Freeman, San Francisco.
59. Cornish N.J. and Shellard E.P.S. (1998) Chaos in quantum cosmology, *Physical Review Letters* **81**, 3571-3574.
60. Zeh H.D. (1970) On the interpretation of measurement in quantum theory, *Foundations of Physics 1*, 69-76.
61. Giulini D., Joos E., Kiefer C., Kupsch J., Stamatesku I.-O. and Zeh H.D. (1996) *Deceherence and the appearance of the classical world in quantum theory*, Springer, Berlin.
62. Kiefer C. (1987) Continuous measurement of minisuperspace variables by higher multipoles, *Classical and Quantum Gravity* **4**, 1369-1382.
63. Paz J.P. and Sinha S. (1992) Decoherence and back reaction in quantum cosmology: Multidimensional minisuperspace examples, *Physical Review* **D 45**, 2823-2842.
64. Kiefer C. (1992) Decoherence in quantum electrodynamics and quantum cosmology, *Physical Review* **D 46**, 1658-1670.
65. Okamura T. (1996) On divergence of decoherence factor in quantum cosmology, *Progress of Theoretical Physics 95*, 565-576.
66. Leibbrandt G. (1975) Introduction to the technique of dimensional regularization, *Reviews of Modern Physics* **47**, 849-876.
67. Duff. M.J. (2000) State of the Unification Address, hep-th/0012249.

NORMAL MODE QUANTIZATION OF RELATIVISTIC SCALAR FIELDS IN AN EINSTEIN ELEVATOR

ARTHUR KOMAR
Syracuse University
Syracuse, New York 13244

ABSTRACT

We present an exact quantization of the Klein-Gordon field in a uniformly accelerated elevator. We then appeal to the Principle of Equivalence in order to infer properties of the related general relativistic quantum field theory. Among our results are: the exact expression for the energy eigenvalues, the appearance of a mass term even when the bare mass of the scalar field vanishes, and perhaps, most interesting, a rigorous proof of the conservation of the number of quanta, contrary to what might be expected from the Unruh effect.

1. Introduction

In a now classic paper, Unruh [1] considered the behavior of a relativistic quantum scalar field as viewed in the frame of reference of a uniformly accelerated observer. By considering global solutions, which necessitated the imposition of boundary conditions on certain characteristic surfaces, he concluded that the observer could interpret his physical situation as being, in effect, in a thermal bath of radiation.

It is not the intention of the present paper to join in the details of the controversy generated by Unruh's result or to cavil about the details of his derivation. Instead we wish to examine a related and important question that seems to be implied by his conclusion and which can be answered directly and precisely: does an observer in a uniformly accelerated elevator in flat Minkowski space find himself to be in a thermal radiation bath? Should that be the case, the Principle of Equivalence would imply that any object in a constant gravitational field, for example, any object in the room in which you are reading at this moment, should be in an Unruh-like thermal radiation bath!

Note that to investigate this modified question, the simplified model that we shall now consider need not be concerned with the more difficult and controversial issue of appropriate boundary conditions of scalar fields on characteristic surfaces. (Vide Narozhny et.al. [2].) We shall require that our scalar field shall vanish on the walls of a uniformly accelerating elevator, a boundary condition that needs the very precise definition to be given below. (Throughout this paper we shall employ natural units where $c = h/2\pi = 1$.)

2. Accelerated Reference Frame

Let us consider a four dimensional Minkowski space initially coordinatized by the standard rectangular variables of an inertial observer t', x', y', z'. The metric in this frame is of course

P.G. Bergmann and V. de Sabbata (eds.),
Advances in the Interplay Between Quantum and Gravity Physics, 243–250.

$$ds^2 = dt'^2 - dx'^2 - dy'^2 - dz'^2 \tag{2.1}$$

Let us now perform a coordinate transformation to the frame of an observer who is constantly accelerating in his instantaneous rest frame in the z′ direction at magnitude g. If we also require that the spatial metric remain rectangular Euclidian and that the observer at the origin of the accelerated frame adjusts his local clock to agree with that of the Minkowski frame observer who is instantaneously at rest with respect to him, we obtain the accelerated frame of reference by performing the coordinate transformation

$$\begin{aligned} z' &= (z + 1/g)\cosh gt - 1/g \\ t' &= (z + 1/g)\sinh gt \\ x' &= x \\ y' &= y \end{aligned} \tag{2.2}$$

The resulting (Riemann-flat) metric of the accelerated frame is found to be

$$ds^2 = \lambda^2 dt^2 - dx^2 - dy^2 - dz^2 \tag{2.3}$$

where

$$\lambda = 1 + gz \tag{2.4}$$

We are now able to define precisely what is to be understood by “the uniformly accelerated elevator” whose volume shall confine the scalar field. It is given in the accelerated frame of reference by the three-dimensional rectangular region, E,

$$0 \le x \le L_1 \qquad 0 \le y \le L_2 \qquad 0 \le z \le L_3 \tag{2.5}$$

3. The Scalar Field

The real Klein-Gordon scalar field is initially best specified by a relativisticly covariant Lagrangian. In this fashion all observers will agree as to the physical situation under consideration. In that spirit let us define the physics of the scalar field φ by stating its covariant Lagrangian density L:

$$L = \tfrac{1}{2} \sqrt{|g^{\alpha\beta}|}\,(g^{\mu\nu} \varphi_{,\mu} \varphi_{,\nu} - m^2 \varphi^2) \tag{3.1}$$

In the metric of the accelerated frame (2.3), the Lagrangian becomes

$$L = \tfrac{1}{2} [\varphi_{,0}\varphi_{,0}/\lambda - \lambda(\nabla\varphi \cdot \nabla\varphi + m^2)] \tag{3.2}$$

The corresponding Euler –Lagrange field equation is found to be

$$\varphi_{,00}/\lambda - \lambda(\nabla^2 \varphi + g\varphi_{,3}/\lambda - m^2\varphi) = 0 \tag{3.3}$$

We shall now proceed to solve this equation in the elevator E. The simplest and most convenient method to solve this equation is to employ the separation of variables. Thus, employing the ansatz

$$\varphi = T(t)\, X(x)\, Y(y)\, Z(z) \tag{3.4}$$

and dividing equation (3.3) by φ, we obtain

$$T''/\lambda T - \lambda(X''/X + Y''/Y + Z''/Z + g\, Z'/\lambda Z - m^2) = 0 \tag{3.5}$$

where the prime superscript denotes derivation with respect to the argument of the respective function. Multiplying this equation by λ, we see that an equation for $T(t)$ easily separates. Designating the separation constant ω^2 we trivially find the most general real solution for $T(t)$

$$T = \alpha \cos \omega t + \beta \sin \omega t \tag{3.6}$$

Substituting (3.6) into (3.5) and dividing by λ, we see that equations for $X(x)$ and $Y(y)$ also separate in the standard familiar manner. Thus, with an eye toward the boundary conditions that X and Y must vanish on the walls of E, we find it convenient to define the separation constants respectively $(n_1\pi/L_1)^2$ and $(n_2\, \pi/L_2)^2$. Thus we easily obtain

$$X = c_1 \sin n_1 x\pi/L_1 \qquad\qquad Y = c_2 \sin n_2 y\pi/L_2 \tag{3.7}$$

where, in order to satisfy the required conditions on E in a non-redundant manner, n_i must be non-negative integers. Finally, substituting (3.6) and (3.7) into (3.5) we obtain the equation that $Z(z)$ must satisfy:

$$Z'' + gZ'/\lambda + (\omega^2/\lambda^2 - K^2)\, Z = 0 \tag{3.8}$$

where

$$K^2 = (n_1\pi/L_1)^2 + (n_2\pi/L_2)^2 + m^2 \tag{3.9}$$

The general solution for (3.8) which vanishes on the floor of E ($z = 0$) is found to be

$$Z = c_3 \exp(-gz/\lambda) \sin [z/\lambda \sqrt{(\omega^2 - g^2/4 - K^2\lambda^2)}] \tag{3.10}$$

The final boundary condition that must be satisfied on E, $Z(L_3) = 0$, is easily seen to yield

$$L_3/(1 + gL_3) \sqrt{[\omega^2 - g^2/4 - K^2(1 + gL_3)^2]} = n_3\pi \tag{3.11}$$

where n_3 is required to be a non-negative integer. Note that, unlike n_1 and n_2, n_3 is not a separation constant. However it will shortly be shown to be a required "quantum number". Solving (3.11) for the separation constant ω we obtain

$$\omega^2 = g^2/4 + (1 + gL_3)^2\; [(n_1\pi/L_1)^2 + (n_2\pi/L_2)^2 + (n_3\pi/L_3)^2 + m^2] \tag{3.12}$$

This relation clearly provides the requisite discrete energy spectrum for the normal modes of our system and, in the limit of vanishing g, becomes the standard expression for the Klein-Gordon spectrum. An unexpected result is that, despite the fact that the z direction is singled out as the direction of acceleration, all three "quantum numbers" n_J enter symmetrically. This property has a curious consequence. If we define

$$\mathcal{M}^2 \equiv g^2/4 + m^2 (1 + gL_3)^2 \qquad (3.13)$$

$$\mathcal{L}_J \equiv L_J/(1 + gL_3) \qquad (3.14)$$

then (3.12) may be written

$$\omega^2 = \mathcal{M}^2 + (n_1\pi/\mathcal{L}_1)^2 + (n_2\pi/\mathcal{L}_2)^2 + (n_3\pi/\mathcal{L}_3)^2 \qquad (3.15)$$

This expression evidently yields the spectrum of a free Klein-Gordon field. We note that according to (3.13) the mass is renormalized in such a fashion that even when the bare mass vanishes (m=0), the effective dressed mass, $\mathcal{M}$, does not. In addition, we see from (3.14) that the density of states is increase by a factor $(1 + gL_3)^3$. It is as if all three edges of the accelerated elevator contracted by the same amount, an unexpected and puzzling result.

4. Normal Mode Pre-quantization

Thus far we have in effect found the spectra of the normal modes of the scalar field φ in a uniformly accelerated elevator. In order to quantize this theory it will be necessary to construct a normed space in which these modes can be shown to be orthonormal. It will then be a relatively straightforward procedure to construct the Fock space for the quantum field theory. Our procedure for field quantization will be closer to the original work of Planck [3] than the more customary contemporary procedure of canonical field quantization, employed for example in Unruh's paper [1].

A virtue of our normal mode quantization procedure is that, once the normal modes have been identified and a Hilbert space in which they are mutually orthonormal is constructed, the coordinate system in which the modes are realized does not affect the subsequent quantization procedure. This contrasts very much with the sensitivity of the closely related canonical quantization procedure to the choice of coordinates. (Indeed, in a much overlooked paper, Uhlhorn [5] has proven many decades ago that the procedure of canonical transformation of phase space coordinates and that of canonical quantization of phase space coordinates only commutes for the symplectic subgroup of the canonical group, that is, only for the linear canonical transformations. Note that the configuration space transformation to the accelerated frame that we have performed, (2.2), is highly nonlinear, as would be the corresponding phase space transformation.)

We have shown in the previous section that the solutions of (3.8) can be enumerated by a triple of integers n_J. It will be more convenient to denote this triple as a single (vector-like object) **N**. Thus we are able to express these solutions by the self-evident notation $Z_N(z)$. We now define, by the following definite integral, what we shall call our *initial* scalar product for these functions of z:

$$(Z_M, Z_N)_I \equiv \int \lambda\, Z_M Z_N \, dz \qquad (4.1)$$

(Throughout this paper all integrals will be understood to be definite, ranging over the respective dimensions of the elevator as given in (2.5).) Let us now define the second order differential operator D^2 as follows:

$$D^2 \equiv -(\partial_3{}^2 + g/\lambda\ \partial_3) \tag{4.2}$$

In terms of this operator the differential equation for Z(z), (3.8), may be written

$$D^2 Z_N = (\omega^2/\lambda^2 - K_N{}^2)\, Z_N \tag{4.3}$$

For future reference it will be important to note that, as a consequence of (3.9), the constant $K_N{}^2$ does not depend on the quantum number n_3, but only on n_1 and n_2. It therefore follows that, for the dimensions of a generic elevator,

$$K_M{}^2 = K_N{}^2 \quad \Leftrightarrow \quad (m_1 - n_1) = (m_2 - n_2) = 0 \tag{4.4}$$

The most important property of our operator D^2 is that it is self-adjoint in the initial scalar product (4.1). That is, it is readily shown by parts integration that

$$(Z_M, D^2 Z_N)_I = (D^2 Z_M, Z_N)_I \tag{4.5}$$

If we now substitute (4.3) into (4.5) we obtain

$$\omega_M{}^2 \int 1/\lambda\ Z_M Z_N\, dz - K_M{}^2 (Z_M, Z_N)_I = \omega_N{}^2 \int 1/\lambda\ Z_M Z_N\, dz - K_N{}^2 (Z_M, Z_N)_I \tag{4.6}$$

Thus, when (4.4) applies, we can conclude in the standard fashion

$$\int 1/\lambda\ Z_M Z_N\, dz = c\delta_{MN} \tag{4.7}$$

Thus we see that, although the operator D^2 is self-adjoint in the space of the *initial* norm, the functions Z_M are orthonormal in quite a different normed space. The result of (4.7) leads us to define what we shall call the *primary* norm:

$$(Z_M, Z_N) \equiv \int 1/\lambda\ Z_M Z_N\, dz \tag{4.8}$$

Finally, if we collect all the expressions that enter into the ansatz (3.4), we see that the general solution of the classical field equation for φ, (3.3), within the elevator E is:

$$\varphi(t, x, y, z) = \Sigma_M [\alpha_M \cos(\omega_M t) + \beta_M \sin(\omega_M t)]\, S_M(x, y, z) \tag{4.9}$$

where we have defined

$$S_M(x, y, z) \equiv C_M \sin(m_1 x\pi/L_1) \sin(m_2 y\pi/L_2)\, Z_M(z) \tag{4.10}$$

We can extend our definition of the primary norm to all three spatial dimensions in the obvious manner:

$$(S_M, S_N) \equiv \iiint 1/\lambda\ S_M(x, y, z)\, S_N(x, y, z)\, dx\, dy\, dz \tag{4.11}$$

In view of the fact that the sine functions are orthogonal in their respective domains and, taking in to consideration the significance of the relations (4.4) and (4.7), it is easy to confirm that, with an appropriate choice of the normalization constants C_M, we obtain for the primary norm:

$$(S_M, S_N) = \delta_{MN} \tag{4.12}$$

The completeness of these basis vectors in our primary vector space enables us to invert (4.12) and thereby obtain in the standard fashion the useful and important relation

$$\Sigma_M\, S_M(\mathbf{x})\, S_M(\mathbf{y}) = \lambda(x_3)\, \delta(\mathbf{x} - \mathbf{y}) \tag{4.13}$$

(Recall that bold face denotes 3-vectors.)

5. Normal Mode Field Quantization

In (4.9) we have exhibited the general time-dependent solution of the scalar field as a linear superposition of normal modes. In this form the quantization of this theory in the Schrödinger representation is essentially trivial; we need only promote the constant coefficients α_M and β_M to algebraic operators satisfying appropriate commutation relations. We shall briefly indicate below an important way to accomplish this; but for the moment, we find to more instructive to employ what we have thus far developed to construct a Heisenberg representation in Fock space.

Let us therefore return to the Lagrangian density (3.2) and construct the associated Hamiltonian density. As per usual we obtain the momentum $\Pi(\mathbf{x})$ canonically conjugate to the field $\varphi(\mathbf{x})$ via

$$\Pi \equiv \partial L/\partial\varphi_{,0} = \varphi_{,0} / \lambda \tag{5.1}$$

The Hamiltonian density is defined as usual:

$$H \equiv \Pi\, \varphi_{,0} - L = \tfrac{1}{2} \lambda\, (\Pi^2 + \nabla\varphi \cdot \nabla\varphi + m^2\, \varphi^2\,) \tag{5.2}$$

The Heisenberg operators are found most conveniently by taking the Cauchy data, that is, the values of our Schrödinger expressions for φ and Π on the initial surface $t = 0$. Thus

$$\varphi(\mathbf{x}) \equiv \Sigma_M\, \alpha_M\, S_M(\mathbf{x}) \tag{5.3a}$$

$$\Pi(\mathbf{x}) \equiv 1/\lambda(x_3)\, \Sigma_M\, \beta_M\, \omega_M\, S_M(\mathbf{x}) \tag{5.3b}$$

Recalling that α_M and β_M are understood at this point to be real constants, it proves convenient to replace them by conjugate pairs of complex numbers by means of the relations

$$\alpha_M \equiv (a_M^\dagger + a_M)/\sqrt{(2\omega_M)} \tag{5.4a}$$

$$\beta_M \equiv i\, (a_M^\dagger - a_M)/\sqrt{(2\omega_M)} \tag{5.4b}$$

Substituting (5.4) into (5.3) we find the canonically conjugate field variables can be written

$$\varphi(\mathbf{x}) = \Sigma_M\, (a_M^\dagger\; + a_M)\, S_M(\mathbf{x}) \,/\, \sqrt{2\omega_M} \tag{5.5a}$$

$$\Pi(\mathbf{x}) = i/\lambda(x_3)\, \Sigma_M\, \sqrt{(\omega_M/2)}\, (a_M^\dagger - a_M)\, S_M(\mathbf{x}) \tag{5.5b}$$

We complete the Heisenberg quantization by promoting constant coefficients a_M and $a_M^\dagger$ to algebraic operators that satisfy the familiar commutation relations of creation and annihilation operators

$$[a_M, a_N^\dagger] = \delta_{MN} \qquad [a_M, a_N] = [a_M^\dagger, a_N^\dagger] = 0 \tag{5.6}$$

Employing (5.6) as well as the completeness relation (4.13), a straightforward computation enables one to derive the Heisenberg algebra for the scalar quantum field theory:

$$[\varphi(\mathbf{x}), \Pi(\mathbf{y})] = i\,\delta(\mathbf{x} - \mathbf{y}) \qquad [\varphi(\mathbf{x}), \varphi(\mathbf{y})] = [\Pi(\mathbf{x}), \Pi(\mathbf{y})] = 0 \tag{5.7}$$

The virtue of employing the creation and annihilation operators in our quantization scheme is that enables the construction of a linear vector space for the realization of quantum field states that can support, in a natural way, a particle interpretation of the theory. In this regard, consider the Hamiltonian of this theory, H, obtained by integrating, at $x_0 = 0$, the Hamiltonian density, H($\mathbf{x}$}, as defined in (5.2), over the entire volume of the elevator E:

$$H \equiv \iiint \mathrm{H}(\mathbf{x})\, d^3\mathbf{x} \tag{5.8}$$

If we substitute (5.5) into (5.8), an intricate computation, employing the completeness relation (4.13), gives

$$H = \tfrac{1}{2} \Sigma_M\, \omega_M\, (a_M^\dagger a_M + a_M a_M^\dagger) \tag{5.9}$$

Or by normal ordering H we can write

$$H = \Sigma_M\, \omega_M\, a_M^\dagger a_M \equiv \Sigma_M\, \omega_M\, N_M \tag{5.10}$$

where N_M has the natural interpretation of the number of particles in the one particle state having the quantum numbers **M**, and the energy of that state is ω_M. The total number of particles is clearly realized by the operator

$$N \equiv \Sigma_M N_M \tag{5.11}$$

Furthermore we trivially observe that

$$[H, N_M] = [H, N] = 0 \tag{5.12}$$

With the above interpretation of our symbols we easily see from (5.12) that the number of particles in every field mode is a constant of the motion. There is no radiation of particles in the accelerating elevator!

6. Principle of Equivalence

The Principle of Equivalence is one of the essential foundation stone of the general theory of relativity. In its weakest form in can be understood to assert that the physical phenomena of nature in the frame of reference of a uniformly accelerating observer cannot be distinguished from those of an observer in a constant gravitational field. Although this principle was formulated for a classical field theory we expect it to be valid for all properly quantized general relativistic theories. It is for this reason that we chose to examine the very the very simple model of this paper.

Granting our premise, an important conclusion of this paper is that the quantum theory of a linear scalar field in a general relativistic background does not require that accelerated observers appear to be immersed in a thermal bath of radiation. The objects that you see about you, at rest in your terrestrial location (which provides, in effect, a frame of reference of a constant gravitational field), should not be expected to be bathed in a gravitationally induced cloud of black body radiation.

Let me hasten to add a caveat emptor. The principle of equivalence is a very local requirement. A true gravitational field involves curvature that cannot be transformed away by choice of a coordinate frame. In such situations it becomes unclear how to state the principle of equivalence when extending our considerations out to finite spatial distances. However quantum theory is highly nonlocal. In our model it is clear that our conclusions depend very critically on the boundary conditions imposed on the walls of the elevator. It was the imposition of those boundary conditions that determined the normal modes to be quantized. Thus our two critical requirements may not be mutually compatible; but in that case it is not clear what is to be understood by a linear quantum field theory in such a context.

We would like to point out at this time the reason why the procedure of field quantization that we employed in this paper need not always agree with that of canonical quantization. The physical assumptions contained in the boundary conditions required for defining a quantum theory are frequently obscured by that formal procedure. These essential boundary conditions are buried in the formal assumption that the field observables are required to be realized as hermitian operators in *some* linear vector space. In the present paper our results were obtained by being very specific as to the physical character of the boundary conditions on the walls of the elevator: the walls were assumed to be fully insulating. Were we to extend our method of quantization to the case of the Maxwell field in the elevator E, the correct normal modes to be quantized would have to be determined by the physical requirements implicit in the work of Planck [3]; that is, the walls of the box are assumed to be conducting rather than insulating. As theoreticians, we are of course free to consider either possibility, but the nature of the physical problem under consideration must really force the set of conditions that is appropriate. Unfortunately, the canonical quantization procedure generally makes an unanalyzed choice that depends solely on the vagaries of the particular coordinate system in which it is being employed. Our procedure for field quantization, which first selects the normal modes that the desired physical boundary conditions require, can therefore be applied in any convenient coordinate system. The essential drawback of our method of quantization is that, since it depends ab initio on the decomposition of the field into the physically appropriate normal modes, it can only be applied to linear field theories.

Of course the question of the quantization of the relativistic theory of gravitation by means of our methods remains open. This is due to (at least) two circumstances: the theory is nonlinear, and there are no gravity shields by which the field can be confined in a finite region, as required for the unambiguous specification of normal modes.

7. Bibliography

1. Unruh, W. G. (1976) Notes on black-hole evaporation, *Physical Review D* **14**, 870-892

2. Narozhny, N. B., Fedotov, A. M., Karnakov, B. M., and Mur, V. D. (1999) Boundary conditions in the Unruh problem, *hep-th*/9906181, 1-31

3. Planck, M. (1914) *The Theory of Heat Radiation*, P. Blakiston Son & Co; (1991) Dover Publications, New York, 81-82

GEOMETRIC ALGEBRA, DIRAC WAVEFUNCTIONS AND BLACK HOLES

A.N. LASENBY AND C.J.L. DORAN
Astrophysics Group
Cavendish Laboratory
Madingley Road
Cambridge CB3 OHE, U.K.

Abstract. In this contribution we describe some applications of *geometric algebra* to the field of black hole physics. Our main focus is on the properties of Dirac wavefunctions around black holes. We show the existence of normalised bound state solutions, with an associated decay rate controlled by an imaginary contribution to the energy eigenvalue. This is attributable to the lack of Hermiticity caused by a black hole singularity. We also give a treatment of the Feynman scattering problem for fermions interacting with black holes that we believe is new, and produces an analogue of the Mott scattering formula for the gravitational case. Throughout, the consistent application of geometric algebra simplifies the mathematical treatment and aids understanding by focusing attention on observable quantities. We finish with a brief review of recent work on the effects of torsion in quadratic theories of gravity. This work demonstrates that a free torsion field can play a significant role in cosmology.

1. Introduction

This contribution provides a broad survey of a number of recent advances in the application of geometric algebra to quantum physics and gravitation. The bulk of the presentation deals with the quantum properties of fermions in a black hole background. We address two related problems. The first question is whether a spectrum of bound state solutions exists around a black hole. The non-relativistic limit of the Dirac equation suggests that this should be the case, and indeed this is borne out by the full general relativistic theory. The relativistic treatment contains a novel feature, however, which complicates the search for bound states. The Hamilto-

P.G. Bergmann and V. de Sabbata (eds.),
Advances in the Interplay Between Quantum and Gravity Physics, 251–283.

nian representing a Dirac fermion in a Schwarzschild black hole background is not Hermitian as the singularity acts as a density sink. This means that the eigenspectrum consists of isolated *complex* energies. The equations are too complicated to admit an analysis in terms of known special functions, so these eigenvalues can only be found by a numerical search. Fortunately, for a range of masses, the real parts of the energies lie close to their non-relativistic values, which simplifies the search process. The imaginary contribution to the energy is responsible for orbital decay and implies that the bound states have finite lifetimes. This provides a novel, physical picture of what happens as a particle falls into a black hole. Some of the questions raised by this work are discussed here.

The second main application is to the scattering problem for black holes. This has been discussed by many authors, both in a classical and quantum context [1]–[6]. Here we tackle the problem in terms of straightforward Feynman propagator theory. Remarkably, this approach does not appear to have been tried before. The approach is suggested naturally by the geometric algebra form of the Dirac equation in a Schwarzschild background, when cast in *Newtonian* gauge form. The name reflects the fact that many equations take on a Newtonian form if expressed in this gauge, though of course all physical predictions are gauge invariant. The result of this approach is that the effects of the black hole can be described by a single vertex factor in momentum space. This factor has the unusual property of vanishing on-shell, so the gravitational cross-section involves a second-order process (in the Newtonian gauge). The calculation can be reproduced in other gauges, and in all cases the cross-section, to leading order in the black hole mass, is

$$\frac{d\sigma}{d\Omega_f} = \frac{(GM)^2}{4\beta^4 \sin^4(\theta/2)} \left(1 + 2\beta^2 - 3\beta^2 \sin^2\frac{\theta}{2} + \beta^4 - \beta^4 \sin^2\frac{\theta}{2}\right), \quad (1)$$

where $\beta = |\boldsymbol{p}|/E$. The cross-section does not depend on the particle mass, which embodies the equivalence principle at the quantum level. The low-velocity limit recovers the standard Rutherford formula. The full result is the gravitational analogue of the Mott formula for the Coulomb scattering cross-section.

2. Geometric Algebra

Throughout this paper we consistently apply the mathematical language of *geometric algebra*. Though this does not affect any of the physical predictions, it does considerably assist in the understanding and solution of the mathematics encountered. The foundations of geometric algebra (GA) were laid in the 19th Century by Hamilton and Grassmann (Fig. 1). Clif-

Figure 1. Hermann Gunther Grassmann (1809–1877), born in Stettin, Germany (now Szczecin, Poland). He worked as a schoolteacher in Germany, and later established a significant reputation in linguistics. The mathematical system he introduced, Grassmann algebra, is now an integral part of differential geometry and theoretical physics.

ford (Fig. 2) then unified their work by showing how Hamilton's quaternion algebra could be included in Grassmann's scheme through the introduction of a new, geometric product. The resulting algebra is known as a *Clifford algebra*, though when used in its intended, geometric role most practitioners prefer Clifford's original name of *geometric algebra*. The algebra was not immediately adopted by physicists, as it became tainted by its association with quaternions. These ran into problems due to their muddled handling of reflections and rotations, though ironically it is the embedding of quaternions within GA which solves all of these difficulties! Clifford algebra was rediscovered by Pauli and Dirac for use in the quantum theory of spin, and then developed further by mathematicians in the 50s and 60s. It was reintroduced in the guise of geometric algebra to physics in the 70s by David Hestenes, who is largely responsible for the modern form of the subject.

At its most fundamental level, geometric algebra is a mathematical language for directly encoding geometric primitives — points, lines, planes, volumes, circles, spheres *etc.* Geometric algebra provides a new product for vectors which generalizes complex analysis to arbitrary dimensions. The key to the development of the algebra was Grassmann's introduction of the *outer product* in the '*Lineale Ausdehnungslehre*' of 1844. This product encodes an oriented plane segment via an antisymmetric 'wedge product' $a \wedge b$. Antisymmetry in encoded mathematically in the statement

$$a \wedge b = -b \wedge a \tag{2}$$

Figure 2. William Kingdon Clifford 1845–1879. Mathematician and philosopher. He died tragically young at the age of only 33.

which implies that $a \wedge a = 0$. For example, suppose that e_1, e_2 are a pair of orthonormal basis vectors in two dimensions. With vectors $a = a_1 e_1 + a_2 e_2$ and $b = b_1 e_1 + b_2 e_2$, we find that

$$\begin{aligned} a \wedge b &= a_1 b_2 e_1 \wedge e_2 + a_2 b_1 e_2 \wedge e_1 \\ &= (a_1 b_2 - b_2 a_1) e_1 \wedge e_2. \end{aligned} \tag{3}$$

The coefficient here is seen to represent the area of the parallelogram defined by the vectors, and the *bivector* term $e_1 \wedge e_2$ defines an orientation. Grassmann extended the outer product to arbitrary dimensions and *grades.* The latter term refers to the number of vectors in each outer product. The total antisymmetry ensures that the outer product of a set of linearly-dependent vectors must vanish. Each step up in grade is therefore a step up in the spatial dimension of the object of interest.

Clifford's great insight was to realise that Grassmann's outer product can be combined with the (symmetric) inner product to define a single product

$$ab = a \cdot b + a \wedge b. \tag{4}$$

Clifford called this the *geometric product.* It satisfies the usual axioms of associativity and distributivity (over addition). Clearly it is *not* commutative, and the separate scalar (inner) and outer products can be obtained from

$$a \cdot b = \tfrac{1}{2}(ab + ba), \qquad a \wedge b = \tfrac{1}{2}(ab - ba). \tag{5}$$

These axioms are sufficient to build up an extremely rich algebra. For example, in two dimensions the algebra is spanned by the set

$$\begin{array}{ccc} 1 & \{e_1, e_2\} & e_1 \wedge e_2 \\ \text{1 scalar} & \text{2 vectors} & \text{1 bivector.} \end{array} \tag{6}$$

As the basis vectors are orthogonal, the bivector $e_1 \wedge e_2$ can be written in any of the equivalent forms:

$$e_1 \wedge e_2 = e_1 e_2 = -e_2 \wedge e_1 = -e_2 e_1. \tag{7}$$

It follows that the square of the bivector satisfies

$$(e_1 \wedge e_2)^2 = (e_1 e_2)(-e_2 e_1) = -e_1 e_2 e_2 e_1 = -1, \tag{8}$$

so the unit bivector squares to minus 1! Furthermore, it is responsible for generating $\pi/2$ rotations in the plane. One can clearly see that the geometric algebra of the Euclidean plane naturally incorporates complex analysis, with the geometric product taking on the role of the complex product.

Similarly, in three dimensions the full algebra is spanned by the set

$$\begin{array}{cccc} 1 & \{\boldsymbol{\sigma}_i\} & \{\boldsymbol{\sigma}_i \wedge \boldsymbol{\sigma}_j\} & I = \boldsymbol{\sigma}_1 \boldsymbol{\sigma}_2 \boldsymbol{\sigma}_3 \\ \text{1 scalar} & \text{3 vectors} & \text{3 bivectors} & \text{1 trivector.} \end{array} \tag{9}$$

(The reason for the notation will emerge shortly.) Each grade encodes a geometrically significant object in three dimensions. The highest grade object, I, is called the *pseudoscalar*. In three dimensions it has negative square, $I^2 = -1$. The pseudoscalar maps vectors to bivectors, and vice-versa. For the bivectors we can write

$$\boldsymbol{\sigma}_1 \boldsymbol{\sigma}_2 = I \boldsymbol{\sigma}_3, \quad \boldsymbol{\sigma}_2 \boldsymbol{\sigma}_3 = I \boldsymbol{\sigma}_1, \quad \boldsymbol{\sigma}_3 \boldsymbol{\sigma}_1 = I \boldsymbol{\sigma}_2. \tag{10}$$

It follows that each bivector squares to -1. If we set

$$\boldsymbol{i} = I \boldsymbol{\sigma}_1, \quad \boldsymbol{j} = -I \boldsymbol{\sigma}_2, \quad \boldsymbol{k} = I \boldsymbol{\sigma}_3 \tag{11}$$

we find that $\boldsymbol{i}, \boldsymbol{j}, \boldsymbol{k}$ satisfy

$$\boldsymbol{i}^2 = \boldsymbol{j}^2 = \boldsymbol{k}^2 = \boldsymbol{ijk} = -1. \tag{12}$$

This are the defining relations of the *quaternion algebra*, which is naturally embedded in the algebra of 3D space. The embedding is important, as it separates the roles of vectors and quaternions (bivectors). This is essential for the correct handling of reflections and rotations.

Further information can be found in a number of introductions to geometric algebra. See for example the papers by Hestenes [7, 8, 9] and

Vold [10, 11] and the books [12, 13, 14, 15]. See also [16] for a complete set of lecture notes and further resources.

3. Spacetime Algebra

The applications in this paper are mainly developed in the geometric algebra of spacetime. This is called the *spacetime algebra* or STA. We let the set $\{\gamma_\mu\}$ denote an orthogonal frame of vectors for spacetime, where $\gamma_0^2 = 1$, and $\gamma_k^2 = -1$ for $k = 1, 2, 3$. The inner product for these is summarised by

$$\gamma_\mu \cdot \gamma_\nu = \tfrac{1}{2}(\gamma_\mu \gamma_\nu + \gamma_\nu \gamma_\mu) = \eta_{\mu\nu} = \mathrm{diag}(+ - - -). \tag{13}$$

These are the defining relations of the Dirac matrix algebra, so the Dirac matrices form a representation for the STA. We do not use the matrix representation directly in calculations, though, as it is invariably slower. For example, one can end up calculating the square of a 4×4 complex matrix, when all this encodes is the result $\gamma_0^2 = 1$. Also, the matrices are naturally defined over the complex field, whereas we are interested in the *real* STA.

Repeated multiplication of the basis vector builds up a complete basis for the STA. This has $2^4 = 16$ elements, and is written

$$\begin{array}{ccccc} 1 & \{\gamma_\mu\} & \{\boldsymbol{\sigma}_k, I\boldsymbol{\sigma}_k\} & \{I\gamma_\mu\} & I \\ \text{scalar} & \text{vectors} & \text{bivectors} & \text{pseudovectors} & \text{pseudoscalar.} \end{array} \tag{14}$$

The bivectors $\boldsymbol{\sigma}_k$ are defined by

$$\boldsymbol{\sigma}_k = \gamma_k \gamma_0. \tag{15}$$

Their algebra is isomorphic to that of Euclidean 3-space, since

$$\boldsymbol{\sigma}_j \cdot \boldsymbol{\sigma}_k = \tfrac{1}{2}(\boldsymbol{\sigma}_j \boldsymbol{\sigma}_k + \boldsymbol{\sigma}_k \boldsymbol{\sigma}_j) = \delta_{jk}. \tag{16}$$

These define a set of basis vectors for the rest frame defined by γ_0. Different timelike vectors define different relative vectors, and hence a different *spacetime split* of the bivectors in the STA. The spacetime pseudoscalar is again denoted I, and is given by

$$I = \boldsymbol{\sigma}_1 \boldsymbol{\sigma}_2 \boldsymbol{\sigma}_3 = \gamma_1 \gamma_0 \gamma_2 \gamma_0 \gamma_3 \gamma_0 = \gamma_0 \gamma_1 \gamma_2 \gamma_3. \tag{17}$$

It follows that $I^2 = -1$ and, in spacetime, I *anticommutes* with all vectors.

For a general vector a we write the spacetime split with γ_0 as

$$a\gamma_0 = a_0 + \boldsymbol{a} \tag{18}$$

where

$$a_0 = a\cdot\gamma_0, \qquad \boldsymbol{a} = a\wedge\gamma_0. \tag{19}$$

Relative vectors are denoted in bold, to distinguish them from spacetime vectors. Of course, in the full STA, relative vectors are treated as bivectors. For example, the momentum p is split according to

$$p\gamma_0 = p\cdot\gamma_0 + p\wedge\gamma_0 = E + \boldsymbol{p}. \tag{20}$$

where E is the energy measured in the γ_0 frame. We immediately recover the Lorentz invariant scalar from

$$p^2 = p\gamma_0\gamma_0 p = (E+\boldsymbol{p})(E-\boldsymbol{p}) = E^2 - \boldsymbol{p}^2. \tag{21}$$

(We use natural units throughout, so $c = \epsilon_0 = \hbar = 1$.) Velocity 4-vectors are invariably normalised so that $v^2 = 1$. For these the relative velocity in the γ_0 frame is defined by

$$\boldsymbol{v} = \frac{v\wedge\gamma_0}{v\cdot\gamma_0}, \tag{22}$$

so that we have

$$v\gamma_0 = (v\cdot\gamma_0)(1+\boldsymbol{v}) = \gamma(1+\boldsymbol{v}) \tag{23}$$

where $\gamma = (1-\boldsymbol{v}^2)^{-1/2}$ is the Lorentz factor.

A general multivector in the STA can contain a mixture of grades. It is useful to have a compact notation to denote the result of projecting out terms of a given grade. We write this as $\langle A\rangle_r$, which returns the grade-r terms in A. For the scalar part ($r = 0$) we drop the subscript and just write

$$\langle A\rangle = \langle A\rangle_0. \tag{24}$$

Multivectors containing terms of only a single grade are called *homogeneous*. For these the inner and outer products extend straightforwardly. For multivectors A_r and B_s of grade r and s respectively, we can write

$$\begin{aligned} A_r\cdot B_s &= \langle A_r B_s\rangle_{|r-s|} \\ A_r\wedge B_s &= \langle A_r B_s\rangle_{r+s}. \end{aligned} \tag{25}$$

In the case where $r = 1$, so that $a = A_r$ is a vector, we have the relation

$$aB_s = a\cdot B_s + a\wedge B_s. \tag{26}$$

The symmetry of both the inner and outer product alternate with increasing grade of B_s,

$$\begin{aligned} a\cdot B_s &= \langle aB_s\rangle_{s-1} = \tfrac{1}{2}(aB_s - (-1)^s B_s a) \\ a\wedge B_s &= \langle aB_s\rangle_{s+1} = \tfrac{1}{2}(aB_s + (-1)^s B_s a). \end{aligned} \tag{27}$$

We adopt the convention that, in the absence of brackets, inner and outer products are performed before geometric products.

An important operation in GA is that of *reversion*. We write the reverse of A as $\tilde{A}$, which reverses all the vector products making up the multivector. This has the property that

$$(AB)\tilde{} = \tilde{B}\tilde{A}. \tag{28}$$

Given a general multivector

$$M = \alpha + a + B + Ib + I\beta, \tag{29}$$

where α and β are scalars, a and b are vectors and B is a bivector, the reverse satisfies

$$\tilde{M} = \alpha + a - B - Ib + I\beta. \tag{30}$$

Lorentz transformations are spacetime rotations and can be performed by the use of a *rotor*. A rotor R is an even-grade element of the STA satisfying

$$R\tilde{R} = \tilde{R}R = 1. \tag{31}$$

A proper orthochronous Lorentz transformation of a vector a can be written

$$a \mapsto a' = Ra\tilde{R}. \tag{32}$$

It can be shown that all proper orthochronous Lorentz transformations can be written in this way. Since R and $-R$ encode the same transformation, rotors form a double cover of the (restricted) Lorentz group. Any rotor can be written in the form

$$R = \pm\exp(-B/2) \tag{33}$$

where B is a bivector. The bivectors form the Lie algebra of the Lorentz group, and directly encode the spacetime plane(s) in which the transformation is performed. The same double-sided transformation law applies to multivectors representing geometric objects. For example, the plane $a \wedge b$ transforms to

$$a\wedge b \mapsto (Ra\tilde{R})\wedge(Rb\tilde{R}) = \tfrac{1}{2}(Ra\tilde{R}Rb\tilde{R} - Rb\tilde{R}Ra\tilde{R}) = R\,a\wedge b\,\tilde{R}. \tag{34}$$

The fact that general multivectors transform in this simple manner is a considerable improvement over matrix-based techniques for carrying out Lorentz transformations.

The main ingredient in field theory in the STA is the *vector derivative* ∇. If x^μ denote the coordinates in each of the γ_μ directions, we write

$$\nabla = \gamma^\mu \frac{\partial}{\partial x^\mu} = \gamma_0 \frac{\partial}{\partial t} + \gamma^i \frac{\partial}{\partial x^i}. \tag{35}$$

The spacetime split of ∇ goes as

$$\nabla\gamma_0 = \partial_t - \boldsymbol{\nabla}, \tag{36}$$

where $\boldsymbol{\nabla} = \boldsymbol{\sigma}_i\partial_i$ is the vector derivative in the relative space picked out by γ_0. The spacetime vector derivative ∇ provides a remarkably compact encoding of the Maxwell equations. We combine the electric and magnetic fields into spacetime bivector $F = \boldsymbol{E} + I\boldsymbol{B}$. The Maxwell equations can then be written as

$$\nabla F = J, \tag{37}$$

where J is the current. This is not merely a cosmetic exercise. The vector derivative ∇ is directly invertible, which provides a number of new techniques for solving the Maxwell equations.

4. Quantum Theory

The fact that the Pauli and Dirac matrix algebras are representations of the geometric algebras of real space and spacetime suggests that GA is a natural tool for studying quantum theory. This is indeed the case. Both Pauli and Dirac spinors can be handled in the real geometric algebras of space and spacetime, and this provides a number of insights into their geometric roles. We start with the case of non-relativistic spinors. A Pauli column spinor ψ^a can be placed in a direct $1 \leftrightarrow 1$ correspondence with an element of the even subalgebra of 3D geometric algebra as follows:

$$\psi^a = \begin{pmatrix} a^0 + ia^3 \\ -a^2 + ia^1 \end{pmatrix} \quad \leftrightarrow \quad \psi = a^0 + a^k I\boldsymbol{\sigma}_k. \tag{38}$$

The actions of the quantum operators $\{\hat{\sigma}_k\}$, and the unit imaginary i, are replaced by the operations

$$\hat{\sigma}_k|\psi\rangle \quad \leftrightarrow \quad \boldsymbol{\sigma}_k\psi\boldsymbol{\sigma}_3 \qquad (k = 1\ldots3) \tag{39}$$

$$i|\psi\rangle \quad \leftrightarrow \quad \psi I\boldsymbol{\sigma}_3. \tag{40}$$

A feature of this scheme is that the complex structure is taken over entirely by the bivector $I\boldsymbol{\sigma}_3$, acting to the right of ψ.

Every calculation that can be performed with the column spinor ψ^a can also be performed with the even element ψ, and in practice the latter approach is usually easier. One reason for this is the natural decomposition of ψ into a density term and a rotor:

$$\psi = \rho^{1/2}R, \tag{41}$$

where

$$\rho = \psi\tilde{\psi}. \tag{42}$$

For example, the spin vector $\boldsymbol{s}$ has components

$$s_k = \langle\psi|\hat{\sigma}_k|\psi\rangle. \tag{43}$$

In terms of GA these become

$$s_k = \langle\tilde{\psi}\boldsymbol{\sigma}_k\psi\boldsymbol{\sigma}_3\rangle = \boldsymbol{\sigma}_k\cdot(\psi\boldsymbol{\sigma}_3\tilde{\psi}). \tag{44}$$

It follows that all of the components of the spin vector can be summarised in the single expression

$$\boldsymbol{s} = \psi\boldsymbol{\sigma}_3\tilde{\psi} = \rho R\boldsymbol{\sigma}_3\tilde{R}. \tag{45}$$

The 3D rotor R is therefore an instruction to rotate the fixed $\boldsymbol{\sigma}_3$ vector onto the observable spin vector. This establishes a natural link with the description of a rotating rigid body [15, 17].

The normalised observables define the unit spin vector

$$\hat{\boldsymbol{s}} = \frac{\psi\boldsymbol{\sigma}_3\tilde{\psi}}{\rho} = R\boldsymbol{\sigma}_3\tilde{R}. \tag{46}$$

This defines a unit vector which is usually represented as a point on the *Bloch sphere*. The quantum density matrix is also easily encoded. For normalised pure states the density matrix is defined by

$$\hat{\rho} = |\psi\rangle\langle\psi|. \tag{47}$$

This maps directly to the equivalent multivector

$$\hat{\rho} = \psi\tfrac{1}{2}(1+\boldsymbol{\sigma}_3)\tilde{\psi} = \tfrac{1}{2}(1+\hat{\boldsymbol{s}}). \tag{48}$$

For mixed states we simply sum the density matrices and normalise, so that

$$\hat{\rho} = \tfrac{1}{2}(1+\boldsymbol{P}), \quad \boldsymbol{P}^2 \leq 1. \tag{49}$$

This is sufficient to encode single particle quantum theory, but what we really need is a version of multiparticle quantum mechanics. Surprisingly, this requires the full, relativistic theory.

A similar construction for Pauli spinors can be applied to Dirac spinors. A Dirac spinor has 8 real components, and these are placed in a $1 \leftrightarrow 1$ correspondence with a even-grade element of the STA as follows:

$$|\psi\rangle = \begin{pmatrix} |\phi\rangle \\ |\eta\rangle \end{pmatrix} \quad \leftrightarrow \quad \psi = \phi + \eta\boldsymbol{\sigma}_3. \tag{50}$$

Here $|\phi\rangle$ and $|\eta\rangle$ are two-component spinors, and ϕ and η are their Pauli-even equivalents. The action of the operators $\{\hat{\gamma}_\mu, \hat{\gamma}_5, i\}$ becomes

$$
\begin{aligned}
\hat{\gamma}_\mu|\psi\rangle &\quad\leftrightarrow\quad \gamma_\mu\psi\gamma_0 \quad (\mu = 0,\ldots,3) \\
i|\psi\rangle &\quad\leftrightarrow\quad \psi I\boldsymbol{\sigma}_3 \\
\hat{\gamma}_5|\psi\rangle &\quad\leftrightarrow\quad \psi\boldsymbol{\sigma}_3.
\end{aligned}
\tag{51}
$$

With these relations one can immediately write down a form of the Dirac equation in the STA as

$$\nabla\psi I\boldsymbol{\sigma}_3 - eA\psi = m\psi\gamma_0, \tag{52}$$

where A is the electromagnetic vector potential and m is the mass. This equation is entirely equivalent to the standard matrix form of the Dirac equation, even though it is written entirely in the real STA.

The main observables in Dirac theory are also simply encoded now. For example, the current is defined by

$$J^\mu = \langle\psi|\hat{\gamma}^\mu|\psi\rangle \quad\leftrightarrow\quad \langle\tilde{\psi}\gamma^\mu\psi\gamma_0\rangle = \gamma^\mu\cdot(\psi\gamma_0\tilde{\psi}). \tag{53}$$

We can therefore reconstitute the vector $J = J^\mu\gamma_\mu$ to obtain

$$J = \psi\gamma_0\tilde{\psi}. \tag{54}$$

The remaining observables have equally simple expressions [18]. Further insight into the role of a spinor is provided by first writing

$$\psi\tilde{\psi} = \rho e^{I\beta} \tag{55}$$

so that we can set

$$\psi = \rho^{1/2}e^{I\beta/2}R. \tag{56}$$

The object R satisfies $R\tilde{R} = 1$, so is a spacetime rotor. We now find that

$$J = \rho R\gamma_0\tilde{R} = \rho v, \tag{57}$$

so the velocity v is obtained from a boost of the γ_0 vector onto the observable. The boost is controlled by the rotor R.

Now suppose we wish to extend to describe an n-particle quantum state. We construct a $4n$-dimensional relativistic configuration space, spanned by the vectors $\{\gamma_\mu^a\}$, where $\mu = 0\ldots3$ labels the spacetime vector, and $a = 1\ldots n$ labels the particle space. Vectors from distinct particle spaces are orthogonal, so we have

$$\gamma_\mu^a\cdot\gamma_\nu^b = \eta_{\mu\nu}\delta^{ab}. \tag{58}$$

With this algebra one can construct a multiparticle Dirac equation, which provides a simple, geometric encoding of Pauli antisymmetrisation [18]. If we now apply a spacetime split in one space we obtain the relative vectors

$$\boldsymbol{\sigma}_j^a = \gamma_j^a \gamma_0^a, \qquad j = 1 \ldots 3, \quad a = 1 \ldots n. \tag{59}$$

Bivectors from distinct spaces *commute*, as can be seen from

$$\begin{aligned} \boldsymbol{\sigma}_i^a \boldsymbol{\sigma}_j^b &= \gamma_i^a \gamma_0^a \gamma_j^b \gamma_0^b \\ &= \gamma_i^a \gamma_j^b \gamma_0^b \gamma_0^a \\ &= \gamma_j^b \gamma_0^b \gamma_i^a \gamma_0^a = \boldsymbol{\sigma}_j^b \boldsymbol{\sigma}_i^a \qquad (a \neq b). \end{aligned} \tag{60}$$

It follows that the geometric product of elements of even grade from different spaces is equivalent to the *tensor product*. This makes it a straightforward exercise to construct multiparticle quantum states, and all that is ever required is the geometric product! This idea is now the basis for a wide range of applications, including quantum information theory and NMR [19, 20, 21].

5. Gravitation

The STA is the geometric algebra of (flat) Lorentzian spacetime. This might appear to pose a problem for its continued application in general relativity, where spacetime becomes curved. This turns out not to be the case. The solution is to work with a gauge theory formulation of gravity. The first satisfactory gauge treatment of gravity was formulated by Kibble in 1961 [22]. The gauge theory approach leads naturally to an extended version of general relativity (GR) known as a spin-torsion theory [23] though, in the absence of macroscopic spin, the equations reduce to those of GR. In [24] the gauge theory treatment was reconsidered in the STA framework. It was shown that gravitation can be developed as a gauge theory built on arbitrary finite transformations. And, unlike Kibble's earlier theory, the combination of the gauge treatment and STA produces a theory which is conceptually and computationally simpler than GR. This theory is known as gauge theory gravitation, or GTG.

The gauge theory of gravitation requires the introduction of two gauge fields. The first is a position-dependent linear function $\bar{\mathsf{h}}(a)$, or $\bar{\mathsf{h}}(a, x)$. This is linear in its vector argument a, and is a general non-linear function of the position vector $x = x^\mu \gamma_\mu$. This gauge field ensures that the equations remain covariant under arbitrary, nonlinear displacement. Suppose that the vector field $J(x)$ is defined by

$$J = \bar{\mathsf{h}}(\nabla \phi). \tag{61}$$

Then, if we replace $\phi(x)$ by $\phi'(x) = \phi(x')$, and transform $\bar{\mathsf{h}}$ accordingly, $J(x)$ transforms simply to $J'(x) = J(x')$. By inserting the $\bar{\mathsf{h}}$ field at various points in STA field equations, we can construct equations which remain covariant under arbitrary displacements.

The second gauge field is denoted $\Omega(a) = \Omega(a,x)$, and is a bivector-valued linear function of its argument a. The position-dependence of $\Omega(a)$ is also generally non-linear. The bivector field $\Omega(a)$ is the connection for the gauge group of Lorentz transformations. Recall that a Lorentz transformation can be written as $a \mapsto Ra\tilde{R}$. Suppose now that all fields are subject to a local rotation generated by the position dependent rotor R. The gauge fields transform as:

$$\bar{\mathsf{h}}(a) \mapsto \bar{\mathsf{h}}'(a) = R\bar{\mathsf{h}}(a)\tilde{R}, \tag{62}$$

and

$$\Omega(a) \mapsto \Omega'(a) = R\Omega(a)\tilde{R} - 2a\cdot\nabla R\,\tilde{R}. \tag{63}$$

The gauge fields ensure that equations can remain covariant under local rotations, as well as arbitrary displacements. For example, the Dirac equation generalises to

$$\bar{\mathsf{h}}(\gamma^\mu)D_\mu\psi I\sigma_3 = m\psi\gamma_0, \tag{64}$$

where

$$D_\mu\psi = \left(\partial_\mu + \tfrac{1}{2}\Omega(\gamma_\mu)\right)\psi. \tag{65}$$

The Dirac wavefunction transforms as $\psi \mapsto R\psi$ under rotations. Observables, such as the current $\psi\gamma_0\tilde{\psi}$, then transform as covariant multivectors.

The gravitational field equations can be expressed in various ways. Suppose we introduce an arbitrary set of coordinates x^μ in the STA, with associated coordinate frame $\{e_\mu\}$ and reciprocal frame $\{e^\mu\}$. From these we define

$$g^\mu = \bar{\mathsf{h}}(e^\mu), \qquad g_\mu = \mathsf{h}^{-1}(\mathsf{e}_\mu). \tag{66}$$

(The overbar denotes the adjoint on the linear function $\mathsf{h}(\mathsf{a})$). In terms of these the metric is defined by

$$g_{\mu\nu} = g_\mu\cdot g_\nu. \tag{67}$$

One can proceed to perform all calculations directly from the metric, but this is by no means the best way to work. Instead it is preferable to work with an expanded set of first-order equations. We define

$$L_a = a\cdot\bar{\mathsf{h}}(\nabla) = a\cdot g^\mu\,\partial_\mu, \quad \omega(a) = a\cdot g^\mu\,\Omega_\mu. \tag{68}$$

Assuming no torsion is present, the $\bar{\mathsf{h}}(a)$ and $\omega(a)$ gauge fields are related by the bracket identity

$$[L_a, L_b] = L_c, \tag{69}$$

where

$$c = L_a b + \omega(a)\cdot b - L_b a - \omega(b)\cdot a. \tag{70}$$

The gravitational field strength is encoded in the Riemann tensor, $\mathcal{R}(a\wedge b)$, which is defined by

$$\mathcal{R}(a\wedge b) = L_a\omega(b) - L_b\omega(a) + \omega(a)\times\omega(b) - \omega(c), \tag{71}$$

with c determined by equation (70). The Ricci and Einstein tensors and the Ricci scalar are defined by

$$\text{Ricci Tensor:} \quad \mathcal{R}(b) = \gamma_\mu\cdot\mathcal{R}(\gamma^\mu\wedge b) \tag{72}$$

$$\text{Ricci Scalar:} \quad \mathcal{R} = \gamma_\mu\cdot\mathcal{R}(\gamma^\mu) \tag{73}$$

$$\text{Einstein Tensor:} \quad \mathcal{G}(a) = \mathcal{R}(a) - \tfrac{1}{2}a\mathcal{R}. \tag{74}$$

6. Spherical Systems

Our analysis of the effects of a spherical black hole starts with the Schwarzschild solution in the following form:

$$ds^2 = dt^2 - \left(dr + \left(\frac{2GM}{r}\right)^{1/2} dt\right)^2 - r^2 d\Omega^2. \tag{75}$$

Here the time coordinate t is the proper time of an observer freely falling from rest at infinity. This simple form of the Schwarzschild solution was first given by Painlevé and Gullstrand (see [25]). The solution is well-behaved at the horizon, and has a number of other attractive properties, but it has been strangely neglected. In terms of the $\bar{\mathsf{h}}$ function, the solution is generated by

$$\bar{\mathsf{h}}(a) = a - \sqrt{2M/r}\, a\cdot e_r\, e_t. \tag{76}$$

We call this particular gauge choice the 'Newtonian gauge', due to its simple Newtonian properties (these are discussed further in [24]).

We use this gauge to put the Dirac equation into Hamiltonian form. The derivation is particularly clear in Cartesian coordinates. The metric is now

$$ds^2 = \eta_{\mu\nu}dx^\mu dx^\nu - \frac{GM}{r}dt^2 - \frac{2}{r}\left(\frac{2GM}{r}\right)^{1/2} a_\mu\, dt\, dx^\mu \tag{77}$$

where $a_\mu = (0, x, y, z)$. The g_μ vectors are now given by

$$g_0 = \gamma_0 + \left(\frac{2GM}{r}\right)^{1/2} \frac{x_i}{r}\gamma_i, \quad g_i = \gamma_i \quad (i = 1, 2, 3) \tag{78}$$

and one can check easily that $g_\mu \cdot g_\nu = g_{\mu\nu}$. The g_μ vectors generate a Dirac equation of the form

$$\nabla\psi I\sigma_3 - \left(\frac{2GM}{r}\right)^{1/2} \gamma_0 \big(\partial_r\psi + 3/(4r)\psi\big) I\sigma_3 = m\psi\gamma_0. \tag{79}$$

The full relativistic wave equation in a black hole background therefore picks up a single interaction term:

$$H_I\psi = \left(\frac{2GM}{r}\right)^{1/2} \hbar \left(\partial_r\psi + 3/(4r)\psi\right) I\sigma_3. \tag{80}$$

(The $\hbar$ has been written explicitly here to emphasise the relation with the standard radial momentum operator $-i\hbar\partial_r$.) This is impressively simple, but there is a subtlety connected with the Hermiticity of H_I. Writing

$$H_I(\psi) = (2M/r)^{1/2} r^{-3/4} \partial_r \big(r^{3/4}\psi\big) I\sigma_3, \tag{81}$$

we find that

$$\begin{aligned} \int d^3x \, \langle \phi^\dagger H_I(\psi) \rangle_S &= \sqrt{2M} \int d\Omega \int_0^\infty r^2 \, dr \, r^{-5/4} \langle \phi^\dagger \partial_r (r^{3/4}\psi) I\sigma_3 \rangle_S \\ &= \int d^3x \, \langle (H_I(\phi)^\dagger \psi \rangle_S + \sqrt{2M} \int d\Omega \left[r^{3/2} \langle \phi^\dagger \psi I\sigma_3 \rangle_S \right]_0^\infty \end{aligned} \tag{82}$$

where $\langle\,\rangle_S$ denotes the projection onto the 'complex' 1 and $I\sigma_3$ terms, and $\phi^\dagger = \gamma_0\tilde{\phi}\gamma_0$. For all normalised states the final term in (82) tends to zero as $r \to \infty$. But it can be shown that wavefunctions tend to the origin as $r^{-3/4}$, so the lower limit is finite and H_I is therefore not (quite) a Hermitian operator. This immediately rules out the existence of normalisable stationary states with constant real energy, and gives us an insight into the nature of wavefunction decay and collapse into a black hole, which will be important below.

6.1. THE SCHRÖDINGER LIMIT

To begin understanding some of the physical properties of this interaction term, we can check that its properties are sensible in the Schrödinger limit. Defining a *reduced* radial wavefunction $U(r)$ via

$$\psi = \frac{U}{r}\Phi(\theta, \phi) \tag{83}$$

one finds that the standard reduction of the Dirac equation to the Schrödinger equation (here in the presence of the interaction term) yields

$$\frac{d^2U}{dr^2} - l(l+1)\frac{U}{r^2} - \frac{2mi}{\hbar}\left(\frac{2GM}{r}\right)^{1/2} r^{1/4}\frac{d}{dr}(Ur^{-1/4}) = -\frac{2mE}{\hbar^2}U \quad (84)$$

Now the spherically symmetric gravitational problem in the Schrödinger case is formally similar to that of a charged nucleus. In both cases we have an underlying inverse square force. In fact we can manipulate the above equation all the way to complete agreement with the Hydrogen atom problem by carrying out a phase transformation as follows. Define the *gravitational Bohr radius*

$$a_0 = \frac{\hbar^2}{GMm^2} \quad (85)$$

and set

$$U = W \exp\left(i(8r/a_0)^{1/2}\right). \quad (86)$$

Then the Schrödinger equation becomes

$$W'' - \left(\frac{l(l+1)}{r} - \frac{2}{a_0 r} - \frac{2mE}{\hbar^2}\right) W = 0 \quad (87)$$

which is now identical to the Hydrogen atom case. We can therefore read off the energy level spectrum:

$$E = -\frac{R_G}{n^2}, \quad (88)$$

where the *gravitational Rydberg* is given by

$$R_G = \frac{m}{2}\left(\frac{GMm}{\hbar}\right)^2. \quad (89)$$

As we shall see later, these energy levels and associated wavefunctions do provide good approximations for the relativistic solutions in cases where the bulk of the probability density lies well outside the black hole horizon. However, we can see that near the black hole horizon, the Schrödinger approach becomes inconsistent. If we consider the Schrödinger current, we find that it is given by

$$\frac{\rho\hbar}{m}\boldsymbol{\nabla}\left(\frac{8r}{a_0}\right)^{1/2} = \rho\left(\frac{2GM}{r}\right)^{1/2}\hat{\boldsymbol{r}} \quad (90)$$

where $\hat{\boldsymbol{r}}$ is a unit vector in the radial direction. This implies a velocity associated with the current of $(2GM/r)^{1/2}$ — the same as that of an observer freely falling from rest at infinity. The problem, of course, is that this becomes equal to c at the horizon, meaning that the Schrödinger approach becomes inconsistent there. Instead, therefore, we must work with the full Dirac theory.

6.2. RELATIVISTIC BOUND STATES

We start by assuming

$$\psi(x) = \phi(\boldsymbol{x})\alpha(t) \tag{91}$$

in equation (79). As usual, the solution of the t-equation is

$$\alpha(t) = \exp(-I\sigma_3 Et), \tag{92}$$

where E is the separation constant. The non-Hermiticity of H_I means that E cannot be purely real if ψ is normalisable. The imaginary part of E is determined by equation (82) and, for suitably normalised states, we find that

$$\mathrm{Im}(E) = -\lim_{r\to 0} 2\pi\sqrt{2M}\langle\phi^\dagger\phi\rangle r^{3/2}. \tag{93}$$

This equation shows that the imaginary part of E is necessarily negative, so the wavefunction decays with time. This is consistent with the fact that the streamlines generated by the conserved current $\psi\gamma_0\tilde{\psi}$ are timelike curves and, once inside the horizon, must ultimately terminate on the origin. However, we can still ask the question of whether a *discrete set* of complex energy eigenvalues exist. This would then give us a set of spectral levels similar to that in a Hydrogen atom, except with the novelty that each state decays with time, making it a type of resonance.

We now show this does in fact happen, but a crucial question before continuing is whether the energy eigenvalues so obtained are physically meaningful. In the GTG approach, this will be the case if we can show that the values obtained are *gauge invariant*. In [24], we show that the radial coordinate r is physically well-defined (e.g. it can be experimentally determined by tidal forces) and hence that the full set of gauge transformations compatible with our setup in the spherically symmetric case are:

$$\begin{aligned} &t \mapsto t + f(r), &&\text{position gauge change}\\ &\psi \mapsto R\psi, &&\text{rotation gauge change}\end{aligned}$$

with the rotor R being a function of $\boldsymbol{x}$ only ($R = R(\boldsymbol{x})$). Suppose that we have a solution in some gauge

$$\psi(x) = \phi(\boldsymbol{x})\exp(-I\sigma_3 Et), \tag{94}$$

where $E = E_r + I\boldsymbol{\sigma}_3 E_i$. We now carry out a gauge transformation of the above form, obtaining

$$\begin{aligned} R\psi(x') &= R\phi(\boldsymbol{x}) \exp(-I\boldsymbol{\sigma}_3 E(t + f(r))) \\ &= R(\boldsymbol{x})\phi(\boldsymbol{x}) \exp(-I\boldsymbol{\sigma}_3 Et) \exp(-I\boldsymbol{\sigma}_3 Ef(r)) \end{aligned}$$

Thus

$$\partial_t R\psi(x') I\boldsymbol{\sigma}_3 = R\psi(x')(E_r + I\boldsymbol{\sigma}_3 E_i) \tag{95}$$

i.e. the new wavefunction is still an eigenfunction of the evolution operator, with the same eigenvalue. This means E_r and E_i are gauge invariant, and hence potentially physically observable.

Proceeding with a solution in the Newtonian gauge, we carry out an angular separation via writing

$$\psi(\boldsymbol{x}, \kappa) = \begin{cases} \psi_l^m u(r) + \hat{\boldsymbol{r}}\psi_l^m v(r) I\boldsymbol{\sigma}_3 & \kappa = l+1 \\ \hat{\boldsymbol{r}}\psi_l^m u(r)\boldsymbol{\sigma}_3 + \psi_l^m iv(r) & \kappa = -(l+1), \end{cases} \tag{96}$$

where κ is a non-zero integer and $u(r)$ and $v(r)$ are complex functions of r (sums of a scalar and an $I\boldsymbol{\sigma}_3$ term). Here the ψ_l^m are *spherical monogenics* (see e.g. [18]), which carry the angular dependency of the wavefunction. Specifically, the unnormalised monogenic ψ_l^m is defined by

$$\psi_l^m = [(l+m+1)P_l^m(\cos\theta) - P_l^{m+1}(\cos\theta) i\sigma_\phi] e^{m\phi I\boldsymbol{\sigma}_3}, \tag{97}$$

where $l \geq 0$, $-(l+1) \leq m \leq l$, and P_l^m are the associated Legendre polynomials.

Substituting (96) into the Dirac equation with the time dependence separated out, and using the properties of the spherical monogenics, we arrive at the coupled radial equations

$$\begin{pmatrix} 1 & -(2M/r)^{1/2} \\ -(2M/r)^{1/2} & 1 \end{pmatrix} \begin{pmatrix} u_1' \\ u_2' \end{pmatrix} = \boldsymbol{A} \begin{pmatrix} u_1 \\ u_2 \end{pmatrix} \tag{98}$$

where

$$\boldsymbol{A} = \begin{pmatrix} \kappa/r & j(E+m) - (2M/r)^{1/2}(4r)^{-1} \\ j(E-m) - (2M/r)^{1/2}(4r)^{-1} & -\kappa/r \end{pmatrix}, \tag{99}$$

u_1 and u_2 are the reduced functions defined by

$$u_1 = ru \qquad u_2 = jrv, \tag{100}$$

and the primes denote differentiation with respect to r. (We employ the abbreviation j for right-sided multiplication by $I\boldsymbol{\sigma}_3$.)

To analyse (98) we first rewrite it in the equivalent form

$$(1-2M/r)\begin{pmatrix}u_1'\\u_2'\end{pmatrix}=\begin{pmatrix}1 & (2M/r)^{1/2}\\(2M/r)^{1/2} & 1\end{pmatrix}\boldsymbol{A}\begin{pmatrix}u_1\\u_2\end{pmatrix}. \tag{101}$$

This makes it clear that the equations have regular singular points at the origin and horizon ($r = 2M$), as well as an irregular singular point at $r=\infty$. To our knowledge, the special function theory required to deal with such equations has not been developed. In the massless case the equations can be manipulated into a second order equation of Heun type [26], but this does not appear to help with the present problem. In the absence of a suitable mathemtical theory, we must either attempt a numerical solution, or look for power series with a limited radius of convergence. We start by considering the latter approach, and look for power-series solutions around the horizon. To this end we introduce the series

$$u_1=\eta^s\sum_{k=0}^{\infty}a_k\eta^k, \qquad u_2=\eta^s\sum_{k=0}^{\infty}b_k\eta^k, \tag{102}$$

where $\eta = r - 2M$. The index s controls the radial dependence of ψ at the horizon, so represents a physical quantity. To find the values that s can take, we substitute (102) into (101) and set $\eta = 0$. This results in the equation

$$\frac{s}{2M}\begin{pmatrix}a_0\\b_0\end{pmatrix}=\begin{pmatrix}1&1\\1&1\end{pmatrix}\begin{pmatrix}\kappa/(2M) & j(E+m)-(8M)^{-1}\\ j(E-m)-(8M)^{-1} & -\kappa/(2M)\end{pmatrix}\begin{pmatrix}a_0\\b_0\end{pmatrix} \tag{103}$$

Rewriting this in terms of a single matrix and setting its determinant to zero yields the two indicial roots

$$s=0 \quad \text{and} \quad s=-\tfrac{1}{2}+4jME. \tag{104}$$

The $s = 0$ solution is entirely sensible — the power series is analytic, and nothing peculiar happens at the horizon. If one calculates the conserved current $\psi\gamma_0\tilde{\psi}$ associated with this solution, one finds it is finite and inward-pointing at the horizon, as one would expect. The second root leads to solutions which are ill-defined at the horizon, and have a discontinuity in the current there. This raises some delicate issues as regards the physical meaning of these solutions, some of which are discussed in Section 8 of [24]. If one disregards these problems then, remarkably, it turns out that the discontinuity can be interpreted as the creation of a net outward flux of particles at the horizon, described by a Fermi-Dirac distribution at a temperature given by the Hawking temperature [27]

$$T=\frac{1}{8\pi Mk_B}. \tag{105}$$

The value of this temperature comes directly from the imaginary part of the second index s in (104). The surprising feature here is that a Fermi-Dirac distribution is obtained without any of the apparatus of quantum field theory. It turns out that repeating this exercise with a field of integer spin (for example a scalar field, or the electromagnetic field) yields the Hawking temperature again, with a flux described by a Bose-Einstein distribution.

This is obviously an interesting and important area to explore more fully, but here we wish to concentrate instead on the question of the existence of bound state solutions. By analogy with the rejection of unnormalisable solutions in the Hydrogen atom problem, we here reject the solution which is singular at the horizon, and ask whether bound state solutions exist composed wholly of the regular solution, with index $s = 0$. Since the wavefunction we are using is described by two complex functions of r (the $u(r)$ and $v(r)$ in equation (96)), there are four real degrees of freedom. Two of these are absorbed when we discard the singular solution, and the remaining two correspond to the amplitude and phase of the regular solution at the horizon. The equations are linear so, without loss of generality, we can fix the parameters to arbitrary values. This means we have no degrees of freedom left in the wavefunction, and can proceed to try to find the (complex) values of energy which lead to normalisable solutions — these will be the eigenfunctions we desire. In the absence of the necessary special function theory, this has been carried out numerically. We hope that one day a more analytic approach may be possible.

The numerical method chosen involved picking trial values for the real and imaginary components of the energy, and then integrating outwards from the horizon as far as possible. What sets the limits to such an integration is that either numerical instability sets in, or the wavefunction density starts diverging exponentially, due to an incorrect (non-eigenvalue) value of E being chosen. The search method involving minimizing the density as a function of the imaginary component of E automatically for each E_r over a grid of values of E_r. Some typical results are shown in Fig. 3. These are for $l = 0$, a black hole mass $M = 1$, and particle mass $m = 0.1$. The y-axis is $\log_{10}$ of the wavefunction density, so a dynamic range in selecting eigenvalues of about 10^4 is displayed here.

The search for eigenvalues is aided by searching near regions in E_r where the non-relativistic theory would predict a bound state. We would expect the non-relativistic energy found in equation (88) to act as a small correction to the rest mass energy mc^2, giving

$$E_n \approx mc^2 - \frac{GMm}{2a_0 n^2} \tag{106}$$

where a_0 is the gravitational Bohr radius (85), and n is the quantum number appropriate in the non-relativistic analysis. For the ground state with $m =$

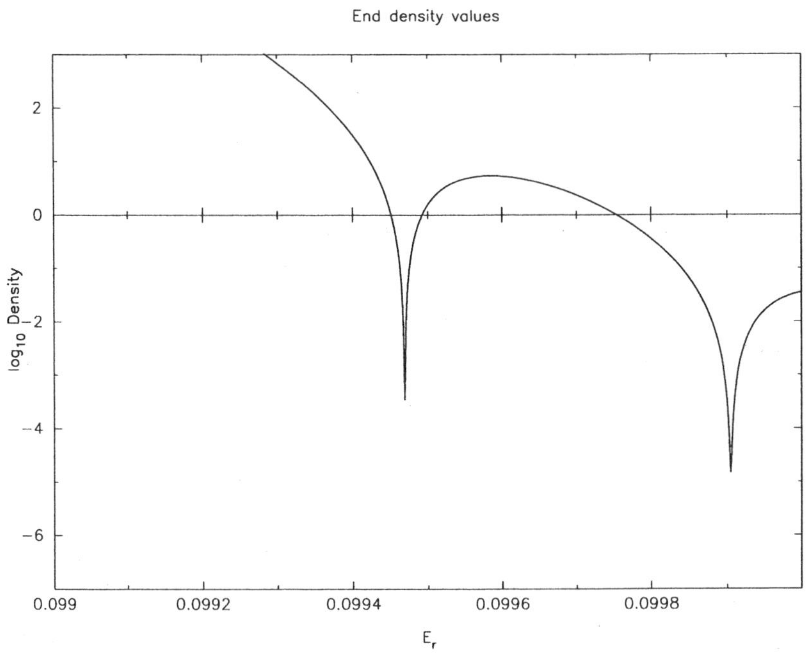

Figure 3. A plot of wavefunction density at a fixed large distance from the black hole, for various values of the real part of the particle energy E_r, with automatic minimization carried out in E_i. The particle mass is 0.1, the angular quantum number $l = 0$ and the black hole mass is 1.0. The y-axis shows $\log_{10}$ of the wavefunction density at a radius $(800\,GM/c^2)$.

0.1, $M = 1$, this predicts $E_r = 0.0995$, indeed very close to where the first dip is seen in Fig. 3. $n = 2$ would yield $E_r = 0.099875$, close to the second dip. The wavefunction density corresponding to this first excited state is shown in Fig. 4.

To establish formally that we have genuine eigenvalues requires integrating out to infinite radial coordinate, and demonstrating that the density integral remains finite. This can be achieved by integrating inwards from infinity and simultaneously outwards from the horizon, and matching at an intermediate point. This process is complicated by the fact that the wavefunction has two types of essential singularity at infinity, which have to be allowed for before series solutions at infinity can be found. One is the $\exp(-\sqrt{m^2 - E^2}\,r)$ behaviour which one expects by analogy with the Dirac solution solution for the Hydrogen atom. The second is an essential singularity in the phase, given by a dependence of the form $\exp(j2E\sqrt{2Mr})$. A term of this kind could be expected from what we found above for the gravitational Schrödinger equation.

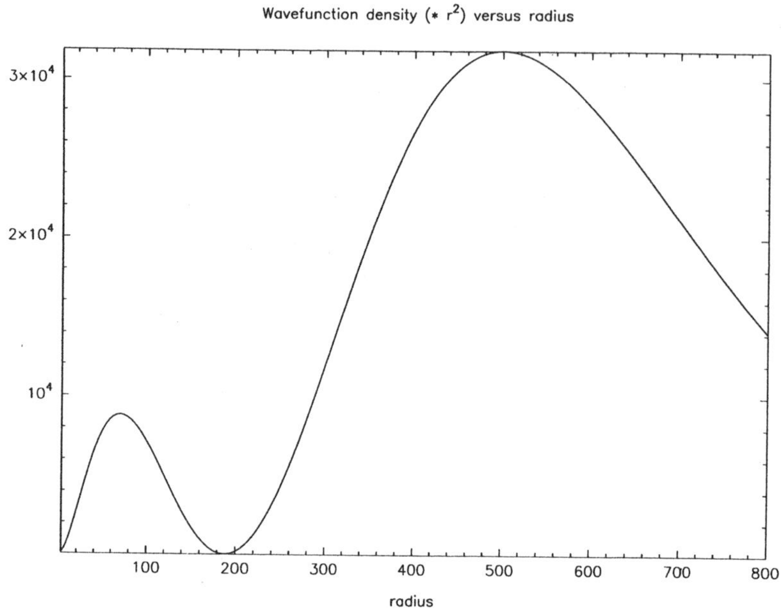

Figure 4. A plot of $r^2\times$ wavefunction density for the first excited state of a fermion bound to a black hole. As in Fig. 3 the particle mass is 0.1, the angular quantum number $l = 0$ and the black hole mass is 1.0.

TABLE 1. Groundstate energies calculated for two values of m and $M = 1$.

m	E_r	E_i
0.1e0	0.099468827746	-2.7870281e-6
0.2e0	0.194834514694	-.000751079389

We have confirmed that this process works, which means we can be confident that our eigenvalues are correct. A factor reinforcing confidence is that all the computations have been repeated in a different gauge (the advanced Eddington-Finkelstein gauge — see below) and exactly the same numerical values were found here as well. Furthermore, one can compare the numerically obtained imaginary component of E with that found by integrating the wavefunction to as large a radius as possible and normalising it, and then using equation (93) at the origin. This also checks out precisely. Some ground state eigenvalues are as shown in Table 1 and for the first excited state in Table 2.

What physical regimes might these solutions be interesting in? We can

TABLE 2. First excited state energies calculated for two values of m and $M = 1$.

m	E_r	E_i
0.1	0.099870235768	-.356282508e-6
0.2	0.19880214433	-.00009960784655

get some idea of the importance of the decay by defining a dimensionless quantity

$$\alpha = \frac{\text{gravitational Bohr radius}}{\text{Schwarzschild radius}/2} \tag{107}$$

We can re-write this relation in the form

$$Mm = \frac{1}{\sqrt{\alpha}} (2.2 \times 10^{-8}\,\text{kg})^2. \tag{108}$$

So, for example, if we take m equal to the electron mass, $m = m_e = 9.1 \times 10^{-31}$ kg, then $M = (1/\sqrt{\alpha})\, 5.3 \times 10^{14}$ kg. Primordial black holes of size 10^{12} to 10^{14} kg say, then lie in an interesting range as regards quantum effects.

The size of α is inversely related to the importance of decay effects. A crude estimate, obtained by working with non-relativistic approximations to the ground state wavefunctions, suggests that (in the near non-relativistic regime) we should find

$$E_i \sim -8\alpha^{-5/2} mc^2 \tag{109}$$

which is roughly borne out by the numerical results. For electrons coupled to primordial black holes, this is therefore of order the 'zitterbewegung' time $\hbar/(m_e c^2)$, (approximately 10^{-21} seconds) which is obviously very fast, and it is not clear how such a decay would manifest itself. Typical questions which should now be faced include:

1. What does the detailed energy spectrum look like?
2. What is the significance of the antiparticle solutions, which have opposite sign of the real part of their energy?
3. Do the energy differences between shells correspond to something observable, given the rapid decay rates?
4. If they do, what is the mechanism of radiation, and what quantum jumps are involved?
5. Are there bound states, for which the expectation value of r lies wholly inside the horizon?

6. What is the extension to the Reissner-Nordstrom and Kerr cases?
7. Can we incorporate multiparticle effects?

The last question is important since a multiparticle approach is necessary to provide a proper link to the Hawking radiation. Equally, in the same way that for the calculation of the Lamb shift it is necessary to evaluate a sum involving bound state energies of the Hydrogen atom, so presumably in the black hole case, vacuum fluctuation effects should be calculated taking proper account of the existence and spectrum of bound states. We are not sure that so far this has been carried out.

7. Propagators and black hole scattering cross-sections

We have considered bound states, so an obvious next topic is fermion scattering by black holes. What we wish to look at is again in the spirit of seeing how far we can get in applying conventional quantum mechanics to a gravitational context, using the added facilities of geometric algebra.

In the Coulomb scattering of an electron by a point charge, the first quantum corrections to the Rutherford formula are embodied in the Mott scattering cross-section. A convenient way to arrive at this is via using the Feynman propagator to find the first order S-matrix linking initial and final states. In a geometric algebra approach, one finds that the S-matrix is replaced by a scaled rotor, which as well as giving the quantum mechanical amplitude, also explicitly embodies the spin transformation properties of the process. In simple cases, this means that the need for spin-sum calculations is obviated, leading to a considerable reduction in the length of the calculations as compared to the conventional approach (see e.g. [18, 28]).

Specifically, let $\psi_i(x)$ be an initial 'input' wavefunction and $\psi_f(x)$ the final 'output' function and Consider the Coulomb scattering problem to first order. What one finds in the GA approach (see [18] for details) is

$$\psi_f(x) = -S_{fi}\psi_i(x)I\boldsymbol{\sigma}_3\frac{(2\pi)^2}{E_f}\delta(E_f - E_i) \tag{110}$$

where E_i and E_f are the initial and final energies and S_{fi} is the scaled rotor just referred to. Clearly it rotates from initial to final momentum and spin states, while its magnitude determines the cross-section via

$$\frac{d\sigma}{d\Omega_f} = S_{fi}\tilde{S}_{fi}. \tag{111}$$

In the case of Coulomb scattering from a nucleus with charge Z one finds explicitly, to first order, that

$$S_{fi} = \frac{Z\alpha}{\boldsymbol{q}^2}(2E + \boldsymbol{q}). \tag{112}$$

Here α is the fine structure constant, $E = E_i = E_f$ is the electron energy and $\boldsymbol{q} = \boldsymbol{p}_f - \boldsymbol{p}_i$ is the change in relative momentum. (Recall the relative momentum is defined by $p\gamma_0 = p{\cdot}\gamma_0 + p\wedge\gamma_0 = E + \boldsymbol{p}$.). This leads to the Mott cross-section

$$\frac{d\sigma}{d\Omega_f} = \frac{Z^2\alpha^2}{\boldsymbol{q}^4}(4E^2 - \boldsymbol{q}^2) = \frac{Z^2\alpha^2}{4\boldsymbol{p}^2\beta^2\sin^4(\theta/2)}\left(1 - \beta^2\sin^2(\theta/2)\right) \qquad (113)$$

where

$$\boldsymbol{q}^2 = (\boldsymbol{p}_f - \boldsymbol{p}_i)^2 = 2\boldsymbol{p}^2(1 - \cos\theta) \quad \text{and} \quad \beta = |\mathbf{p}|/E. \qquad (114)$$

Now the interaction Hamiltonian for the Coulomb potential is the simple scalar operator

$$H_C\psi = \frac{Ze^2}{4\pi\epsilon_0 r}\psi \qquad (115)$$

while for an electron interacting with a black hole using the Newtonian gauge we have seen the interaction Hamiltonian has the form

$$H_I\psi = \left(\frac{2GM}{r}\right)^{1/2}\hbar\,(\partial_r\psi + 3/(4r)\psi)\, I\boldsymbol{\sigma}_3 \qquad (116)$$

The essential part of this is another scalar operator, but this time with a derivative in r present in it. The simplicity of this form suggests we ask whether we can employ the same Feynman propagator techniques we used in the Coulomb calculation to obtain an analogue of the Mott scattering formula in the gravitational case. Along the way we would find the gravitational analogue of the 'rotor' S_{fi}, which would again have the benefit of obviating the need for spin sums.

The key part of the derivation will be finding the momentum space representation of H_I. We may write schematically

$$\mathcal{H}_I(\boldsymbol{p}_2, \boldsymbol{p}_1) = \int d^3x e^{-j\boldsymbol{p}_2\cdot\boldsymbol{x}}\mathcal{H}_I e^{j\boldsymbol{p}_1\cdot\boldsymbol{x}} \qquad (117)$$

which yields

$$\mathcal{H}_I(\boldsymbol{p}_2, \boldsymbol{p}_1) = \frac{j3\pi^{3/2}(GM)^{1/2}}{|\boldsymbol{p}_2 - \boldsymbol{p}_1|^{7/2}}(\boldsymbol{p}_2^2 - \boldsymbol{p}_1^2). \qquad (118)$$

On multiplying this by γ_0 we obtain the desrired vertex factor. This has the unusual property of vanishing on-shell, when $|\boldsymbol{p}_1| = |\boldsymbol{p}_2|$. (Energy is conserved throughout since the interaction Hamiltonian is independent of time). In retrospect, this is as we might expect, since otherwise we would have the matrix element going as $\sqrt{M}$ in leading order, rather than as M, which is what what we would expect on the basis of the analogy with Z in the Coulomb case.

While this makes sense, it means we have to go to second order in our calculations, which is a perhaps unfortunate feature of the Newtonian gauge. It is thus of interest to find an alternative gauge in which the desired scattering result can emerge at first order. If this answer agrees with that from the second order Newtonian gauge answer, this would be a powerful indication that both are right and that the result we are calculating is physically meaningful (gauge invariant).

The suitable gauge in which to attempt this is the advanced Eddington Finkelstein (AEF) gauge. This has as its h-function

$$\bar{\mathsf{h}}(a) = a + \frac{M}{r}\, a{\cdot}n\, n \quad \text{where} \quad n = \gamma_0 - \hat{r} \tag{119}$$

and $\hat{r}$ is a unit spacelike vector in the radial direction, so that n is null. This h-function has a simple form, and the gravitational effects of the black hole enter linearly in M, as we want, rather than proportional to $\sqrt{M}$ as in the Newtonian gauge case. However, unlike the latter case, the interaction Hamiltonian is no longer a simple scalar operator on ψ, but has a multivector structure, not analogous to the Coulomb case. Specifically, the Dirac equation in this gauge is

$$\nabla\psi I\boldsymbol{\sigma}_3 + \frac{GM}{r}(\gamma_0 - \hat{r})\left(\frac{\partial}{\partial t} - \frac{\partial}{\partial r} - \frac{1}{2r}\right)\psi I\boldsymbol{\sigma}_3 = m\psi\gamma_0 \tag{120}$$

We can still apply the Feynman techniques even in this case, however, by using the following route. For the general Dirac equation in a gravitational field,

$$D\psi I\boldsymbol{\sigma}_3 - m\psi\gamma_0 = 0 \tag{121}$$

let us rewrite it instead as

$$\nabla\psi i\sigma_3 - m\psi\gamma_0 = \nabla\psi i\sigma_3 - D\psi i\sigma_3 \tag{122}$$

This appears to be a trivial rewriting, but means that we can view the right hand side as an interaction term which can be solved for using the free-particle Feynman propagator appropriate to the left hand side. This is quite a powerful technique, and will apply wherever there are asymptotically free in and out states. Some details of the evaluation of the matrix element in this case are discussed in [6], but here we simply state the result:

$$S_{fi} = \frac{GM}{\boldsymbol{q}^2}\big(E(2E + \boldsymbol{q}) + \boldsymbol{p}^2 + \boldsymbol{p}_f\boldsymbol{p}_i\big). \tag{123}$$

This is the scaled rotor which transforms between initial and final states for a fermion scattering from a black hole, to first order in M, calculated using the AEF gauge. This is to be compared and contrasted with the equivalent Coulomb matrix element (112). It is similar in the first term, though with an extra factor of E, and then contains two extra terms. We note immediately that the presence of the extra E in the first term is a manifestation of the equivalence principle! Without it, the cross-section would depend explicitly on the particle mass m. With it, the cross-section is a function of the particle velocity only.

Before continuing to discuss the cross-section coming from this matrix element, we have to satisfy ourselves that it is physically meaningful. Here we have the advantage that we can verify that we obtain the same result using the Newtonian gauge to *second* order. The details are given in [6] but it turns out that indeed the result is identical to (123), which helps verify gauge invariance. A further argument in favour of gauge invariance is that the similar computation in the Coulomb case is definitely gauge invariant — if we replace the electromagnetic potential A with $A - \nabla\alpha(r)$, so that $A \mapsto A + \hat{r}\, d\alpha/dr$, one finds that (on-shell) the matrix element S_{fi}, and therefore the cross-section, is unaffected. A final piece of evidence is that we can be certain that in the AEF gauge we will not obtain an effect at order $M^{3/2}$, since the perturbation series will involve whole powers of M. This predicts that the *third* order effect in the Newtonian gauge should vanish. This can be shown to be indeed the case by explicit computation in the Newtonian gauge. We therefore take it that our result is physically meaningful, although an explicit proof of gauge invariance (ideally to all orders in M) would of course be desirable.

The cross-section following from (112) is

$$\frac{d\sigma}{d\Omega_f} = \frac{(GM)^2}{4\beta^4 \sin^4(\theta/2)} \left(1 + 2\beta^2 - 3\beta^2 \sin^2\frac{\theta}{2} + \beta^4 - \beta^4 \sin^2\frac{\theta}{2}\right) \tag{124}$$

and is plotted for a particular case in Figure 5. It is possible to check our result against the classical limit for black hole scattering by comparing with the result of Collins, Delbourgo and Williams in [1]. They were able to obtain an explicit series solution for the classical cross-section in the limit of small θ. The first term in their series is of order M^2 and is (translating to our notation)

$$\frac{d\sigma}{d\Omega} = \left(\frac{2GM(2\gamma^2 - 1)}{\theta^2(\gamma^2 - 1)}\right)^2 \tag{125}$$

where γ is the Lorentz factor $(1 - \beta^2)^{-1/2}$. This agrees precisely with the first term in the small angle expansion of our result, equation (124). Their second term

$$\frac{3\pi(GM)^2(5\gamma^2 - 1)}{4\theta^3(\gamma^2 - 1)} \tag{126}$$

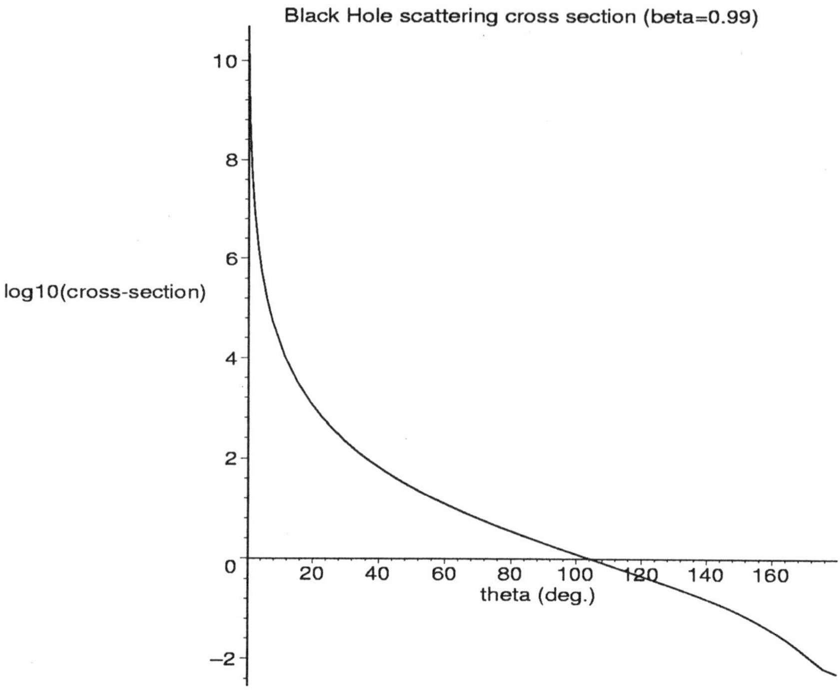

Figure 5. Plot of the black hole scattering cross-section formula (124) for a high speed particle ($\beta = 0.99$), and with $M = 1$.

is also of order M^2, and therefore might be expected to agree with us also. However, our second term is

$$-\frac{(GM)^2\left(4\gamma^4 - 7\gamma^2 + 1\right)}{3\theta^2\left(\gamma^2 - 1\right)^2} \tag{127}$$

which as well as being quite different numerically, is also different as regards its order in θ, going as θ^{-2} rather than θ^{-3}. We cannot explain this discrepancy, and indeed it seems strange that an odd power of θ could be obtained at all, even in the classical result.

The massless limit $m \mapsto 0$ is also well-defined and leads to the simple formula

$$\frac{d\sigma}{d\Omega} = \frac{(GM)^2 \cos^2(\theta/2)}{\sin^4(\theta/2)}. \tag{128}$$

Again, the low-angle limit recovers the classical formula for the bending of light. This result also predicts zero amplitude in the backward direction, $\theta = \pi$. Null geodesics produce a significant flux in the backward direction, and the fact that zero is predicted here is a diffraction effect for neutrinos which goes beyond the predictions of geometric optics. A similar prediction of zero back-scattering for neutrinos was made in [3]. A more detailed analysis of the cross section in the backward direction also reveals a large 'glory' scattering [3, 5]. In the geometric optics limit this is attributable to multiple orbits, and in the quantum description the glory scattering is described by higher-order terms in GM. To describe these effects in the present scheme requires extending to higher order in perturbation theory. This is currently under investigation.

Extending to higher orders also raises the question of the convergence of the iterative scheme proposed here. This is not a straightforward issue to address as there is no dimensionless coupling constant in the problem. Also, it is not clear whether higher-order quantum terms should still be expected to obey the equivalence principle. One can easily formulate desirable criteria for convergence, such as $GME < 1$ or $GMEv < 1$, but these are too restrictive, given that the low angle formula we arrive at is expected to be valid for all masses and velocities. It would appear that the only way to investigate convergence is to compute the next order terms in the perturbation series directly.

This work should also have clarified the importance of working consistently to the correct order in M. This is particularly clear in the Schwarzschild gauge, where the interaction term contains factors which go as $1-(1-2GM/r)^{1/2}$. An iterative scheme based on this gauge choice should expand out the vertex factor as a series in M, and then keep all of the terms up to the desired order. Such a scheme is workable, but has the disadvantage of introducing new vertex terms at each order in the series solution. It is straightforward to confirm that such a scheme reproduces our result for the fermion cross section, to lowest order.

The next steps include the following:

1. Extension of the above results to second order in the AEF gauge (and perhaps 4th order in Newtonian gauge, to verify gauge invariance at order M^2). Potential new features which may emerge include the 'glory' back scattering effect, mentioned above ([5]).
2. Extension to the Riessner-Nordstrom and Kerr cases. The interaction with the angular momentum in the Kerr case will be particularly interesting, but the computations currently look difficult.
3. Another vertex which can be tried is interaction with a photon, in the presence of the background gravitational field, in order to calculate the gravitational equivalent of bremsstrahlung. This could shed some light on the long-standing problem of the radiation caused by a freely falling electron.

8. Riemann-squared theory and torsion effects in the early universe

We would like finally to discuss some progress in applying GA techniques to higher-order Lagrangian field theories. Our group has already looked at the effects of torsion in gauge theory gravity, and found a new solution for the Dirac field coupled self-consistently to gravity in which torsion was important [29, 30]. We have also considered a GA approach to quadratic gravitational Lagrangians and to topological invariants, in [31]. Here, we wish to highlight some recent work which may be important in application to the early universe.

The higher-order Lagrangian theory which is perhaps best motivated in analogy with other gauge theories is one where the Lagrangian is given by the same expression as would be used in any Yang-Mills theory — the field strength tensor squared. Written in the GA approach this is a term in

$$\mathcal{L}_2 = \mathcal{R}(e^\mu \wedge e^\nu)\cdot\mathcal{R}(e_\nu \wedge e_\mu). \tag{129}$$

This has been explored several times and is known to lead to spherically symmetric metrics of Schwarzschild-like form but with

$$\left(1 - \frac{2M}{r} + \kappa r^2\right) \quad \text{in place of} \quad \left(1 - \frac{2M}{r}\right) \tag{130}$$

If one demands asymptotic flatness, then $\kappa = 0$ and one is back with the standard Schwarzschild metric, which means that the theory passes all the standard solar system tests.

A key feature of this type of theory, which makes it particularly attractive, is that it satisfies *scale covariance* as well as position and rotation gauge covariance. It thus encompasses all the symmetries one might expect to be present. A further feature is that it includes torsion. This is the case even if there is no source of quantum spin, which is a significant difference from theories using the using a Ricci scalar Lagrangian for the gravitational field. For these the torsion is generated entirely by matter. In cosmology, for example, we can consider coupling self-consistently with a Dirac field, which is the form of torsion considered in [29] and [30].

Working with the Lagrangian $\mathcal{L}_2$ we can derive the field equations appropriate for cosmology. In what follows we have assumed spatial flatness. The equations yields a remarkable result. Writing

$$A = 2H^2 + \dot{H} - \frac{1}{4}Q^2 \quad \text{and} \quad B = \frac{3}{2}QH + \frac{1}{2}\dot{Q} \tag{131}$$

where H is the Hubble parameter and Q is a measure of the 'free-space' torsion (both real), we find that it is possible to write the evolution equations

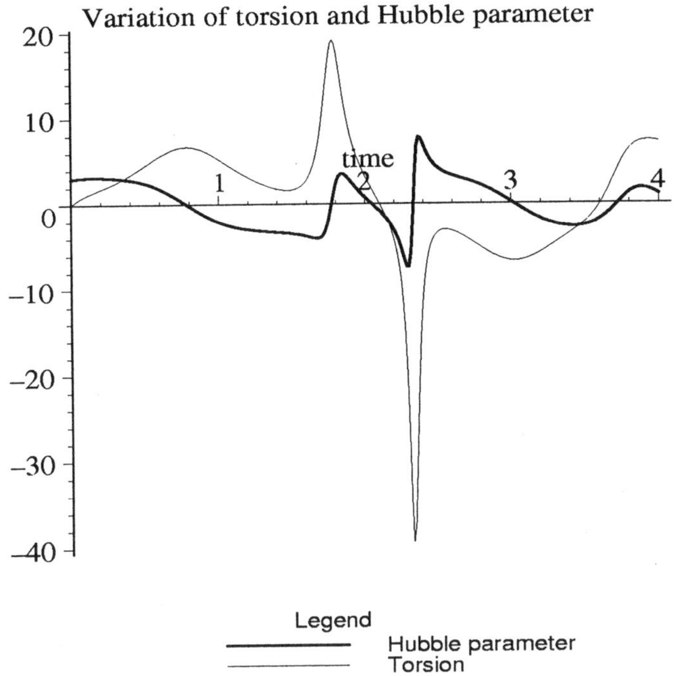

Figure 6. Illustrative plot of variation of Hubble parameter (H) and and free-space torsion (Q) with Q starting at small values. In this case, the matter density is zero at all times, so it is only the torsion field providing the dynamics.

together jointly in a complex form:

$$\frac{d}{dt}(A + iB) = iQ(A + iB) \tag{132}$$

Thus the modulus of the quantity $(A + iB)$ is constant, but its phase gets driven round in a potentially chaotic fashion by the torsion, Q. These are the full equations of the cosmology. They reduce to equations people have studied before in Riemann squared cosmology for $Q = 0$, but it is not clear whether the above behaviour is known for the torsion case.

If one computes numerically some typical results, one finds that the regular behaviour of H is disturbed by the torsion Q which, provided it is not precisely zero initially, builds up and then forces a rapid change in H. Q then declines in value again, before another episode of building up. A typical example is shown in Figure 6. This is actually computed for zero matter density, and shows that the addition of torsion by itself is quite sufficient to give some interesting dynamics. A problem for the matter sector of any scale invariant theory, is that the matter stress-energy tensor has to have zero trace. This precludes using normal matter, and means that such cosmologies are probably unrealistic. One of the main current cosmological problems, however, is how to generate a period of inflation in the early universe, and then a hugely smaller effective cosmological constant today. It is certainly interesting that torsion can give somewhat chaotic behaviour of the type illustrated, with very large swings of the Hubble parameter, unrelated to the underlying matter field.

Acknowledgements

ANL would like to thank Venzo de Sabbata for the opportunity of attending the meeting in Erice and contributing to this volume. CJLD is supported by the EPSRC.

References

1. P.A. Collins, R. Delbourgo, and R. M. Williams. On the elastic Schwarzschild scattering cross section. *J. Phys. A*, 6:161–169, 1973.
2. N. Sanchez. Elastic scattering of waves by a black hole. *Phys. Rev. D*, 18(6):1798–1804, 1978.
3. J.A.H. Futterman, F.A. Handler, and R.A. Matzner. *Scattering from Black Holes.* Cambridge University Press, 1988.
4. S. Chandrasekhar. *The Mathematical Theory of Black Holes.* Oxford University Press, 1983.
5. N. Andersson and B.P. Jensen. Scattering by black holes. In R. Pike and P. Sabatier, editors, *Scattering.* Academic Press, 2000. See also gr-qc/0011025.
6. C.J.L. Doran and A.N. Lasenby. Perturbation theory calculation of the black hole elastic scattering cross section. To appear in: *Physical Review D* (gr-qc/0106039), 2001.
7. D. Hestenes. Vectors, spinors, and complex numbers in classical and quantum physics. *Am. J. Phys.*, 39:1013, 1971.
8. D. Hestenes. Proper particle mechanics. *J. Math. Phys.*, 15(10):1768, 1974.
9. D. Hestenes. Observables, operators, and complex numbers in the Dirac theory. *J. Math. Phys.*, 16(3):556, 1975.
10. T.G. Vold. An introduction to geometric algebra with an application to rigid body mechanics. *Am. J. Phys.*, 61(6):491, 1993.
11. T.G. Vold. An introduction to geometric calculus and its application to electrodynamics. *Am. J. Phys.*, 61(6):505, 1993.
12. D. Hestenes and G. Sobczyk. *Clifford Algebra to Geometric Calculus.* Reidel, Dordrecht, 1984.
13. D. Hestenes. *New Foundations for Classical Mechanics (Second Edition).* Kluwer Academic Publishers, Dordrecht, 1999.
14. C.J.L Doran and A.N. Lasenby. *Geometric Algebra for Physicists.* Cambridge University Press, 2001. In press.
15. C.J.L. Doran, A.N. Lasenby, S.F. Gull, and J. Lasenby. Lectures in geometric algebra. In W.E. Baylis, editor, *Clifford (Geometric) Algebras*, pages 65–236. Birkhauser, Boston, 1996.
16. C.J.L. Doran and A.N. Lasenby. Physical applications of geometric algebra. Lecture notes available from `http://www.mrao.cam.ac.uk/~clifford`.
17. C.J.L. Doran, A.N. Lasenby, and S.F. Gull. States and operators in the spacetime algebra. *Found. Phys.*, 23(9):1239, 1993.

18. C.J.L Doran, A.N. Lasenby, S.F. Gull, S.S. Somaroo, and A.D. Challinor. Spacetime algebra and electron physics. *Adv. Imag. & Elect. Phys.*, 95:271, 1996.
19. S.S. Somaroo, A.N. Lasenby, and C.J.L. Doran. Geometric algebra and the causal approach to multiparticle quantum mechanics. *J. Math. Phys.*, 40(7):3327–3340, 1999.
20. T.F. Havel and C.J.L. Doran. Geometric algebra in quantum information processing. quant-ph/0004031, *AMS Contemporary Math series*, to appear.
21. S. Somaroo, D.G. Cory, and T.F. Havel. Expressing the operations of quantum computing in multiparticle geometric algebra. *Phys. Lett. A*, 240:1–7, 1998.
22. T.W.B. Kibble. Lorentz invariance and the gravitational field. *J. Math. Phys.*, 2(3):212, 1961.
23. F.W. Hehl, P. von der Heyde, G.D. Kerlick, and J.M. Nester. General relativity with spin and torsion: Foundations and prospects. *Rev. Mod. Phys.*, 48:393, 1976.
24. A.N. Lasenby, C.J.L. Doran, and S.F. Gull. Gravity, gauge theories and geometric algebra. *Phil. Trans. R. Soc. Lond. A*, 356:487–582, 1998.
25. K. Martel and E. Poisson. Regular coordinate systems for Schwarzschild and other spherical spacetimes. *Am. J. Phys.*, 69(4):476, 2001.
26. A. Ronveaux (ed.). *Heun's Differential Equations.* Oxford University Press, 1995.
27. S. W. Hawking. Black hole explosion? *Nature*, 248:30, 1974.
28. A.M. Lewis, C.J.L. Doran, and A.N. Lasenby. Electron scattering without spin sums. *Int. J. Theor. Phys.*, 40(1), 2001.
29. C.J.L Doran, A.N. Lasenby, A.D. Challinor, and S.F Gull. Effects of spin-torsion in gauge theory gravity. *J. Math. Phys.*, 39(6):3303, 1998.
30. A.D. Challinor, A.N. Lasenby, C.J.L Doran, and S.F Gull. Massive, non-ghost solutions for the Dirac field coupled self-consistently to gravity. *General Rel. Grav.*, 29:1527, 1997.
31. A.M. Lewis, C.J.L. Doran, and A.N. Lasenby. Quadratic Lagrangians and topology in gauge theory gravity. *Gen. Rel. Grav.*, 32(1):161, 2000.

MODERN TRENDS IN MULTIDIMENSIONAL GRAVITY AND COSMOLOGY

V.R. GAVRILOV
Russian Gravitational Society, 3-1 M. Ulyanovoy Str., Moscow 117313, Russia
Moscow State Technical University, 2-nd Bauman Str. 5, Moscow 107005, Russia

AND

V.N. MELNIKOV
Center for Gravitation and Fundamental Metrology, VNIIMS, 3-1 M. Ulyanovoy Str., Moscow 117313, Russia
Institute of Gravitation and Cosmology, PFUR, Michlukho-Maklaya Str. 6, Moscow 117198, Russia

Abstract. The role of multidimensional gravitational models with different matter sources in solving the basic problems of cosmology is stressed. The choice, nature, classification and precision of determination of fundamental physical constants are described. The problem of their temporal variations is also discussed, temporal and range variations of G in particular. A need for their more precise measurements is pointed out. The novel multipurpose space project SEE, aimed for measuring G and its stability in space and time 3-4 orders better than at present, may answer many important questions posed by gravitation, cosmology and unified theories. A short review of multidimensional integrable gravitational and cosmological models is given. As an example, D-dimensional cosmological model describing the evolution of a perfect fluid with negative pressure (x-fluid) and a fluid possessing both shear and bulk viscosity in n Ricci-flat spaces is investigated. The second equations of state are chosen in some special form of metric dependence of the shear and bulk viscosity coefficients. The equations of motion are integrated and the dynamical properties of the exact solutions are studied. The possibility is shown to resolve the cosmic coincidence problem when the x-fluid plays role of quintessence and the viscous fluid is used as a cold dark matter.

P.G. Bergmann and V. de Sabbata (eds.),
Advances in the Interplay Between Quantum and Gravity Physics, 285–315.

1. Introduction

The necessity of studying multidimensional models of gravitation and cosmology [1, 2] is motivated by several reasons. First, the main trend of modern physics is the unification of all known fundamental physical interactions: electromagnetic, weak, strong and gravitational ones. During the recent decades there has been a significant progress in unifying weak and electromagnetic interactions, some more modest achievements in GUT, supersymmetric, string and superstring theories.

Now, theories with membranes, p-branes and more vague M- and F-theories are being created and studied. Having no definite successful theory of unification now, it is desirable to study the common features of these theories and their applications to solving basic problems of modern gravity and cosmology. Moreover, if we really believe in unified theories, the early stages of the Universe evolution and black hole physics, as unique superhigh energy regions, are the most proper and natural arena for them.

Second, multidimensional gravitational models, as well as scalar-tensor theories of gravity, are theoretical frameworks for describing possible temporal and range variations of fundamental physical constants [3, 4, 5, 6].

Lastly, applying multidimensional gravitational models to basic problems of modern cosmology and black hole physics, we hope to find answers to such long-standing problems as singular or nonsingular initial states, creation of the Universe, creation of matter and its entropy, acceleration, cosmological constant, origin of inflation and specific scalar fields which may be necessary for its realization, isotropization and graceful exit problems, stability and nature of fundamental constants [4], possible number of extra dimensions, their stable compactification etc.

Bearing in mind that multidimensional gravitational models are certain generalizations of general relativity which is tested reliably for weak fields up to 0.001 and partially in strong fields (binary pulsars), it is quite natural to inquire about their possible observational or experimental windows. From what we already know, among these windows are:

– possible deviations from the Newton and Coulomb laws, or new interactions,

– possible variations of the effective gravitational constant with a time rate smaller than the Hubble one,

– possible existence of monopole modes in gravitational waves,

– different behaviour of strong field objects, such as multidimensional black holes, wormholes and p-branes,

– standard cosmological tests etc.

Since modern cosmology has already become a unique laboratory for testing standard unified models of physical interactions at energies that are far beyond the level of the existing and future man-made accelerators and other installations on Earth, there exists a possibility of using cosmological and astrophysical data for discriminating between future unified schemes.

As no accepted unified model exists, in our approach we adopt simple, but general from the point of view of number of dimensions, models based on multidimensional Einstein equations with or without sources of different nature: cosmological constant, perfect and viscous fluids, scalar and electromagnetic fields, their possible interactions, dilaton and moduli fields, fields of antisymmetric forms (related to p-branes) etc.

Our program's main objective was and is to obtain exact self-consistent solutions (integrable models) for these models and then to analyze them in cosmological, spherically and axially symmetric cases. In our view this is a natural and most reliable way to study highly nonlinear systems. It is done mainly within Riemannian geometry. Some simple models in integrable Weyl geometry and with torsion were studied as well.

Here we dwell mainly upon some problems of fundamental physical constants, the gravitational constant in particular, upon the SEE project shortly and exact solutions with a mixture of a perfect fluid with negative pressure and a fluid possessing both shear and bulk viscosity within a multidimensional cosmology as an example.

2. Fundamental physical constants

2.1. In any physical theory we meet constants which characterize the stability properties of different types of matter: of objects, processes, classes of processes and so on. These constants are important because they arise independently in different situations and have the same value, at any rate within accuracies we have gained nowadays. That is why they are called fundamental physical constants (FPC) [3]. It is impossible to define strictly this notion. It is because the constants, mainly dimensional, are present in definite physical theories. In the process of scientific progress some theories are replaced by more general ones with their own constants, some relations between old and new constants arise. So, we may talk not about an absolute choice of FPC, but only about a choice corresponding to the present state of the physical sciences.

Really, before the creation of the electroweak interaction theory and some Grand Unification Models, it was considered that this *choice* is as follows:

$$c,\ \hbar,\ \alpha,\ G_F,\ g_s,\ m_p\ (\text{or } m_e),\ G,\ H,\ \rho,\ \Lambda,\ k,\ I, \tag{1}$$

where α, G_F, g_s and G are constants of electromagnetic, weak, strong and gravitational interactions, H, ρ and Λ are cosmological parameters (the Hubble constant, mean density of the Universe and cosmological constant),

k and I are the Boltzmann constant and the mechanical equivalent of heat which play the role of conversion factors between temperature on the one hand, energy and mechanical units on the other. After adoption in 1983 of a new definition of the meter ($\lambda = ct$ or $\ell = ct$) this role is partially played also by the speed of light c. It is now also a conversion factor between units of time (frequency) and length, it is defined with the absolute (null) accuracy.

Now, when the theory of electroweak interactions has a firm experimental basis and we have some good models of strong interactions, a more preferable choice is as follows:

$$\hbar,\ (c),\ e,\ m_e,\ \theta_w,\ G_F,\ \theta_c,\ \Lambda_{QCD},\ G,\ H,\ \rho,\ \Lambda,\ k,\ I \qquad (2)$$

and, possibly, three angles of Kobayashi-Maskawa — θ_2, θ_3 and δ. Here θ_w is the Weinberg angle, θ_c is the Cabibbo angle and Λ_{QCD} is a cut-off parameter of quantum chromodynamics. Of course, if a theory of four known now interactions will be created (M-, F-or other), then we will probably have another choice. As we see, the macro constants remain the same, though in some unified models, i.e. in multidimensional ones, they may be related in some manner. From the point of view of these unified models the above mentioned ones are low energy constants.

All these constants are known with different *accuracies*. The most precisely defined constant was and remain the speed of light c: its accuracy was 10^{-10} and now it is defined with the null accuracy. Atomic constants, e, $\hbar$, m and others are determined with errors $10^{-6} \div 10^{-8}$, G up to 10^{-4} or even worse, θ_w — up to 10%; the accuracy of H is also about 10%. An even worse situation is now with other cosmological parameters (FPC): mean density estimations vary within an order of magnitude; for Λ we have now data that its corresponding density exceeds the matter density (0.7 of the total mass).

As to the *nature* of the FPC, we may mention several approaches. One of the first hypotheses belongs to J.A. Wheeler: in each cycle of the Universe evolution the FPC arise anew along with physical laws which govern this evolution. Thus, the nature of the FPC and physical laws are connected with the origin and evolution of our Universe.

A less global approach to the nature of dimensional constants suggests that they are needed to make physical relations dimensionless or they are measures of asymptotic states. Really, the speed of light appears in relativistic theories in factors like v/c, at the same time velocities of usual bodies are smaller than c, so it plays also the role of an asymptotic limit. The same sense have some other FPC: $\hbar$ is the minimal quantum of action, e is the minimal observable charge (if we do not take into account quarks which are not observable in a free state) etc.

Finally, FPC or their combinations may be considered as natural scales determining the basic units. While the earlier basic units were chosen more or less arbitrarily, i.e., the second, meter and kilogram, now the first two are based on stable (quantum) phenomena. Their stability is believed to be ensured by the physical laws which include FPC.

Another interesting problem, which is under discussion, is why the FPC have values in a very narrow range necessary for supporting life (stability of atoms, stars lifetime etc.). There exist several possible but far from being convincing explanations [19]. First, that it is a good luck, no matter how improbable is the set of FPC. Second, that life may exist in other forms and for another FPC set, of which we do not know. Third, that all possibilities for FPC sets exist in some universe. And the last but not the least: that there is some cosmic fine tuning of FPC: some unknown physical processes bringing FPC to their present values in a long-time evolution, cycles etc.

An exact knowledge of FPC and precision measurements are necessary for testing main physical theories, extention of our knowledge of nature and, in the long run, for practical applications of fundamental theories. Within this, such theoretical problems arise:

1) development of models for confrontation of theory with experiment in critical situations (i.e. for verification of GR, QED, QCD, GUT or other unified models);

2) setting limits for spatial and temporal variations of FPC.

As to a *classification* of FPC, we may set them now into four groups according to their generality:

1) Universal constants such as $\hbar$, which divides all phenomena into quantum and nonquantum ones (micro- and macro-worlds) and to a certain extent c, which divides all motions into relativistic and non-relativistic ones;

2) constants of interactions like α, θ_w, Λ_{QCD} and G;

3) constants of elementary constituencies of matter like m_e, m_w, m_x, etc., and

4) transformation multipliers such as k, I and partially c.

Of course, this division into classes is not absolute. Many constants move from one class to another. For example, e was a charge of a particular object – electron, class 3, then it became a characteristic of class 2 (electromagnetic interaction, $\alpha = \frac{e^2}{\hbar c}$ in combination with $\hbar$ and c); the speed of light c has been in nearly all classes: from 3 it moved into 1, then also into 4. Some of the constants ceased to be fundamental (i.e. densities, magnetic moments, etc.) as they are calculated via other FPC.

As to the *number* of FPC, there are two opposite tendencies: the number of "old" FPC is usually diminishing when a new, more general theory is created, but at the same time new fields of science arise, new processes are discovered in which new constants appear. So, in the long run we may come to some minimal choice which is characterized by one or several FPC, maybe connected with the so-called Planck parameters — combinations of c, $\hbar$ and G:

$$L = \left(\frac{\hbar G}{c^3}\right)^{1/2} \sim 10^{-33}\ \mathrm{cm}, \ m_L = \left(\frac{c\hbar}{2G}\right)^{1/2} \sim 10^{-5}\ \mathrm{g}, \ \tau_L = \frac{L}{c} \sim 10^{-43}\ \mathrm{s}.$$

The role of these parameters is important since m_L characterizes the energy of unification of four known fundamental interactions: strong, weak, electromagnetic and gravitational ones, and L is a scale where the classical notions of space-time loose their meaning.

2.2. The problem of the gravitational constant G measurement and its stability is a part of a rapidly developing field, called gravitational-relativistic metrology (GRM). It has appeared due to the growth of measurement technology precision, spread of measurements over large scales and a tendency to the unification of fundamental physical interaction [6], where main problems arise and are concentrated on the gravitational interaction.

The main subjects of GRM are:

- general relativistic models for different astronomical scales: Earth, Solar System, galaxies, cluster of galaxies, cosmology - for time transfer, VLBI, space dynamics, relativistic astrometry etc.(pioneering works were done in Russia by Arifov and Kadyev, Brumberg in 60's);

- development of generalized gravitational theories and unified models for testing their effects in experiments;

- fundamental physical constants, G in particular, and their stability in space and time;

- fundamental cosmological parameters as fundamental constants: cosmological models studies, measurements and observations;

- gravitational waves (detectors, sources...);

- basic standards (clocks) and other modern precision devices (atomic and neutron interferometry, atomic force spectroscopy etc.) in fundamental gravitational experiments, especially in space...

There are three problems related to G, which origin lies mainly in unified models predictions:

1) absolute G measurements,

2) possible time variations of G,

3) possible range variations of G – non-Newtonian, or new interactions.

Absolute measurements of G. There are many laboratory determinations of G with errors of the order 10^{-3} and only 4 on the level of 10^{-4}. They are (in 10^{-11} m^3 kg^{-1} s^{-2}):

1. Facy and Pontikis, France	1972 —	6,6714 ± 0.0006
2. Sagitov et al., Russia	1979 —	6,6745 ± 0.0008
3. Luther and Towler, USA	1982 —	6,6726 ± 0.0005
4. Karagioz, Russia	1988 —	6,6731 ± 0.0004

From this table it is evident that the first three experiments contradict each other (the results do not overlap within their accuracies). And only the fourth experiment is in accord with the third one.

The official CODATA value of 1986

$$G = (6,67259 \pm 0.00085) \cdot 10^{-11} \cdot m^3 \cdot kg^{-1} \cdot s^{-2} \tag{3}$$

is based on the Luther and Towler determination. But after very precise measurements of G in Germany and New Zealand the situation became more vague. Their results deviate from the official CODATA value by more than 600 ppm.

As it may be seen from the Cavendish conference data [26], the results of 7 groups may agree with each other only on the level 10^{-3} and this is the official uncertainty from 1998. The most recent and precise G-measurement [27] diverge also from the CODATA value of 1986.

This means that either the limit of terrestrial accuracies has been reached or we have some new physics entering the measurement procedure [6]. The first means that, maybe we should turn to space experiments to measure G [18], and second means that a more thorough study of theories generalizing Einstein's general relativity or unified theories is necessary.

There exist also some satellite determinations of G (namely $G \cdot M_{\text{Earth}}$) on the level of 10^{-9} and several less precise geophysical determinations in mines.

The precise knowledge of G is necessary, first of all, as it is a FPC; next, for the evaluation of mass of the Earth, of planets, their mean density and, finally, for construction of Earth models; for transition from mechanical to electromagnetic units and back; for evaluation of other constants through relations between them given by unified theories; for finding new possible types of interactions and geophysical effects; for some practical applications like increasing of modern gradiometers precision, as they demand a calibration by a gravitational field of a standard body depending on G: high accuracy of their calibration (10^{-5} - 10^{-6}) requires the same accuracy of G. (I am indebted to Dr.N.Kolosnitsyn for this last remark.)

The knowledge of constants values has not only a fundamental meaning but also a metrological one. The modern system of standards is based mainly on stable physical phenomena. So, the stability of constants plays a crucial role. As all physical laws were established and tested during the last 2-3 centuries in experiments on the Earth and in the near space, i.e. at a rather short space and time intervals in comparison with the radius and age of the Universe, the possibility of slow *variations* of constants (i.e. with the rate of the evolution of the Universe or slower) cannot be excluded a priori.

So, the assumption of absolute stability of constants is an extrapolation and each time we must test it.

2.3. *Time Variations of G.* The problem of variations of FPC arose with the attempts to explain the relations between micro- and macro-world phenomena. Dirac was the first to introduce (1937) the so-called "Large Numbers Hypothesis" which relates some known very big (or very small) numbers with the dimensionless age of the Universe $T \sim 10^{40}$ (age of the Universe in seconds 10^{17}, divided by the characteristic elementary particle time 10^{-23} seconds). He suggested (after Milne in 1935) that the ratio of the gravitational to strong interaction strengths, $Gm_p^2/\hbar c \sim 10^{-40}$, is inversely proportional to the age of the Universe: $Gm_p^2/\hbar c \sim T^{-1}$. Then, as the age varies, some constants or their combinations must vary as well. Atomic constants seemed to Dirac to be more stable, so he chose the variation of G as T^{-1}.

After the original *Dirac hypothesis* some new ones appeared and also some generalized *theories* of gravitation admitting the variations of an effective gravitational coupling. We may single out three stages in the development of this field:

1. Study of theories and hypotheses with variations of FPC, their predictions and confrontation with experiments (1937-1977).
2. Creation of theories admitting variations of an effective gravitational constant in a particular system of units, analyses of experimental and observational data within these theories [9, 3] (1977-present).
3. Analyses of FPC variations within unified models [6, 4, 1] (present).

Within the development of the first stage from the analysis of the whole set of existed astronomical, astrophysical, geophysical and laboratory data, a conclusion was made [9, 10] that variations of atomic constants are excluded, but variations of the effective gravitational constant in the atomic system of units do not contradict the available experimental data on the level $10^{-11} \div 10^{-12}$year^{-1}. Moreover, in [11, 9, 10] the conception was worked out that variations of constants are not absolute but depend on the system of measurements (choice of standards, units and devices using this or that fundamental interaction). Each fundamental interaction through dynamics, described by the corresponding theory, defines the system of units and the system of basic standards.

Earlier reviews of some hypotheses on variations of FPC and experimental tests can be found in [3, 4].

Following Dyson (1972), we can introduce dimensionless combinations of micro- and macro-constants:

$$\alpha = e^2/\hbar c = 7,3 \cdot 10^{-3}, \qquad \gamma = Gm^2/\hbar c = 5 \cdot 10^{-39},$$
$$\beta = G_F\ m^2 c/\hbar^3 = 9 \cdot 10^6, \qquad \delta = H\hbar/mc^2 = 10^{-42},$$
$$\varepsilon = \rho G/H^2 = 2 \cdot 10^{-3}, \qquad t = T/(e^2/mc^3) \approx 10^{40}$$

We see that α, β and ε are of order 1 and γ and δ are of the order 10^{-40}. Nearly all existing hypotheses on variations of FPC may be represented as follows:

Hyposesis 1 (standard):

α, β, γ are constant, $\delta \sim t^{-1}$, $\varepsilon \sim t$.

Here we have no variations of G while δ and ε are determined by cosmological solutions.

Hyposesis 2 (Dirac):

α, β, ε are constant, $\gamma \sim t^{-1}$, $\delta \sim t^{-1}$.

Then $\dot{G}/G = 5 \cdot 10^{-11}$year^{-1} if the age of the Universe is taken to be $T = 2{\cdot}10^{10}$ years.

Hyposesis 3 (Gamow):

$\gamma/\alpha = Gm^2/e^2 \sim 10^{-37}$, so e^2 or α are varied, but not $G, \beta, \gamma; \varepsilon =$ const, $\alpha \sim t^{-1}$, $\delta \sim t^{-1}$.

Then $\dot{\alpha}/\alpha = 10^{-10}$ year^{-1}.

Hyposesis 4 (Teller):

trying to account also for deviations of α from 1, he suggested $\alpha^{-1} = \ln \gamma^{-1}$.

Then β, ε are constants, $\gamma \sim t^{-1}$, $\alpha \sim (\ln t)^{-1}$, $\delta \sim t^{-1}$

$$\dot{\alpha}/\alpha = 5 \cdot 10^{-13} year^{-1}. \qquad (4)$$

The same relation for α and γ was used also by Landau, DeWitt, Staniukovich, Terasawa and others, but in approaches other than Teller's.

Some other variants may be also possible, e.g. the Brans-Dicke theory with $G \sim t^{-r}$, $\rho \sim t^{r-2}$, $r = [2 + 3\omega/2]^{-1}$, a combination of Gamow's and Brans-Dicke etc. [3].

2.4. There are different astronomical, geophysical and laboratory *data* on possible variations of FPC.

Astrophysical data:

a) from comparison of fine structure ($\sim \alpha^2$) and relativistic fine structure ($\sim \alpha^4$) shifts in spectra of radio galaxies, Bahcall and Schmidt (1967) obtained

$$|\dot{\alpha}/\alpha| \leq 2 \cdot 10^{-12} \text{ year}^{-1}; \qquad (5)$$

b) comparing lines in optical ($\sim Ry = me^4/\hbar^2$) and radio bands of the same sources in galaxies Baum and Florentin-Nielsen (1976) got the estimate

$$|\dot{\alpha}/\alpha| \leq 10^{-14} \text{ year}^{-1}; \tag{6}$$

c) from observations of superfine structure in H-absorption lines of the distant radiosource Wolf et al. (1976) obtained that

$$|\alpha^2(m_e/m_p)g_p| < 2 \cdot 10^{-14} ; \tag{7}$$

From these data it is seen that Hyposeses 3 and 4 are excluded. Recent data only strengthen this conclusion. Comparing the data from absorption lines of atomic and molecular transition spectra in high redshifts QSO's, Varshalovich and Potekhin, Russia, [20] obtained for $z = 2.8 - 3.1$:

$$|\dot{\alpha}/\alpha| \leq 1.6 \cdot 10^{-14} \text{ year}^{-1} \tag{8}$$

and Drinkwater et al. [22]:

$$|\dot{\alpha}/\alpha| \leq 10^{-15} \text{ year}^{-1} \text{ for } z = 0.25 \tag{9}$$

and

$$|\dot{\alpha}/\alpha| \leq 5 \cdot 10^{-16} \text{ year}^{-1} \text{ for } z = 0.68 \tag{10}$$

for a model with zero deceleration parameter and $H = 75\text{km} \cdot \text{s}^{-1} \cdot \text{Mpc}^{-1}$.

The same conclusion is made on the bases of *geophysical data*. Indeed,

a) α-decay of $U_{238} \rightarrow Pb_{208}$. Knowing abundances of U_{238} and P_{238} in rocks and independently the age of these rocks, one obtains the limit

$$|\dot{\alpha}/\alpha| \leq 2 \cdot 10^{-13} \text{ year}^{-1}; \tag{11}$$

b) from spontaneous fission of U_{238} such an estimation was made:

$$|\dot{\alpha}/\alpha| \leq 2,3 \cdot 10^{-13} \text{ year}^{-1}. \tag{12}$$

c) finally, β-decay of Re_{187} to Os_{187} gave:

$$|\dot{\alpha}/\alpha| \leq 5 \cdot 10^{-15} \text{ year}^{-1} \tag{13}$$

We must point out that all astronomical and geophysical estimations are strongly model-dependent. So, of course, it is always desirable to have *laboratory tests* of variations of FPC.

a) Such a test was first done by the Russian group in the Committee for Standards (Kolosnitsyn, 1975). Comparing rates of two different types of clocks, one based on a Cs standard and another on a beam molecular generator, they found that

$$|\dot{\alpha}/\alpha| \leq 10^{-10} \text{ year}^{-1}. \tag{14}$$

b) More recent data were obtained by J. Prestage et al. [23] by comparing mercury and H-maser clocks. Their result is

$$|\dot{\alpha}/\alpha| \leq 3.7 \cdot 10^{-14} \text{ year}^{-1}. \tag{15}$$

All these limits were placed on the fine structure constant variations. From the analysis of decay rates of K_{40} and Re_{187}, a limit on possible variations of the weak interaction constant was obtained (see approach for variations of β, e.g. in [12])

$$|\dot{\beta}/\beta| \leq 10^{-10} \text{ year}^{-1}. \tag{16}$$

But the most strict data were obtained by A. Schlyachter in 1976 (Russia) from an analysis of the ancient natural nuclear reactor data in Gabon, Oklo, because the event took place $2 \cdot 10^9$ years ago. They are the following:

$$\begin{aligned} \dot{G}_s/G_s| &< 5 \cdot 10^{-19} \text{ year}^{-1}, \\ |\dot{\alpha}/\alpha| &< 10^{-17} \text{ year}^{-1}, \\ |\dot{G}_F/G_F| &< 2 \cdot 10^{-12} \text{ year}^{-1}. \end{aligned} \tag{17}$$

Quite recently Damour and Dyson [21] repeated this analysis in more detail and gave more cautious results:

$$|\dot{\alpha}/\alpha| \leq 5 \cdot 10^{-17} \text{ year}^{-1}, \ |\dot{G}_F/G_F| < 10^{-11} \text{ year}^{-1}. \tag{18}$$

So, we really see that all existing hyposeses with variations of atomic constants are excluded. Quite recently some data on possible variations on α appeared, but they are not confirmed.

2.5. Now we still have no unified theory of all four interactions. So it is possible to construct systems of measurements based on any of these four interactions. But practically it is done now on the basis of the mostly worked out theory — on electrodynamics (more precisely on QED). Of course, it may be done also on the basis of the gravitational interaction (as it was partially earlier). Then, different units of basic physical quantities arise based on dynamics of the given interaction, i.e. the atomic (electromagnetic) second, defined via frequency of atomic transitions or the gravitational second defined by the mean Earth motion around the Sun (ephemeris time).

It does not follow from anything that these two seconds are always synchronized in time and space. So, in principal they may evolve relative to each other, for example at the rate of the evolution of the Universe or at some slower rate.

That is why, in general, variations of the gravitational constant are possible in the atomic system of units (c, $\hbar$, m are constant) and masses of all particles — in the gravitational system of units (G, $\hbar$, c are constant by definition). Practically we can test only the first variant since the modern basic standards are defined in the atomic system of measurements. Possible variations of FPC must be tested experimentally but for this it is necessary to have the corresponding theories admitting such variations and their certain effects.

Mathematically these systems of measurement may be realized as conformally related metric forms. Arbitrary conformal transformations give us a transition to an arbitrary system of measurements.

We know that scalar-tensor and multidimensional theories are corresponding frameworks for these variations. So, one of the ways to describe variable gravitational coupling is the introduction of a *scalar field* as an additional variable of the gravitational interaction. It may be done by different means (e.g. Jordan, Brans-Dicke, Canuto and others). We have suggested a variant of gravitational theory with a conformal scalar field (Higgs-type field [13, 3]) where Einstein's general relativity may be considered as a result of spontaneous symmetry breaking of conformal symmetry (Domokos, 1976) [3]. In our variant spontaneous symmetry breaking of the global gauge invariance leads to a nonsingular cosmology [14]. Besides, we may get variations of the effective gravitational constant in the atomic system of units when m, c, $\hbar$ are constant and variations of all masses in the gravitational system of units (G, c, $\hbar$ are constant). It is done on the basis of approximate [15] and exact cosmological solutions with local inhomogeneity [16].

The effective gravitational constant is calculated using the equations of motions. Post-Newtonian expansion is also used in order to confront the theory with existing experimental data. Among the post-Newtonian parameters the parameter f describing variations of G is included. It is defined as

$$\frac{1}{GM}\frac{d(GM)}{dt} = fH. \tag{19}$$

According to Hellings' data [17] from the Viking mission,

$$\tilde{\gamma} - 1 = (-1.2 \pm 1.6) \cdot 10^3, \qquad f = (4 \pm 8) \cdot 10^{-2}. \tag{20}$$

In the theory with a conformal Higgs field [15, 16] we obtained the following relation between f and $\tilde{\gamma}$: $f = 4(\tilde{\gamma} - 1)$. Using Hellings' data for $\tilde{\gamma}$, we can calculate in our variant f and compare it with f from [17]. Then we get $f = (-9,6 \pm 12,8) \cdot 10^{-3}$ which agrees with (20) within its accuracy.

We used here only Hellings' data on variations of G. But the situation with experiment and observations is not so simple. Along with [17], there are some other data [3, 4]:

1. Reasenberg's estimates (1987) of the same Viking mission as in [17]:

$$|\dot{G}/G| < (0 \pm 2) \cdot 10^{-11}\ \text{year}^{-1} \tag{21}$$

2. Hellings' result in the same form is

$$|\dot{G}/G| < (2 \pm 4) \cdot 10^{-12} \text{ year}^{-1}. \tag{22}$$

3. A result from nucleosynthesis (Acceta et al., 1992):

$$|\dot{G}/G| < (\pm 0.9) \cdot 10^{-12} year^{-1}. \tag{23}$$

4. E.V.Pitjeva's result, Russia (1997), based on satellites and planets motion:

$$|\dot{G}/G| < (0 \pm 2) \cdot 10^{-12} year^{-1} \tag{24}$$

As we see, there is a vivid contradiction in these results. As to other experimental or observational data, the results are rather inconclusive. The most reliable ones are based on lunar laser ranging (Muller et al, 1993 and Williams et al, 1996). They are not better than 10^{-12} per year. Here, once more we see that there is a need for corresponding theoretical and experimental studies. Probably, future space missions like Earth SEE-satellite [18] or missions to other planets and lunar laser ranging will be a decisive step in solving the problem of temporal variations of G and determining the fates of different theories which predict them, since the greater is the time interval between successive measurements and, of course, the more precise they are, the more stringent results will be obtained.

As we saw, different theoretical schemes lead to temporal variations of the effective gravitational constant:

1. Empirical models and theories of Dirac type, where G is replaced by $G(t)$.
2. Numerous scalar-tensor theories of Jordan-Brans-Dicke type where G depending on the scalar field $\sigma(t)$ appears.
3. Gravitational theories with a conformal scalar field arising in different approaches [3, 24].
4. Multidimensional unified theories in which there are dilaton fields and effective scalar fields appearing in our 4-dimensional spacetime from additional dimensions [25, 1]. They may help also in solving the problem of a variable cosmological constant from Planckian to present values.

As was shown in [4, 25, 1] temporal variations of FPC are connected with each other in *multidimensional models* of unification of interactions. So, experimental tests on $\dot{\alpha}/\alpha$ may at the same time be used for estimation of $\dot{G}/G$ and vice versa. Moreover, variations of G are related also to the cosmological parameters ρ, Ω and q which gives opportunities of raising the precision of their determination.

As variations of FPC are closely connected with the behaviour of internal scale factors, it is a direct probe of properties of extra dimensions and the corresponding theories [7, 8, 1].

2.6. *Non-Newtonian interactions, or range variations of G.* Nearly all modified theories of gravity and unified theories predict also some deviations from the Newton law (inverse square law, ISL) or composition-dependent violations of the Equivalence Principle (EP) due to appearance of new possible massive particles (partners) [4]. Experimental data exclude the existence of these particles at nearly all ranges except less than *millimeter* and also at *meters and hundreds of meters* ranges. The most recent result in the range of 20 to 500 m was obtained by Achilli et al. using an energy storage plant experiment with gravimeters. They found a positive result for the deviation from the Newton law with the Yukawa potential strength α between 0.13 and 0.25. Of course, these results need to be verified in other independent experiments, probably in space ones [18].

In the Einstein theory G is a true constant. But, if we think that G may vary with time, then, from a relativistic point of view, it may vary with distance as well. In GR massless gravitons are mediators of the gravitational interaction, they obey second-order differential equations and interact with matter with a constant strength G. If any of these requirements is violated, we come in general to deviations from the Newton law with range (or to generalization of GR).

In [5] we analyzed several classes of such theories:

1. Theories with massive gravitons like bimetric ones or theories with a Λ-term.

2. Theories with an effective gravitational constant like the general scalar-tensor ones.

3. Theories with torsion.

4. Theories with higher derivatives (4th-order equations etc.), where massive modes appear leading to short-range additional forces.

5. More elaborated theories with other mediators besides gravitons (partners), like supergravity, superstrings, M-theory etc.

6. Theories with nonlinearities induced by any known physical interactions (Born-Infeld etc.)

7. Phenomenological models where the detailed mechanism of deviation is not known (fifth or other force).

In all these theories some effective or real masses appear leading to Yukawa-type deviation from the Newton law, characterized by strength and range.

There exist some model-dependant estimations of these forces. The most well-known one belongs to Scherk (1979) from supergravity where the graviton is accompanied by a spin-1 partner (graviphoton) leading to an additional repulsion. Other models were suggested by Moody and Wilczek (1984) – introduction of a pseudo-scalar particle – leading to an additional attraction between macro-bodies with the range $2 \cdot 10^{-4}$ cm $< \lambda <$ 20 cm and strength α from 1 to 10^{-10} in this range. Another supersymmetric model was elaborated by Fayet (1986, 1990), where a spin-1 partner of a massive graviton gives an additional repulsion in the range of the order 10^3 km and α of the order 10^{-13}.

A scalar field to adjust Λ was introduced also by S. Weinberg in 1989, with a mass smaller than 10^{-3}eV$/c^2$, or a range greater than 0.1 mm. One more variant was suggested by Peccei, Sola and Wetterich (1987) leading to additional attraction with a range smaller than 10 km. Some p-brane models also predict non-Newtonian additional interactions in the mm range, what is intensively discussed nowadays. About PPN parameters for multi-dimensional models with p-branes see below.

2.7. SEE - Project

We saw that there are three problems connected with G. There is a promising new multi-purpose space experiment SEE - Satellite Energy Exchange [18], which addresses all these problems and may be more effective in solving them than other laboratory or space experiments.

This experiment is based on a limited 3-body problem of celestial mechanics: small and large masses in a drag-free satellite and the Earth. Unique horse-shoe orbits, which are effectively one-dimensional, are used in it.

The aims of the SEE-project are to measure: Inverse Square law (ISL) and Equivalence Principle (EP) at ranges of meters and the Earth radius, G-dot and the absolute value of G with unprecedented accuracies.

We studied some aspects of the SEE-project [28] :

1. Wide range of trajectories with the aim of finding optimal ones:
- circular in spherical field;
- the same plus Earth quadrupole modes;
- elliptic with eccentricity less than 0.05.

2. Estimations of other celestial bodies influence.

3. Estimation of relative influence of trajectories to changes in G and α.

4. Modelling measurement procedures for G and α by different methods, for different ranges and for different satellite altitudes: optimal - 1500 km, ISS free flying platform - 500 km and also for 3000 km.

5. Estimations of some sources of errors:
- radial oscillations of the shepherd's surface;
- longitudal oscillations of the capsule;
- transversal oscillations of the calsule;
- shepherd's nonsphericity;

- limits on the quadrupole moment of the shepherd;
- limits on admissible charges and time scales of charging by high energy particles etc.

6. Error budgets for G, G-dot and $G(r)$.

The general conclusion is that the SEE-project may really improve our knowledge of these values by 3-4 orders better than we have nowadays.

3. Multidimensional Viscous Models

Herein, we assume that the cosmic fluid (the source of the gravitational field at early stages) is viscous, which might simulate high energy physics processes (such as the particles creation). The effects related to viscosity in 4-dimensional Universe were studied through different viewpoints. Before developing the multidimensional model let us briefly discuss (extensive review of the subject was given by Gron [29]) the main trends in 4-dimensional cosmology with viscous fluid as a source.

First, Misner [30] considered neutrino viscosity as a mechanism of reducing the anisotropy in the Early Universe. Stewart [31] and Collins and Stewart [32] proved that it is possible only if initial anisotropies are small enough. Another series of papers which concerns the production of entropy in the viscous Universe was started by Weinberg [33]. Both isotropization and production of entropy during lepton era in models of Bianchi types I,V were considered by Klimek [34]. Caderni and Fabbri [35] calculated coefficients of shear and bulk viscosity in plasma and lepton eras within the model of Bianchi type I. The next approach is connected with obtaining singularity free viscous solutions. The first nonsingular solution was obtained by Murphy [36] within the flat Friedman-Robertson-Walker (FRW) model with fluid possessing a bulk viscosity. However, Belinsky and Khalatnikov [37, 38] showed that this solution corresponds to the very peculiar choice of parameters and is unstable with respect to the anisotropy perturbations. Other nonsingular solutions with bulk viscosity were obtained by Novello and Araujo [39], Romero [40], Oliveira and Salim [41].

The crucial feature of each viscous cosmological model is assuming of the so called "second equation of state", which provides us with the viscosity coefficients dependence on time. Further we denote by ζ and η the bulk and shear viscosity coefficients, correspondingly. Murphy [36] integrated the 4-dimensional flat FRW model with bulk viscosity by assuming $\zeta \sim \rho$ (as second equation of state), where ρ is the density of the viscous fluid. Belinsky and Khalatnikov [37, 38] studied the behavior of this model as well as homogeneous anisotropic models of Bianchi types I and IX by means of qualitative methods with more general second equations of state $\zeta, \eta \sim \rho^{\nu}$, where ν is constant. Lukacs [42] integrated the homogeneous and isotropic 4-dimensional model with a viscous pressureless fluid and a second equation of state given by $\zeta \sim$ [scale factor]$^{-1}$. A curvature-dependent bulk viscosity was studied in multidimensional cosmology by Wolf [43]. Motta and Tomimura [44] studied a 4-dimensional inhomogeneous cosmology with a bulk viscosity coefficient which depends on the metric. In our previous papers [45] exact solutions in multidimensional models with bulk viscosity were obtained and their properties were studied for the following type of the second equation of state: $\zeta, \eta \sim$ [volume of the Universe]$^{-1}$.

Within multidimensional cosmology one assumes that the Universe had a higher dimension at a very early stage of its evolution, and that quantum processes have been responsible for the (topological) partition of the space, which provides us at present with the usual 3-dimensional (*external*) space, in addition to *internal* space(s). The manifold which accounts for such a multidimensional spacetime has the following topology $\mathbb{M} = \mathbb{R} \times M_1 \times \ldots \times M_n$, where $\mathbb{R}$ stands for the cosmic time axis, and the product with one part of manifolds $M_1, \ldots, M_n$ gives the external space, when the remaining part stands for internal spaces. The classical stage of the evolution is governed by the multidimensional version of Einstein's equations. According to the classical description, the Riemann curvatures of spaces $M_1, \ldots, M_n$ are assumed to be constant (Einstein spaces). One easily understands that the use of extra dimensions (for the physical spacetime) can be a sensible scenario only at primordial epochs. Hence, it is clear that a reduction process (called *dynamical compactification* of additional dimensions) is required before such an epoch, to make the internal spaces contracting themself down to unobservable sizes.

The most probable candidate for the so called quintessence matter responsible for the current phase of the accelerated expansion of the Universe is a Λ term or more generally an exotic x-fluid - perfect fluid with negative pressure satisfying a linear barotropic equation of state (see, for instance, [47] and refs. therein). To describe the present stage of evolution this x-fluid is to be added to the normal matter, which mainly consists of cold dark matter. For instance, ΛCDM model [48],[49] describes the flat FRW Universe filled with a mixture of quintessence represented by the cosmological constant Λ and cold dark matter in the form of pressureless perfect fluid (dust).

However, the flat FRW cosmologies with a x-fluid and a normal perfect fluid are not free from some difficulties. One of them is the so called coincidence problem [50]: to explain why the quintessence density and the normal matter density are comparable today, one has to tune their initial ratio very carefully. The problem may be ameliorated by replacement the x-fluid by the so called quintessence scalar field - homogeneous scalar field Q slowly rolling down with some potential $V(Q)$ [50]-[52]. For instance, the potential $V(Q) = M^{4+\alpha}Q^{-\alpha}$, $\alpha > 0$, leads to the so called "tracker" solution, which "attracts" solutions to the equations of motion before the present stage for very wide range of initial conditions. Another way for resolving the problem was proposed in [53]. The idea is to use a fluid with bulk viscosity in combination with a quintessence scalar field.

Herein, we show that the cosmic coincidence problem may be resolved by using a x-fluid as the quintessence and a viscous fluid as the normal matter. Viscosity of the normal matter can be explained by its own multicomponent structure. As is known, a mixture of different fluids admits a description as a single viscous fluid. Other possible cosmological origins of viscosity were discussed, for instance, in [53], [54]. We adopt the "second equations of state" in the form of some special metric dependence of the bulk and shear viscosity coefficients. These "second equations of state" generalize the so called "linear dissipative regime" in FRW world model, when the bulk viscosity coefficient is linearly proportional to the Hubble parameter. It should be noted, that the choice of the equations of state crucially affects on the dynamical properties of the cosmological model and needs a physical substantiation. In other words, one has separate a stage of the cosmological evolution and find the most appropriate from the physical viewpoint equations of state. Here we adopt another way. Following many authors (see, for instance, the papers [55],[56] on viscosity-driven inflation) we rather analyze the cosmological consequences of some specific equations of state. Moreover, taking into account the possible multidimensionality of our world, we don't restrict ourselves to 4-dimensional world model. From the very beginning we study D-dimensional homogeneous anisotropic cosmology, which allows to describe the dynamical compactification of the extra dimensions (see, for instance, our previous paper [46] on viscous cosmology).

We assume the following metric

$$\mathrm{d}s^2 = -\mathrm{e}^{2\gamma(t)}\mathrm{d}t^2 + \sum_{i=1}^{n} \exp[2x^i(t)]\mathrm{d}s_i^2, \tag{25}$$

on the D-dimensional space-time manifold

$$\mathbb{M} = \mathbb{R} \times M_1 \times \ldots \times M_n, \tag{26}$$

where $\mathrm{d}s_i^2$ is the metric of the Ricci-flat factor space M_i of dimension d_i, $\gamma(t)$ and $x^i(t)$ are scalar functions of the cosmic time t. $a_i \equiv \exp[x^i]$ is the scale factor of the space M_i and the function $\gamma(t)$ determines a time gauge. The synchronous time t_s is defined by the equation $\mathrm{d}t_s = \exp[\gamma(t)]\mathrm{d}t$.

We consider a source of gravitational field in the form of 2-component cosmic fluid. The first one is a perfect fluid with a density $\rho^{(1)}(t)$ and a pressure $p^{(1)}(t)$. The second component is supposed to be a viscous fluid. It is characterized by a density $\rho^{(2)}(t)$, a pressure $p^{(2)}(t)$, a bulk viscosity coefficient $\zeta(t)$ and a shear viscosity coefficient $\eta(t)$. The overall energy-momentum tensor of the cosmic fluid reads

$$T^{\mu}_{\nu} = \left(\rho^{(1)} + \rho^{(2)}\right) u^{\mu} u_{\nu} + \left(p^{(1)} + p^{(2)} - \zeta\theta\right) P^{\mu}_{\nu} - 2\eta\sigma^{\mu}_{\nu}, \tag{27}$$

where u^{μ} is the D-dimensional velocity of the fluid, $\theta = u^{\mu}_{;\mu}$ denotes the scalar expansion, $P^{\mu}_{\nu} = \delta^{\mu}_{\nu} + u^{\mu} u_{\nu}$ is the projector on the $(D-1)$-dimensional space orthogonal to u^{μ}, $\sigma^{\mu}_{\nu} = \frac{1}{2}\left(u_{\alpha;\beta} + u_{\beta;\alpha}\right) P^{\alpha\mu} P^{\beta}_{\nu} - (D-1)^{-1}\theta P^{\mu}_{\nu}$ is the traceless shear tensor and $\mu, \nu = 0, 1, \ldots, D-1$.

By assuming the comoving observer condition $u^{\mu} = \delta^{\mu}_{0}\, \mathrm{e}^{-\gamma(t)}$, the overall energy-momentum tensor may be written as

$$T^{\mu}_{\nu} = T^{\mu(1)}_{\nu} + T^{\mu(2)}_{\nu}, \tag{28}$$

where

$$\left(T^{\mu(1)}_{\nu}\right) = \mathrm{diag}(-\rho^{(1)}, p^{(1)}, \ldots, p^{(1)}), \tag{29}$$

$$\left(T^{\mu(2)}_{\nu}\right) = \mathrm{diag}(-\rho^{(2)}, \tilde{p}^{(2)}_{1}\delta^{k_1}_{l_1}, \ldots, \tilde{p}^{(2)}_{n}\delta^{k_n}_{l_n}), \tag{30}$$

$k_i, l_i = 1, \ldots, d_i$ for $i = 1, \ldots, n$. Here $\tilde{p}^{(2)}_{i}$ denotes the effective pressure including the dissipative contribution of the viscous fluid in the factor space described by the manifold M_i. It reads

$$\tilde{p}^{(2)}_{i} = p^{(2)} - \mathrm{e}^{-\gamma}\left[\zeta\dot{\gamma}_0 + 2\eta\left(\dot{x}^{i} - \frac{\dot{\gamma}_0}{D-1}\right)\right], \tag{31}$$

where

$$\gamma_0 = \sum_{i=1}^{n} d_i x^{i}. \tag{32}$$

Furthermore, we assume that the barotropic equations of state hold

$$p^{(\alpha)} = \left(1 - h^{(\alpha)}\right)\rho^{(\alpha)}, \quad \alpha = 1, 2, \tag{33}$$

where the $h^{(\alpha)}$ are constants such that $h^{(1)} \neq h^{(2)}$.

The Einstein equations $R^{\mu}_{\nu} - \frac{1}{2}\delta^{\mu}_{\nu}R = \kappa^2 T^{\mu}_{\nu}$, where κ^2 is the gravitational constant, can be written as $R^{\mu}_{\nu} = \kappa^2(T^{\mu}_{\nu} - [T/(D-2)]\delta^{\mu}_{\nu})$. Further, the equations $R^{0}_{0} - \frac{1}{2}\delta^{0}_{0}R = \kappa^2 T^{0}_{0}$ and $R^{a}_{b} = \kappa^2(T^{a}_{b} - [T/(D-2)]\delta^{a}_{b})$, where $a, b = 1, \ldots, D$, give the following equations of motion

$$\dot{\gamma}_0^2 - \sum_{i=1}^{n} d_i(\dot{x}^{i})^2 = 2\kappa^2\, \mathrm{e}^{2\gamma}\left(\rho^{(1)} + \rho^{(2)}\right), \tag{34}$$

$$\ddot{x}^i + (\dot{\gamma}_0 - \dot{\gamma})\dot{x}^i = \kappa^2 \mathrm{e}^{\gamma}\left[\frac{\mathrm{e}^{\gamma}}{D-2}\sum_{\alpha=1}^{2} h^{(\alpha)}\rho^{(\alpha)} + \frac{\zeta}{D-2}\dot{\gamma}_0 - 2\eta\left(\dot{x}^i - \frac{\dot{\gamma}_0}{D-1}\right)\right], \quad (35)$$

$i = 1, \ldots, n$.

The energy conservation law $\nabla_\mu T_0^{\mu(2)} = 0$ for a viscous fluid described by a tensor given by (30) reads

$$\dot{\rho}^{(2)} + \sum_{i=1}^{n} d_i \dot{x}^i \left(\rho^{(2)} + \tilde{p}_i^{(2)}\right) = 0. \quad (36)$$

Due to the constraint $\nabla_\mu T_0^\mu = 0$ for the overall energy-momentum tensor given by equation (27) the similar energy conservation law is valid for the perfect fluid

$$\dot{\rho}^{(1)} + \left(\rho^{(1)} + p^{(1)}\right)\sum_{i=1}^{n} d_i \dot{x}^i = 0. \quad (37)$$

Taking into account equation (33), ones easily integrates equation (37). The result is

$$\rho^{(1)} = A e^{(h^{(1)}-2)\gamma_0}, \quad (38)$$

where A is a positive constant. We note that the contribution of the perfect fluid component with $h^{(2)} = 2$ to the overall energy momentum tensor is equivalent to the presence of Λ-term with $\Lambda = \kappa^2 A$.

Furthermore, by using equations (34) and (38) the presence of the densities $\rho^{(1)}$ and $\rho^{(2)}$ in equations (35) can be cancelled. Thus, we obtain the main governing set of equations

$$\ddot{x}^i + (\dot{\gamma}_0 - \dot{\gamma})\dot{x}^i = \kappa^2 A \frac{h^{(1)} - h^{(2)}}{D-2} \mathrm{e}^{[h^{(1)}-2]\gamma_0 + 2\gamma}$$
$$+ \frac{h^{(2)}}{2(D-2)}\left(\dot{\gamma}_0^2 - \sum_{i=1}^{n} d_i (\dot{x}^i)^2\right) + \kappa^2 \mathrm{e}^{\gamma}\left[\frac{\zeta}{D-2}\dot{\gamma}_0 - 2\eta\left(\dot{x}^i - \frac{\dot{\gamma}_0}{D-1}\right)\right]. \quad (39)$$

We use an integration procedure which is based on the n-dimensional Minkowsky-like geometry. Let $\mathbb{R}^n$ be the real vector space and $\mathbf{e}_1, \ldots, \mathbf{e}_n$ be the canonical basis in $\mathbb{R}^n$ (i.e. $\mathbf{e}_1 = (1, 0, \ldots, 0)$ etc). Let us define a symmetrical bilinear form $\langle , \rangle$ on $\mathbb{R}^n$ by

$$\langle \mathbf{e}_i, \mathbf{e}_j \rangle = \delta_{ij} d_j - d_i d_j. \quad (40)$$

Such a form is non-degenerate and has the pseudo-Euclidean signature $(-, +, \ldots, +)$ [57] . With this in mind, a vector $\mathbf{y} \in \mathbb{R}^n$ is timelike, spacelike or isotropic respectively, if $\langle \mathbf{y}, \mathbf{y} \rangle$ takes negative, positive or null values respectively and two vectors $\mathbf{y}$ and $\mathbf{z}$ are orthogonal if $\langle \mathbf{y}, \mathbf{z} \rangle = 0$. Hereafter, we use the following vectors

$$\mathbf{x} = x^1(t)\mathbf{e}_1 + \ldots + x^n(t)\mathbf{e}_n, \tag{41}$$

$$\mathbf{u} = u^1\mathbf{e}_1 + \ldots + u^n\mathbf{e}_n, \; u^i = \frac{-1}{D-2}, \; u_i = d_i, \tag{42}$$

where the covariant coordinates u_i of the vector $\mathbf{u}$ are introduced by the usual way. Moreover, we obtain

$$\langle \dot{\mathbf{x}}, \dot{\mathbf{x}} \rangle = \sum_{i=1}^{n} d_i (\dot{x}^i)^2 - \dot{\gamma}_0^2, \; \langle \mathbf{u}, \mathbf{x} \rangle = \gamma_0, \; \langle \mathbf{u}, \mathbf{u} \rangle = -\frac{D-1}{D-2}. \tag{43}$$

Thus, using equations (41)-(43) we rewrite the main governing set of equations in the following vector form

$$\begin{aligned} \ddot{\mathbf{x}} + (\dot{\gamma}_0 - \dot{\gamma})\dot{\mathbf{x}} &= \left[\kappa^2 A \left(h^{(2)} - h^{(1)} \right) \mathrm{e}^{[h^{(1)}-2]\gamma_0 + 2\gamma} + \frac{h^{(2)}}{2} \langle \dot{\mathbf{x}}, \dot{\mathbf{x}} \rangle \right] \mathbf{u} \\ &- \kappa^2 \mathrm{e}^{\gamma} \left[\left(\zeta + \frac{D-2}{D-1} 2\eta \right) \dot{\gamma}_0 \mathbf{u} + 2\eta \dot{\mathbf{x}} \right]. \end{aligned} \tag{44}$$

Moreover, the density $\rho^{(2)}$ can be expressed as

$$\rho^{(2)} = -\frac{\langle \dot{\mathbf{x}}, \dot{\mathbf{x}} \rangle}{2\kappa^2} \mathrm{e}^{-2\gamma} - \rho^{(1)}. \tag{45}$$

Now we summarize thermodynamics principles. The first law of thermodynamics applied to the viscous fluid reads

$$T \mathrm{d}S = \mathrm{d} \left(\rho^{(2)} V \right) + p^{(2)} \mathrm{d}V, \tag{46}$$

where V stands for a fluid volume in the whole space $M_1 \times \ldots \times M_n$, S is an entropy in the volume V and T is a temperature of the viscous fluid. By assuming that the baryon particle number N_{B} in the volume V is conserved, equation (46) transforms to

$$nT\dot{s} = \dot{\rho}^{(2)} + \left(\rho^{(2)} + p^{(2)} \right) \sum_{i=1}^{n} d_i \dot{x}^i \tag{47}$$

where $s = S/N_{\mathrm{B}}$ and $n = N_{\mathrm{B}}/V$ stands for the entropy per baryon and the baryon number density. The comparison between equations (36) and (47) gives the variation rate of entropy per baryon

$$nT\dot{s} = \sum_{i=1}^{n} d_i \dot{x}^i \left(p^{(2)} - \tilde{p}_i^{(2)} \right) = \mathrm{e}^{-\gamma} \left[\left(\zeta + \frac{D-2}{D-1} 2\eta \right) \dot{\gamma}_0^2 + 2\eta \langle \dot{\mathbf{x}}, \dot{\mathbf{x}} \rangle \right]. \tag{48}$$

To integrate equation (44) one needs a second set equations of state, involving the bulk viscosity coefficient ζ and the shear viscosity coefficient η. Herein, we suppose

$$\zeta = \frac{\zeta_0}{\kappa^2}\frac{D-2}{D-1}\dot{\gamma}_0\,\mathrm{e}^{-\gamma}, \quad \eta = \frac{\eta_0}{2\kappa^2}\dot{\gamma}_0\,\mathrm{e}^{-\gamma}, \tag{49}$$

where $\zeta_0 \geq 0$ and η_0 are constants. When the cosmological model is isotropic, i.e. $\dot{x}^i = \dot{\gamma}_0/(D-1)$, $i = 1, \ldots, n$, and the shear viscosity is not significant, the expression $H \equiv \dot{\gamma}_0\,\mathrm{e}^{-\gamma}/(D-1)$ is the Hubble parameter.Then we get from equation (49): $\zeta = \frac{D-2}{\kappa^2}\zeta_0 H$, i.e. the bulk viscosity coefficient is linearly proportional to the Hubble parameter. Such kind of the second equation of state describes the so called "linear dissipative regime" in the FRW world model (see, for instance, [53]). As we study an anisotropic cosmological model in general, we must involve a shear viscosity as well as a bulk one. So, we propose equations (49) for the anisotropic model as a simplest generalization of the second equation of state describing the linear dissipative regime.

In order to integrate equation (44), we use the orthogonal basis

$$\frac{\mathbf{u}}{\langle \mathbf{u}, \mathbf{u}\rangle}, \mathbf{f}_2, \ldots, \mathbf{f}_n \in \mathbb{R}^n, \tag{50}$$

where the vector $\mathbf{u}$ was introduced by equation (42) The orthogonality property reads

$$\langle \mathbf{u}, \mathbf{f}_j\rangle = 0, \ \langle \mathbf{f}_j, \mathbf{f}_k\rangle = \delta_{jk}, \quad (j, k = 2, \ldots, n). \tag{51}$$

Let us note that the basis vectors $\mathbf{f}_2, \ldots, \mathbf{f}_n$ are space-like, since they are orthogonal to the time-like vector $\mathbf{u}$. The vector $\mathbf{x} \in \mathbb{R}^n$ decomposes as follows

$$\mathbf{x} = \gamma_0 \frac{\mathbf{u}}{\langle \mathbf{u}, \mathbf{u}\rangle} + \sum_{j=2}^{n} \langle \mathbf{x}, \mathbf{f}_j\rangle \mathbf{f}_j. \tag{52}$$

Hence, under the above assumptions equation (44) reads in the terms of coordinates in such basis as follows

$$\ddot{\gamma}_0 + \left[\left(1 - \frac{h^{(2)}}{2} - \zeta_0\right)\dot{\gamma}_0 - \dot{\gamma}\right]\dot{\gamma}_0$$
$$= \frac{D-1}{D-2}\left[\kappa^2 A\left(h^{(1)} - h^{(2)}\right)\mathrm{e}^{[h^{(1)}-2]\gamma_0 + 2\gamma} - \frac{h^{(2)}}{2}\sum_{j=2}^{n}\langle \dot{\mathbf{x}}, \mathbf{f}_j\rangle^2\right], \tag{53}$$

$$\langle \ddot{\mathbf{x}}, \mathbf{f}_j\rangle + [(1+\eta_0)\dot{\gamma}_0 - \dot{\gamma}]\,\langle \dot{\mathbf{x}}, \mathbf{f}_j\rangle = 0 \quad (j = 2, \ldots, n). \tag{54}$$

The integration of equation (54) gives

$$\langle \dot{\mathbf{x}}, \mathbf{f}_j \rangle = p^j \, \mathrm{e}^{\gamma-(1+\eta_0)\gamma_0}, \tag{55}$$

where p^j is an arbitrary constant.

Further we determine the time gauge by

$$\gamma = k\gamma_0, \tag{56}$$

where k is a constant.

By substituting the functions $\langle \dot{\mathbf{x}}, \mathbf{f}_j \rangle$ into equation (53) we obtain the following integrable by quadrature ordinary differential equation

$$\ddot{\gamma}_0 - \left(k - 1 + \frac{h^{(2)}}{2} + \zeta_0 \right) \dot{\gamma}_0^2$$

$$= \frac{D-2}{D-1} \mathrm{e}^{2(k-1)\gamma_0} \left[\kappa^2 A \left(h^{(1)} - h^{(2)} \right) \mathrm{e}^{h^{(1)}\gamma_0} - \frac{h^{(2)}}{2} \sum_{j=2}^{n} \left(p^j \right)^2 \mathrm{e}^{-2\eta_0\gamma_0} \right]. \tag{57}$$

In what follows we accept the agreement

$$\mathrm{d}t > 0, \tag{58}$$

i.e. the cosmic time increases during the evolution.

It should be noted that the solutions of equation (57) corresponding to different sets of the parameters $h^{(1)}, h^{(2)}, \zeta_0$ and η_0 may lead to nonsatisfactory (from the physical viewpoint) cosmological evolutions. Further, we study only ones, which satisfy the following *consistency condition*: neither the density $\rho^{(2)}(t)$ nor the variation rate of entropy $\dot{s}(t)$ have negative values on any time interval.

3.1. THE ISOTROPIC MODEL

The *isotropic model* is described by the metric given by equation (25) with

$$a \equiv \mathrm{e}^{x^i} = \mathrm{e}^{\gamma_0/(D-1)}, \ i = 1, \ldots, n. \tag{59}$$

The scale factor a of the whole isotropically evolving space $M_1 \times \ldots \times M_n$ can be obtained by integration of equation (57) with

$$p^j = 0, \ j = 2, \ldots, n. \tag{60}$$

It can be proven that the above mentioned *consistency condition* leads to the following constraint

$$h^{(1)} - h^{(2)} - 2\zeta_0 > 0. \tag{61}$$

At first we present the special solution to equation (57). The special solution describes the asymptotical behaviour of all solutions at late time. It is the steady state solution

$$a \sim \exp\left[\sqrt{\frac{2\Lambda(1+\varepsilon_0)}{(D-1)(D-2)}}t_s\right], \; \rho^{(1)} = A \equiv \Lambda/\kappa^2, \tag{62}$$

for $h^{(1)} = 2$ and shows a power-law behaviour

$$a \sim t_s^{2/[(2-h^{(1)})(D-1)]}, \; \rho^{(1)} \sim t_s^{-2} \tag{63}$$

for $h^{(1)} \neq 2$, where t_s is the synchronous time. Moreover, we present the deceleration parameter

$$q \equiv -\frac{a(\ddot{a} - \gamma\dot{a})}{\dot{a}^2} = -1 + \frac{D-1}{2}\left(2 - h^{(1)}\right), \tag{64}$$

the density ratio

$$\rho^{(2)}/\rho^{(1)} = \varepsilon_0 \equiv \frac{2\zeta_0}{h^{(1)} - h^{(2)} - 2\zeta_0} \tag{65}$$

and the overall pressure

$$p^{(1)} + \tilde{p}^{(2)} = \left(1 - h^{(1)}\right)(1+\varepsilon_0)\rho^{(1)}. \tag{66}$$

The variation rate of entropy is positive (if $\zeta_0 > 0$) and $nT\dot{s} \sim \rho^{(1)}$.

To obtain a general solution to equation (57) in the isotropic case we suppose that the parameter k specifying the time gauge by condition (56) is

$$k = 1 - h^{(1)}/2. \tag{67}$$

(We note that the time t becomes synchronous if the perfect fluid component appears as the Λ-term, i.e. $h^{(1)} = 2$). This yields that equation (57) under the condition, given by equation (60), is integrable by elementary methods. The result is

$$a = a_0\left(\frac{\tau^2}{1-\tau^2}\right)^{1/[(D-1)(h^{(1)}-h^{(2)}-2\zeta_0)]}, \tag{68}$$

$$\rho^{(1)} = Aa^{(D-1)[h^{(1)}-2]}, \tag{69}$$

$$q = -1 + \frac{D-1}{2}\left[2 - h^{(1)} + \left(h^{(1)} - h^{(2)} - 2\zeta_0\right)(1-\tau^2)\right], \tag{70}$$

$$\rho^{(2)}/\rho^{(1)} = (1+\varepsilon_0)/\tau^2 - 1, \tag{71}$$

$$p^{(1)} + \tilde{p}^{(2)} = (1+\varepsilon_0)\left[\left(1 - h^{(2)} - 2\zeta_0\right)\frac{1-\tau^2}{\tau^2} + 1 - h^{(1)}\right]\rho^{(1)}, \tag{72}$$

where we introduced the following variable

$$\tau = \tanh\left[\left(h^{(1)} - h^{(2)}\right)\sqrt{\frac{\kappa^2(D-1)A}{2(D-2)(1+\varepsilon_0)}}(t-t_0)\right], \quad t > t_0. \tag{73}$$

Now we analyze the role of viscosity in this isotropic 2-component model using the obtained exact solution. Main features of the model are the following. Under the weak energy condition ($\rho^{(1)}+\rho^{(2)} \geq 0$, $\rho^{(1)}+\rho^{(2)}+p^{(1)}+\tilde{p}^{(2)} \geq 0$), which leads to the following restriction on the parameters $2 \geq h^{(1)} > h^{(2)} + 2\zeta_0$, the Universe expands eternally from the initial singularity. Near the singularity we obtain in the main order $a \sim t_s^{2/[(2-h^{(2)}-2\zeta_0)(D-1)]}$ There is only the nonsingular solution given by equation (62).

All solutions describe the accelerated expansion at least on the late phase of evolution if $h^{(1)} > 2(D-2)/(D-1)$. The solutions given by (68) describe the period of decelerated expansion if $h^{(2)} + 2\zeta_0 < 2(D-2)/(D-1)$. We note that the cosmic deceleration phase is important within the mechanism of the clumping of matter into galaxies (see, for instance, [48]). Under the assumption $2 \geq h^{(1)} > 2(D-2)/(D-1) > h^{(2)} + 2\zeta_0$ the decelerated expansion takes place during the time interval (t_0, t^*) and $\Delta t = t^* - t_0$ is given by

$$\Delta t = \frac{\operatorname{arcosh}\sqrt{1 + \frac{2(D-2)/(D-1) - h^{(2)} - 2\zeta_0}{h^{(1)} - 2(D-2)/(D-1)}}}{\sqrt{\kappa^2 \frac{A(D-1)}{2(D-2)}\left(h^{(1)} - h^{(2)}\right)\left(h^{(1)} - 2\frac{D-2}{D-1}\right)\left[1 + \frac{2(D-2)/(D-1) - h^{(2)} - 2\zeta_0}{h^{(1)} - 2(D-2)/(D-1)}\right]}}$$

This equation shows that introduction of the bulk viscosity reduces the phase of the decelerated expansion.

The density ratio given by (65) and (71) exhibits the following property

$$\lim_{t\to+\infty}\rho^{(2)}/\rho^{(1)} = \varepsilon_0 \equiv \frac{2\zeta_0}{h^{(1)} - h^{(2)} - 2\zeta_0}. \tag{74}$$

So, the bulk viscosity allows to resolve the coincidence problem which appears in the nonviscous 2-component model.

3.2. THE ANISOTROPIC MODEL

Now we consider the general anisotropic behaviour of the model. Once integrating equation (57) we get the first integral of the form $\dot{\gamma}_0 = F(\gamma_0, C)$, where F is some function and C is an arbitrary constant. Substituting $\dot{\gamma}_0$ and the functions $\langle \dot{\mathbf{x}}, \mathbf{f}_j \rangle$ given by equation (55) into equations (45) and (48) we express $\rho^{(2)}$ and $nT\dot{s}$ via γ_0. The subsequent analysis of the expressions for γ_0, $\rho^{(2)}$ and $nT\dot{s}$ shows the presence of solutions with physically nonsatisfactory behaviour near the initial singularity. To exclude such solutions we put the following restrictions on the parameters

$$2 \geq h^{(1)} > h^{(2)} + 2\zeta_0 \geq -2\eta_0 \geq 2\zeta_0, \; -2\eta_0 \geq h^{(2)} \geq 0. \tag{75}$$

These restrictions guarantee the following properties for all solutions: the density $\rho^{(2)}$ and the variation rate of entropy $\dot{s}$ are positive during the evolution, which starts at the initial singularity of the Kasner type and proceeds eternally to the subsequent isotropic expansion; the density ratio $\rho^{(2)}/\rho^{(1)}$ tends to the nonzero constant as $t \to +\infty$ (see equation (74)).

To study these behaviour in detail let us obtain the exact solution. We note that the equation $\dot{\gamma}_0 = F(\gamma_0, C)$ is integrable by quadrature for arbitrary parameters $h^{(1)}, h^{(2)}, \zeta_0$ and η_0. To express the exact solution in elementary functions we put the following relation on the parameters

$$h^{(1)} - 2h^{(2)} - 2\eta_0 - 4\zeta_0 = 0. \tag{76}$$

For the nonviscous model ($\eta_0 = \zeta_0 = 0$) the relation reads $h^{(1)} = 2h^{(2)}$. The latter corresponds, for instance, to the so called ΛCDM cosmological model with Λ-term ($h^{(1)} = 2$) and dust ($h^{(2)} = 1$). We remind that the parameters obey the inequalities given by formula (75). Comparing equation (76) and formula (75), one gets $h^{(2)} > 0$.

Now we start the integration procedure in the time gauge defined by equations (56) and (67) . The integration of equation (57) gives

$$e^{\beta\gamma_0(t)} = C_0 \frac{\tau(\sin\alpha + \tau\cos\alpha)}{1-\tau^2}, \tag{77}$$

where we cancelled the constant $\sum_{j=2}^{n}(p^j)^2$ by introducing the following constants

$$C_0 \geq \sqrt{\frac{2h^{(2)}}{\kappa^2 A(h^{(1)} - h^{(2)})} \sum_{j=2}^{n} (p^j)^2},$$

$$\alpha = \arcsin \frac{\sqrt{\frac{2h^{(2)}}{\kappa^2 A(h^{(1)} - h^{(2)})} \sum_{j=2}^{n}(p^j)^2}}{C_0} \in [0, \pi/2], \tag{78}$$

the parameter β is defined as follows

$$\beta = h^{(1)} - h^{(2)} - 2\zeta_0 > 0. \tag{79}$$

The variable τ was introduced by equation (73). Substituting equation (77) into equation (55) and taking into account equations (56) and (67), we obtain by integration

$$\langle \mathbf{x}, \mathbf{f}_j \rangle = \sqrt{\frac{(D-2)\beta}{(D-1)h^{(2)}\sum_{j=2}^{n}(p^j)^2}}\, p^j \ln\left[\frac{\tau}{\sin\alpha + \tau\cos\alpha}\right]^{1/\beta} + q^j, \qquad (80)$$

where q^j are arbitrary constants. Substituting equations (77) and (80) into the decomposition given by equation (52), we get

$$\begin{aligned} \mathbf{x} &= \ln\left[C_0 \frac{(\sin\alpha + \tau\cos\alpha)^2}{1-\tau^2}\right]^{1/\beta} \frac{\mathbf{u}}{\langle \mathbf{u}, \mathbf{u} \rangle} \\ &+ \ln\left[\frac{\tau}{\tau(\sin\alpha + \tau\cos\alpha)}\right]^{1/\beta} \mathbf{r} + \mathbf{s}, \end{aligned} \qquad (81)$$

where the vectors $\mathbf{r} \in \mathbb{R}^n$ and $\mathbf{s} \in \mathbb{R}^n$ are defined as follows

$$\mathbf{r} \equiv \sum_{i=1}^{n} r^i \mathbf{e}_i = \sqrt{\frac{(D-2)\beta}{(D-1)h^{(2)}\sum_{j=2}^{n}(p^j)^2}} \sum_{j=2}^{n} p^j \mathbf{f}_j + \frac{\mathbf{u}}{\langle \mathbf{u}, \mathbf{u} \rangle},$$

$$\mathbf{s} \equiv \sum_{i=1}^{n} s^i \mathbf{e}_i = \sum_{j=2}^{n} q^j \mathbf{f}_j. \qquad (82)$$

Owing to the orthogonality property given by equation (51) the coordinates r^i and s^i of these vectors in the canonical basis $\mathbf{e}_1, \ldots, \mathbf{e}_n$ satisfy the following constraints

$$\sum_{i=1}^{n} d_i \left(r^i\right)^2 = \langle \mathbf{r}, \mathbf{r} \rangle - \sum_{i,j=1}^{n} d_i d_j r^i r^j = \frac{(D-2)\beta}{(D-1)h^{(2)}} + \frac{1}{D-1}, \qquad (83)$$

$$\sum_{i=1}^{n} d_i r^i = \langle \mathbf{u}, \mathbf{r} \rangle = 1, \; \sum_{i=1}^{n} d_i s^i = \langle \mathbf{u}, \mathbf{s} \rangle = 0. \qquad (84)$$

The constants r^i may be called Kasner-like parameters, because of the existence of the constraints given by equation (83). We remind, that the coordinates of the vector x in the canonical basis are the logarithms of the scale factors $a_i \equiv \exp[x^i]$.

Finally, we present the exact solution:

$$\begin{aligned} ds^2 = &- \left[C_0 \frac{\tau(\sin\alpha + \tau\cos\alpha)}{1-\tau^2}\right]^{(2-h^{(1)})/\beta} dt^2 \\ &+ \left[C_0 \frac{(\sin\alpha + \tau\cos\alpha)^2}{1-\tau^2}\right]^{2/[(D-1)\beta]} \sum_{i=1}^{n} \left[\frac{\tau}{\sin\alpha + \tau\cos\alpha}\right]^{2r^i/\beta} e^{2s^i} ds_i^2, \end{aligned} \qquad (85)$$

$$\rho^{(1)} = A\left[C_0\frac{\tau(\sin\alpha+\tau\cos\alpha)}{1-\tau^2}\right]^{-(2-h^{(1)})/\beta}, \tag{86}$$

$$\rho^{(2)}/\rho^{(1)} = (1+\varepsilon_0)F_1^2(\tau)\left[1-\frac{\beta}{h^{(2)}}F_2^2(\tau)\right]-1, \tag{87}$$

$$nT\dot{s} = \sqrt{8\kappa^2 A\frac{D-1}{D-2}(1+\varepsilon_o)^3F_1^3(\tau)}\left[\zeta_0+\frac{\eta_0\beta}{h^{(2)}}F_2^2(\tau)\right]\rho^{(1)}, \tag{88}$$

where

$$F_1(\tau) = \frac{\frac{1}{2}(1+\tau^2)\sin\alpha+\tau\cos\alpha}{\tau(\sin\alpha+\tau\cos\alpha)}, \quad F_2(\tau) = \frac{(1-\tau^2)\sin\alpha}{(1+\tau^2)\sin\alpha+2\tau\cos\alpha}. \tag{89}$$

The solution has the following integration constants: $C_0 > 0$, $\alpha \in [0, \pi/2]$, t_0, $r^1, \ldots, r^n$, $s^1, \ldots, s^n$. The constants r^i and s^i satisfy the constraints given by equations (83) and (84). Then the number of free integration constants is $2n$ as required. The limit for $\alpha \to +0$ of this exact solution is the isotropic solution obtained in section 3.2.

Before we start the studying of the obtained exact solution near the initial singularity let us remind the multi-dimensional generalization of the well-known *Kasner solution* [58]. It reads (for the synchronous time t_s) as follows

$$\mathrm{d}s^2 = -\mathrm{d}t_s^2 + \sum_{i=1}^{n} A_i t_s^{2\varepsilon^i}\mathrm{d}s_i^2. \tag{90}$$

Such a metric describes the evolution of a vacuum model under consideration. The Kasner parameters ε^i satisfy the constraints

$$\sum_{i=1}^{n} d_i\varepsilon^i = 1, \quad \sum_{i=1}^{n} d_i\left(\varepsilon^i\right)^2 = 1. \tag{91}$$

The generalized Kasner solution describes the contraction of some spaces from the set $M_1, \ldots, M_n$ and the expansion for the other ones. According to equation (91), the number of either contracting or expanding spaces depends on n (the total number of spaces) and $d_{i=1,n}$ (their dimensions).

We note that the constraints given by equation (83) for r^i coincide with these constraints for ε^i when $\zeta_0 = \eta_0 = 0$ and $h^{(1)} = 2h^{(2)}$, i.e. the Kasner-like parameters r^i become exactly Kasner parameters ε^i in the absence of viscosity. Therefore, if the parameters ζ_0 and η_0 are small enough, then the model describes a behaviour of the Kasner type as $\tau \to +0$, i.e. towards to the initial singularity. However, too strong viscosity suppress a behaviour of the Kasner type. It can be shown that if the parameters ζ_0, η_0 are large enough, then the model describes expansion of all factor spaces $M_1, \ldots, M_n$ near the initial singularity. Consider as an example the 4-dimensional Bianchi I-type model (the corresponding manifold is $\mathbb{M}^4 = \mathbb{R} \times M_1^1 \times M_2^1 \times M_3^1$) with Λ-term ($h^{(1)} = 2$) and radiation ($h^{(2)} = 2/3$). Near the initial singularity the obtained metric has the form (90) for $n = 3$, where instead of the Kasner parameters $\varepsilon^1, \varepsilon^2, \varepsilon^3$ one obtains the following parameters

$$\epsilon^1 = \frac{1 + 2\sqrt{2 - 3\zeta_0}\sin\varphi_0}{2(2 - 3\zeta_0)}, \tag{92}$$

$$\epsilon^2 = \frac{1 - \sqrt{2 - 3\zeta_0}\sin\varphi_0 + \sqrt{3}\sqrt{2 - 3\zeta_0}\cos\varphi_0}{2(2 - 3\zeta_0)}, \tag{93}$$

$$\epsilon^2 = \frac{1 - \sqrt{2 - 3\zeta_0}\sin\varphi_0 - \sqrt{3}\sqrt{2 - 3\zeta_0}\cos\varphi_0}{2(2 - 3\zeta_0)}. \tag{94}$$

Here φ_0 is an arbitrary constant and $\zeta_0 \in [1/3, 2/3)$ due to the restrictions (75) and the relation (76). It is easy to see, that if $\zeta_0 \in (7/12, 2/3)$, all parameters ϵ^i, $i = 1, 2, 3$ are positive, i.e. the model describes the expansion for each of three spatial directions.

One can prove that the final stage of the evolution ($\tau \to 1 - 0$) exhibits the isotropic expansion. The asymptotical behaviour of the model for $\tau \to 1 - 0$ is described by the exact solution given by equations (62)-(66).

We have integrated the D-dimensional homogeneous anisotropic cosmological model, assuming that the Universe is filled by a mixture of a perfect fluid with negative pressure (x-fluid) and a viscous fluid. The x-fluid is considered as a quintessence matter responsible for the accelerated expansion and the viscous fluid describes the normal matter, which probably possesses viscosity owing to its mixture content. The exact solution, describing the isotropic expansion, is considered separately. In this case we deal simply with the flat FRW model for $D = 4$. The bulk viscosity coefficient is supposed to be linearly proportional to the Hubble parameter. Using the exact solution, we have shown that the ratio of the normal matter density to the quintessence density tends to some nonzero constant as $t \to +\infty$. The constant depends on the parameters from the equations of state. This suggest the resolving to the cosmic coincidence problem, arising for the corresponding nonviscous model, when the ratio vanishes at large time. However, the bulk viscosity reduces the phase of the decelerated expansion, which is important within the mechanism of the matter clumping into galaxies. Generalizing the "second equations of state" to anisotropic case, we have obtained the general anisotropic solution to the equations of motion. It describes the Kasner-like behaviour near the initial singularity and the isotropic eternal expansion at the final stage of evolution. However, if the viscosity is strong enough, the Kasner-like behaviour may be suppressed.

References

1. Melnikov, V.N. (1993) *Multidimensional Classical and Quantum Cosmology and Gravitation. Exact Solutions and Variations of Constants*, CBPF-NF-051/93, Rio de Janeiro;
   Melnikov, V.N. (1994) in: *Cosmology and Gravitation*, ed. M. Novello, Editions Frontieres, Singapore, 1994, p. 147.
2. Melnikov, V.N. (1995) *Multidimensional Cosmology and Gravitation*, CBPF-MO-002/95, Rio de Janeiro;
   Melnikov, V.N. (1996) In: *Cosmology and Gravitation. II*, ed. M. Novello, Editions Frontieres, Singapore, p. 465.
3. Staniukovich, K.P. and Melnikov, V.N. (1983) *Hydrodynamics, Fields and Constants in the Theory of Gravitation*, Energoatomizdat, Moscow (in Russian).
4. Melnikov, V.N. (1994) *Int. J. Theor. Phys.* **33**, 1569.
5. de Sabbata, V., Melnikov, V.N. and Pronin, P.I. (1992) *Prog. Theor. Phys.* **88**, 623.
6. Melnikov, V.N. (1988) In: *Gravitational Measurements, Fundamental Metrology and Constants*, eds. V. de Sabbata and V.N. Melnikov, Kluwer Academic Publ., Dordtrecht, p. 283.
7. Ivashchuk, V.D. and Melnikov, V.N. (1988) *Nuovo Cimento* **B 102**, 131.
8. Bronnikov, K.A., Ivashchuk, V.D. and Melnikov, V.N. (1988) *Nuovo Cimento* **B 102**, 209.
9. Melnikov, V.N. and Staniukovich, K.P. (1978) In: *Problems of Gravitation and Elementary Particle Theory*, Atomizdat, Moscow, v. 9, p. 3 (in Russian).
10. Zaitsev, N.A. and Melnikov, V.N. (1979) In: *Problems of Gravitation and Elementary Particle Theory*, Atomizdat, Moscow, v. 10, p. 131 (in Russian).
11. Canuto, V. et al. (1977) *Phys. Rev.* **D 16**, 1643.
12. Novello, M. and Rotelli, P. (1972) *J. Phys.* **A 5**, 1488.
13. Bronnikov, K.A., Melnikov, V.N. and Staniukovich, K.P. (1968) ITP-68-69, Kiev.
14. Melnikov, V.N. (1979) *Dokl. Acad. Nauk* **246**, N 6, 1351;
    Melnikov, V.N. and Orlov, S.V. (1979) *Phys.Lett.* **A 70**, N4, 263.
15. Melnikov, V.N. and Radynov, A.G. (1984) In: *Problems of Gravitation and Elementary Particle Theory*, Atomizdat, Moscow, v. 14, p. 73 (in Russian).
16. Melnikov, V.N. and Radynov, A.G. (1985) In: *On Relativity Theory*, Singapore, WS, v. 2, p. 196.
17. Hellings R. (1983) *Phys. Rev. Lett.* **51**, 1609.
18. Sanders, A. and Deeds, W. (1992) *Phys. Rev.* **D 46**, 480.
19. Barrow, J.D., gr-qc/9711084.
20. Varshalovich, D.A. and Potekhin, A.Y. (1995) *Space Sci. Rev.* **74**, 259; (1996) *Pis'ma Atron. Zh.* **22**, 3; (1996) *Astron. Lett.*, **22**, 6.
21. Damour, T. and Dyson, F. (1996) *Nucl. Phys.* **B 480**, 37.
22. Drinkwater, M. et al. (1999) *MNRAS*, November, astro-ph/9711290.
23. Prestage, J.D. et al. (1995) *Phys. Rev. Lett.* **74**, 3511.
24. Melnikov, V.N. et al. (1985) *GRG* **17**, 63.
25. Melnikov, V.N. (1991) In: *Results of Science and Technology. Gravitation and Cosmology*, ed. V.N. Melnikov, **1**, p. 49 (in Russian).
26. (1999) *Measurement Science and Technology* **10**.
27. Gundlach, J.H.and Merkowitz, S.M., gr-qc/0006043.

28. Alexeev, A.D., Bronnikov, K.A., Kolosnitsyn, N.I., Konstantinov, M.Yu. Melnikov, V.N. and Radynov, A.G. (1993) *Izmeritelnaya tekhnika* N 8, p. 6; N 9, p. 3; N 10, p. 6; (1994) N 1, p. 3.; (1994) *Int. J. Mod. Phys.* **D 3**, N 4, 773.
Antonyuk, P.N., Bronnikov, K.A. and Melnikov, V.N. (1994) *Astron.Lett.* **20**, N 1, 59;
Bronnikov, K.A., Konstantinov, M.Yu. and Melnikov, V.N. (1997) *Grav. Cosm.* **3**, N 4(12), 293;
Alexeev, A.D. et al. (1997) *Izmerit. Tekhnika* N 10, p. 3; (1999) *Grav. Cosm.* **5**, N 1(17), 67.
Sanders, A et al. (1997) *Grav. Cosm.* **3**, 287; (1999) *Meas.Sci.Technol.* **10**, 514; (2000) *Class.Quant.Grav.* **17**, N 1.
Melnikov, V.N. (2000) *Science in Russia*, N 6, November.
29. Gron, O. (1990) *Astrophys. Space Sci.* **173**, 191.
30. Misner, C.W. (1968) *Astrophys. J.* **151**, 431.
31. Stewart, J.M. (1968) *Astrophys. Lett.* **2**, 133.
32. Collins, C.B. and Stewart, J.M. (1971) *Monthly Notices Roy. Astron. Soc.* **151**, 419.
33. Weinberg, S. (1971) *Astrophys. J.* **168**, 175.
34. Klimek, Z. (1976) *Nuovo Cimento* **B 35**, 249.
35. Caderni, N. and Fabbri, R. (1978) *Nuovo Cimento* **B 44**, 228.
36. Murphy, G. (1973) *Phys. Rev.* **D 8**, 4231.
37. Belinsky, V.A, and Khalatnikov, I.M. (1975) *Soviet Phys. JETP Letters* **21**, 99.
38. Belinsky, V.A. and Khalatnikov, I.M. (1976) *Soviet Phys. JETP* **42**, 205.
39. Novello, M. and Araujo, T.A. (1988) *Phys. Rev.* **D 22**, 260.
40. Romero, C. (1988) *Rev.Bras. de Fisica* **18**, 75.
41. Oliveira, H.P. and Salim, J.M. (1988) *Acta Phys. Pol.* **B 19**, 649.
42. Lukacs, B. (1976) *Gen. Rel. Gravit.* **7**, 635.
43. Wolf, C. (1989) *Phys. Scripta* **40**, 9.
44. Motta, D. and Tomimura, N. (1992) *Astrophys. J.* **401**, 437.
45. Gavrilov, V.R., Melnikov, V.N. and Novello, M. (1995) *Grav. Cosm.* **1**, 149; (1996) *Grav. Cosm.* **4(8)**, 325.
46. Gavrilov, V.R., Melnikov, V.N. and Triay, R. (1997) *Class. Quantum Grav.* **14**, 2203.
47. Waga, I. and Miceli, A. *Cosmological constraints from lensing statistics and supernova on the cosmic equation of state*, astro-ph/9811460.
48. Johri, V.B. *Constraints over cosmological constant and quintessence fields in an accelerating Universe*, astro-ph/00070079.
49. Bludman, S. and Roos, M. *Vacuum energy: cosmological constant or quintessence*, astro-ph/0004328.
50. Zlatev, I., Wang, L. and Steinhardt, P.J. *Quintessence, cosmic coincidence and the cosmological constant*, astro-ph/9807002.
51. de Ritis, R. and Marino, A.A. *Effective cosmological "constant" and quintessence*, astro-ph/007128.
52. Brax, P. and Martin, J. *High energy physics and quintessence*, astro-ph/0005449.
53. Chimento, L.S., Jakubi, A.S. and Pavón, D. *Enlarged Q-matter cosmology*, astro-ph/005070.
54. Brevik, I. and Heen, L.T. (1994) *Astrophys. Space Sci.* **219**, 99.
55. Zakari, M. and Jou, D. (1993) *Phys. Rev.* **D 48**, 1597.
56. Maartens, R. and Méndez, V. (1997) *Phys. Rev.* **D 55**, 1937.
57. Ivashchuk, V.D., Melnikov, V.N. and Zhuk, A.I.(1989) *Nuov. Cim.* **B 104**, 575.
58. Ivashchuk, V.D. (1992) *Phys. Lett.* **A 170**, 16.

QUANTUM SYSTEMS IN WEAK GRAVITATIONAL FIELDS

G. PAPINI
Department of Physics, University of Regina
Regina, Sask. S4S 0A2, Canada

International Institute for Advanced Scientific Studies
Vietri sul Mare (SA), Italy

AND

Canadian Institute for Theoretical Astrophysics
University of Toronto, On., Canada

1. Introduction

Fully covariant wave equations predict the existence of a class of inertial-gravitational effects that can be tested experimentally. In these equations inertia and gravity appear as external classical fields, but, by conforming to general relativity, provide very valuable information on how Einstein's views carry through in the world of the quantum. Experiments already confirm that inertia and Newtonian gravity affect quantum particles in ways that are fully consistent with general relativity down to distances of $\sim 10^{-4}cm$ for superconducting electrons [1] and of $\sim 10^{-8}cm$ for neutrons [2, 3, 4]. Other aspects of the interaction of gravity with quantum systems are just beginning to be investigated.

Gravitational-inertial fields in the laboratory are weak and remain so in the cosmos for most astrophysical sources. These are the fields considered here. They are adequately described by the weak field approximation (WFA).

Gravitational-inertial fields affect particle wave functions in a variety of ways. They induce quantum phases that afford a unified treatment of interferometry and gyroscopy. They interact with particle spins giving rise to a number of significant effects. They finally shift energy levels in particle spectra [5]. While it still is difficult to predict when direct measurements will become possible in the latter case, rapid experimental advances in particle interferometry [6, 7, 8] require that quantum phases be derived

P.G. Bergmann and V. de Sabbata (eds.),
Advances in the Interplay Between Quantum and Gravity Physics, 317–338.

with precision. This will be done below for Schroedinger, Klein-Gordon, Maxwell and Dirac equations. Large, sensitive interferometers hold great promise in many of these investigations. They can play a role in testing general relativity.

Spin-inertia and spin-gravity interactions are the subject of numerous theoretical [9, 10, 11, 12, 13, 14, 15, 16] and experimental efforts [17, 18, 19, 20, 21]. At the same time precise Earth-bound and near space experimental tests of fundamental theories require that inertial effects be identified with great accuracy. It is shown below that spin-rotation coupling is particularly important in precise tests of fundamental theories and in certain types of neutrino oscillations. Surprisingly, particle accelerators may be also called to play a role in these investigations [22].

2. Wave equations

The quantum phases induced by inertia and gravity are derived in this section for Schroedinger, Klein-Gordon, Maxwell and Dirac equations. Some applications are given in Section 3.

2.1. THE SCHROEDINGER EQUATION

Starting from the action principle ($\hbar = c = 1$)

$$S = -m\int ds = -m\int \sqrt{g_{\mu\nu}\dot{x}^{\mu}\dot{x}^{\nu}}dx^{0}, \tag{1}$$

where $\dot{x}^{\mu} = dx^{\mu}/dx^{0}$, one arrives at the Lagrangian

$$L = -m(g_{ij}\dot{x}^{i}\dot{x}^{j} + 2g_{i0}\dot{x}^{i} + g_{00})^{1/2}, \tag{2}$$

where $i, j = 1, 2, 3$. From L one obtains

$$p_i = \partial L/\partial \dot{x}^{i} = -m(g_{ij}\dot{x}^{j} + g_{i0})(g_{lk}\dot{x}^{l}\dot{x}^{k} + 2g_{k0}\dot{x}^{k} + g_{00})^{-1/2}. \tag{3}$$

Substituting (3) into $H = p_i\dot{x}^{i} - L$ one finds

$$H = m(g_{i0}\dot{x}^{i} + g_{i0})(g_{lk}\dot{x}^{l}\dot{x}^{k} + 2g_{k0}\dot{x}^{k} + g_{00})^{-1/2}. \tag{4}$$

In the WFA $g_{\mu\nu} \simeq \eta_{\mu\nu} + \gamma_{\mu\nu}, g^{\mu\nu} \simeq \eta^{\mu\nu} - \gamma^{\mu\nu}, g^{ij}g_{jk} \simeq \delta^{i}_{k}$. From (3) one then gets

$$g^{ij}p_j \simeq -m(\dot{x}^{i} + g^{ij}g_{j0})(g_{lk}\dot{x}^{l}\dot{x}^{k} + 2g_{k0}\dot{x}^{k} + g_{00})^{-1/2}. \tag{5}$$

Eq.(5) can be solved for $-m\dot{x}^{i}/(g_{lk}\dot{x}^{l}\dot{x}^{k} + 2g_{k0}\dot{x}^{k} + g_{00})^{-1/2}$ and gives

$$\dot{x^{j}} = -\frac{1}{m}(g_{lk}\dot{x}^{l}\dot{x}^{k} + 2g_{k0}\dot{x}^{k} + g_{00})^{1/2}g^{jk}p_k - g^{jk}g_{k0}. \tag{6}$$

On using Eq.(6), one finds

$$g_{lk}\dot{x}^l\dot{x}^k + 2g_{k0}\dot{x}^k + g_{00} = (g_{00} - g^{il}g_{i0}g_{l0})/(1 - 1/m^2 g^{lk}p_l p_k). \tag{7}$$

By substituting (6) and (7) into (4), one obtains

$$H \simeq \sqrt{p^2 + m^2}(1 + 1/2\gamma_{00}) + 1/2\gamma^{ij}p_i p_j/\sqrt{p^2 + m^2} - p^l\gamma_{l0}. \tag{8}$$

In the presence of electromagnetic fields and in the low velocity limit, the Hamiltonian (8) leads to the Schroedinger equation [23]

$$i\partial\psi(x)/\partial t = [1/2m(p_i - eA_i + m\gamma_{0i})^2 - eA_0 + 1/2m\gamma_{00}]\psi(x). \tag{9}$$

The WFA does not fix the reference frame entirely. The transformations $x_\mu \rightarrow x_\mu + \xi_\mu$ are still allowed and lead to the "gauge" transformations $\gamma_{\mu\nu} \rightarrow \gamma_{\mu\nu} - \xi_{\mu,\nu} - \xi_{\nu,\mu}$. In the stationary case the transformations $\gamma_{00} \rightarrow \gamma_{00}, \gamma_{0i} \rightarrow \gamma_{0i} - \xi_{0i}$ leave Eq.(9) invariant. Returning to normal units, the solution of the Schroedinger equation is in this case

$$\psi(x) = exp\left\{imc/\hbar\int^x \gamma_{0i}dx^i - ie/c\hbar\int^x A_i dx^i\right\}\psi_0(x), \tag{10}$$

where ψ_0 is the solution of the field-free Schroedinger equation. If the electron-lattice interaction is added to Eq.(9), then the resulting equation can be applied to the study of BCS superconductors in weak stationary gravitational fields [24, 25]. This is desirable because BCS superconductors behave in many respects as non viscous fluids. They also exhibit quantization on a macroscopic scale and appear ideally suited to magnify small physical effects.

2.2. THE KLEIN-GORDON EQUATION

A well-known form of the the fully covariant Klein-Gordon equation is

$$(g^{\mu\nu}\nabla_\mu\nabla_\nu - m^2)\Phi(x) = 0, \tag{11}$$

where ∇_μ represents covariant differentiation. To first order in the WFA, Eq.(11) becomes

$$[(\eta^{\mu\nu} - \gamma^{\mu\nu})\partial_\mu\partial_\nu - (\gamma^{\alpha\mu} - 1/2\gamma^\sigma_\sigma\eta^{\alpha\mu})_{,\mu}\,\partial_\alpha]\phi(x) = 0. \tag{12}$$

Eq.(12) has the exact solution [26, 27]

$$\Phi(x) = exp\{-i\Phi_g\}\,\phi_0(x) = (1 - i\Phi_g)\phi_0(x), \tag{13}$$

where $\phi_0(x)$ is the solution of the field-free equation in Minkowski space, and

$$\begin{aligned} i\Phi_g\phi_0 &= [\frac{1}{4}\int_P^x dz^\lambda(\gamma_{\alpha\lambda,\beta}(z) - \gamma_{\beta\lambda,\alpha}(z))[(x^\alpha - z^\alpha)\partial^\beta - (x^\beta - z^\beta)\partial^\alpha] - \\ &\quad \frac{1}{2}\int_P^x dz^\lambda \gamma_{\alpha\lambda}(z)\partial^\alpha]\phi_0. \end{aligned} \tag{14}$$

Eq.(14) is related to Berry's phase [28]. It is easy to prove by direct substitution that (13) is a solution of (12). In fact

$$\begin{aligned} i\partial_\mu(\Phi_g\phi_0) &= \frac{1}{4}\int_P^x dz^\lambda(\gamma_{\alpha\lambda,\beta}(z) - \gamma_{\beta\lambda,\alpha}(z))[\delta_\mu^\lambda\partial^\beta - \delta_\mu^\beta\partial^\alpha]\phi_0(x) + \\ &\quad \frac{1}{4}\int_P^x dz^\lambda(\gamma_{\alpha\lambda,\beta}(z) - \gamma_{\beta\lambda,\alpha}(z))[(x^\alpha - z^\alpha)\partial^\beta - \\ &\quad (x^\beta - z^\beta)\partial^\alpha]\partial_\mu\phi_0(x) - \frac{1}{2}\int_P^x dz^\lambda\gamma_{\alpha\lambda}(z)\partial^\alpha\partial_\mu\phi_0(x) - \\ &\quad \frac{1}{2}\gamma_{\alpha\mu}(x)\partial^\alpha\phi_0(x), \end{aligned} \tag{15}$$

from which one gets

$$i\partial_\mu\partial^\mu(\Phi_g\phi_0) = im^2\Phi_g\phi_0 - \gamma_{\mu\alpha}\partial^\mu\partial^\alpha\phi_0 - (\gamma^{\beta\mu} - \frac{1}{2}\gamma^\sigma_\sigma\eta^{\beta\mu})_{,\mu}\,\partial_\beta\phi_0. \tag{16}$$

The result is proven by substituting (16) into (12). For a closed path in space-time one finds

$$i\Delta\Phi_g\phi_0 = \frac{1}{4}\oint R_{\mu\nu\alpha\beta}L^{\alpha\beta}d\tau^{\mu\nu}\phi_0, \tag{17}$$

where $L^{\alpha\beta}$is the angular momentum of the particle of mass m and $R_{\mu\nu\alpha\beta}$ is the linearized Riemann tensor. The result found is therefore manifestly gauge invariant.

Unlike the case of the Schroedinger equation discussed above, the gravitational fields considered in this section need not be stationary.

Since Eq.(13) is also a solution of the Landau-Ginzburg equation [26], the present results may be applied to the description of charged and neutral superfluids and Bose-Einstein condensates.

Applications of (13) to the detection of gravitational waves can be found in the literature [26, 29].

2.3. MAXWELL EQUATIONS

Consider now Maxwell equations

$$\nabla_\nu\nabla^\nu A_\mu - R_{\mu\sigma}A^\sigma = 0, \tag{18}$$

where the electromagnetic field A_μ satisfies the condition$\nabla_\mu A^\mu = 0$. If the second term in Eq.(18) is negligible, then Maxwell equations in the WFA are

$$\nabla_\nu\nabla^\nu A_\mu \simeq (\eta^{\sigma\alpha}-\gamma^{\sigma\alpha})A_{\mu,\alpha\sigma}+R_{\mu\sigma}A^\sigma-(\gamma_{\sigma\mu,\nu}+\gamma_{\sigma\nu,\mu}-\gamma_{\mu\nu,\sigma})A^{\sigma,\nu} = 0, \quad (19)$$

where use has been made of the Lanczos-DeDonder gauge condition

$$\gamma_{\alpha\nu},^{\nu} - \frac{1}{2}\gamma^{\sigma}_{\sigma},_{\alpha} = 0. \quad (20)$$

Eq.(19) has the solution [27]

$$\begin{aligned} A_\mu(x) &= a_\mu(x) - \frac{1}{4}\int_P^x dz^\lambda(\gamma_{\alpha\lambda,\beta}(z) - \gamma_{\beta\lambda,\alpha}(z))[(x^\alpha - z^\alpha)\partial^\beta a_\mu(x) - \\ &(x^\beta - z^\beta)\partial^\alpha a_\mu(x)] + \frac{1}{2}\int_P^x dz^\lambda\gamma_{\alpha\lambda}(z)\partial^\alpha a_\mu(x) + \\ &\frac{1}{2}\int_P^x dz^\lambda(\gamma_{\beta\mu,\lambda}(z) + \gamma_{\beta\lambda,\mu}(z) - \gamma_{\mu\lambda,\beta}(z))a^\beta(x), \end{aligned} \quad (21)$$

where $\partial_\nu\partial^\nu a_\mu = 0$ and $\partial^\nu a_\nu = 0$. Eq.(21) can also be written in the form $A_\mu = exp(-i\xi)a_\mu$, where

$$\begin{aligned} \xi &= -\frac{1}{4}\int_P^x dz^\lambda(\gamma_{\alpha\lambda,\beta}(z) - \gamma_{\beta\lambda,\alpha}(z))J^{\alpha\beta} + \\ &\frac{1}{2}\int_P^x dz^\lambda\gamma_{\alpha\lambda}(z)k^\alpha - \frac{1}{2}\int_P^x dz^\lambda\gamma_{\alpha\beta,\lambda}(z)T^{\alpha\beta}, \end{aligned} \quad (22)$$

$J^{\alpha\beta} = L^{\alpha\beta} + S^{\alpha\beta}$ is the total angular momentum, $(S^{\alpha\beta})^{\mu\nu} = -i(g^{\mu\alpha}g^{\nu\beta} - g^{\mu\beta}g^{\nu\alpha})$ is the spin-1 operator, $(T^{\alpha\beta})^{\mu\nu} \equiv -i\frac{1}{2}(g^{\mu\alpha}g^{\nu\beta} + g^{\mu\beta}g^{\nu\alpha})$ and k^α is the momentum of the free photon. All spin effects are therefore contained in the $S^{\alpha\beta}$ and $T^{\alpha\beta}$ terms. For a closed path one can again find Eq.(17).

2.4. THE GENERALLY COVARIANT DIRAC EQUATION

Some of the most precise experiments in physics involve spin-1/2 particles. They are very versatile tools that can be used in a variety of experimental situations and energy ranges while still retaining essentially a non-classical behaviour. Within the context of general relativity, De Oliveira and Tiomno [30] and Peres [31] conducted comprehensive studies of the fully covariant Dirac equation. The latter takes the form

$$[i\gamma^\mu(x)D_\mu - m]\Psi(x) = 0, \quad (23)$$

where $D_\mu = \nabla_\mu + i\Gamma_\mu$, $\{\gamma^\mu(x), \gamma^\nu(x)\} = 2g^{\mu\nu}(x)$, and the spin connection Γ^μ is defined by $D_\mu\gamma_\nu(x) = \nabla_\mu\gamma_\nu(x) + i[\Gamma_\mu(x), \gamma_\nu(x)] = 0$. By using the definitions $\Psi(x) = S\tilde{\Psi}(x)$, $S = exp(-i\int_P^x dz^\lambda \Gamma_\lambda(z))$ and $\tilde{\gamma}^\mu(x) = S^{-1}\gamma^\mu(x)S$, in (23) one finds

$$[i\tilde{\gamma}^\mu(x)\nabla_\mu - m]\tilde{\Psi} = 0. \tag{24}$$

By substituting $\tilde{\Psi} = [-i\tilde{\gamma}^\alpha(x)\nabla_\alpha - m]\psi'$ into (24) , one obtains

$$(g^{\mu\nu}\nabla_\mu\nabla_\nu + m^2)\psi' = 0 \tag{25}$$

which, as shown above, has the WFA solution $\psi' = exp(-i\Phi_g)\psi_0$, where ψ_0 is a solution of the Dirac equation in Minkowski space. It is again possible to show that for a closed path the total phase difference suffered by the Dirac wave function is gauge invariant and is given by $-\frac{1}{4}\oint R_{\mu\nu\alpha\beta}J^{\alpha\beta}d\tau^{\mu\nu}$, where the total angular momentum is now $J^{\alpha\beta} = L^{\alpha\beta} + \sigma^{\alpha\beta}$, $\sigma^{\alpha\beta} = -\frac{1}{2}[\gamma^\alpha, \gamma^\beta]$ and γ^β represents a usual, constant Dirac matrix [27].

3. Applications

Several applications of solutions (10) and (13)to superconductors, gyroscopy and interferometry can now be discussed.

3.1. SUPERCONDUCTORS

By comparing the Schroedinger equation for superconductors in electromagnetic fields with (9) one can immediately draw the following conclusions [24, 25].

i)$\vec{\nabla}(A_0 - \frac{1}{2}\frac{mc^2}{e}\gamma_{00}) = 0$. This means that the gravitational field generates an electric field inside the superconductor, contrary to the gravity-free case ($\gamma_{00} = 0$) that yields $\vec{E} = 0$. In principle one could therefore detect a gravitational field by means of the electric field it produces inside the superconductor. If the field is Newtonian, then $\vec{E} = \frac{mg}{e}$ which is the field Schiff and Barnhill [32] predicted gravity would produce inside normal conductors.

ii)$B_i + \frac{mc^2}{e}\varepsilon_{ijk}\partial^j\gamma^{0i} = 0$ well inside the superdonductor where B_i is known to vanish in the absence of gravitational fields.

iii) The total flux $\oint(A_i - \frac{mc^2}{e}\gamma_{0i})dx^i = n\frac{\hbar c}{2e}$ is quantized, rather than just the flux of B_i. This again signifies that γ_{0i} could be measured if the magnetic field it generates were sufficiently large. When the superconductor rotates $\gamma_{0i} = (\frac{\vec{\omega}\times\vec{r}}{c})_i$, one finds $\vec{B} = \frac{2mc}{e}\vec{\omega}$ which is the London moment of rotating superconductors. This result offers tangible evidence that inertia interacts with a quantum system in ways that are compatible with Einstein's views down to lengths of the order of $10^{-4}cm$.

These conclusions only apply to *stationary* gravitational fields. Other examples of gravity-induced electric and magnetic fields are discussed in the literature [25].

3.2. ROTATION

Consider for simplicity a square interferometer ABCD of side l in the (xy)-plane, rotating with angular velocity ω about the z-axis. The emission and interference of spinless particles of mass m take place at A and C respectively. Using the metric

$$ds^2 = (1-\omega^2\frac{x^2+y^2}{c^2})(dx^0)^2 + \frac{2\omega}{c}(ydx - xdy)dx^0 - dx^2 - dy^2 - dz^2, \quad (26)$$

and indicating by $\wp_1$ the path ABC and by $\wp_2$ the path ADC, the non-vanishing contributions to $i\Phi_g\phi_0$ given by Eq.(14) are

$$\begin{aligned}
\Delta\chi &= -\frac{1}{2}\int_{A,\wp_1}^{C} dz^\lambda \gamma_{\alpha\lambda}(z)k^\alpha + \frac{1}{2}\int_{A,\wp_2}^{C} dz^\lambda\gamma_{\alpha\lambda}(z)k^\alpha = \\
& -\frac{1}{2}\int (dz^0\gamma_{10}k^1 + dz^0\gamma_{20}k^2 + dz^1\gamma_{01}k^0 + dz^2\gamma_{02}k^0) = \\
& \frac{\omega}{2c}[-\int_{0,\wp_1}^{\frac{lc}{v}} dz^0 yk^1 + \int_{\frac{lc}{v}}^{\frac{2lc}{v}} dz^0 xk^2 - \int_{0,\wp_2}^{\frac{lc}{v}} dz^0 xk^2 + \\
& \int_{\frac{lc}{v}\wp_2}^{\frac{2lc}{v}} dz^0 yk^1] - k^0[\int_{0,\wp_1}^{l} ydx - \int_{0,\wp_1}^{l} xdy + \int_{0,\wp_2}^{l} xdy - \int_{0,\wp_2}^{l} ydx] = \\
& \frac{\omega l^2}{c}(k\frac{c}{v} + k^0). \qquad (27)
\end{aligned}$$

For non-relativistic particles $k^0 \sim \frac{mc}{\hbar}(1+\frac{v^2}{2c^2})$, $k \sim \frac{mv}{\hbar}$ and the result is $\Delta\chi \sim \frac{2ml\omega^2}{\hbar}(1+\frac{v^2}{8c^2})$. The first term agrees with the results of several relativistic and non-relativistic approximations. In general one obtains from Eq.(14)

$$\Delta\chi = (\frac{2m}{\hbar} + \frac{\hbar k^2}{2mc^2})\vec{\omega}\cdot\vec{a}, \quad (28)$$

where $\vec{a}$ represents the area of the interferometer oriented along its normal [26]. It therefore appears that gyroscopy is completely controlled by the quantum phase (28). One also finds that $\frac{(\Delta\chi)_{ph}}{(\Delta\chi)_{part}} = \frac{\lambda_c}{(\lambda)_{ph}}$, where λ_c is the Compton wavelength of the particle circulating in the interferometer. This ratio indicates that particle interferometers are more sensitive than photon interferometers for particle masses $m > \frac{h\nu_{ph}}{c^2}$.

On applying (22) to photons, one finds that the time integral part of ξ yields

$$-\frac{1}{4}\int_P^x dz^0(\gamma_{\alpha 0,\beta}-\gamma_{\beta 0,\alpha})S^{\alpha\beta}-\frac{1}{2}\int_P^x dz^0\gamma_{\alpha\beta,0}T^{\alpha\beta} = -\frac{1}{2}\int_P^x dz^0\gamma_{i0,j}S^{ij}=\int dt\omega S_z \tag{29}$$

which represents the spin-rotation coupling , or Mashhoon effect, for photons [13, 14, 27].

3.3. GRAVITATIONAL RED-SHIFT

Two light sources of the same frequency are at distances r_A and r_B from the origin at the initial time x_1^0. They are compared at r_A at the later time x_2^0. Neglecting spin effects, the phase difference can be simply obtained from (14) using the closed space-time path in the (r, x^0)-plane with vertices at (r_A, x_1^0), (r_B, x_1^0), (r_B, x_2^0), (r_A, x_2^0). The gravitational field is represented by $\gamma_{00}(r) = 2\varphi(r)$, where $\varphi(r)$ is the Newtonian potential. One finds

$$\begin{aligned}\Delta\chi \;=\; & \frac{1}{2}\int_{x_1^0}^{x_2^0} dz^0[\gamma_{\alpha 0,\beta}(r_B)-\gamma_{\beta 0,\alpha}(r_B)](x^\alpha - z^\alpha)k^\beta + \\ & \frac{1}{2}\int_{x_2^0}^{x_1^0} dz^0\gamma_{\alpha 0,\beta}(r_A)(x^\alpha - z^\alpha)k^\beta - \frac{1}{2}\int_{x_1^0}^{x_2^0} dz^0\gamma_{\alpha 0}(r_B)k^\alpha - \\ & \frac{1}{2}\int_{x_2^0}^{x_1^0} dz^0\gamma_{\alpha 0}(r_A)k^\alpha = \\ & -\frac{k^0}{2}(x_2^0 - x_1^0)[\gamma_{00}(r_B)-\gamma_{00}(r_A)] - \\ & \frac{k^0}{4}(x_2^0 - x_1^0)^2[\gamma_{00,1}(r_B)+\gamma_{00,1}(r_A)].\end{aligned} \tag{30}$$

The first term gives the usual red-shift formula $(\frac{\Delta\nu}{\nu})_1 = -\frac{1}{c^2}[\varphi(r_B)-\varphi(r_A)]$. The second term yields the additional correction $(\frac{\Delta\nu}{\nu})_2 = -\frac{x_2^0-x_1^0}{4}[\gamma_{00,1}(r_B)+\gamma_{00,1}(r_A)]$. In an experiment of the type carried out by Pound and Rebka the ratio of the two terms is $(\frac{\Delta\nu}{\nu})_2/(\frac{\Delta\nu}{\nu})_1 \simeq \frac{2l}{R_\oplus}$, where $l = r_B - r_A$. The second term $(\frac{\Delta\nu}{\nu})_2$ should therefore be measurable for sufficiently high values of l.

3.4. SCHWARZSCHILD METRIC

If Earth is assumed perfectly spherical and homogeneous and rotation is neglected, then its gravitational field can be described by the Schwarzschild metric[33]

$$ds^2 = (1 - \frac{2M_\oplus}{r})(dx^0)^2 - (1 - \frac{2M_\oplus}{r})^{-1}dr^2 - r^2 d\theta^2 - r^2 \sin\theta^2 d\varphi^2. \quad (31)$$

Assuming, for simplicity that a square interferometer of side l is placed in a vertical plane at one of the poles and that particle emission and interference occur at opposite corners, one finds

$$\Delta\chi = \frac{GM_\oplus l^2 m}{R_\oplus^2 \hbar v}\left(1 + \frac{3v^2}{2c^2} - \frac{3l}{2R_\oplus} - \frac{3v^2 l}{4c^2 R_\oplus}\right). \quad (32)$$

The first term in Eq.(32) is the term observed with a neutron interferometer in the well known COW experiment [2]. If $v \sim 10^{-5}c$, then the De Broglie's wavelength for neutrons is $\sim 10^{-8}cm$. General relativity appears therefore to be valid down to lengths of this order of magnitude. The ratio of the terms on the r.h.s. of (32) is $1 : 10^{-10} : 10^{-7} : 10^{-17}$. The last term is extremely small and may be neglected. Of the remaining terms, the second represents a special relativistic correction, which is smaller than the general relativity effect represented by the third term. The ratio of the latter to the magnitude of the Sagnac effect also is $\sim 10^{-7}$ and appears difficult to observe at present.

3.5. LENSE-THIRRING FIELD OF EARTH

The non-vanishing components of $\gamma_{\mu\nu}$ are in this instance [34]

$$\begin{aligned}\gamma_{00} &= \gamma_{11} = \gamma_{22} = \gamma_{33} = \frac{2GM_\oplus}{c^2 r} \\ \gamma_{01} &= \frac{4GM_\oplus \omega a^2 (y + y')}{5c^3 r^3}, \gamma_{02} = \frac{4GM_\oplus \omega a^2 (x + x')}{5c^3 r^3},\end{aligned} \quad (33)$$

where $r^2 = (x+x')^2 + (y+y')^2 + (z+z')^2$, Earth is again assumed spherical and homogeneous, ω its angular velocity about the z'-axis, and (x', y', z') are the coordinates at the point A at which the interferometer beam is split in the coordinate system with origin at the centre of the Earth. Interference occurs at the opposite vertex C. The frame z^μ has origin at A, is at rest relative to z'^μ and the plane of the interferometer is chosen for simplicity to coincide with the (x, y)-plane and parallel to the (x', y')-plane. The time at which the particle beam is split at A is $z^0 = z'^0 = 0$ [35]. If, in particular, A coincides with a pole, then $\Delta\chi = \frac{2G}{c^2 R^3} J_\oplus \frac{ml^2}{\hbar}$, where $J_\oplus = \frac{2M_\oplus R_\oplus^2 \omega}{5}$ is the angular momentum of Earth. Taking into account that the precession frequency of a gyroscope in orbit is $\Omega = \frac{GJ_\oplus}{2c^2 R_\oplus^3}$, one can also write $\Delta\chi = \Omega\Pi$, where $\Pi = \frac{4ml^2}{\hbar}$ replaces the period of a satellite in the classical calculation. Its value, $\Pi \sim 1.4 \times 10^8 s$ for neutron interferometers with $l \sim 10^2 cm$, is rather high and yields $\Delta\chi \sim 10^{-7} rad$. This suggests that the development and use of heavy particle interferometers would be particularly advantageous in attempts to measure the Lense-Thirring effect.

4. Helicity precession of fermions

Consider the line element $ds^2 = g_{\mu\nu}(x)dx^\mu dx^\nu$ and the set of tangent vectors $\vec{e}_\mu = \partial_\mu \vec{P}$ that forms the coordinate basis that spans the manifold $g_{\mu\nu}(x) = \partial_\mu \vec{P} \cdot \partial_\nu \vec{P} \equiv \vec{e}_\mu \cdot \vec{e}_\nu$. The principle of equivalence ensures the existence of an orthonormal tetrad frame $\vec{e}_{\hat{\mu}} = \partial_{\hat{\mu}} \vec{P}$ such that for a local tangent space defined at any given point of space-time $\eta_{\hat{\mu}\hat{\nu}} = \vec{e}_{\hat{\mu}} \cdot \vec{e}_{\hat{\nu}}$. The principle underlying the tetrad formalism therefore requires that for a sufficiently small region of space-time $\vec{e}_\mu$ be mapped onto $\vec{e}_{\hat{\mu}}$ using a set of projection functions $e^\nu_{\hat{\mu}}$ and their inverses $e^\nu_{\hat{\mu}}$ such that

$$\begin{aligned} \vec{e}_{\hat{\mu}} &= e^\nu_{\hat{\mu}} \vec{e}_\nu, \vec{e}_\mu = e^{\hat{\nu}}_\mu, \eta_{\hat{\mu}\hat{\nu}} = e^\mu_{\hat{\mu}} e^\nu_{\hat{\nu}} g_{\mu\nu}(x), \\ e^\nu_{\hat{\mu}} e^{\hat{\alpha}}_\nu &= \delta^{\hat{\alpha}}_{\hat{\mu}}, e^{\hat{\nu}}_\mu e^\alpha_{\hat{\nu}} = \delta^\alpha_\mu . \end{aligned} \tag{34}$$

When $\vec{e}_\mu$ refers to an observer with acceleration $\vec{a}$ rotating with angular velocity $\vec{\omega}$, one finds [9]

$$ds^2 = [(1 + \vec{a} \cdot \vec{x})^2 + (\vec{\omega} \cdot \vec{x})^2 - \omega^2 x^2] dx_0^2 - 2dx_0 d\vec{x} \cdot (\vec{\omega} \times \vec{x}) - d\vec{x} \cdot d\vec{x}, \tag{35}$$

while

$$\begin{aligned} \vec{e}_{\hat{0}} &= (1 + \vec{a} \cdot \vec{x})^{-1} [\vec{e}_0 - (\vec{\omega} \times \vec{x})^k \vec{e}_k], \vec{e}_{\hat{i}} = \vec{e}_i, \\ e^0_{\hat{0}} &= (1 + \vec{a} \cdot \vec{x})^{-1}, e^k_{\hat{0}} = -(1 + \vec{a} \cdot \vec{x})^{-1} (\vec{\omega} \times \vec{x}^k), \\ e^{\hat{0}}_0 &= 1 + \vec{a} \cdot \vec{x}, e^{\hat{k}}_0 = (\vec{\omega} \times \vec{x})^k, e^{\hat{0}}_i = 0, e^{\hat{k}}_i = \delta^k_i . \end{aligned} \tag{36}$$

Also, from $D_\mu \gamma_\nu(x) = 0$ and $\gamma_\mu(x) = e^{\hat{\mu}}_\mu \gamma_{\hat{\mu}}$, where $\gamma_{\hat{\mu}}$ represents the usual Dirac matrices, one finds $\Gamma_\mu(x) = \frac{1}{4}\sigma^{\hat{\alpha}\hat{\beta}} \Gamma_{\hat{\alpha}\hat{\beta}\hat{\mu}} e^{\hat{\mu}}_\mu$. The Ricci coefficients are $\Gamma_{\hat{\nu}\hat{\alpha}\hat{\beta}} = \frac{1}{2}(C_{\hat{\nu}\hat{\alpha}\hat{\beta}} + C_{\hat{\alpha}\hat{\beta}\hat{\nu}} - C_{\hat{\beta}\hat{\nu}\hat{\alpha}})$ and $C_{\hat{\nu}\hat{\alpha}\hat{\beta}} = \eta_{\hat{\mu}\hat{\nu}} e^\alpha_{\hat{\alpha}} e^\beta_{\hat{\beta}} (\partial_\alpha e^{\hat{\mu}}_\beta - \partial^{\hat{\mu}}_\beta)$. It also follows that $\Gamma_0 = -\frac{1}{2} a_i \sigma^{0i} - \frac{1}{2}\vec{\omega} \cdot \vec{\sigma}, \Gamma_i = 0$, with $\sigma^{0i} = \frac{i}{2}[\gamma^0, \gamma^i]$. The Hamiltonian is obtained by isolating the time derivative in the Dirac equation. The risult is

$$\begin{aligned} H &= \vec{\alpha} \cdot \vec{p} + m\beta + V(x) \\ V(x) &= \frac{1}{2}[(\vec{a} \cdot \vec{x})(\vec{p} \cdot \vec{\alpha}) + (\vec{p} \cdot \vec{\alpha})(\vec{a} \cdot \vec{x})] + m\beta(\vec{a} \cdot \vec{x}) - \vec{\omega} \cdot (\vec{L} + \frac{\vec{\sigma}}{2}), \end{aligned} \tag{37}$$

where $\vec{L}$ is the orbital angular momentum and $\vec{\sigma}$ are the usual Pauli matrices. The first three terms in $V(x)$ represent relativistic energy-momentum effects. The term $-\vec{\omega}\cdot\vec{L}$ is a Sagnac-type effect. The last term, $-\frac{1}{2}\vec{\omega}\cdot\vec{\sigma}$, is the spin-rotation coupling, or Mashhoon effect. The non-relativistic effects can be obtained by applying three successive Foldy-Wouthysen transformations to H. One obtains to lowest order

$$H = m\beta + \beta\frac{p^2}{2m} + \beta m(\vec{a}\cdot\vec{x}) + \frac{\beta}{2m}\vec{p}(\vec{a}\cdot\vec{x})\cdot\vec{p} - \vec{\omega}\cdot(\vec{L} + \frac{\sigma}{2}) + \frac{1}{4m}\vec{\sigma}\cdot(\vec{a}\times\vec{p}). \quad (38)$$

The third term in Eq.(38) is the energy-momentum effect observed by Bonse and Wroblewski [4]. The term $-\vec{\omega}\cdot\vec{L}$ was predicted by Page [36] and observed by Werner and collaborators [3]. The term $-\vec{\omega}\cdot\frac{\vec{\sigma}}{2}$ was found by Mashhoon. Hehl and Ni [9] re-derived all terms and also predicted the existence of the fourth term (a kinetic energy effect) and of the last term (spin-orbit coupling). Equations (37) and (38) can also be obtained by isolating the quantum phase in the wave function. The spin-rotation coupling term deserves a few comments. As discussed by Mashhoon, the effect violates the hypothesis of locality, namely that an accelerated observer is locally equivalent to an instantaneously comoving observer. This hypothesis is valid for classical point-like particles and optical rays and is widely used in relativity. The effect also violates the equivalence principle because it does not couple universally to matter [37]. No direct experimental verification of the Mashhoon effect has so far been reported, though the data given in [19] can be re-interpreted as due to the coupling of Earth's rotation to the nuclear spins of mercury. The effect is also consistent with a small depolarization of electrons in storage rings [38]. It is shown below that it plays an essential role in measurements of the anomalous magnetic moment, or $g-2$ factor, of the muon.

4.1. SPIN-ROTATION COUPLING IN MUON G-2 EXPERIMENTS

Precise measurements of the $g-2$ factor involve muons in a storage ring consisting of a vacuum tube, a few meters in diameter, in a uniform, vertical magnetic field $\vec{B}$. Muons on equilibrium orbits within a small fraction of the maximum momentum are almost completely polarized with spin vectors pointing in the direction of motion. As the muons decay, the highest energy electrons with spin almost parallel to the momentum, are projected forward in the muon rest frame and are detected around the ring. Their angular distribution does therefore reflect the precession of the muon spin along the cyclotron orbits [39, 40]. Let us start from the covariant Dirac equation (23). It is convenient to use the chiral representation for the usual Dirac matrices

$$\gamma^0 = \beta = \begin{pmatrix} 0 & -I \\ -I & 0 \end{pmatrix}, \gamma^i = \begin{pmatrix} 0 & \sigma^i \\ -\sigma^i & 0 \end{pmatrix}, \alpha^i = \begin{pmatrix} \sigma^i & 0 \\ 0 & -\sigma^i \end{pmatrix},$$
$$\sigma^{0i} = i\begin{pmatrix} \sigma^i & 0 \\ 0 & -\sigma^i \end{pmatrix}, \sigma^{ij} = \epsilon^{ij}_k \begin{pmatrix} \sigma^k & 0 \\ 0 & \sigma^k \end{pmatrix}, \gamma^5 = \begin{pmatrix} I & o \\ 0 & -I \end{pmatrix}. \qquad (39)$$

One must now add to the the Hamiltonian the effect of a magnetic field $\vec{B}$ on the total (magnetic plus anomalous) magnetic moment of the particle. Assuming for simplicity that all quantities in H are time-independent and referring them to a left-handed tern of axes comoving with the particle in the x_3-direction and rotating in the x_2-direction, one finds

$$H = \alpha^3 p_3 + m\beta + \frac{1}{2}[-a_1 R(\alpha^3 p_3) - (\alpha^3 p_3) a_1 R] + \beta m a_1 R - \vec{\omega}\cdot\vec{L} - \frac{1}{2}\omega_2\sigma^2 + \mu B\sigma^2 \equiv H_0 + H', \qquad (40)$$

where $B_2 = -B, \mu = (1 + \frac{g-2}{2})\mu_0$, $\mu_0 = \frac{e\hbar}{2mc}$ is the Bohr magneton, $H' = -\frac{1}{2}\omega_2\sigma^2 + \mu\beta\sigma^2$ and R is the radius of the muon's orbit. Electric fields used to stabilize the orbits and stray radial electric fields can also affect the muon spin. Their effects can be cancelled by choosing an appropriate muon momentum and will be neglected in what follows. Before decay the muon states can be represented as

$$|\psi(t)>= a(t)|\psi_+ > + b(t)|\psi_- >, \qquad (41)$$

where $|\psi_+ >$ and $|\psi_- >$ are the right and left helicity states of H_0. Substituting (41) into the Schroedinger equation $i\frac{\partial}{\partial t}|\psi(t)>= H|\psi(t)>$, one obtains

$$i\frac{\partial}{\partial t}\begin{pmatrix} a \\ b \end{pmatrix} = \begin{pmatrix} <\psi_+|(H_0 + H')|\psi_+> & <\psi_+|H'|\psi_-> \\ <\psi_-|H'|\psi_+> & <\psi_-|(H_0 + H')|\psi_-> \end{pmatrix}\begin{pmatrix} a \\ b \end{pmatrix}$$
$$= \begin{pmatrix} E - i\frac{\Gamma}{2} & i(\frac{\omega_2}{2} - \mu B) \\ -i(\frac{\omega_2}{2} - \mu B) & E - i\frac{\Gamma}{2} \end{pmatrix} \equiv M\begin{pmatrix} a \\ b \end{pmatrix}, \qquad (42)$$

where Γ represents the width of the muon. Notice that the spin-rotation coupling is off diagonal in (42). This is a clear indication that the Mashhoon effect violates the equivalence principle [37]. The matrix M can be diagonalized. Its eigenvalues are $h_1 = E - i\frac{\Gamma}{2} + (\frac{\omega_2}{2}) - \mu B), h_2 = E - i\frac{\Gamma}{2} - (\frac{\omega_2}{2} - \mu B)$, with the corresponding eigenvectors

$$|\psi_1 >= \frac{1}{\sqrt{2}}[i|\psi_+ > + |\psi_- >]; |\psi_2 >= \frac{1}{\sqrt{2}}[-i|\psi_+ > + |\psi_- >]. \qquad (43)$$

The solution of Eq.(42) is therefore

$$|\psi(t)> \;=\; \frac{1}{\sqrt{2}}(e^{-ih_1t}|\psi_1> + e^{-ih_2t}\psi_2>) =$$
$$\frac{1}{2}[(ie^{-ih_1t} - ie^{-ih_2t})|\psi_+> + (e^{-ih_1t} + e^{-ih_2t})|\psi_->], \quad (44)$$

where $|\psi(0)>=|\psi_->$. The spin-flip probability is

$$P_{\psi_-\to\psi_+} = |<\psi_+|\psi>|^2 = \frac{e^{-\Gamma t}}{2}[1-\cos(2\mu B-\omega_2)t], \quad (45)$$

where the Γ-term accounts for the observed exponential decrease in electron counts due to the loss of muons by radioactive decay [40]. The spin-rotation contribution to$P_{\psi\to\psi_+}$ is represented by ω_2 which is the cyclotron angular velocity $\frac{eB}{m}$. The spin-flip angular frequency is then

$$\Omega = 2\mu B - \omega_2 = (1+\frac{g-2}{2})\frac{eB}{m} - \frac{eB}{m} = \frac{g-2}{2}\frac{eB}{m}, \quad (46)$$

which is precisely the observed modulation frequency of the electron counts [41](see also Fig. 19 of Ref.[40]). This result is independent of the value of the anomalous magnetic moment of the particle. It is therefore the Mashhoon effect that gives prominence to the $g-2$ term in Ω by exactly cancelling, in $2\mu B$, the much larger contribution μ_0 that comes from fermions with no anomalous magnetic moment [42].

It is perhaps odd that spin-rotation coupling as such has almost gone unnoticed for such a long time. It is however significant that its effect is observed in an experiment that has already provided crucial tests of quantum electrodynamics and a test of Einstein's time-dilation formula to better than a 0.1 percent accuracy. Recent versions of the experiment [43, 44, 45] have improved the accuracy of the measurements from 270 ppm to 1.3 ppm. This bodes well for the detection of effects involving spin, inertia and electromagnetic fields, or inertial fields to higher order.

5. Neutrino Oscillations

Neutrino beams produced in weak interactions may be considered as a superposition of different mass eigenstates. As a beam propagates, different components of the beam evolve differently so that the probability of finding different eigenstates in the beam varies with time. The consequences of this can be explored in a number of cases.

5.1. NEUTRINO HELICITY OSCILLATIONS

Let us consider a beam of high energy neutrinos. If the neutrino source rotates, the effective Hamiltonian for the mass eigenstates can be written as $H_e = (p^2 + m^2)^{\frac{1}{2}} + \Gamma_0 \approx p + \frac{m^2}{2E} - \frac{1}{2}\vec{\omega} \cdot \vec{\sigma}$. For simplicity, consider a one-generation of neutrino that can now be written as a superposition of ν_L and ν_R in the form

$$|\nu(t) >= a_L(t)|\nu_L > + b_R(t)|\nu_R > . \qquad (47)$$

It is well known that the standard model contemplates only the existence of ν_L, while ν_R is considered sterile and therefore unobservable. Strictly speaking one should consider the helicity states $\nu_\pm$(that are mass eigenstates) in (47), however at high energies $\nu_L \simeq \nu_-, \nu_R \simeq \nu_+$. Assuming that $m_1 \neq m_2$, taking $p_1 \sim p_2$ along the x_3-axis and substituting (47) into the Schroedinger equation that corresponds to H_e, one obtains

$$i\frac{\partial}{\partial t}\begin{pmatrix} a_L \\ b_R \end{pmatrix} = \begin{pmatrix} p + \frac{m_1^2}{2E} & -\frac{\omega_1}{2} - i\frac{\omega_2}{2} \\ -\frac{\omega_1}{2} + i\frac{\omega_2}{2} & p + \frac{m_2^2}{2} \end{pmatrix}\begin{pmatrix} a_L \\ b_R \end{pmatrix} \equiv M_{12}\begin{pmatrix} a_L \\ b_R \end{pmatrix}. \qquad (48)$$

The eigenvalues of M_{12} are

$$k_\mp = p + \frac{m_1^2 + m_2^2}{4E} \mp [(\frac{\Delta m^2}{2E})^2 + \omega_\perp^2]^{\frac{1}{2}}, \qquad (49)$$

where $\Delta m^2 \equiv m_1^2 - m_2^2$, and $\omega_\perp^2 \equiv \omega_1^2 + \omega_2^2$. The eigenvectors are

$$|\nu_1 >= b_1[\eta_1|\nu_L > +|\nu_R >], |\nu_2 >= b_2[\eta_2|\nu_L > +|\nu_R >], \qquad (50)$$

where

$$\eta_1 = \frac{\omega_1 + i\omega_2}{\Omega + \frac{\Delta m^2}{2E}}, \eta_2 = \frac{\omega_1 + i\omega_2}{-\Omega + \frac{\Delta m^2}{2E}}, \qquad (51)$$

$$|b_1|^2 = \frac{1}{1 + |\eta_1|^2}, |b_2|^2 = \frac{1}{1 + |\eta_2|^2}, \qquad (52)$$

and

$$\Omega = \left[\left(\frac{\Delta m^2}{2E}\right)^2 + \omega_\perp^2\right]^{\frac{1}{2}}. \qquad (53)$$

One therefore finds

$$\begin{aligned} |\nu(t) > &= \frac{b_1}{\eta_1 - \eta_2} exp\left[-i\left(p + \frac{m_1^2 + m_2^2}{4E}\right)t\right] \times \\ &\left[\left(e^{i\frac{\Omega}{2}t}\eta_1 - e^{-i\frac{\Omega}{2}t}\eta_2\right)\nu_L + 2i\sin(\frac{\Omega t}{2})\nu_R\right], \end{aligned} \qquad (54)$$

where the initial condition is $\nu(0) = \nu_L$. One obtains the transition probability

$$P_{\nu_L \to \nu_R} = | < \nu_R|\nu(t) > |^2 = \frac{\omega_\perp^2}{2\Omega^2}[1 - \cos(\Omega t)]. \qquad (55)$$

If the neutrinos have mass, then the magnitude of the transition probability becomes appreciable if

$$\omega_\perp \geq \frac{\Delta m^2}{2E}. \tag{56}$$

Unlike the flavour oscillations generated by the MSW-mechanism [46, 47] that require $\Delta m^2 \neq 0$, helicity oscillations can occur also when $m_1 = m_2$ and $m_1 = m_2 = 0$. They are interesting because ν_R's, if they exist, do not interact with matter and would therefore provide an energy dissipation mechanism with possible astrophysical implications. The conversion rate that $\nu_L \to \nu_R$ is not large for galaxies and white dwarfs. Assume in fact that $\omega_\perp \gg \Delta m^2/2E$ and that the beam of neutrinos consists of $N_L(0)$ particles at $z = 0$. One immediately obtains from (55) that the relative numbers of particles at $z = 0$ are [48]

$$N_L(z) = N_L(0)\cos^2\left(\frac{\omega_\perp z}{2c}\right), N_R(z) = N_L(0)\sin^2\left(\frac{\omega_\perp z}{2c}\right). \tag{57}$$

One then obtains from (57) $N_R \sim 10^{-6} N_L(0)$ for galaxies of typical size L such that $\omega_\perp L \sim 200 km/s$. Similarly, for white dwarfs $\omega_\perp \sim 1.0 s^{-1}$ and one finds $N_R \sim 10^{-4} N_L(0)$. In the case of the Sun $\omega_\perp \sim 7.3 \times 10^{-5} - 2.4 \times 10^{-6} s^{-1}$ and the conversion rate peaks at distances $L \sim 10^{15} - 4 \times 10^{16} cm$, well in excess of the average Sun-Earth distance. Helicity oscillations could not therefore explain the solar neutrino puzzle without additional assumptions about the Sun's structure [49]. For neutron stars, however, the dynamics of the star could be affected by this cooling mechanism. In fact neutrinos diffuse out of a canonical neutron star in a time 1 to $10 s$, during which they travel a distance $3 \times 10^9 cm$ between collisions. At distances $L \sim 5 \times 10^6 cm$ (the star's radius) the conversion rate is $N_R \sim 0.5 N_0$. Even higher cooling rates may occur at higher rotational speeds and prevent the formation of a pulsar. These results do not require the existence of a magnetic moment for the neutrino (which would also require some mass). Its effect could be taken into account by adding the term $\vec{\mu} \cdot \vec{B}$ to H_e. In all instances considered, however, magnetic spin-flip rates of magnitude comparable to those discussed would require neutrino magnetic moments vastly in excess of the value $10^{-19}\mu_0 \left(\frac{m_\nu}{1eV}\right)$ predicted by the standard $SU(2) \times U(1)$ electroweak theory [50].

5.2. HELICITY OSCILLATIONS IN A MEDIUM

The behaviour of neutrinos in a medium is modified by a potential V. When this is taken into account, the effective Hamiltonian becomes (after subtracting from the diagonal terms a common factor which contributes only to the overall phase)

$$H = p + \frac{m^2}{2E} - V - \frac{1}{2}\vec{\omega}\cdot\vec{\sigma}. \tag{58}$$

For simplicity, consider again a one-generation of neutrino and assume that V is constant. Applying the digonalization procedure of the previous section to the new Hamilltonian, leads to the transition probability

$$P_{\nu_L \to \nu_R} = \frac{\omega_\perp^2}{2\Omega'^2}[1 - \cos(\Omega' t)], \tag{59}$$

where $\Omega' = [(V + \frac{\Delta m^2}{2E})^2 + \omega_\perp^2]^{\frac{1}{2}}$. One finds from (59) that spin-flip transitions are strongly suppressed when $V + \frac{\Delta m^2}{2E} > \omega_\perp$ and only the ν_L component is present in the beam. If $\omega_\perp > V + \frac{\Delta m^2}{2E}$, then the ν_L flux has effective modulation. Resonance occurs at $V = \frac{-\Delta m^2}{2E}$. Consider now the rotating core of a supernova. In this case V can be relatively large, of the order of several electron volts and corresponds to the interaction of neutrinos with the particles of the medium. For right-handed neutrinos V vanishes. Assuming that the star does not radiate more energy as ν_R's than as ν_L's, one finds $L_{\nu_L} \sim L_{\nu_R} \sim 5 \times 10^{53} erg/s$. As the star collapses, spin-rotation coupling acts on both ν_L and ν_R. The ν_L's become trapped and leak toward the exterior ($l \sim 1.5 \times 10^7 cm$), while their interaction with matter is $V \sim 14(\rho/\rho_c)eV$ and increases therefore with the medium's density, which at the core is $\rho_c \sim 4 \times 10^{14} g/cm^3$. The ν_R's escape. As ρ increases, the transition $\nu_L \to \nu_R$ is inhibited (off resonance). One also finds $\frac{\Delta m^2}{2E} < \omega_\perp$ when $\Delta m^2 < 10^{-5} eV^2, E \sim 10 MeV, \omega_\perp \sim 6 \times 10^3 s^{-1}$. It then follows from (59) that

$$L_{\nu_L} \sim L_{\nu_R} \sin^2(\frac{\omega_\perp l}{2c}), \tag{60}$$

where $\frac{\omega_\perp l}{2c} \sim \frac{\pi}{2}$. In the time $\frac{l}{c} \sim 5 \times 10^{-4} s$, the energy associated with the $\nu_R \to \nu_L$ conversion is $\sim 2.5 \times 10^{50} erg$ which is just the missing energy required to blow up the mantle of the collapsing star [48].

5.3. NEUTRINO FLAVOUR OSCILLATIONS

Consider a beam of neutrinos of fixed energy E emitted at point (r_A, t_A) of the (r,t)-plane. Assume also that the particles are in a weak flavour eigenstate that is a linear superposition of mass eigenstates m_1 and m_2, with $m_1 \neq m_2$. It is argued in the literature [51, 52] that if interference is observed at the same space-time point (r_B, t_B), then the lighter component must have left the source at a later time $\Delta t = \frac{r_B - r_A}{v_1} - \frac{r_B - r_A}{v_2}$, where v_1 and v_2 are the velocities of the eigenstates of masses m_1 and m_2 respectively.

Because of the difference in travel time Δt, gravity induced neutrino flavour oscillations will ensue even though gravity couples universally to matter. Ignoring spin contributions, the phase difference of the two mass eigenstates can be calculated from (14) in a completely gauge invariant way. Assume the neutrinos propagate in a gravitational field described by the Schwarzschild metric. When the closed space-time path in (14) is extended to the triangle (r_A, t_A), (r_B, t_B), $(r_A, t_A + \Delta t)$, one obtains

$$\begin{aligned}
&(i\Delta\Phi_g\phi_0)_{m_1} - (i\Delta\Phi_g\phi_0)_{m_2} = \\
&\frac{r_g E}{2}[-\frac{v_1\Delta t}{2} - \frac{1}{v_1}\ln(\frac{r_B}{r_A + v_1\Delta t}) + (-v_1 + \frac{1}{v_2} + v_2)\ln\frac{r_B}{r_A}] \simeq \\
&\frac{r_g E}{2}(\frac{1}{v_2} - \frac{1}{v_1} + v_2 - v_1)\ln\frac{r_B}{r_a},
\end{aligned} \tag{61}$$

where the approximation $v_1\Delta t \ll r_A$ has been used in deriving the last result. On using the equation $1/v = E/p$ and the approximations $v \sim 1 - \frac{m^2}{2E^2} - \frac{m^4}{8E^4}$, $1/v \sim 1 + \frac{m^2}{2E^2} + \frac{3m^4}{8E^4}$, one arrives at the final result

$$\begin{aligned}
&(i\Delta\Phi_g\phi_0)_{m_1} - (i\Delta\Phi_g\phi_0)_{m_2} \simeq \\
&\frac{MGc^5}{4\hbar E^3}(m_2^4 - m_1^4)\ln\frac{r_B}{r_A} = \\
&1.37 \times 10^{-19}(\frac{M}{M_\odot})(\frac{\Delta m^4}{eV^4})(\frac{MeV}{E})^3\ln\frac{r_B}{r_A}.
\end{aligned} \tag{62}$$

The effect therefore exists, but is extremely small in typical astrophysical applications.

Torsion-induced neutrino oscillations have been considered by de Sabbata and collaborators [53, 54].-

5.4. THE EQUIVALENCE PRINCIPLE AND NEUTRINO OSCILLATIONS

Gravitational fields can not generate neutrino oscillations if gravity couples universally to matter. As first pointed out by Gasperini [55], violations of the equivalence principle could in principle affect the behaviour of neutrinos and be tested in experiments on neutrino oscillations [56]. Consider a spinless particle in a Newtonian gravitational field. Its Hamiltonian in the WFA is

$$H = (1 - \gamma_{00})^{\frac{1}{2}}(p^2 - \gamma^{ij}p_ip_j + m^2)^{\frac{1}{2}} - p_i\gamma^{i0}, \tag{63}$$

which, in the simple case of a Newtonian potential, becomes

$$H \sim p + \frac{m^2}{2p} - \frac{1}{2}p\gamma_{00}, \tag{64}$$

where $\gamma_{00} = 2\alpha\varphi(r)$ and $\alpha = 1$ if the principle of equivalence is not violated. Deviations from the equivalence principle are parameterized by assuming that $\alpha \neq 1$ and takes different values for different neutrino mass eigenstates. Assume the mass eigenstates ν_1, ν_2 are related to the weak eigenstates by the transformation $\begin{pmatrix} \nu_1 \\ \nu_2 \end{pmatrix} = U \begin{pmatrix} \nu_e \\ \nu_\mu \end{pmatrix}$, where the unitary matrix $U = \begin{pmatrix} \cos\theta & -\sin\theta \\ \sin\theta & \cos\theta \end{pmatrix}$ represents mixing in the two-generation case. The weak eigenstates then evolve according to the equation

$$i\frac{\partial}{\partial t}\begin{pmatrix} \nu_e \\ \nu_\mu \end{pmatrix} = U^\dagger(p + \frac{m^2}{2E} - \alpha E)U\begin{pmatrix} \nu_e \\ \nu_\mu \end{pmatrix}, \tag{65}$$

where $m^2 = \begin{pmatrix} m_1^2 & 0 \\ 0 & m_2^2 \end{pmatrix}$, and $\alpha = \begin{pmatrix} \alpha_1 & 0 \\ 0 & \alpha_2 \end{pmatrix}$. One therefore finds

$$i\frac{\partial}{\partial t}\begin{pmatrix} \nu_e \\ \nu_\mu \end{pmatrix} = M_W\begin{pmatrix} \nu_e \\ \nu_\mu \end{pmatrix}, \tag{66}$$

$$M_W = \begin{pmatrix} E + (\frac{m_1^2}{2E} - E\varphi\alpha_1)\cos^2\theta + (\frac{m_2^2}{2E} - E\varphi\alpha_2)\sin^2\theta & (\frac{\Delta m^2}{4E} - E\varphi\frac{\Delta\alpha}{2})\sin(2\theta) \\ (\frac{\Delta m^2}{4E} - E\varphi\frac{\Delta\alpha}{2})\sin(2\theta) & E + (\frac{m_1^2}{2E} - E\varphi\alpha_1)\sin^2\theta + (\frac{m_2^2}{2E} - E\varphi\alpha_2)\cos^2\theta \end{pmatrix}. \tag{67}$$

Since the overall phase is unobservable, subtracting the constant $E + (m_1^2 - E\varphi\alpha_1)\sin^2\theta + (m_2^2 - E\varphi\alpha_2)\cos^2\theta$ from the diagonal terms of M_W does not affect oscillations in which only the relative phases of the mass eigenstates are involved. This leads to the equation

$$i\frac{\partial}{\partial t}\begin{pmatrix} \nu_e \\ \nu_\mu \end{pmatrix} = \frac{1}{2}(\frac{\Delta m^2}{2E} - E\varphi\Delta\alpha)\begin{pmatrix} -2\cos(2\theta) & \sin(2\theta) \\ \sin(2\theta) & 0 \end{pmatrix}\begin{pmatrix} \nu_e \\ \nu_\mu \end{pmatrix}. \tag{68}$$

The solutions of (68) are

$$\begin{aligned} \nu_e(t) &= C_1 e^{-i\omega t} + C_2 e^{i\omega t} \\ \nu_\mu(t) &= D_1 e^{-i\omega t} + D_2 e^{i\omega t}, \end{aligned} \tag{69}$$

with the condition $|\nu_e(t)|^2 + |\nu_\mu(t)|^2 = 1$. One finds $\omega = \frac{\Delta m^2}{4E} - E\varphi\frac{\Delta\alpha}{2}$. The initial condition $\nu_e(0) = 1$ is also used to determine the constants in (69).

One finds $C_1 = \sin^2\theta, C_2 = \cos^2\theta, D_1 = \sin\theta\cos\theta, D_2 = -\sin\vartheta\cos\theta$. The transition probability therefore is

$$P_{\nu_e \to \nu_\mu} = \sin^2(2\theta)\sin^2(\omega t). \tag{70}$$

In the absence of gravity, $\alpha_1 = \alpha_2 = 0$, flavour oscillations in vacuum occur according to the MSW mechanism and are driven by Δm^2. The MSW oscillations take place also when $\alpha_1 = \alpha_2 \neq 0$. On the other hand, when gravity is present and $\alpha_1 \neq \alpha_2 \neq 0$, flavour oscillations occur not only if $\Delta m^2 \neq 0$, but also when $\Delta m^2 = 0$, with either $m_1 = m_2$ or $m_1 = m_2 = 0$ [57]. The charged-current interactions of ν_e's with electrons in a star can also be taken into account by introducing the additional potential energy $\sqrt{2}G_F N_e(r) \leq 10^{-12}eV$. For the Sun $N_e(r) = N_0 exp(-10.54\frac{r}{R_\odot})cm^{-3}$, where N_0 is the number of electrons at its centre [58]. Assuming $\Delta m^2 = 0$ for simplicity, the equations of motion become in this case

$$i\left(\frac{\partial}{\partial t}\right)\begin{pmatrix}\nu_e\\ \nu_\mu\end{pmatrix} = \begin{pmatrix}\sqrt{2}G_F N_e(r) - \Delta\alpha E\varphi\cos(2\theta) & \frac{\Delta\alpha}{2}E\varphi\sin(2\theta)\\ \frac{\Delta\alpha}{2}E\varphi\sin(2\theta) & 0\end{pmatrix}\begin{pmatrix}\nu_e\\ \nu_\mu\end{pmatrix}.$$

The resonance condition $\sqrt{2}G_F N_e(r) = E\varphi\Delta\alpha\cos(2\theta)$ is satisfied only when $\Delta\alpha < 0$ because $\varphi < 0$.

6. Summary

These lectures have dealt with non-relativistic and relativistic wave equations in weak, external gravitational and inertial fields. Only two fundamental aspects of the interaction have been considered: the generation of quantum phases and spin-gravity coupling.

As shown in Section 2, quantum phases can be calculated exactly to first order in the field and in a manifestly covariant way for Klein-Gordon, Maxwell and Dirac equations. They can then be tested in experiments of increasing accuracy.

The behaviour of quantum systems is consistent with that predicted by general relativity, intended as a theory of both gravity and inertia, down to distances $\sim 10^{-8}cm$. This is borne out of measurements on superconducting electrons ($\sim 10^{-4}cm$) and on neutrons ($\sim 10^{-8}cm$) which are not tests of general relativity *per se*, but confirm that the behaviour of inertia and Newtonian gravity is that predicted by wave equations that satisfy the principle of general covariance. Atomic and molecular interferometers will push this limit down to $10^{-9} - 10^{-11}cm$ and perhaps lead to new tests of general relativity. Prime candidates are in this regard a correction term to the gravitational red-shift of photons, a (general) relativistic correction to the gravitational field of Earth, and the Lense-Thirring effect of Earth. The last two experiments may require use of a near space laboratory to obviate the effect of rotation in the first instance and, on the contrary, to sense it in the latter case.

The Mashhoon effect offers interesting insights into the interaction of inertia-gravity with spin. Spin is, of course, a quantum degree of freedom *par excellence*. Spin-rotation coupling plays a fundamental role in precision measurements of the anomalous magnetic moment of muons. These extend the validity of the fully covariant Dirac equation down to distances comparable with the muon's wave-length, or $\sim 2 \times 10^{-13} cm$.

Rotational inertia does not couple universally to matter and does therefore violate the weak equivalence principle. This generates particle helicity oscillations that may play a role in some astrophysical processes.

Other violations of the equivalence principle may play a role in neutrino oscillations. Even small violations, $\Delta\alpha \sim 10^{-14}$, could be amplified by a concomitant gravitational field. The resulting oscillations would then become comparable in magnitude with those due to the MSW effect.

More serious violations of the equivalence principle would invalidate the principle of general covariance and render problematic the use of covariant wave equations.

7. References

1. Hildebrandt, A.F., Saffren, M.M., 1965, *Proc. 9th Int. Conf. on Low-Temp. Phys. Pt. A*, p. 459; Hendricks, J.B. and Rorschach, H.E. Jr. *ibid.*, p. 466; Bol, M. and Fairbank, W.M. *ibid.*, p. 471; Zimmerman, J.E. and Mercereau, J.E., 1965, *Phys. Rev. Lett.* **14**, p. 887.
2. Colella, R., Overhauser, A.W. and Werner, S.A., 1975, *Phys. Rev. Lett.* **34**, 1472.
3. Werner, S.A., Staudenman, J-L. and Colella, R., 1979, *Phys. Rev. Lett.* **42**, 1103.
4. Bonse, V. and Wroblewski, T., 1983, *Phys. Rev. Lett.* **51**, 1401.
5. Parker, Leonard, 1980, *Phys. Rev. D* **22**, 1922.
6. Opat, G.I., 1983, *Proc. 3rd M. Grossmann Meeting in General Relativity*, edited by H. Ning, Science Press, North-Holland, Amsterdam.
7. Werner, S.A. and Kaiser, H., 1990, *Quantum Mechanics in Curved Space-Time*, edited by J. Audretsch and V. de Sabbata, Plenum Press, New York, p. 1.
8. Riehle, F., Kister, Th., Witte, A., Helmcke, J., Bordé,Ch. J., 1991, *Phys. Rev. Lett.* **67**, 177; Bordé, Ch.J., 1989, *Phys. Lett. A* **140**, 10; Bordé, Ch.J. et al. *Phys. Lett. A* **188**, 187; Bordé, Ch.J., 1997, in *Atom Interferometry*, edited by P. Berman, Academic Press, London.
9. Hehl, F.W. and Ni, W.-T., 1990, *Phys. Rev. D* **42**, 2045.
10. Hehl, F.W. and Lämmerzahl, C., 1992, *Relativistic Gravity Research*, adited by J. Ehlers and G. Schäfer, Springer, Berlin.
11. Audretsch, J. and Lämmerzahl, C., 1992, *Appl. Phys. B* **54**, 351.
12. Huang, J.C., 1994, *Ann. der Physik* **3**, 53.
13. Mashhoon, B., 1990, *Phys. Lett. A* **143**, 176; **145**, 147 ; 1992, *Phys. Rev. Lett.* **68**, 3812.
14. Mashhoon, B., 1988, *Phys. Rev. Lett.* **61**, 2639.
15. Singh, D. and Papini, G., 2000, *Nuovo Cimento B* **115**, 223.

16. Obukhov, Yuri N., 2001, *Phys. Rev. Lett.* **86**, 192.
17. Ni, W.-T., 1992, *Proc. of the* 6th *Marcel Grossmann Meeting on General Relativity,* edited by H. Sato and T. Nakamura, World Scientific, Singapore, p. 356; Pan, S.-S., Ni, W.-T.and S.-C. Chen, *ibid.* p. 364; Jen, T.-H., Ni, W.-T., Pan, S.-S. and Wang, S.-L.,*ibid.* p. 489; Daniels, J. M. and Ni, W.-T., *ibid.* p. 1629.
18. Silverman, M.P., 1991, *Phys. Lett. A* **152**, 133 ; 1992, *Nuovo Cimento D* **14**, 857.
19. Venema, B. J., Majumder, P. K., Lamoreaux, S. K., Heckel, B. R. and Fortson, E. N., 1992, *Phys. Rev. Lett.* **68**, 135.
20. Wineland, D. J., Bollinger, J. J., Heinzen, D. J., Itano, W. M. and Raizen, M. G., 1991, *Phys. Rev. Lett.* **67**, 1735.
21. Ritter, R. C., Winkler, L. I. and Gillies, G. T., 1993, *Phys. Rev. Lett.* **70**, 701.
22. Cai, Y.Q., Lloyd, D.G. and Papini, G., 1993, *Phys. Lett. A* **178**, 2.
23. Papini, G., 1967, *Nuovo Cimento B***52**, 136.
24. DeWitt, B.S., 1966, *Phys. Rev. Lett.* **16**, 1092.
25. Papini, G., 1966, *Nuovo Cimento* **45**, 66; 1966, *Phys. Lett.* **23**, 418; 1967, Phys. Lett. A **24**, 32; 1969, *Nuovo Cimento B* **63**, 549.
26. Cai, Y.Q., Papini, G., 1989, *Class. Quantum Grav.* **6**, 407.
27. Cai, Y.Q., Papini, G., 1991, *Phys. Rev. Lett.* **66**, 1259; 1992, **68**, 3811.
28. Cai, Y.Q., Papini, G., 1989, *Modern Physics Letters A* **4**, 1143; 1990, *Class. Quantum Grav.* **7**, 269; 1990, *Gen. Rel. Grav.* **22**, 259.
29. Thompson, R. and Papini, G., 1993, *Proc. of the* 5th *Canadian Conference on General Relativity and Relativistic Astrophysics*, edited by R.B. Mann and R.G. McLenaghan, World Scientific, Singapore, p. 459.
30. De Oliveira, C.G. and Tiomno, J. 1962, *Nuovo Cimento* **24**, 672.
31. Peres, A. 1962, *Suppl. Nuovo Cimento* **24**, 389.
32. Schiff, L.I. and Barnhill, M.V., 1966, *Phys. Rev.* **151**, 1067.
33. Fang, L.Z. and Ruffini, R., 1983, *Concepts in Relativistic Astrophysics*, World Scientific, Singapore.
34. Lense, J. and Thirring, H., 1918, *Z. Phys.* **19**, 156; English translation: Mashhoon, B., Hehl, F.W. and Theiss, D.S., 1984, *Gen. Rel. Grav.* **16**, 711.
35. Cai, Y.Q.and Papini, G., 1991, in *Modern Problems of Theoretical Physics*, edited by P.I. Pronin and Yu. N. Obukhov, World Scientific, Singapore.
36. Page, L.A., 1975, *Phys. Rev. Lett.* **35**, 543.
37. Mashhoon, B., 2000, *Class. Quantum Grav.* **17**, 2399.
38. Bell, J.S. and Leinaas, J., 1987, *Nucl. Phys. B* **284**, 488.
39. Bailey, J. *et al.*, 1979, *Nucl. Phys.* **B150**, 1.
40. Farley, F.J.M. and Picasso, E., 1990, *Advanced Series in High-Energy Physics, vol. 7, Quantum Electrodynamics*, edited by T. Kinoshita, World Scientific, Singapore, p. 479.
41. Farley, F.J. and Picasso, E., 1979, *Ann. Rev. Nucl. Part. Sci.* **29**, 243.
42. Papini, G., Lambiase, G., 2001, *arXiv:gr-qc/0106066.*
43. Carey, R.M. *et al.*, 1999, Muon (g-2) Collaboration, *Phys. Rev. Lett.* **82**, 1632.
44. Brown, H.N. *et al.*, 2000, Muon (g-2) Collaboration, *Phys. Rev. D* **62**, 091101.
45. Brown, H.N. *et al.*, 2001, Muon (g-2) Collaboration, *Phys. Rev. Lett.* **86**, 2227.
46. Wolfenstein, L., 1978, *Phys. Rev. D* **17**, 2369.
47. Mikheyev, S.P. and Smirnov, A.Yu., 1985, *Sov. J. Nucl. Phys.* **42**, 913; 1986, *Nuovo Cimento C* **9**, 17.
48. Papini, G., *Proc of the* 5th *Canadian Conference on General Relativity and Relativistic Astrophysics*, edited by R.B. Mann and R.G. McLenaghan, World Scientific, Singapore, p. 10.
49. de Sabbata, V. and Sivaram, C., 1991, *Nuovo Cimento A* **104**, 1577. These author consider the possibility that large magnetic fields exist in the innermost region of the Sun's core.
50. de Sabbata, V. and Sivaram, C., 1990, *Nuovo Cimento B* **105**, 1181. The authors show that higher magnetic moments may occur as a consequence of the Kaluza-Klein

theory.
51. Ahluwalia, D.V. and Burgard, C., 1998, *Phys. Rev. D* **57**, 4274.
52. Bhattacharya, T., Habib, S. and Mottola, E., 1999, *Phys. Rev. D* **59**,067301.
53. de Sabbata, V. and Gasperini, M., 1981, *Nuovo Cimento A* **65**, 479.
54. de Sabbata, V. and Sivaram, C. 1989, *Annalen der Physik Leipzig* **46**, 8.
55. Gasperini, M., 1988, *Phys. Rev. D* **38**, 2635; 1989, *Phys. Rev. D* **39**, 3606.
56. Halprin, A. and Leung, C.N., 1991, *Phys. Rev. Lett.* **67**, 1833.
57. Butler, M.N., Nozawa, S., Malaney, R.A., Boothroyd, A.I., 1993, *Phys. Rev. D* **47**, 2615.
58. Bahcall, J.N. and Ulrich, R.K., 1988, *Rev. Mod. Phys.* **60**, 297.

LABORATORY SEARCHES FOR EFFECTS BEYOND CLASSICAL GRAVITY

R.C.RITTER and G.T. GILLIES
University of Virginia
Department of Physics
382 McCormick Road
Charlottesville, VA, U.S.

KEYWORDS/ABSTRACT: gravity / experiments / relativity / strings / spin / extra dimensions / torsion pendulums / polarized / scales / Planck / fundamental interactions / weak bosons

Advances in the past few years in theoretical gravity and particle physics suggest radical changes in our understanding of gravity, and have directed physicists' attention to new types of gravitational experiments in the laboratory. From among these, this review describes recent tests of the inverse square law at submillimeter range, and anomalous (non-magnetic) spin coupling experiments in the laboratory.

1.1 INTRODUCTION

An important theme at several of the past International Schools of Cosmology and Gravitation held in Erice has been the investigation of the interaction of gravitation with other fields or phenomena, and specifically the consideration of observable effects that might be predicted from such interactions. Both the 15th and the 17th Courses of the School have been focused particularly on these concepts, with the former exploring the role of spin and torsion in gravity, and the latter concentrating on the interplay between quantum and gravity physics.

A rich variety of gravitational effects can be discussed within these contexts, including the possibilities of time variation of the Newtonian gravitational constant, breakdowns of the inverse square law of gravity at different scales of distance, temperature-dependent gravity, spin-couplings to the gravitational force, and several forms of violation of the weak equivalence principle and the universality of free fall. By their very nature, several such suggested phenomena are measurable or inferable only from astronomical or cosmological observations, while others can be sought in high precision laboratory experiments. Of particular interest in recent times have been those observations and experiments which provide insight into the exactness of general relativity and which simultaneously allow physicists to probe the extra dimensions predicted by string theory. Many significant technological advances have made it possible to design instrumentation systems that for the first time allow terrestrial studies not only of "laboratory astrophysics and particle physics" but also of "laboratory high energy and string physics." A very important thing about these exciting possibilities is that also for the first time, experimental studies in this area are now motivated strongly by key theoretical developments that call specifically for them. Thus, as we shall see below, one

P.G. Bergmann and V. de Sabbata (eds.),
Advances in the Interplay Between Quantum and Gravity Physics, 339–352.

can meaningfully suggest experiments in areas of string theory which were previously thought to be testable only within the context of quantum-gravity gedanken experiments, eg., on black holes.

With both the theoretical and experimental efforts in these areas developing rapidly, it seemed timely to attempt to capture the status of the field as it emerges by including a review on this subject here in the Proceedings of the 17th Course. Other reviews that also deal with searches for exotic gravitational phenomena and/or theoretical developments that are relevant in this area include the following. Gillies [1] discusses the efforts aimed at determining the absolute value of the Newtonian gravitational constant and the attempts to experimentally uncover variations in it with respect to time and temperature, as well as measurements of several different categories of breakdown of the inverse square law. Mashhoon [2] has written a very useful review of the role of and tests for spin in gravity, and Carlip [3] has provided a thorough description of the many recent developments in theories of quantum gravity.

We open the present paper with a discussion of the theoretical motivations for new tests of the inverse square law at very short scales of distance, and we then consider the various experimental efforts that are attempting to examine sub-millimeter gravitational forces. We continue with an overview of the physical foundations that underlie searches for a spin-dependence in the gravitational force, and end with some commentary on and intercomparison of the various experiments that have been reported to date on this topic, including our own which is active and ongoing.

2.1 THE INVERSE SQUARE LAW AT SUBMILLIMETER RANGE - MOTIVATION

Some years after its exciting revival, string theory has started to provide explicit predictions for new experimental tests of gravity. In 1966 Witten [4] and Lykken [5] proposed that there might be internal dimensions of the strings large compared with the usually assumed Planck length. This was taken up in a specific concept by N. Arkani-Hamad, S. Dimopoulus, and G. Dvali "ADD" [6], who suggest that there are two energy scales in nature, the electroweak scale m_{EW}~1 TeV, and the Planck scale $M_{Pl} = G_N^{-1/2} \sim 10^{16}$ TeV. These would arise from a difference between the strings associated with gravity and those associated with the other three forces. It is considered that this can be responsible for the well known hierarchy problem—the apparent huge relative weakness of gravity from existing measurement. At the same time this new concept would predict a related effect—a modification of the inverse square law, at short distances.

Gravity is treated as an intrinsically different force in string theory. While electromagnetism and the strong and weak nuclear forces are coupled by spin-1 gauge bosons (the photon, the gluon, and the heavy weak boson, respectively) gravity is coupled by the spin-2 graviton. In string theory these two types of couplings behave differently, but each can be associated with at least some of the extra dimensions of a 10-dimension string theory. Ordinarily, strings are presumed to be compactified into loops about 10^{-33} cm diameter, the Planck length. Strings associated with spin-1 bosons are treated as open strings which are stuck on our 4-dimensional membrane. The spin-2 graviton, however, is a closed-loop string and is free to sample other dimensions.

These differences are used [6] to modify the inverse square force law of the Newtonian gravitational potential V_N:

$$V_N(r) = G\, m_a m_b / r. \qquad (1)$$

The G in this interaction potential between two (point) masses separated a distance r is the value measured in ordinary experiments and was assigned the value 6.673(10) x 10^{-11} $m^3kg^{-1}s^{-2}$ in the 1998 CODATA recommendation [7]. Ordinary experiments here mean those in the laboratory at ranges down to about one cm, and astronomical observations out to solar system dimensions.

The new form for gravitational potential suggested by ADD has terms for added dimensions labelled by n,

$$V(r) = G_n'(m_a m_b/r)(1/R^n). \qquad (2)$$

Here G_n' is a new, much stronger coupling constant for the "true" gravitational interaction. It is assumed here that $R << r$. When $n > 0$ the potential terms are assumed to associate with added dimensions that gravity can act in, and which thereby are responsible for the strength loss apparent in the G of equation (1).

ADD assume that all forces unify intrinsically at some distance $L_{ew} \sim 10^{-17}$ cm, that is at the electroweak interaction length, and that G_n' is the fundamental coupling strength of gravity. According to the theory, this has been masked in the past by the limited range of existing measurements in gravity, unlike in particle experiments on the other forces, which probe down to ~10^{-17} cm.

To see consequences of this, we recall the Planck mass $m_{Pl} = (hc/G)^{1/2}$ and Planck length $L_{Pl} = (hG/c^3)^{1/2}$. In a set of units for which $h = c = 1$,

$$G = L_{Pl}^2 = m_{Pl}^{-2}. \qquad (3)$$

A generalized Gauss' Law will yield, for the new coupling strength

$$G_n' = L_{ew}^{2+n} \qquad (4)$$

This with equations (1) and (2) yields a relation between the two strength scales,

$$G = G_n'/R^n = L_{ew}^2(L_{ew}/R)^n. \qquad (5)$$

Equations (3) and (5) link the range R and the electroweak length scale,

$$R^n = L_{ew}^n(L_{ew}/L_{Pl})^2 = L_{ew}^n \text{ x } (10^{16})^2 \qquad (6)$$

For $n = 1$, this yields $R = 10^{15}$ cm, a value that would conflict with gravitational dynamics of the solar system. But for $n = 2$, it provides $R = 0.1$ cm, the previously untested value in the laboratory. It this is correct, at some small distance < 1 cm we might expect a deviation from the inverse square law due to the added dimension(s).

Others have added to the theoretical study of the violation of the inverse square law at small distances. Among these, Kehagias and Sfetsos [8] provide expanded theoretical predictions for deviations from the inverse square law, based on extra dimensions. These include compactification on toroids, spheres and other manifolds and point to what they consider to be more realistic expectations. Results are expressed in terms of an added interaction term of Yukawa form. For an attracting mass M this has the form for its potential,

$$V(r) = (GM/r)(1 + \alpha e^{-r/\lambda}). \qquad (7)$$

Predictions of these theories depend on the manifold on which compactification could take place. For an n-torus, $\alpha = 2n$, and for an n-sphere, $\alpha = n + 1$. Figure 1 shows the coupling strength, α^2, predicted by these theories, along with the dilaton [9], and the experimental bounds considered the best at that writing.

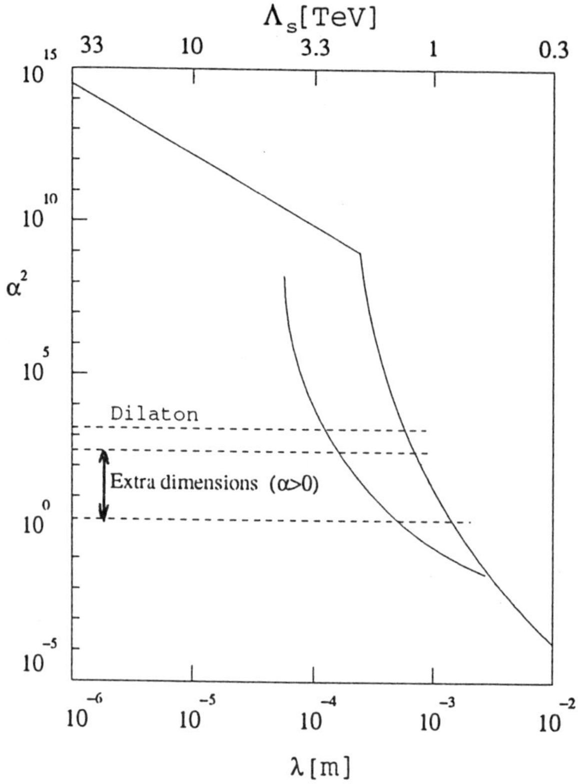

Figure 1. Strength of Yukawa term predicted by [8] for interactions involving extra space-time dimensions

2.2 EXPERIMENTAL TESTS OF THE INVERSE SQUARE LAW AT SUBMILLIMETER RANGE

Previous to ADD, experimenters had been motivated on more general grounds to test the inverse square law at smaller inter-mass distances. The presence of Casimir forces was an inhibiting and limiting factor in these earlier experiments. This force is given, for flat parallel plates of separation r, as

$$F_c = \pi \hbar c A/(480\, r^4), \tag{8}$$

where A is their area. At some separations r less than a millimeter, the patch effect can be an even greater electrical interfering effect. Most experimenters now use thin electrostatic shields to reduce or remove these effects. Without a similar electrical problem, Hoskins et al [10], in a Gauss' Law type of experiment at Irvine, had in 1985 achieved a null result for departure from the inverse square law with sensitivity α less than unity for a range of about 1to 2 mm and greater. Mitrofanov et al [11], in 1988 provided results which, extended to about 0.1 mm, but exceeded $\alpha = 1$ at about 0.5 mm range.

These are difficult experiments, and the number of them is much fewer than other gravity experiments at greater ranges. Nevertheless, the impetus of the theoretical papers cited above, along with others, caused experimenters in a few more laboratories to commence experiments at sub-millimeter distances. Long et al at the University of Colorado had published [12] plans for an experiment using a vibrating plate mass. Price, of that group, discussed progress in that experiment at an American Physical Society meeting in Long Beach [13]. There Kapitulnik of Stanford [14] told of plans using a different version of that method. During an informal extension of that meeting session Adelburger of the University of Washington "Eot-Wash" group announced progress with a sub-millimeter test of the inverse square law using a rotating Cavendish balance method. At that same meeting, Paik at the University of Maryland discussed plans for an experiment using superconducting differential gradiometer technology. Three of these experiments will be discussed in the following sections.

2.2.1. *University of Washington*

This is the only experiment for which submillimeter limits have been published at a strength significantly below unity. Hoyle et al [15] use a rotating torsion balance of the type refined in many experiments over the years at that institution. In this experiment a seldom used trick was employed: the use of "negative mass" of holes in gravitationally interacting disks as a way of enhancing the sensitivity with small mass separation. Figure 2 shows the sensing and attracting elements of the torsion pendulum.

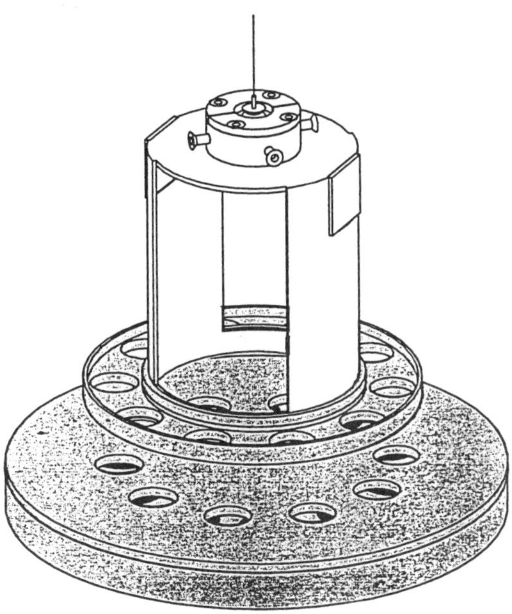

Figure 2. Torsion pendulum of University of Washington inverse square law experiment.

An aluminum disk with ten equally spaced holes, suspended by a fiber, senses the gravitational torques. Below it, two copper attractor disks are rotated at 17 times the period of the fiber-mounted disk, and the fiber twist observed with an autocollimator. The two attractor disks each had ten holes of size which, combined with an 18 degree shift of the lower, provided an attracting potential aimed at reducing a Newtonian signal relative to a short range interaction. A thin beryllium-copper membrane between the attractor and pendulum disks provided electrical shielding. Precision machining and symmetry considerations in analysis provided additional sensitivity in analysis for a short range signal. Capacitance measurements of the attractor – pendulum separation, which was made as low as 0.218 μm.

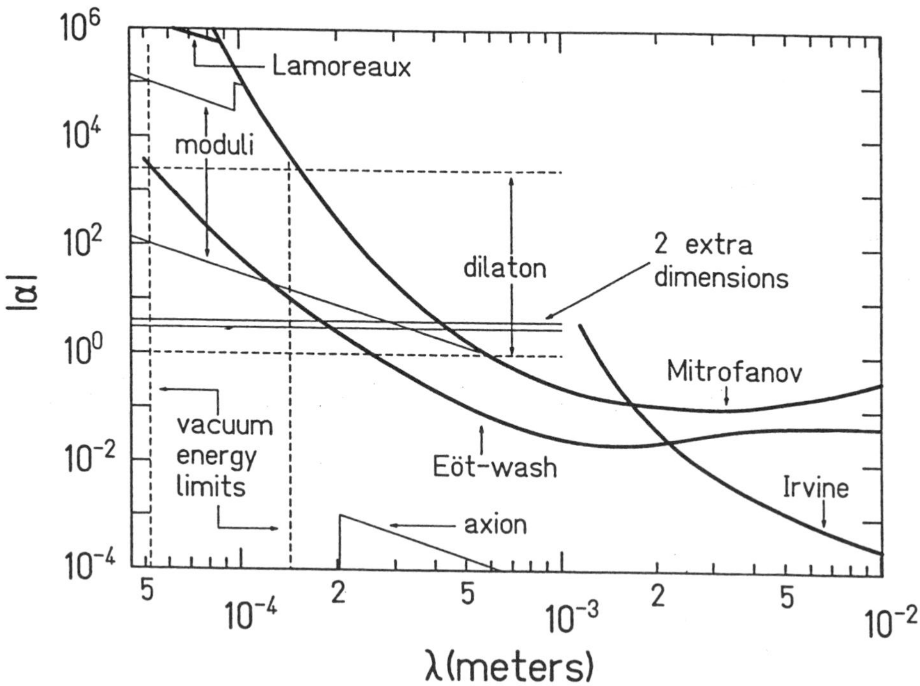

Figure 3 is a plot of the range-dependence of limits on α achieved for a non-newtonian signal from this experiment, along with those from earlier measurements at other laboratories. The double line is for n = 2, with a range of values of α predicted by [8] for toroidal and spherical compactifications. The dilaton prediction is from [9].

The Eot-Wash experimental limit of $\alpha = 0.02$ at about 1 mm range can be used to place a limit on an n = 2 scale factor M* (the limit replacement of $M_{ew} = 1/L_{ew}$ of equation (6)). In this case

$$R = (1/M^*)(m_{Pl}/M^*)^{2/n} \tag{9}$$

For n = 2, solving for M*:

$$M^* = (m_{Pl}/R)^{1/2} \tag{10}$$

And, in ordinary units

$$M^* = (\hbar m_{Pl}/Rc)^{1/2} \quad (11)$$

From this, M* > 3.5 TeV is found to be the scale factor limit implied by this experiment.

2.2.2 *University of Colorado*

This experiment [12,13] is planned for two stages. The first stage is a room temperature experiment, aimed at serious testing of the vibrating mass concept. Figure 4 shows the elements of the room temperature device. A thin flat attractor (source) mass vibrates at 1 kHz near a flat detecting mass, with a stiff conducting shield between the two to reduce electrical and magnetic interaction between them. They expect to achieve a useable separation of about 0.1 mm. After series of preliminary runs the experimenters have found a number of background effects. These include pressure effects, vibration from the attractor motion, Casimir forces, and electrical and magnetic forces. Ultimately they believe the experiment will be thermal noise limited for gaps this size.

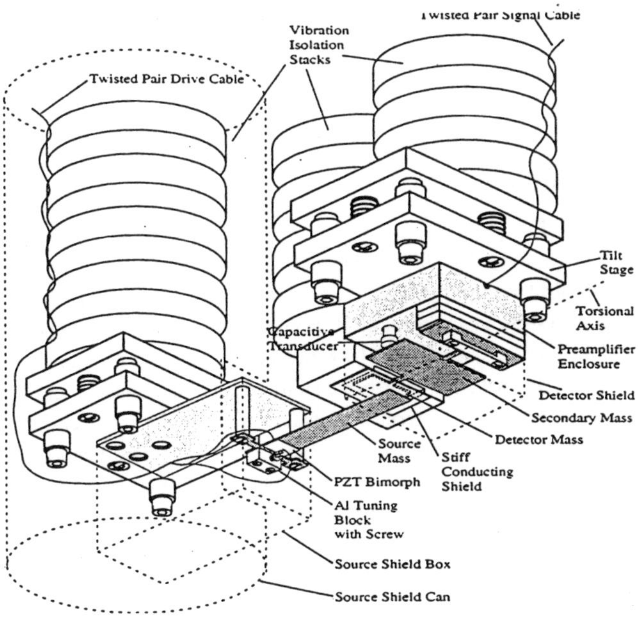

Figure 4. Source and sensing elements on mounting -- Unversity of Colorado experiment.

A limit for α of about unity is anticipated for the room temperature experiment, at a range down to approximately 0.1 mm. A subsequent experiment at 4 K would be expected to reduce noise and improve the Q of the instrument, along with appropriate material changes. A projected limit for the 4K experiment is about two orders of magnitude lower than the room temperature. Since thermal noise falls as the square root of the temperature, this prediction must include other improvements. Figure 5 shows the predictions of the cooled version of this experiment, along with limits, achieved and proposed, of the other two experiments.

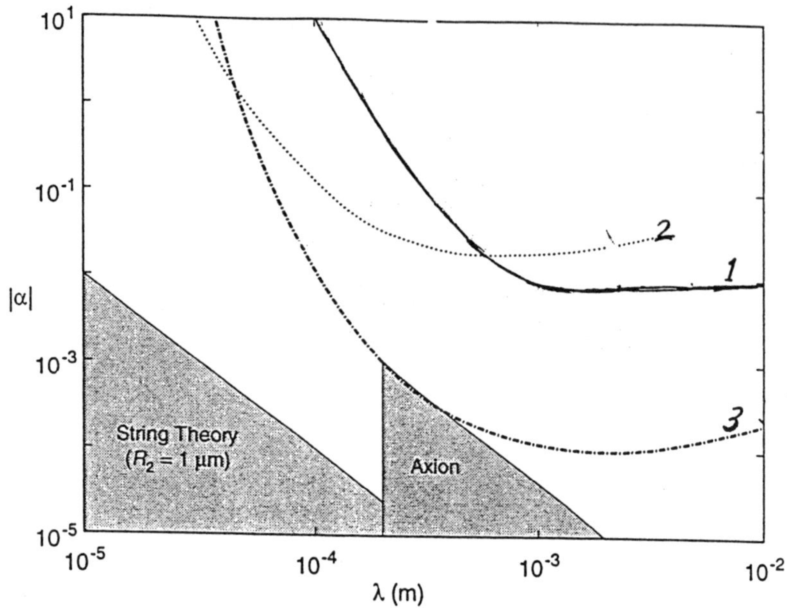

Figure 5. Sensitivity of 3 experiments: 1 - Washington; 2- Colorado (proposed); 3- Maryland (proposed)

2.2.3 *University of Maryland*

A group led by H. Paik of Maryland has been developing superconducting gravity gradient technology for many years, and has published extensively on superconducting accelerometry [16]. With this technology they have published stringent limits for the dependence of G on range. Presently the method has been used for a test of the inverse square law [17]. Now, a new version will search for sub-millimeter departure [13]. Figure 6 shows the spatially symmetrized arrangement of the levitated sensing coils and the source masses. The circuit diagram of the differential gravimeter consists of the two sensing masses coupled with superconducting turns, some of which levitate the masses and some of which read out to SQUIDs a relative positional change of the two masses. As used in this experiment, the coupled pair of source masses, one a plate and one a ring, is vibrated at low frequency, ~ 0.1 Hz, to generate appropriately different fields at the two test masses. Synchronous detection of the summed sensor mass signals provides a null test of Newtonian gravity. With highly accurate machining and many design features to eliminate standard sensitive gravitational interferences, the experiment is predicted to have a sensitivity of $\alpha \sim 10^{-4}$ at about 1 mm as shown in Figure 5.

3.1 EXPERIMENTAL TESTS OF ANOMALOUS SPIN DEPENDENCE IN GRAVITY

A review of the such tests and their motivations was presented at this school in 1997 [18]. In this section we will update and expand that information. A major motivation for such experiments is the interest in torsion in space-time and coupling to it that has been a major component of this school at Erice.

Two references among many on this major space-time question are Datta, de Sabbata and Ronchetti [19] and Yu X [20, and this school].

Other motivations include fundamental interaction symmetry violation considerations. Early work by Leitner and Okubo [21] with a heuristic gravitational potential having an added $\boldsymbol{\sigma}\cdot\mathbf{r}$ term initiated such work. Later, several authors included other symmetry violating terms, leading to the more generalized CPT form,

$$U_{CPT}(r) = U_o(r)\{1 + A_1\,\boldsymbol{\sigma}\cdot\mathbf{r} + A_2\,\boldsymbol{\sigma}\cdot(\mathbf{v}/c) + A_3\,\mathbf{r}\cdot[(\mathbf{v}/c)\times\boldsymbol{\sigma}]\} \qquad (12)$$

Here we will be concerned with the first added term of this expression.

Moody and Wilczek [22] provided an explicit theory for testing the effects of spin on gravity in their work on fundamental fermions weakly coupled by very light bosons having long interaction range (e.g., of mass ~ 10^{-5} eV). They provided one especially significant insight for experiments, namely the notion of the coherent action of huge numbers of polarized electrons in a macroscopic mass. The importance of this can be gauged by the weakness of the interaction we are interested in. For nucleons the dimensionless coupling is

$$g_s^2 \sim (m_N/m_{Planck})^2 \sim 10^{-38}. \qquad (13)$$

Two 1-gram masses separated by one cm provide a gravitational force,

$$F \sim (6 \times 10^{23})^2\,(m_N/m_{Planck})^2\,\hbar c/1\text{cm} = 6.7 \times 10^{-8}\ \text{dyne} \qquad (14)$$

3.1.1 *The Moody-Wilczek Formalism*

For a pure space-like momentum transfer $\vec{q}$ on shell: $p_f = p + q/2$, $p_i = p - q/2$ in momentum space, they find the two types of vertices:

Scalar:

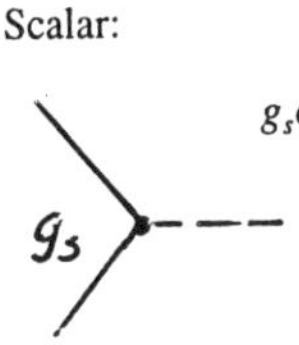

$$g_s\phi(q)\bar{\psi}(p_f)\psi(p_i) = g_s\phi(q)[(p^\mu p_\mu/M^2)\bar{\psi}(p_f)\psi(p_i) - i(p_\mu q_\nu/2M^2)\bar{\psi}(p_f)\sigma^{\mu\nu}\psi(p_i)] \qquad (15)$$

and,

Pseudoscalar: *polarized mass*

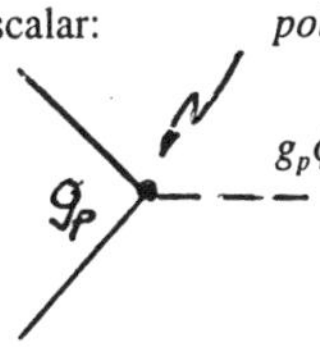

$$g_p\phi(q)\bar{\psi}(p_f)i\gamma_5\psi(p_i) = g_p\phi(q)(q^\mu/2M)\bar{\psi}(p_f)i\gamma_5\gamma_\mu\psi(p_i) \qquad (16)$$

In the inverse Born approximation the potential is:

$$V(r) = \int d^3q/(2\pi)^3\,(\text{vertex 1})\,(\text{vertex 2})\,e^{i\mathbf{q}\cdot\mathbf{r}}/(q^2 + m_\phi^2), \qquad (17)$$

in the non-relativistic limit.

This yields three types of single-boson coupling between fermions:

MONOPOLE-MONOPOLE (two scalar vertices)

$$V(r) = -g_s^1 g_s^2 e^{-m\phi r}/4\pi r \qquad (18)$$

(Experiments in the inverse square law and composition dependence)

MONOPOLE-DIPOLE (vertex 1- scalar, vertex 2 - pseudoscalar)

$$V(r) = (g_s^1 g_p^2)(\hat{\sigma}_2 \cdot \hat{r})/(8\pi M_2)[m_\phi/r + 1/r^2]\, e^{-m\phi r} \qquad (19)$$

(Searches for axions)

DIPOLE-DIPOLE (two pseudoscalar vertices)

$$V(r) = [g_p^1 g_p^2/(16\pi M_1 M_2)][tensor\ form]\, e^{-m\phi r} \qquad (20)$$

Practical Problem: Because of the unnatural parity associated with it, each pseudoscalar vertex has an inhibition of ~ 10^{12} to 10^{13} relative to a scalar vertex. That is,

$$g_p^2 \sim 10^{-12.5}\, g_s g_p \sim 10^{-12.5}\, g_s^2 \qquad (21)$$

How strong are the spin interactions between electrons? Comparing with magnetic interactions:

$$V_{mag} = \mu_e \hat{\sigma}_1 \cdot \mu_e \hat{\sigma}_2 / r : \qquad V_{spin} = \alpha_s \mu_e \hat{\sigma}_1 \cdot \mu_e \hat{\sigma}_2 / r \qquad (22)$$

Experimental limits on these dipole-dipole electron spin interactions were presented at this school in 1997 [18], and there have been no significant new experiments of this type since that time. A partial listing illustrates the history of this experimental effort:

Ramsey and Weinland 1979 [23]	$\alpha_s < 5 \times 10^{-3}$
Anselm 1983 [24]	$\alpha_s < 10^{-8}$
Graham & Newman 1985 [25]	$\alpha_s < 3 \times 10^{-10}$
Ritter et al 1990 [26]	$\alpha_s < 9 \times 10^{-12}$
Ni et al 1994 [27]	$\alpha_s\, s < 1.2 \times 10^{-14}$

It should be remarked that the range tested in these experiments was some few cm, except for the Ramsey experiment which studied the proton-proton interaction in the hydrogen molecule.

3.1.2 *Experiments of the* $\boldsymbol{\sigma \cdot r}$ *Type*

A more interesting case is the monopole-dipole, $\boldsymbol{\sigma \cdot r}$ interactions which are P- and T- violating. Four experiments will be used to represent the several methods and best limits on these interactions between electrons. Recall that this monopole-dipole interaction has one scalar and one pseudoscalar vertex, and the latter is the particle which is polarized. (However, we shall see a case where the polarization is induced, but not present in the initial state.)

3.1.3 University of Virginia

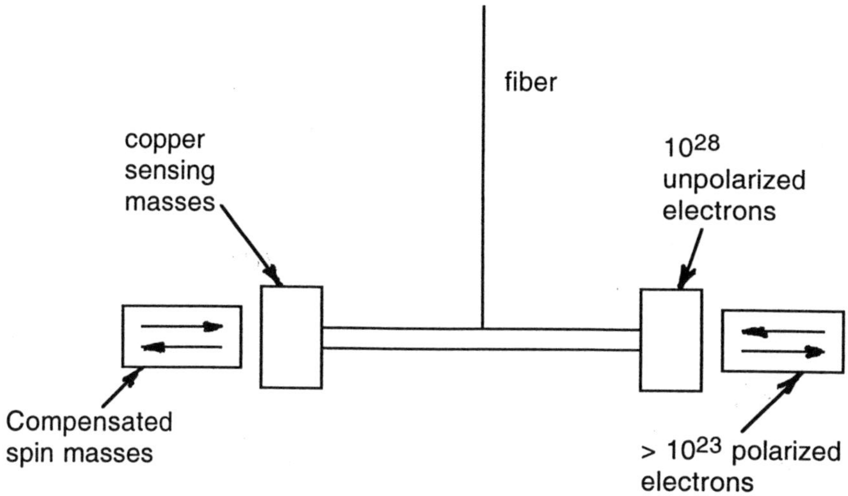

Figure 6. University of Virginia $\boldsymbol{\sigma} \cdot \mathbf{r}$ experiment.

This experiment used a classical torsion pendulum [18,28] but employed source masses having ~ 10^{23} polarized electrons interacting with unpolarized copper sensing masses containing ~ 10^{28} baryons and similar numbers of electrons. Figure 6 diagrams the arrangement of the two copper masses on a fiber in an evacuated chamber. The spin polarized masses were located outside the chamber, each pointing inwards toward a copper mass. Every 3 hours the polarized masses were reversed to invert the sign of the interaction. After 8 months of running, the experimental results were collected and analyzed for a limit on the difference of the spin in – spin out arrangements. From 6-dimensional integration over the dimensions of the two masses in each pair, a result was obtained for the dimensionless coupling limit,

$$g_s g_p / 4\pi\hbar c < 3 \times 10^{-27} \tag{23}$$

The sensitive range of this experiment was limited to $\lambda > 3$ cm. As such it was able to provide an early limit on the existence of the axion in a part of the Turner astrophysical window (10^{-6} eV < axion mass < 10^{-3} eV). The limit as a function of interaction range is shown in Figure 7 along with limits from other experiments.

Amherst College

The NMR method was used in this experiment [29], which evaluated the splitting of magnetic sublevels in the atomic ground state of ^{199}Hg upon the repositioning of two 475 kg masses which provided a local source of gravity. The purpose of the local sources was to perform the test at smaller effective ranges than the NMR methods [30] which used the earth as a source mass. This test was able to improve over the existing limit for some ranges as shown in Figure 7. The asymptotic limit achieved was

$$g_s g_p / 4\pi\hbar c < 2.3 \times 10^{-29} \tag{24}$$

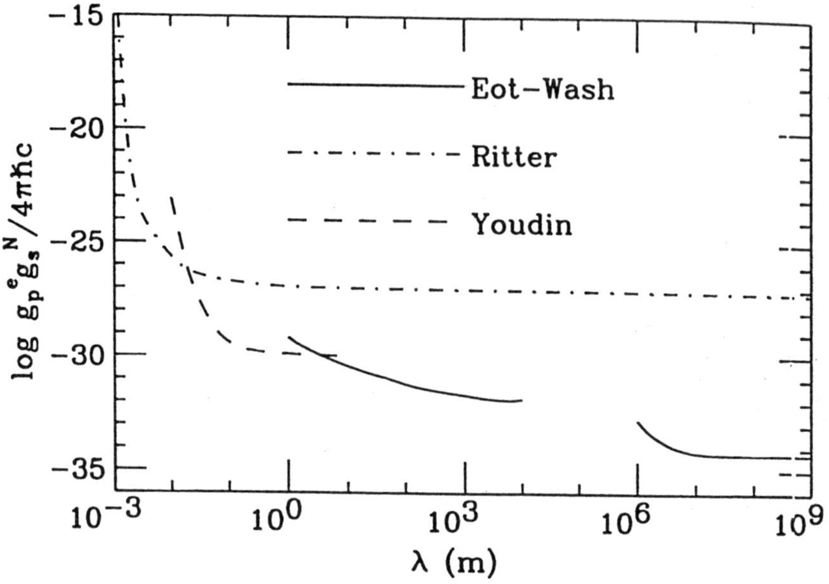

Figure 7. Limits on dimensionless coupling constants: 3 experiments testing **σ · r** interaction.

Hsinchu Taiwan

This very clever experiment [31] uses no initially polarized material. Instead, an unpolarized copper source mass rotates around a magnetically shielded magnetometer, having a paramagnetic salt as the sensing material. The hypothetical **σ · r** interaction would penetrate the magnetic shield and cause a rotating magnetic moment in the salt. A synchronously detected SQUID reads any induced signal from a pickup coil around the salt. This method is somewhat like that of Vorobyov and Gitarts [32], which rotated a permanent magnet source around a shielded piece of permeable iron for the same purpose. The problem in that experiment is that the permeability of the sensing material required to provide the claimed limit was many orders of magnitude below what had ever been measured for iron. The asymptotic limit of interactdion strength (λ > 10 cm) achieved in the Taiwan experiment was

$$g_s g_p/4\pi\hbar c < 8 \times 10^{-30} \qquad (25)$$

University of Washington

This is the most sensitive experiment on spin interactions thus far, as can be seen in Figure 7. In this apparatus the sensing pendulum [33] consists of stacked octagonal rings, each of which uses a combination of Alnico 5 and SmCo magnets magnetized to the same field. Consequently orbital magnetization by the Sm ions compensates the spin alignment so that there is a very small remaining magnetic moment which is easily shielded These provided a net spin alignment estimated at 8 x 10^{22} electrons. As it rotates, this spin aligned pendulum would couple with external fixed masses, The sosurce masses considered are local: 1 to > 1000 m, and distant: 10^6 to 10^9 meters. These provide two asymptotic limits,

$$g_s g_p / 4\pi\hbar c < 10^{-32} \quad (26)$$

and

$$g_s g_p / 4\pi\hbar c < 10^{-34} \quad (27)$$

respectively.

Recently Bluhm and Kostelecky [34] have analyzed the data of this experiment and fit it to a potential angle-varying signal torque which would follow from Lorentz- and CPT-violating spin couplings. The perturbative Lorentz-violating Lagrangian employed was

$$L = -a^e_\mu \overline{\psi}\gamma^\mu\psi - b^e_\mu \overline{\psi}\gamma_5\gamma^\mu\psi - 1/2\, H^e_{\mu\nu}\overline{\psi}\sigma^{\mu\nu}\psi + 1/2\, ic^e_{\mu\nu}\overline{\psi}\gamma^\mu D^\nu y + 1/2\, id^e_{\mu\nu}\overline{\psi}\gamma_5\gamma^\mu D^\nu\psi \quad (28)$$

with $\hbar = c = 1$, and ψ the electron field. Rotation of the pendulum spin vectors would cause it to perceive a torque in response to coupling to a fixed spatial axis. Rotation of the Earth further turns the pendulum axis relative to the fixed spatial axis. Appropriate trigonometric connections then compare the angles with the torque read on the spins in the experiment. This analysis places a value on the parameter representing the CPT-violating component of the torque of (1.4 +/- 0.8) x 10^{-28} GeV. This is to be considered with studies of clock-comparison experiments and cyclotron frequencies of trapped particles as analyzed by these authors [35].

References

1. Gillies, G.T., Rep. Prog. Phys., **60**, 151 (1997).
2. Mashoon, B., Class. Quant. Gravit., **17**, 2399 (2000).
3. Carlip, S., Prog. Phys., **64**, 885 (2001).
4. Witten, E., Nucl.Phys. **B471**, 135 (1996).
5. Lykken, J.D. Phys. Rev. **D54**, 3693 (1996).
6. Arkani-Hamad, N,, Dimopoulus, S.,and Dvali, G., Phys. Lett. **B429**, 264 (1998); Phys. Rev.Lett. **84**, 586 (2000).
7. Mohr, P.J. and Taylor, B.N., Revs. Mod Phys., **72**, 351 (2000).
8. Kehagias, A. and Sfetsos, K., hep-ph/9905417 (1999).
9. Kaplan, D.B. and Wise, M.B., hep-ph/0008116 (2000).
10. Hoskins, J.K., Newman, R.D., Spero, R. and Schultz, J., Phys. Rev. **D32**, 3804 (1985).
11. Mitrofanov, V.P. and Ponomareva, O.I., Sov. Phys. JETP **67**, 1963 (1988).
12. Long, G.C., Chang, H. and Price, J., Nucl. Phys. **B529**, 23 (1999).
13. Bull. Am. Phys. Soc. **45**, No. 2, Session V12, 146 (2000).
14. Bull. Am. Phys. Soc. **45**, No. 2, Session V12, 146, 3rd paper (2000).
15. Hoyle, C.D., Schmidt, U., Heckel, B.R., Adelberger, E.G., Gundlach, J.H., Kapner,, D.J. and Swanson, H.E., Phys. Rev. Lett. **86**, 1418 (2001).
16. Paik, H.J. and Chan, J.A., Phys. Rev. **D35**, 3572 (1987).
17. Paik, H.J., Phys. Rev. Lett. **70**, 1195 (1993).
18. Ritter, R.C., Gillies, G.T. and Winkler, L.I., in *Spin in Gravity*, p 199, World Scientific, Singapore, New Jersey, London and Hong Kong, (1997).
19. Datta, B.K., de Sabbata, V. and Ronchetti, L., Il Nuovo Cimento, **113B**, 711 (1998).
20. Yu, X., Astro. Space Sci. **154**, 32 (1989); This School (2001).
21. Leitner, J. and Okubo, S., Phys. Rev. **136**, B1542 (1964).
22. Moody, J.E. and Wilczek, F., Phys. Rev. **D30**, 130 (1984).
23. Ramsey, N.R., Physica **96A**, 185 (1979).

24. Ansel'm, A., Pis'ma Zh. Eksp. Teor. Fiz., **35**, 266 (1982).
25. Newman, R.D., in Proc. *Third Marcel Grossman Meeting on General Relativity*, p 1497, North-Holland, Amsterdam (1983).
26. Ritter, R.C., Goldblum, C.E., Ni, W.-T., Gillies, G.T. and Speake, C.C., Phys. Rev. **D42**, 977 (1990).
27. Ni, W.-T., Chui, T.C.P., Pan, S.-S., and Cheng, B.-Y., Physica **194B**, 153 (1994).
28. Ritter, R.C., Winkler, L.I. and Gillies, G.T., Phys. Rev. Lett. **70**, 701 (1993).
29. Youdin, A.N., Krause, D., Jagannathan, K., Hunter, L.R. and Lamoreaux, S.K. Phys. Rev. Lett. **77**, 2170 (1996).
30. Venema, B.J., Majumder, P.K., Lamoreaux, S.K., Heckel, B.R. and Fortson, E.N., Phys. Rev. Lett. **68**, 135 (1992).
31. Ni, W.-T, Chinese Jour. Phys. 34, 962 (1996); in Proc *Seveventh Marcel Grossman Meeting on General Relativity,* p 1625, World Scientific, Singapore (1996).
32. Vorobyov, P.V. and Gitarts, Y.I., Phys. Lett. **B208**, 146 (1988).
33. Heckel, B.R., Adelberger, E., Baessler, S., Gundlach, J., Harris, M., Hoyle, C., Sharp, A., Smith, G., and Swanson, E., *Advances in Space Research*, 32nd COSPAR Sciednt. Assembly (1998).
34. Bluhm, R. and Kostelecky, V.A., Phys. Rev. Lett. **84**, 1381 (2000).
35. See ref. cited in [34].

PROSPECTS FOR A TEST OR MEASUREMENT OF *(G-dot)/G* BY A SEE (SATELLITE ENERGY EXCHANGE) MISSION

A. J. SANDERS
Department of Physics and Astronomy,
University of Tennessee,
Knoxville, TN 37996-1200, U.S.A., and
Engineering Physics and Technology Division,
Oak Ridge National Laboratory
2360 Cherohala Boulevard
Knoxville, TN, 37932, U.S.A.
E-mail: ASanders@utk.edu

G. T. GILLIES
Department of Physics, University of Virginia
Charlottesville, VA 22904-4714, U.S.A.
E-mail: gtg@virginia.edu

ABSTRACT

Gravity remains the missing link in unification theory. The question of whether the gravitational constant G is truly constant or whether it might be time-varying is of particular importance to modern theories of gravitation and, hence, to efforts to achieve a satisfactory unification theory. Few things could do more to invigorate physics than the uncontestable discovery that one of the "constants" of nature is in fact time-varying. A SEE (Satellite Energy Exchange) mission is expected to be capable of detecting apparent seasonal variation and secular change in G at cosmologically significant levels. The experimental strategy is basically to observe the orbital period of the Shepherd mass, and it also entails combining the data from GRACE and SEE missions.

1. Introduction

A thorough understanding of the gravitational force—especially a satisfactory quantum theory of gravity—is the missing link in efforts to achieve a satisfactory unification theory. Most of the current promising approaches–including string theories, p-brane theories, and supergravity—seek to incorporate gravity at a fundamental level. Although a number of different theoretical schemes have been proposed, a lack of precise experimental evidence presently makes it almost impossible to assess the validity of alternative schemes.

A striking feature of recent theories of quantum gravity and string theory is that

P.G. Bergmann and V. de Sabbata (eds.),
Advances in the Interplay Between Quantum and Gravity Physics, 353–363.

they cannot retain a constant *G*, but rather require various secular rates of change. It is evident that Einstein's picture of gravity—*i.e.*, the curvature of spacetime—is central to the concept of renormalization gauge theory, and the scaling behavior of fields and coupling "constants" near critical-point phase transitions (Schmidhuber, 1997). However, it is also the case that gravity need not be synonymous with general relativity if new fields, besides the metric, are introduced, such as the dilaton field. The assumption that the "coupling constant" *G* is actually constant is not consistent with these theories. Their predictions of *(G-dot)/G* are typically a few parts in 10^{11}/yr or somewhat less.

The orbital motion of an ultra-drag-free satellite, such as the large test body of the SEE (Satellite Energy Exchange) satellite [Sanders & Deeds, 1992 and 1993, and Sanders *et al.*, 1993, and Sanders & Gillies 1998a], known as the "Shepherd," may possibly provide the best test for time variation of the gravitational constant *G* at the level of parts in 10^{14}—two or three orders of magnitude beyond the predictions of current theories [Gillies, 1997]. Thus, the very precise experimental data to be provided by a SEE mission augurs for significant advances in gravitation theory, with concordant implications for unification theory.

A test of *(G-dot)/G* is one of the very few ways of discriminating among various theories (see, for example, Marciano, 1984; Bronnikov, Ivashchuk & Melnikov, 1988; Melnikov, 1994; Drinkwater *et al.* 1999; and Ivashchuk & Melnikov, 2000). There has never been a credible laboratory measurement of *(G-dot)/G* (using test masses in a controlled situation) at cosmologically interesting levels of precision. Attempts have been made to do so, but all have fallen far short of the presently foreseen requirements [Gillies, 1997]. The current experimental—viz., observational—evidence regarding *(G-dot)/G* is open to interpretation, as discussed below. SEE, on the other hand, will provide a *controlled* experiment and very fine accuracy, which may discriminate among various possible theories.

It would be very significant scientifically to discover that a fundamental "constant" of Nature is not constant. This would certainly be invigorating for the development of new theories, most of which do allow for time variation of *G* and other fundamental "constants." Of course, a finding of non-zero *(G-dot)/G* would require modification of general relativity, since it assumes a constant value of *G*. More broadly, this would clearly mark the boundaries where general relativity is valid, and specify the onset of new physics.

2. Background

The fundamental question of the variability of *G* has taken on substantial urgency in recent years. Dirac's conjecture about variation of the fundamental constants, summarized in his "Large Numbers Hypothesis," opened the door to initial speculations in this area, and his original concept of two metrics (one for "mechanical," *i.e.*, orbital processes and another for "atomic" processes) still echoes today in the theories discussing extra dimensions.

By roughly mid-century, the scalar-tensor theories of gravity were essentially the first to make quantitative predictions of a non-zero *(G-dot)/G*. The essential feature was that the gravitational field would have a secularly-varying component as well as a modulation of the nominal value of *G* as the Earth orbited the Sun each year. Such theories, particularly that of Brans and Dicke, were hotly debated within the context of general relativity, and eventually gave way to increasingly precise experimental confirmations of general relativistic predictions. However, perhaps the most striking thing about the debate at this point was that a non-zero *(G-dot)/G* lay outside of the predictions of general relativity; hence, if some evidence for it could be discovered, then Einsteinian geometrodynamics would be incomplete, at best.

While that particular fact motivated much of the work on searches for a non-zero *(G-dot)/G* which took place over the latter half of the 1900s, it was the advent of string theories and the successor brane theories that eventually provided a modern predictive basis for the existence of such a phenomenon. The compactification of the extra dimensions that arise in such theories leads to the possibility of scenario-dependent non-zero values for *(G-dot)/G*; hence any *bona fide* observation would open a window onto substantial new areas of physics that might have direct impact on the formulation of a quantized theory of gravity.

Of course, there were also many other motivations for measuring *(G-dot)/G* that have been important as well. These include interest in understanding pulsar dynamics, attempts to study the evolution of coupling constants due to the presence of a dilaton field, refinement of primordial nucleosynthesis models, analysis of standard stellar evolution models, and attempts to improve the understanding of white-dwarf luminosity functions. Very interesting results for *(G-dot)/G* have been obtained from efforts in all of these areas, but with the values for it essentially consistent with *(G-dot)/G* = 0. All of these approaches to the topic are reviewed by Gillies (1997); see also Chiba (2001).

Most recently, a new motivation for the measurement of *(G-dot)/G* has arisen within the context of attempts to reveal the presence of a dynamical background energy in the universe, the so-called "dark energy" or quintessence." Chiba (1999), for instance, has pointed out that a dynamical coupling of the quintessence field to the gravitational field can give rise to a *(G-dot)/G* effect, and he has used the existing experimental values to constrain the size of such a coupling.

Although there have been perhaps a dozen laboratory experiments proposed to measure *(G-dot)/G*, none of them have been successfully carried out at cosmologically significant levels of interest. The early work at Princeton by several students of Dicke who used precision pendulums to study this problem provide the best, but only weak, constraints at the level of *(G-dot)/G* $< 4\times10^{-8}$ per year. The observational evidence provided by lunar laser ranging, pulsar spindown studies and radar ranging to the Viking Mars landers is much more stringent, with typical limits of $10^{-12} < (G\text{-}dot)/G < 10^{-11}$ per year or slightly less. In certain particular cases, orbital dynamics studies of solar system bodies place even tighter limits on the effect, but these are based largely on inference, not direct measurement. All of the existing data through 1997 are reviewed by Gillies (1997).

3. Experimental method for determining *(G-dot)/G* on the SEE mission

The experimental approach to measuring *(G-dot)/G* by a SEE mission is described in a previous article [Sanders *et al.*, 2000]. The main idea is to use the orbital period as a clock running in comparison with atomic clocks. Thus, unless the period between successive equatorial crossings at the ascending (or descending) node is constant except for various known and/or characterizeable perturbations, *G* will be found to be changing (strictly speaking, we can look only for changes in the product M_EG). Sanders *et al.* (2000) showed that the accuracy with which *(G-dot)/G* may be measured is limited by (1) available position resolution if the observation time is less than one year and (2) accuracy in accounting for perturbations if the observation time is greater than one year. The capability of time measurement will not be a limiting factor in measuring *(G-dot)/G*, assuming the next generation of atomic clocks is available. However, with the atomic clocks that are now in use as standards, stability would be a severe problem for observation times longer than a few months.

Great caution is required in satellite design to make the Shepherd as nearly drag-free as possible. Observable perturbations on the Shepherd due to the satellite itself are analyzed. Techniques for effecting automatic cancellation of unobservably small intra-satellite perturbations are expected to be very beneficial. In fact, it is shown that along-track acceleration may be about $\sim 10^{-18}$ g if averaged over periods of several months, provided that the capsule is periodically "flipped" and that the location of the Shepherd with respect to the capsule is carefully varied [Sanders *et al.*, 2000].

Newtonian orbital perturbations due to the Earth's potential must be taken into account. Time variation of the Earth's geopotential is presently a major limitation to the use of any orbital motion to test for secular variation of *G*. The new NASA geopotential mission, GRACE, which is scheduled for launch in November 2001, will do much to alleviate the problem of the Earth's potential as an error source in *(G-dot)/G* at both short and long observation periods. The key capability of GRACE is the ability to determine very accurately the time variation of the coefficients of the Earth's field on a time scale of weeks to months.

For completeness, we note that one of the goals of the SEE mission is to search for anisotropy of space, which requires measuring *G-dot* with observation periods of a few months. This goal is not sensitive to errors in the geopotential [Sanders *et al.*, 2000], because for observation periods shorter than one year the error in *(G-dot)/G* is dominated by the error in tracking accuracy (see Figure 1).

GRACES's capabilities will provide virtually all the information about time variation of the Earth's potential with sufficient accuracy for a measurement of *(G-dot)/G* by a SEE mission at a cosmologically significant level. However, GRACE cannot measure the Earth's $J_{2,0}$ harmonic (the equatorial "bulge") adequately. The reason is as follows. Heating and cooling of the GRACE satellites occur cyclically at the frequency of the orbital motion, ν, albeit not sinusoidally, and they also occur at the harmonic frequency 2ν. These heating effects are very difficult to disentangle from the orbital perturbations due to the $J_{2,0}$ term in the Earth's potential, which also occur at frequency 2ν. Consequently, the accuracy with which GRACE can measure $J_{2,0}$ is far less than for other harmonics. We suggest that the data from the GRACE and SEE missions be combined to eliminate this ambiguity and thus to both (1) refine our knowledge of the Earth's time-varying geopotential and (2) reduce the major source of error for an orbital test of secular change in *G*. Our proposed strategy is essentially to use the rate at which the orbital plane precesses to infer $J_{2,0}$. The orbital precession rate may be estimated to first order from $J_{2,0}$ alone:

$$\Omega\text{-dot} = -(3/2)\, J_{2,0}\, (GM_E / R_E{}^3)^{3/2} \times (a/R_E)^{-7/2} \times \cos I$$

Here *I* is the inclination of the orbit. See Sanders & Gillies, 1996a.

The full expression for the precession rate will obviously involves other harmonics, chiefly the other even-degree zonal harmonics. Thus, it would not normally be possible to extract the value of $J_{2,0}$ *per se* from the orbital precession rate, since the effects of $J_{2,0}$ are entangled with those of the other harmonics. However, since GRACE will provide the values of these harmonics with extreme accuracy, we expect to be able to subtract their effects and thus isolate the effect of $J_{2,0}$, and so allow for a very precise determination of this coefficient.

4. Expected results

An estimated error budget for *G*-dot on a SEE mission is given in Sanders, Alexeev *et al.* (2000) and is reproduced in Table 1 below. The authors were not aware at the time of publication of the difficulties which GRACE would have in measuring $J_{2,0}$. In the present paper we consider whether the previous estimates of the unmodeled forces remain valid, given the impact of the difficulty with $J_{2,0}$. We find that the answer is a qualified "yes."

Table 1
Error Budget for *G*-dot
One-Year and Four-Year Observation Periods

Error Source	Average Force ($x10^{-17}$ N)	$\delta(G\text{-dot}/G)$ ($x10^{-15}$) 1 yr	4 yrs	Brief Comments
Tracking error	NA	15.6	2.0	GPS/SLR accuracy = 1 cm
Timekeeping error	NA	1	1	Future atomic clocks
Blackbody radiation	10.0	8.6	4.3	$\Delta\Theta < 0.1$ mK
Electrostatic forces	<15	<10	<5	Surface potential < 6.4 mV
Lorentz forces	small	zero	zero	Perpendicular to velocity
Earth's field	<1.4	<0.9	<0.5	With GRACE or equivalent
Capsule mass defects	22.2	15	7.4	Many defects ~10 mg
Gravity of particle	<0.22	<0.15	<0.08	Newton's 3rd law
Shepherd's moments	small	small?	small?	Not evaluated yet
Outgassing jets	small	small	small	Obviate by baking
Total	**NA**	**26**	**14**	

To avoid any misunderstanding, we emphasize that the following discussion applies only to the issue of *secular* change in *G* and not to a possible seasonally-occurring apparent variation of *G* (perhaps of the type predicted previously by scalar-tensor theories), which might result from anisotropy of space. This is because errors during short-term (< 1 yr) observation of *G* are dominated by the tracking error of the along-track position rather than unmodeled forces and therefore are not affected by the $J_{2,0}$ problem. Thus, the capability of a SEE mission to detect anisotropy of space would not be vitiated by difficulty in measuring $J_{2,0}$.

Measurement resolution considerations: Since $J_{2,0}$ influences the orbital period at a level of less than one part in 1000, it will be necessary to determine $J_{2,0}$ to about a few parts in 10^{11} in order to determine *(G-dot)/G* to one part in 10^{14}. Two facts essentially determine the prospect of achieving this resolution: We will have centimeter-level resolution of the position of the Shepherd, and we will have a large number of data points to use in statistical analysis. Regarding the latter, we will have ≈ 4600 independent measurements per year (the number of orbital revolutions in one year), with the implication that the statistical uncertainty in the subsequent measured quantities will scale as roughly $(4600)^{1/2} = 68$. For instance, the circumference of the orbit is 50,000 km, or $5x10^9$ cm. Since $68x(5x10^9) = 3x10^{11}$ it follows that the resolution-induced error in a measurement of $J_{2,0}$ from nodal precession would be about 1 part in 10^{11} with an observation time of one year. Thus, the resolution does not appear likely to be a barrier to measuring $J_{2,0}$ with the

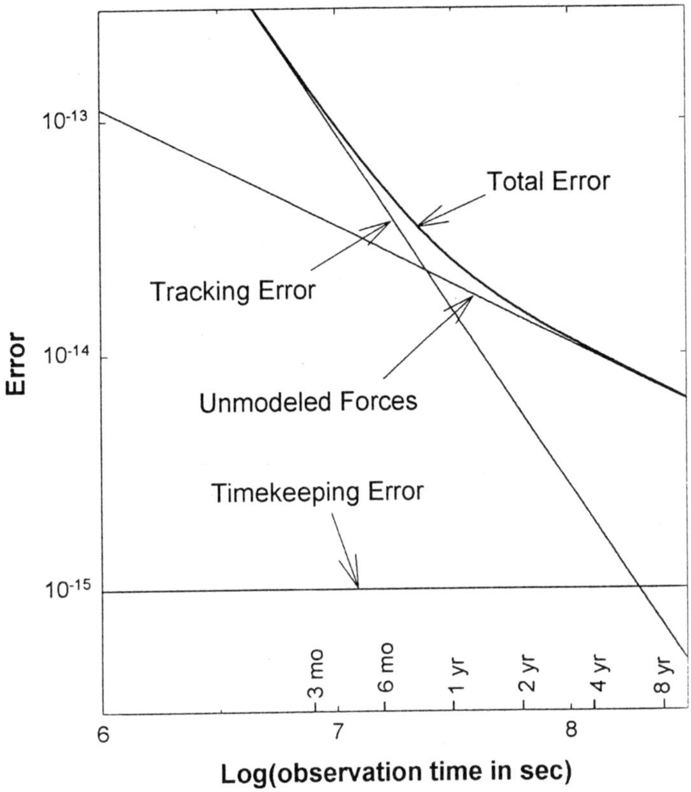

Figure 1. Error Budget for $\dot{G}/G$ by Class of Source.

accuracy of about 1 part in 10^{11}.

Perturbations: In Sanders, Alexeev *et al.* (2000) the perturbations of chief concern are those which entail forces that act largely in parallel with the orbital velocities ("along track" forces). Such forces affect the orbital energy and therefore the period of the test bodies, thus mimicking a change in *G*. In the present paper, however, our focus is on the "cross-track" forces on the Shepherd, since these forces alter its lateral position and therefore the precession rate of the orbital plane, which in turn is used to determine of $J_{2,0}$. In the remainder of this section we examine the cross-track forces on the SEE Shepherd—those that are horizontal and approximately perpendicular to the orbital velocity.

The average cross-track acceleration on a SEE test body due to $J_{2,0}$ is 3.6×10^{-3} m/s^2, so the average cross-track force on the 200-kg Shepherd is 0.7 N. The objective is to measure $J_{2,0}$ to within 1 part in 10^{11}, so it is necessary to show that the unmodeled portion of the various cross-track perturbation forces is at least 11 orders of magnitude smaller than the cross-track force due to $J_{2,0}$. Thus, the question is whether all cross-track perturbations can be accounted for to within $\approx 10^{-11}$ N. Heuristic considerations suggest that this is likely—that the transverse forces on the SEE Shepherd are sufficiently small that they probably will not vitiate a measurement of $J_{2,0}$ at the required accuracy.

The other main cross-track forces on the SEE Shepherd are the Lorenz force, the electrostatic force due to the image charge in the capsule wall, and the cross-track component of the residual gravitational force due to the capsule.

The maximum size of the Lorenz force, $q\mathbf{v}\times\mathbf{B}$ (in SI units), is about 3×10^{-14} N. This is 13 orders of magnitude below the cross-track force due to $J_{2,0}$. (Here we have taken the surface potential of the Shepherd to be 6.4 mV [Sanders, Alexeev *et al.*, 2000], which implies a charge of $q=1.3 \times 10^{-13}$ C; the orbital velocity is about 7.3 km/s, and the Earth's magnetic field is taken as 0.33 gauss.) Thus, the Lorenz force is negligible.

The image-charge electrostatic force on the Shepherd is roughly $(1/4\pi\varepsilon_0)(q^2/r^2) = 6 \times 10^{-16}$ N. This is 15 orders of magnitude smaller than the cross-track force due to $J_{2,0}$ and therefore negligible. (Here we have taken r=50 cm as the distance of the Shepherd from its image charge.)

The gravitational force due to a 1–gram mass defect in the capsule wall is $GMm/r^2 = 2 \times 10^{-10}$ N if the defect is 25 cm away. In Sanders & Gillies (1996a, 1996b, 1997, and 1998b) we show that mass defects in the capsule wall can be detected and accounted for at a level much smaller than 1 gram—certainly 0.1 gram, and possibly 0.01 gram. Consequently, it seems plausible that the unmodeled cross-track gravitational force due to the residual gravitational field of the capsule will be at least

11 orders of magnitude below the cross-track force due to $J_{2,0}$. This would be sufficient.

Thus, neither resolution considerations nor perturbations due to cross-track forces seem likely to spoil a determination of $J_{2,0}$ at the level of 1 part in 10^{11}, given precise determination of the Earth's other harmonics by a GRACE mission or the equivalent. Therefore, a determination of (*G*-dot)/*G* at 1 part in 10^{14} with observation times longer than 1 year may be possible.

5. Summary

Observation of the orbital period of the Shepherd in a SEE mission is expected to be capable of detecting apparent seasonal fluctuation of *G* at the level of about a part in 10^{13} and measuring the secular (long-term) rate of change, *(G-dot)/G*, at about the level of 10^{-14}/yr. Both capabilities require precise knowledge of the Earth's gravitational field. The GRACE mission or an equivalent is capable of supplying all necessary information about the Earth's field except a precise measurement of the harmonic coefficient $J_{2,0}$. It might seem that the difficulty which GRACE has with $J_{2,0}$ raises questions about whether SEE can achieve its accuracy goal for measuring the secular change in *G*. However, we have shown in the present paper that, by combining data from GRACE and SEE missions, one might obtain a precise measurement of $J_{2,0}$ from the precession rate of the Shepherd's orbital plane. This would complete the set of coefficients of the Earth's harmonics and allow a determination of *(G-dot)/G* to about 10^{-14}/yr.

Acknowledgments

We are pleased to acknowledge the encouragement of Venzo de Sabbata, the hospitality of the E. Majorana Center, the support by and interest of NATO Division des Afffaires scientifiques et de l'Environnement,, ESTEC, NASA Fundamental Physics Program, NASA Marshall Space Flight Center, and NASA Jet Propulsion Laboratory.

References

K.A. Bronnikov, V.D. Ivashchuk and V.N. Melnikov; *Nuovo Cim. B* **102**, 209 (1988).

T. Chiba, "Quintessence, the gravitational constant, and gravity," *Phys. Rev. D* **60**, 083508 (1999).

T. Chiba, "Constancy of the Constants of Nature," October 2001, gr-qc/0110118 .

M. Drinkwater *et al.* MNRAS, 1999, November 1999, astro-ph/9711290.

G.T. Gillies, "The Newtonian gravitational constant: Recent measurements and related studies"; *Reports on Progress in Physics* **60**, 151-225 (1997).

V.D. Ivashchuk and V.N. Melnikov. In : Lecture Notes in Physics, v. 537. *Mathematical and Quantum Aspects of Relativity and Cosmology.* Springer, 2000, p. 214.

V.D. Ivashchuk and V.N. Melnikov, "Multidimensional p-brane Cosmological Models and Billiard Behaviour," In *Proc. 2nd ICRA Network Workshop "The Chaotic Universe"*, World Scientific, Singapore (2000).

W.J. Marcino, "Time variation of the fundamental 'constants' and Kaluza-Klein theories," *Phys. Rev. Lett.* **52**, 489-491 (1984).

V.N. Melnikov, "Fundamental physical constants and their stability," *Int. J. Theor. Phys.* **33** (No. 7), 1569-1579 (1994).

A.J. Sanders, A.D. Alexeev, S.W. Allison, K.A. Bronnikov, J.W. Campbell, M.R. Cates, T.A. Corcovilos, D.D. Earl, T. Gadfort, G.T. Gillies, M.J. Harris, N.I. Kolosnitsyn, M.Yu. Konstantinov, V.N. Melnikov, R.J. Newby, R.G. Schunk, and L.L. Smalley, "Project SEE (Satellite Energy Exchange): Proposal for Space-based Gravitational Measurements," at the conference The Gravitational Constant: Theory and Experiment 200 Years after Cavendish, November 23-24, 1998, London; published in *Meas. Sci. Technol.*, **10** (No. 6), 514-524 (1999).

A.J. Sanders, A.D. Alexeev, S.W. Allison, V. Antonov, K.A. Bronnikov, J.W. Campbell, M.R. Cates, T.A. Corcovilos, D.D. Earl, T. Gadfort, G.T. Gillies, M.J. Harris, N.I. Kolosnitsyn, V.N. Melnikov, R.J. Newby, R.G. Schunk, and L.L. Smalley, "Project SEE (Satellite Energy Exchange): an international effort to develop a space-based mission for precise measurements of gravitation". *Class. Quant. Grav.* **17**, 2331-2346 (2000).

Alvin J. Sanders and W. Edward Deeds, "Proposed New Determination of the Gravitational Constant *G* and Tests of Newtonian Gravitation," *Phys. Rev. D.* **46** (No. 2), 489-505 (1992).

Alvin J. Sanders and W. Edward Deeds, "Reply to 'Perturbative Forces in the Proposed Satellite Energy Exchange Experiment'," *Phys. Rev. D.* **47** (No.8), 3660-3661 (1993).

Alvin J. Sanders, W. Edward Deeds and George T. Gillies, "Proposed New Space-

Based Method for More Accurate Gravitational Measurements," in *The Earth and the Universe: Festschrift für Prof. Hans-Jürgen Treder* [retiring as editor of *Annalen der Physik*], edited by Wolfgang Schröder (Bremen, 1993).

Alvin J. Sanders and George T. Gillies, "A Comparative Study of Proposals for Space-Based Determination of the Gravitational Constant *G*," *Rivista del Nuovo Cimento*, **19** (No. 2), 1-54 (1996).

Alvin J. Sanders and George T. Gillies, "Comparison of Proposals for Space-Based Determination of *G*," *Gravitation and Cosmology* **2** (No. 1) 57-60 (1996) (in English and Russian).

A.J. Sanders and G. T. Gillies, "A comparative survey of proposals for space-based determination of the gravitational constant *G*"; *Proc. SPIE* **3116**, 88-96, Small Spacecraft, Space Environments, and Instrumentation Technologies, Oct. 1997.

A.J. Sanders and G.T. Gillies, "Project SEE (Satellite Energy Exchange): Proposed Space-Based Method for More Accurate Gravitational Measurements," in Bergmann, P. G., de Sabbata, V., Gillies, G. T., and Pronin, P. I., eds., *Spin in Gravity: Is it Possible to Give an Experimental Basis to Torsion?* (International School of Cosmology and Gravitation XVth Course), The Science and Culture Series - Physics, No. 16 (World Scientific, Singapore, 1998), pp. 225-230.

A.J. Sanders and G.T. Gillies, "A Comparative Survey of Proposals for Space-Based Determination of the Gravitational Constant *G*," *Loc. cit.*, pp. 231-234.

C. Schmidhuber, "On water, steam, and string theory," Institut für Theoretische Physik, Universität Bern, Sidlerstr. 5, 3012 Bern, Schweiz, preprint (1997).

LAGRANGIAN DESCRIPTION OF DUALITY–SYMMETRIC FIELDS

D. SOROKIN
Institute for Theoretical Physics, NSC KIPT, Kharkov, Ukraine
and
INFN, Sezione Di Padova Via F. Marzolo, 8,
35131 Padova, Italia

We review the construction, group– and field–theoretical features of actions which describe the dynamics of fields obeying duality symmetry and discuss physical models where these fields play an important part.

1. Introduction

In this lecture I would like to discuss peculiarities of the Lagrangian description of field theories which exhibit duality symmetries.

The problem of the Lagrangian description of fields which possess duality properties has a rather long history. A classical example of duality symmetry is the electric–magnetic duality of the free Maxwell equations

$$\nabla \times \vec{E} = -\frac{\partial \vec{B}}{\partial t}, \qquad \nabla \vec{E} = 0, \tag{1}$$

$$\nabla \times \vec{B} = \frac{\partial \vec{E}}{\partial t}, \qquad \nabla \vec{B} = 0. \tag{2}$$

The equations (1) and (2) are invariant under the interchange of the electric $\vec{E}$ and magnetic $\vec{B}$ field strength

$$\vec{E} \to \vec{B}, \qquad \vec{B} \to -\vec{E}. \tag{3}$$

If the Maxwell field is coupled to an electric current the first pair (1) of the Maxwell equations acquires the right hand side

$$\nabla \times \vec{E} = -\frac{\partial \vec{B}}{\partial t}, \qquad \nabla \vec{E} = \vec{j}_e, \tag{4}$$

P.G. Bergmann and V. de Sabbata (eds.),
Advances in the Interplay Between Quantum and Gravity Physics, 365–385.

while the second pair (2) remains intact. Coupling to the electrically charged matter breaks the electric–magnetic duality (3) of the Maxwell equations.

Assuming that the electric–magnetic duality is a fundamental symmetry of Nature Dirac conjectured [1] the existence of magnetically charged particles, monopoles, whose current would contribute to the right hand side of the second pare of the Maxwell equations

$$\nabla \times \vec{B} = \frac{\partial \vec{E}}{\partial t}, \qquad \nabla \vec{B} = \vec{j}_g . \tag{5}$$

Then eqs. (4) and (5) are invariant under (3) if $\vec{j}_e \to \vec{j}_g$, $\vec{j}_g \to -\vec{j}_e$.

To describe monopoles on an equal footing with electrically charged particles one should formulate the electromagnetodynamics in a duality symmetric way. This can be done by introducing two Maxwell potentials $A^1_m(x)$ (m=0,1,2,3) and $A^2_m(x)$ (which correspond, respectively, to the electric and magnetic field strength) whose field strengths $F_{mn} = \partial_m A^I_n - \partial_n A^I_m$ (I=1,2) are related by the Hodge duality

$$F^1_{mn} = \frac{1}{2}\varepsilon_{mnlp}F^{2lp}, \quad F^2_{mn} = -\frac{1}{2}\varepsilon_{mnlp}F^{1lp}, \quad \text{or} \quad F^I_{mn} = \frac{1}{2}\epsilon^{IJ}\varepsilon_{mnlp}F^{Jlp}, \tag{6}$$

where $\epsilon^{IJ} = -\epsilon^{JI}$, $\epsilon^{12} = 1$.

Eq. (6) is an example of the duality relation which implies that among the fields $A^1_m(x)$ and $A^2_m(x)$ only one is independent and is associated with the Maxwell field. For instance, if we choose $A^1_m(x)$ as the independent field, then taking the derivative of (6) we get the Maxwell equations for these field

$$\partial_m F^{1mn} = 0, \quad \partial_m {}^*F^{1mn} = 0 . \tag{7}$$

In more detail we shall consider the duality symmetric formulation of Maxwell theory in Section 3, and here I would just like to point out that though the Maxwell equations are invariant under the duality transformations (3) the Maxwell action is not. It changes the sign, as one can see from the definition of the Maxwell action

$$S = -\frac{1}{4}\int d^4x F_{mn}F^{mn} = \frac{1}{2}\int d^4x \left(\vec{E}^2 - \vec{B}^2\right). \tag{8}$$

Thus we encounter the problem of constructing a Lagrangian formulation of duality–symmetric theories which would respect the duality properties of the fields.

The present interest to self–dual or duality–symmetric fields, which are also called chiral p–form fields, or simply chiral bosons, is mainly related to the fact that they appear as part of the spectrum of string theories and M–theory and play an important role in establishing various duality relations between perturbative and non–perturbative sectors of string/M–theory.

Let us give the most notable examples of chiral fields.

In two–dimensional space–time ($D = 2$) there are chiral scalars which live on the worldvolume of the heterotic string and satisfy the chirality condition

$$\partial_m \phi(x) = \epsilon_{mn} \partial^n \phi(x) . \tag{9}$$

In $D = 4$ we have the electromagnetic duality of the Maxwell field (3), (6).

$D = 6$ supergravities and a five–brane (which is a fundamental extended object of eleven–dimensional M–theory) have a chiral two–form field $A_{mn}(x)$ with the self–dual field strength

$$3\partial_{[l} A_{mn]} = F_{lmn} = \frac{1}{3!} \epsilon_{lmnpqr} F^{pqr}, \quad \text{or} \quad F_{(3)} = {}^* F_{(3)}. \tag{10}$$

$D = 10$ type IIB supergravity and superstring have a chiral four–form field $A_{(4)}(x)$ with the self–dual five–form field strength

$$dA_{(4)} = F_{(5)} = {}^* F_{(5)}. \tag{11}$$

The knowledge of the details of the dynamics, interactions and quantum properties of these fields might allow one to gain a deeper insight into the structure of string theory. This is why the construction of actions for chiral bosons has attracted great deal of attention. And here a problem arises that duality and Lorentz symmetry hardly like to coexist in one and the same action.

Several approaches have been developed to handle this problem. They can be divided in two categories. The first ones sacrifice manifest Lorentz invariance in favor of duality. They are economic in the sense that the construction of the duality–symmetric actions does not require the use of auxiliary fields. To the second category belong approaches in which the Lorentz covariance is restored by a proper introduction of auxiliary fields.

The first non–manifestly Lorentz–invariant formulation of the duality symmetric action for Maxwell theory interacting with dyons was constructed by Zwanziger in 1971 [2]. An alternative action was proposed by Deser and Teitelboim 1976 [3] and rediscovered by Schwarz and Sen in [4].

In 1987 Floreanini and Jackiw [5] and Henneaux and Teitelboim [6] proposed actions for self–dual fields in space–times of dimension $D = 4p+2$ ($p = 0, 1, 2, \cdots \infty$). Particular examples of these fields have been discussed above (see eqs. (9)–(11)).

In 1990 Tseytlin considered a duality–symmetric action for a string [7] which made T–duality of string theory manifest. Finally Schwarz and Sen [4] constructed duality symmetric actions for dual tensor fields in any space–time dimensions.

The drawback of the above mentioned formulations is that they are not manifestly Lorentz invariant. This causes problems with consistent coupling the chiral fields to other fields and in particular to gravity, and also with their quantization and the proof of the Lorentz invariance of the quantum theory. These unsatisfactory features stimulated the search for a covariant formulation of the actions for chiral fields. Several approaches to the construction of covariant duality–symmetric actions have been proposed.

The first covariant formulation was proposed by Siegel [8] and its modification was considered by Kavalov and Mkrtchyan [9]. In this approach the 'square' of the self–duality condition enters the action with a tensorial Lagrange multiplier. In $D = 2$ there is enough local symmetries for the Lagrange multiplier to be an auxiliary field, while in $D > 2$ the question remains whether or not the Lagrange multiplier has propagating degrees of freedom. So the Siegel formulation works well in $D = 2$, but its quantization leads to an anomalous theory which requires additional modification of the Siegel action [10].

To solve the problem of the consistent covariant quantization of the chiral bosons in $d = 2$ MacClain, Wu and Yu [11] and Wotzasek [12] proposed a covariant formulation of the chiral boson action which involves infinite number of Lagrange multiplier fields. This formulation has been generalized to higher space–time dimensions in [13]. It is worth noting that in [14] this formulation was derived from a $D = 10$ superstring field theory. Its drawback is that the infinite sum over Lagrange multiplier terms requires a regularization whose choice is well justified by group–theoretical arguments only in $D = 2$ [11].

An alternative covariant formulation of the chiral field actions was proposed in [15]. This formulation, which is a 'direct' covariantization of the actions of [3, 4], uses only one auxiliar field which enters the action in a specific non–polynomial way. Such a construction is applicable to all known theories with (self) dual fields in space–time with Lorentz signature of an arbitrary dimension. Its relation to the 'infinite field' formulation was demonstrated in [16], and in [17] (see also [18]) it has been found how via dualization of the auxiliary scalar field the non–covariant actions of [3, 4] are related to that of Zwanziger [2]. This formulation is also well suited for quantization as has been demonstrated for various models in [16, 19]–[22].

In this lecture I would like to explain basic features of the covariant Lagrangian description of chiral fields based on the formulation with the single auxiliary field and to show how it is related to other non–covariant and covariant formulations. In Section 2 we shall start the consideration with the example (10) of a chiral two–form field in $D = 6$. This is an instructive example which comprises all the peculiarities of the Lagrangian formulation of duality symmetric theories and, on the one hand side is related to $D = 6$

supergravity models and the M–theory 5–brane and on the other hand its dimensional reduction down to $D = 4$ gives 'for free' the duality–symmetric formulation of Maxwell theory, which we shall demonstrate in Section 4. In Section 5 we shall consider the coupling of the duality–symmetric Maxwell theory to electric and magnetic sources, and to fermionic fields in a supersymmetric fashion. In Conclusion we shall mention other developments and applications of the covariant Lagrangian formulation, in particular, the coupling of chiral fields to gravity and the construction of the worldvolume action for the M5–brane.

2. The chiral two–form field in $D = 6$

2.1. NON–COVARIANT LAGRANGIAN FORMULATION

We start with explaining why the construction of an action for chiral fields encounters problems. In the case under consideration the field strength of the chiral two–form field $A_{mn}(x)$ satisfies the self–duality condition

$$\mathcal{F}_{lmn} \equiv F_{lmn} - \frac{1}{3!}\epsilon_{lmnpqr}F^{pqr} = 0, \tag{12}$$

which is actually a first–order differential equation of motion of the field $A_{mn}(x)$. Taking the second derivative of (12) we get the conventional second–order differential equation for the tensorial gauge field

$$\partial_l F^{lmn} = 0\,. \tag{13}$$

Thus the chiral field satisfying the self–duality condition is a *particular* irreducible solution of the eq. (13). This field has twice less (i.e. three) physical degrees of freedom in comparison with a generic $D = 6$ two–form gauge field satisfying (13).

We would like to find an action which would produce (12) as an equation of motion. This turns out to be a non–trivial problem because (12) is a first order differential equation, while all the conventional bosonic fields obey second–order differential equations for which the action principle is well known. Eq. (13) is derived taking the $A_{(2)}$–field variation of the following action

$$S = -\int d^6x \frac{1}{6} F_{lmn}F^{lmn}\,. \tag{14}$$

The question is whether it is possible to modify (14) (**without the use of auxiliary fields**) in such a way that it would produce the self–duality condition and will thus describe solely a single chiral field[1]. The answer is

[1] One may wonder whether the self–duality condition can be added to an action with a Lagrange multiplier. We shell give the answer to this question in Section 5.

positive if one sacrifices the manifest Lorentz invariance of the construction. The prescription is to add to the action (14) a term quadratic in components of the anti–self–dual field strength $\mathcal{F}_{lmn}$ defined on the left hand side of (12). Note that $\mathcal{F}_{lmn}$ is anti–self–dual by construction, i.e.

$$\mathcal{F}_{(3)} \equiv -{}^{*}\mathcal{F}_{(3)} \qquad (15)$$

and hence its square is identically zero

$$\mathcal{F}_{lmn}\mathcal{F}^{lmn} = 3\mathcal{F}_{0ij}\mathcal{F}^{0ij} + \mathcal{F}_{ijk}\mathcal{F}^{ijk} \equiv 0, \qquad (16)$$

where we have decomposed the sum in time–like and space–like components, $i, j, k = 1, \cdots, 5$ being spatial indices.

For the modification of the action (14) we take the time–like term of eq. (16) and write the following action

$$S = -\int d^6x \left[\frac{1}{6}F_{lmn}F^{lmn} + \frac{1}{2}\mathcal{F}_{0ij}\mathcal{F}^{0ij}\right]. \qquad (17)$$

The variation of this action with respect to A_{mn} produces the self–duality condition (12) and thus describes the dynamics of the chiral field as we shall see in a minute. But first let us note that in a different but algebraically equivalent form the action (17) was constructed in [6]. We should also note that instead of singling out in (16) the time–like component term we might equally well to single out any of the spatial components in $\mathcal{F}_{lmn}$ and use its square for the construction of the chiral field action. This implies that though the Lorentz invariance of (17) is explicitly broken down to its spatial subgroup $SO(5)$, the action should possess a non–manifest space–time symmetry. There is indeed such a symmetry under the following transformations of A_{lm} [2]

$$\delta A_{ij} = \left[x^0 v^k \partial_k A_{ij} + (v^k x^k)\partial_0 A_{ij}\right] - (v^k x^k)\mathcal{F}_{0ij}\,, \qquad (18)$$

where v^k is a constant parameter which can be associated with the spatial velocity of the corresponding coordinate system (Lorentz boost). Indeed the terms in the square brackets are nothing but the standard Lorentz boosts of the tensor field, and the last term in (18) vanishes on the mass shall if the self–duality condition (12) is satisfied.

The action is invariant under Abelian gauge transformations of A_{mn}

$$\delta A_{lm} = \partial_{[l}\varphi_{m]} \qquad (19)$$

[2]For simplicity we present the space–time boosts in the gauge $A_{0i} = 0$ associated with the local symmetry $\delta A_{lm} = \partial_{[l}\varphi_{m]}$ of the action (17).

which reduce **15** degrees of freedom of the two–rank tensor down to **6** transverse physical degrees of freedom. As we have mentioned the chiral field has twice less number of physical degrees of freedom, i.e. **3**. Therefore one may expect that, if the action (17) describes the chiral field, there should be additional symmetry which reduces the six degrees of freedom down to three. The action (17) indeed has a local symmetry which shifts in an arbitrary way the time component of A_{mn}

$$\delta A_{0i} = \phi_i(x). \tag{20}$$

It is this symmetry which is responsible for the self–duality of the A_{mn} field strength. To show this consider the equations of motion of A_{mn} obtained by varying the action (17). Note that the variation of (17) with respect to A_{0i} does not produce any non–trivial equations of motion, since because of the local symmetry (20) A_{0i} is a pure gauge degree of freedom. So we should consider only the A_{ij} equation of motion which has the form of a Bianchi identity

$$\frac{\delta S}{\delta A_{ij}} = \varepsilon^{ijkfg}\partial_k \mathcal{F}_{fg0} = 0. \tag{21}$$

From (21) it follows that (at least locally)

$$\mathcal{F}_{ij0} = \partial_{[i}\tilde{\phi}_{j]}. \tag{22}$$

Now notice that under (20) $\mathcal{F}_{ij0}$ transforms in a similar fashion, i.e.

$$\delta \mathcal{F}_{ij0} = \partial_{[i}\phi_{j]}. \tag{23}$$

Thus if we choose $\phi_j = -\tilde{\phi}_j$ we can gauge fix

$$\mathcal{F}_{ij0} = \partial_{[i}\tilde{\phi}_{j]} = 0. \tag{24}$$

Since $\mathcal{F}_{lmn}$ is anti–self–dual (see (15)) eq. (24) means that also $\mathcal{F}_{ijk} = 0$ and, hence

$$\mathcal{F}_{lmn} = F_{lmn} - {}^*F_{lmn} = 0. \tag{25}$$

In this way we have got the Lorentz covariant self–duality condition and has just proved that the action (17) describes the dynamics of a single chiral two–form field. Let us now try to covariantize this action.

2.2. COVARIANT ACTION FOR THE CHIRAL FIELD

The idea of the covariantization is based on an assumption that the non–covariant action (17) is a gauge fixed form of a covariant action which possess an additional symmetry whose gauge fixing breaks Lorentz invariance

[15, 16]. A natural guess on the possible modification of the non–covariant term is to replace $\mathcal{F}^{0ij}$ by $u^l(x)\mathcal{F}_{lmn}$, where $u^l(x)$ is an auxiliary vector field which, by use of a local symmetry can be put equal to the time component of the unit matrix, i.e. $u^l(x) = \delta^l_0$. It turns out that such a modification is indeed possible if $u_l(x)$ is the normalized derivative of a scalar field a(x)

$$u_l = \frac{1}{\sqrt{\partial_m a(x)\partial^m a(x)}}\partial_l a(x). \tag{26}$$

The covariant chiral field action has the following form

$$S = -\int d^6x \left[\frac{1}{6}F_{lmn}F^{lmn} + \frac{1}{2(\partial a)^2}\partial^l a\mathcal{F}_{lmn}\mathcal{F}^{mnp}\partial_p a\right]. \tag{27}$$

This action possesses the following symmetries:

i)Lorentz invariance (by construction);

ii) the local Abelian gauge symmetry (19);

iii) local symmetry which generalizes the 'self–duality' symmetry (20)

$$\delta A_{mn} = \partial_{[m}a(x)\phi_{n]}(x); \tag{28}$$

iv) a new local symmetry which ensures the scalar field $a(x)$ to be an auxiliary field

$$\delta a(x) = \varphi(x), \quad \delta A_{mn} = \frac{\varphi(x)}{2(\partial a)^2}\mathcal{F}^{mnp}\partial_p a\,. \tag{29}$$

From eq. (29) we see that $a(x)$ can be gauged away everywhere except the light–like direction $\partial_l a\partial^l a = 0$, which is a singular gauge. This turns out to be enough for the $a(x)$ to be a non–propagating field. For instance, we can choose the gauge where

$$a(x) = x^0, \tag{30}$$

i.e. coincides with the time coordinate, then $u_l = \delta^0_l$ and the action (27) reduces to the non–covariant action (17).

It is instructive to show how the modified non–manifest space–time symmetry (18) arises as a result of such a gauge fixing. The gauge condition (30) remains intact under the following combined Lorentz and a(x)–field (29) transformation

$$\bar{\delta}a(x) = \delta_{Lorentz}a + \delta_\phi a = 0, \tag{31}$$

if the parameter $\phi(x) = -v^k x^k$, where v^k is the parameter of Lorentz boosts under which the time coordinate transforms as $\delta x^0 = v^k x^k$.

Since the gauge field A_{mn} transforms nontrivially under (29) the combined transformation $\bar{\delta}A_{mn} = \delta_L A_{mn} + \delta_\phi A_{mn}$ gives rise the unconventional space–time transformations (18) which form a symmetry of the gauge fixed action (17). We have thus clarified the origin of this peculiar symmetry.

Let us consider now the equations of motion of A_{mn} and $a(x)$ which one derives from the covariant action (27).

The equation of motion of A_{mn} are the Lorentz covariant counterpart of eqs. (21)

$$\varepsilon^{mnlpqr}\partial_l\left[\frac{1}{(\partial a)^2}\partial_p a(F-{}^*F)_{qrs}\partial^s a\right]=0\,. \tag{32}$$

As one can verify (due to the local symmetry (28)) the general solution of this equation reduces to the self–duality condition (25). Thus the scalar field $a(x)$ does not appear in the final expressions, its equation of motion being identically satisfied when eq. (32) holds, which is the consequence of the local symmetry (29) [15, 16]. This confirms that $a(x)$ is completely auxiliary and does not propagate. We have thus constructed a consistent Lorentz covariant action for the description of the dynamics of the chiral field.

3. Dimensional reduction to Maxwell theory in $D = 4$

We now demonstrate how one can get the duality symmetric formulation of Maxwell electrodynamics by performing the dimensional reduction of the $D = 6$ action (27) down to $D = 4$.

The dimensional reduction procedure basically consists in the following. Let the fields $A_{\hat m\hat n}(x)$[3] and $a(x)$ do not depend on x^4 and x^5. Thus we effectively deal with a four dimensional theory. The $D = 6$ Lorentz group $SO(1,5)$ gets broken down to the $D = 4$ Lorentz group $SO(1,3)$ times an internal Abelian group $SO(2)$ which is associated with the duality symmetry group

$$SO(1,5)\quad\rightarrow\quad SO(1,3)\times SO(2)\,. \tag{33}$$

The $D = 6$ indices split into $D = 4$ indices $m = 0,1,2,3$ and the $SO(2)$ index $I = 1,2$ which correspond to the 4th and 5th 'reduced' spatial coordinate of the $D = 6$ theory ($\hat m = (m,I)$).

Thus in $D = 4$ the gauge field $A_{\hat m\hat n}(x)$ splits into the scalar $A_{IJ} = \epsilon_{IJ}\Phi(x)$ and the second rank tensor A_{mn} which on the mass shell, in virtue of (25), are related by the "axion" duality [23]

$$\partial_m\Phi=\frac{1}{6}\varepsilon_{mnlp}\partial^n A^{lp}, \tag{34}$$

[3]In this Section I put "hats" on the $D = 6$ indices $\hat m = 0,1,\cdots,5$ while those without hats will correspond to $D = 4$.

and into two $D = 4$ vector fields $A^I_m(x)$ whose field strengths $F^I_{mn} = \partial_m A^I_n - \partial_n A^I_m$ are connected by the duality relation (6) which also follows from (25).

In what follows, for simplicity, we shall put the scalar and the tensor field equal to zero. Then the $D = 6$ chiral field action (27) reduces to the $D = 4$ duality symmetric action for the Maxwell theory

$$S = -\int d^4x \left[\frac{1}{8}(F^I_{mn})^2 - \frac{1}{4(\partial a)^2}\partial^l a \mathcal{F}^I_{lm}\mathcal{F}^{Imp}\partial_p a\right], \tag{35}$$

where $\mathcal{F}^I_{mn} = F^I_{mn} - \epsilon^{IJ}{}^*F^J_{mn}$.

The action (35) is invariant under the following local transformations:

i) two $U(1)$ gauge symmetries

$$\delta A^I_m = \partial_m \varphi^I(x), \tag{36}$$

ii) the "duality" symmetry

$$\delta A^I_m = \partial_m a(x)\phi^I(x), \quad \delta a(x) = 0, \tag{37}$$

iii) the a–field symmetry

$$\delta a(x) = (\partial a)^2\varphi(x), \quad A^I_m = \varphi(x)\mathcal{F}^I_{mn}\partial^n a. \tag{38}$$

Using (38) we can put $\partial_m a(x) = \delta^0_m$, or $\partial_m a(x) = \delta^3_m$. Then the action (35) reduces to the non–covariant actions of [3, 4]. Let us note that the Zwanziger action [2] differs from that of [3, 4] by the sign of the second term in (35) (when we choose the gauge $\partial_m a(x) = \delta^3_m$). To get the Zwanziger action we should dualize the scalar field $a(x)$ into a two–form tensor field $\Lambda_{mn}(x)$ and fix gauge symmetries associated with this dual auxiliary field. Such a dualization, which was found and described in [17], relates the non–covariant actions of [2] and [3, 4].

It is worth noting that as the Hamiltonian analysis shows (see [16]) the duality–symmetric formulation of Maxwell theory based on the action (35) as well as the formulation of the $D = 6$ chiral field obeys only first–class constraints associated with the local symmetries (36)–(38). The presence of the first–class constraints only is a merit of this formulation, which simplifies the quantization of the theory.

The constraints which generate the transformations (36)–(38), respectively have the following form

$$P^0_I = 0, \qquad \partial_i P^i_I = 0, \quad (i = 1, 2, 3), \tag{39}$$

$$\Pi^{-i}_I \partial_i a(x) = 0, \tag{40}$$

$$\Lambda \equiv \epsilon^{IJ}\varepsilon_{ijk}\Pi_I^{-i}\Pi_J^{-j}\partial^k a + P_{a(x)}(\partial_i a)^2 = 0, \tag{41}$$

where $P_I^m = (P_I^0, P_I^i)$ and $P_{a(x)}$ are canonical momenta of the fields A_m^I and $a(x)$ obtained by varying the action (37) with respect to the velocities $\partial_0 A_m^I$ and $\partial_0 a(x)$, and

$$\Pi_I^{\pm i} = P_I^i \pm \frac{1}{2}\epsilon_{IJ}\varepsilon^{ijk}F_{jk}^J. \tag{42}$$

One can check that under the Poisson brackets the constraints (39)–(41) form a closed algebra.

The canonical Hamiltonian which can be derived from the Lagrangian (37) is quadratic in "self–dual" momenta Π_I^{+i} (while the constraints depend on Π_I^{-i})

$$H_0 = \frac{1}{4}\int d^3x \Pi_I^{+i}\Pi_I^{+i}. \tag{43}$$

The fact that H_0 depends only on Π_I^{+i} and not on Π_I^{-i} (i.e. only on the half of the gauge field momenta) implies from somewhat different point of view that the physical modes of the model indeed correspond to a single Abelian gauge field.

Note also that H_0 does not depend on $a(x)$ which confirms the auxiliary nature of this field. But, of course, due to Dirac, the full Hamiltonian should include also Lagrange multiplier terms with the constraints (39)–(41) which involve $a(x)$.

From the above consideration we can conclude that the action (35) indeed describes Maxwell theory in a duality–symmetric way, at least classically. The quantization of this action has been carried out in [35, 22] and demonstrated that the quantum theory is equivalent to the standard quantum electrodynamics.

We now turn to the consideration of coupling the action (35) to matter fields.

4. Coupling of duality–symmetric Maxwell theory to matter

4.1. COUPLING TO DYONS

In a duality–symmetric Hamiltonian formulation the coupling of electrically and magnetically charged particles to Maxwell fields was considerd by Schwinger [24]. We shall consider this coupling in the Lagrangian formulation based on the action (35). The construction below can be regarded as the dimensionally reduced construction of Medina and Berkovits [25] who considered coupling of the $D = 6$ chiral field action (27) to matter. Their construction is the covariantization of that of [26].

Let $j_I^m(x)$ be two currents of charged matter which we assume to be "minimally" coupled to the gauge fields $A_m^I(x)$

$$S_{couple} = \int d^4x\, A_m^I\, j_I^m. \tag{44}$$

We would like to add the coupling term to the action (35) in such a way that the local symmetries (36)–(38) are not spoiled. The prescription of how to do this is actually based on the original idea of Dirac on how to couple monoples to the electro–magnetic field [1]. For this one should consider a so called Dirac string which is associated with the magnetic flux in– or outgoing from the magnetic source. We introduce a two–form tensor $G_{mn}^I(x)$ which has the support on the worldsheet of the Dirac string. By definition $G_{mn}^I(x)$ is a (non)local solution of the equation

$$\partial_m\, G^{Imn}(x) = j^{In} \tag{45}$$

which can be formally written as

$$G_{mn}^I = \partial_{[m}^{-1} j_{n]}^I. \tag{46}$$

Thus $G_{mn}^I(x)$ is determined by the current.

We use $G_{mn}^I(x)$ to construct generalized field strengths of the gauge fields as follows

$$\hat{F}_{mn}^I = F_{mn}^I - {}^*G_{mn}^J \epsilon^{IJ} \tag{47}$$

and to replace F_{mn}^I with $\hat{F}_{mn}^I$ in the action (35). The action for the coupled system takes the form

$$S = -\int d^4x \left[\frac{1}{8}(\hat{F}_{mn}^I)^2 - \frac{1}{4(\partial a)^2}\partial^l a \hat{\mathcal{F}}_{lm}^I \hat{\mathcal{F}}^{Imp}\partial_p a + A_m^I\, j_I^m\right]. \tag{48}$$

One can directly check, using eqs. (45) and (47), that this action is invariant under the transformations (36)–(38). Therefore, the equations of motion of the fields A_m^I, as in the free Maxwell field case, result in the duality relation for the generalized field strengths

$$\hat{F}_{mn}^I = \frac{1}{2}\epsilon^{IJ}\varepsilon_{mnlp}\hat{F}^{Jlp}. \tag{49}$$

Taking a derivative of (49) and using eq. (45) we get the Maxwell equations couple to electrically and magnetically charged sources (dyons) a la Dirac

$$\partial_m\, \hat{F}^{Imn} = j^{In}, \quad \rightarrow \quad \partial_m \hat{F}^{1mn} = j^{1n}, \quad \partial_m {}^*\hat{F}^{1mn} = j^{2n}. \tag{50}$$

From (50) we see that j^1_m and j^2_m can be associated with electric and magnetic current, respectively. This concludes the construction of the classical theory of electromagentodynamics.

The quantum theory of dyons coupled to the Maxwell field has been developed by Lechner and Marchetti [35]. They have shown that the results of quantization do not depend either on the scalar field $a(x)$ or on the choice of the Dirac string thus confirming the consistency of the construction based on the covariant action (48).

4.2. COUPLING TO NEUTRAL FERMIONS AND SUPERSYMMETRY

If we add to the action (35) the free Dirac action for a Majorana fermion field $\Psi_\alpha(x)$ the action

$$S = -\int d^4x \left[\frac{1}{8}(F^I_{mn})^2 - \frac{1}{4(\partial a)^2}\partial^l a \mathcal{F}^I_{lm}\mathcal{F}^{Imp}\partial_p a + i\bar{\Psi}\gamma^m\partial_m\Psi\right] \tag{51}$$

becomes supersymmetric [4, 15] under the following transformations of the fields

$$\delta a(x) = 0, \quad \delta A^I_m = i\bar{\Psi}\gamma_m\epsilon^I,$$

$$\delta\Psi = \frac{1}{8}F^{Imn}\gamma_m\gamma_n\epsilon^I - \frac{1}{4(\partial a)^2}\epsilon^{IJ}\partial_p a\mathcal{F}^{Jp[m}\partial^{n]}a\gamma_m\gamma_n\epsilon^I, \tag{52}$$

where ϵ^I $(I = 1, 2)$ are constant Grassmann Majorana spinor parameters subject to the duality relation $\epsilon^I = i\gamma_5\epsilon^{IJ}\epsilon^J$. Thus, only one of ϵ^I is independent and we deal with $N = 1$ supersymmetry.

On the mass shell (upon eliminating A^2_m) eqs. (52) reduce to the conventional $N = 1$ supersymmetry transformations of the vector supermultiplet (A^1_m, Ψ_α)

$$\delta A^1_m = i\bar{\Psi}\gamma_m\epsilon^1, \quad \delta\Psi = \frac{1}{4}F^{1mn}\gamma_m\gamma_n\epsilon^1. \tag{53}$$

Note that $a(x)$ is an $N = 1$ supersymmetry singlet. This unusual feature of the present model implies that the $N = 1$ supersymmetry algebra is not the standard super Poincare algebra. It has the following form

$$\{Q_\alpha, Q_\beta\} = 2i\gamma^m_{\alpha\beta}\,(P_m - \frac{1}{(\partial a)^2}\Lambda\partial_m a), \tag{54}$$

where Q_α is the supercharge, P_m is the translation generator and Λ is the generator of the local transformations (38) introduced in (41). Such a form of the supersymmetry algebra ensures that the action of the anticommutator of the supercharges on the field $a(x)$ is zero, which is in accordance with the invariance of $a(x)$ under supersymmetry.

5. Relation to other covariant methods of the description of chiral fields

We now show how the formulation of the dynamics of the chiral fields considered above is formally related to other approaches. We shall consider again the case of the $D = 6$ chiral field.

5.1. INFINITE FIELD APPROACH

Let us try to get the self–duality condition (12) by adding to the action (14) the Lagrange multiplier term

$$S = -\int d^6 \left[\frac{1}{6} F_{lmn} F^{lmn} - \Lambda_1^{lmn} (F_{lmn} - {}^*F_{lmn}) \right], \tag{55}$$

where $\Lambda^{lmn}(x)$ is an anti–self–dual Lagrange multiplier. Then varying (55) with respect to $\Lambda_1^{lmn}(x)$ we get the self duality condition for the chiral field A_{mn}. But varying (55) with respect to A_{mn} and taking into account the self–duality condition (12) we get the dynamical equation of motion for the anti–self–dual $\Lambda_1^{lmn}(x)$

$$\partial_l \Lambda_1^{lmn}(x) = 0, \quad \rightarrow \quad \Lambda^1_{lmn} \equiv -{}^*\Lambda^1_{lmn} = \partial_{[l} B_{mn]}, \tag{56}$$

where $B_{mn}(x)$ is an anti–chiral field whose field strength is $\Lambda_1^{lmn}(x)$. Thus the model based on the action (55) describes two, chiral and anti–chiral, fields A_{mn} and B_{mn} while our aim has been to find the formulation for the single chiral field. To cure the situation we might try to add to (4) one more Lagrange multiplier term

$$S_2 = \int d^6 x \, \Lambda_1^{lmn} \Lambda_{2lmn}, \tag{57}$$

where Λ_{2lmn} is a self–dual Lagrange multiplier which "kills" Λ_1^{lmn}. But one can easily verify that Λ_{2lmn} has propagating degrees of freedom which should be eliminated by adding another Lagrange multiplier term. This procedure, in general, cannot be stopped at any step and we arrive at the action with the infinite number of terms and Lagrange multiplier fields

$$S = -\int d^6 \left[\frac{1}{6} F_{lmn} F^{lmn} + \Lambda_1^{lmn} (F_{lmn} - {}^*F_{lmn}) + \sum_{n=0}^{\infty} (-1)^n \Lambda_{(n+1)}^{lmn} \Lambda_{lmn}^{(n+2)} \right], \tag{58}$$

where $\Lambda_{(n+1)} = (-1)^{n*} \Lambda_{(n+1)}$. The formulation of the dynamics of the single chiral field $A_{mn}(x)$ based on the action (58) has been developed in [13, 14]. To be consistent it requires a regularization of the infinite sum.

As we have mentioned when $\Lambda_{(n)}$ are generic it is not possible to stop the sum at an $(N+1)$–th step. But we can "cut the tail" at $(N+1)$ if we replace in (58) the generic $\Lambda^{lmn}_{(N)} + \Lambda^{lmn}_{(N+1)}$ with $\partial^{[l}a\Lambda_0^{mn]p}\partial_p a$, where Λ_0 is an arbitrary 3–form. Then we can eliminate all remaining Λ_n ($n = 0, 1, \cdots, N-1$) by solving their algebraic equations of motion and arrive at the action (27). We have thus demonstrated the relation between the two covariant Lagrangian descriptions of the chiral field [16].

5.2. THE SIEGEL ACTION

The action for the description of the chiral field proposed by Siegel [8] has the following form

$$S = -\int d^6 \left[\frac{1}{6}F_{lmn}F^{lmn} + \Lambda_{mn}\mathcal{F}^{mpq}\mathcal{F}^n{}_{pq}\right], \tag{59}$$

where Λ_{mn} is a Lagrange multiplier. It can be shown that the variation of the action (59) with respect to Λ_{mn} gives rise to the self–duality condition (12). However, there are no enough local symmetries to gauge away degrees of freedom contained in Λ_{mn} if $D > 2$ [8]. In this sense the Siegel formulation in $D > 2$ is not self consistent. We can make it consistent if we formally put

$$\Lambda_{mn} = \frac{1}{2}\frac{\partial_m a(x)\partial_n a(x)}{(\partial a)^2}$$

and substitute this form of Λ_{mn} into (59). We thus again arrive at the action (27).

We should stress that in spite of the formal relation of the chiral field description based on the action (27) to the formulations (58) and (59) there is a substantial difference of the former from the latter two. In the formulation (27) the self–duality condition arises as a consequence of the *dynamical* equations of motion of the chiral field A_{mn}, while in (58) and (59) it appears as a "constraint" multiplied by the Lagrange multipliers. This difference shows up in the quantization of the chiral bosons and makes quantization simpler in the $a(x)$–field approach (see, for instance, [16, 35]).

Another advantage of the $a(x)$–field formulation is that this is the only formulation which one has managed to generalize to describe non–linear self–interaction of chiral fields, which is required, for instance, for the description of the worldvolume dynamics of the M5–brane. We shall briefly consider this generalization in the next Section.

6. Coupling to gravity and non–linear self–interaction of chiral fields

An advantage of the use of the Lorentz covariant formulation is that it can be directly generalized to the description of the chiral fields in a gravitational background. For this one should only replace the Minkowski metric with the metric $g_{mn}(x)$ of a curved space–time. In addition, in $D = 6$ (and upon the dimensional reduction also in $D = 4$) one can generalize the action (27) to describe a Dirac–Born–Infeld–like self–interaction of the chiral field A_{mn} in a gravitational background. The action has the following form [27]

$$S = -\int d^6x \left[\sqrt{-\det\left(g_{mn} + i\frac{\partial^p a}{(\partial a)^2}\,{}^*F_{pmn}\right)} - \frac{\sqrt{-g}}{4(\partial a)^2}\partial_l a^* F^{lmn} F_{mnp}\partial^p a \right]. \tag{60}$$

It possesses all the symmetries (19), (28) and (29) of the action (27), the only difference being in the form of the transformations (29) where $\mathcal{F}^{mnp}\partial_p a$ gets replaced by

$$\mathcal{V}^{mn} \equiv F^{mnp}\partial_p a - 2\sqrt{-\frac{(\partial a)^2}{g}}\,\frac{\delta\sqrt{-\det(g_{pq} + i\frac{\partial^r a}{(\partial a)^2}\,{}^*F_{rpq})}}{\delta(\frac{\partial^r a}{(\partial a)^2}\,{}^*F_{rpq})}. \tag{61}$$

(The definition of $\mathcal{V}^{mn}$ is such that in the linearized approximation it coincides with $\mathcal{F}^{mnp}\partial_p a$.)

The equation of motion of A_{mn} which one derives from the action (60) is

$$\mathcal{V}^{mn} = 0.$$

This is the non–linear generalization of the self–duality condition (25). Though (61) contains the derivative of the auxiliary field $a(x)$ it can be shown [28] that the generalized self–duality condition can be rewritten in such a way that the dependence on $a(x)$ disappears.

If we choose the metric $g_{mn}(x)$ in (60) to be the induced metric of a six–dimensional worldvolume (parametrized by x^m) embedded into an eleven–dimensional superspace (parametrized by coordinates Y^M $(M = 0, 1, \cdots, 10)$)

$$g_{mn}(x) = \partial_m Z^M(x)\partial_n Z^N(x) G_{MN}(Y),$$

where G_{MN} is a $D = 11$ metric, the action (60) becomes the action for a 5–brane propagating in $D = 11$ gravitational background and carrying the chiral gauge field $A_{mn}(x)$ on its worldvolume [27]. This is a bosonic counterpart of the supersymmetric 5–brane of M–theory for which the complete action was constructed in [29]. When the field $a(x)$ is gauge fixed such that

$\partial_m a = \delta_m^0$ or $\partial_m a(x) = \delta_m^5$, the action (60) reduces to the non covariant formulation of the self-interacting $D = 6$ chiral field and the M5-brane developed in [30].

7. Conclusion

In this lecture we have considered main principles of the construction and particular features of the covariant Lagrangian formulation of the theory of chiral fields which uses the auxiliary field $a(x)$. This formulation has proved to be useful in studying problems of different theories where duality-symmetric and self-dual fields appear and in particular in studying various aspects of string/M-theory associated with the existence of the M5-brane.

In conclusion, let me give an overview of some papers where the $a(x)$-field approach has been developed and applied for the solution of classical and quantum problems.

The basic superalgebra of M-theory has been derived from the M5-brane action in [31]. The M5-brane Hamiltonian has been constructed and analyzed in [32]. A similar Hamiltonian (in a quadratic approximation [13, 16]) has been quantized on a T^6-torus [33] where it has been demonstrated that the quantum theory on T^6 of the chiral field is modular invariant.

The duality-symmetric formulation of $D = 11$ supergravity and its coupling to the M5-brane was constructed and analyzed in [34, 35]. The problem of quantum anomalies of the M5-brane action has been studied in [37, 35].

The M5-brane action has been used for studying aspects of $AdS_7 \times S^4$ correspondence in [36]. Other aspects of the M5-brane and related topics have been considered in [38].

$D = 6$ supergravities with two-form chiral fields were studied in [39].

The action for type IIB D=10 supergravity containing the four-form chiral field (11) has been constructed in [40]. It has been used for the analysis of the correspondence between the type IIB superstring on $AdS_5 \times S^5$ and $N = 4$ super-Yang-Mills theory in [41].

The quantization of chiral fields in the $a(x)$-field approach has been discussed by several authors. In [10, 37, 42] anomalies associated with chiral fields have been analyzed. The BRST quantization has been carried out in [19, 21] and a finitness of self-dual models in $D = 2, 4, 6$ has been analyzed in [22]. The quantum field theory of dyons has been developed in [35].

As has been shown the final results of quantization (carried out on topologically trivial manifolds, or tori) do not depend on the field $a(x)$ which confirms its auxiliary nature. An interesting problem for further investigation would be to study properties of the chiral-field action when it is quantized on a topologically nontrivial manifolds.

Acknowledgements. The author is grateful to Norma Sanchez and Venzo de Sabbata for the invitation to give this lecture and for hospitality at the Ettore Majorana Centre for Scientific Culture in Erice. This work was partially supported by the European Commission TMR Programme ERBFMPX-CT96-0045 to which the author is associated with the University of Padua, by the Grant N 2.51.1/52-F5/1795-98 of the Ukraine Ministry of Science and Technology and by the INTAS Research Project N 2000-254.

References

1. Dirac, P. A. (1931) Quantised Singularities In The Electromagnetic Field, *Proc. Roy. Soc. Lond.* A **133**, 60.
   Dirac, P. A. (1948) The Theory Of Magnetic Poles, *Phys. Rev.* **74**, 817 .
2. Zwanziger, D. (1971) Local Lagrangian Quantum Field Theory Of Electric And Magnetic Charges, *Phys. Rev.* D **3**, 880.
3. Deser, S. and Teitelboim, C. (1976) Duality Transformations Of Abelian And Non-abelian Gauge Fields, *Phys. Rev.* D **13**, 1592.
4. Schwarz, J. H. and Sen, A. (1994) Duality symmetric actions, *Nucl. Phys.* B **411**, 35.
5. Floreanini, R. and Jackiw, R. (1987) Selfdual Fields As Charge Density Solitons, *Phys. Rev. Lett.* **59**, 1873.
6. Henneaux, M. and Teitelboim, C. (1988) Dynamics Of Chiral (Selfdual) P Forms, *Phys. Lett.* B **206**, 650.
7. Tseytlin, A. A. (1990) Duality Symmetric Formulation Of String World Sheet Dynamics, *Phys. Lett.* B **242**, 163.
   Tseytlin, A. A. (1991) Duality symmetric closed string theory and interacting chiral scalars, *Nucl. Phys.* B **350**, 395.
8. Siegel, W. (1984) Manifest Lorentz Invariance Sometimes Requires Nonlinearity, *Nucl. Phys.* B **238**, 307.
9. Kavalov, A. R. and Mkrtchian, R. L. (1987) Lagrangian Of The Selfduality Equation And D = 10, N=2b Supergravity, *Sov. J. Nucl. Phys.* **46**, 728.
10. Imbimbo, C. and Schwimmer, A. (1987) The Lagrangian Formulation Of Chiral Scalars, *Phys. Lett.* B **193**, 455.
    Hull, C. M. (1988) Covariant Quantization Of Chiral Bosons And Anomaly Cancellation, *Phys. Lett.* B **206**, 234.
    Labastida, J. M. and Pernici, M. (1988) On The Brst Quantization Of Chiral Bosons, *Nucl. Phys.* B **297**, 557. Mezincescu, L. and Nepomechie, R. I. (1988) Critical Dimensions For Chiral Bosons, *Phys. Rev.* D **37**, 3067.
11. McClain, B. , Yu, F. and Wu, Y. S. (1990) Covariant Quantization Of Chiral Bosons And Osp(1,1|2) Symmetry, *Nucl. Phys.* B **343**, 689.
12. Wotzasek, C. (1991) The Wess-Zumino term for chiral bosons, *Phys. Rev. Lett.* **66**, 129.
13. Martin, I. and Restuccia, A. (1994) Duality symmetric actions and canonical quantization, *Phys. Lett.* B **323**, 311.
    Bengtsson, I. and Kleppe, A. (1997) On chiral p-forms, *Int. J. Mod. Phys.* A **12**, 3397.
14. Berkovits, N. (1996) Manifest electromagnetic duality in closed superstring field theory, *Phys. Lett.* B **388**, 743.
    Berkovits, N. (1997) Local actions with electric and magnetic sources, *Phys. Lett.* B **395**, 28.
    Berkovits, N. (1997) Super-Maxwell actions with manifest duality, *Phys. Lett.* B **398**, 79.

15. Pasti, P. , Sorokin, D. and Tonin, M. (1995) Note on manifest Lorentz and general coordinate invariance in duality symmetric models, *Phys. Lett.* B **352**, 59.
Pasti, P. , Sorokin, D. and Tonin, M. (1995) Duality symmetric actions with manifest space-time symmetries, *Phys. Rev.* D **52**, 4277 (1995).
Pasti, P. , Sorokin, D. and Tonin, M. (1995) Space-Time Symmetries in Duality Symmetric Models, hep-th/9509052.
16. Pasti, P. , Sorokin, D. and Tonin, M. (1997) On Lorentz invariant actions for chiral p-forms, *Phys. Rev.* D **55**, 6292 (1997).
17. Maznytsia, A., Preitschopf, C. R. and Sorokin, D. (1999) Duality of self-dual actions, *Nucl. Phys. B* **539**, 438.
Maznytsia, A., Preitschopf, C. R. and Sorokin, D. (1999) Dual actions for chiral bosons, hep-th/9808049.
18. Miao, Y. and Muller-Kirsten, H. J. (2000) Self-duality of various chiral boson actions, *Phys. Rev.* D **62**, 045014.
Miao, Y. , Manvelyan, R and Muller-Kirsten, H. J. (2000) Self-duality beyond chiral p-form actions, *Phys. Lett.* B **482**, 264.
Miao, Y. , Muller-Kirsten, H. J. and Park, D. K. (2001) Constructing doubly self-dual chiral p-form actions in D = 2(p+1) spacetime dimensions, hep-th/0106197.
19. Van Den Broeck, C. and Van Hoof, K. (1999) Batalin-Vilkovisky gauge-fixing of a chiral two-form in six dimensions, *Class. Quant. Grav.* **16**, 4011.
20. Lechner, K. and Marchetti, P. A. (2000) Duality-invariant quantum field theories of charges and monopoles, *Nucl. Phys.* B **569**, 529.
Lechner, K. and Marchetti, P. A. (2000) Spin-statistics transmutation in relativistic quantum field theories of dyons, *JHEP* **0012**, 028.
21. Bekaert, X. and Cucu, S. (2001) Antifield BRST quantization of duality-symmetric Maxwell theory, *JHEP* **0101**, 015.
22. Del Cima, O. M. ,Piguet, O. and Sarandy, M. S. (2001) Finiteness of PST self-dual models, *Nucl. Phys.* B **600**, 387 [hep-th/0012067].
23. Ogievetsky, V. I. and Polubarinov, I. V. (1967) The Notoph And Its Possible Interactions, *Sov. J. Nucl. Phys.* **4**, 156.
Kalb, M. and Ramond, P. (1974) Classical Direct Interstring Action, *Phys. Rev.* D **9**, 2273.
24. Schwinger, J. (1966) Magnetic Charge And Quantum Field Theory, *Phys. Rev.* **144**, 1087.
Schwinger, J. (1975) Magnetic Charge And The Charge Quantization Condition, *Phys. Rev.* D **12**, 3105.
25. Medina, R. and Berkovits, N. (1997) Pasti-Sorokin-Tonin actions in the presence of sources, *Phys. Rev.* D **56**, 6388.
26. Deser, S. Gomberoff, A. Henneaux, M. and Teitelboim, C. (1997) Duality, self-duality, sources and charge quantization in abelian N-form theories, *Phys. Lett.* B **400**, 80.
27. Pasti, P. , Sorokin, D. and Tonin, M. (1997) Covariant action for a D = 11 five-brane with the chiral field, *Phys. Lett.* B **398**, 41.
28. Howe, P. S. Sezgin, E. and West, P. C. (1997) The six-dimensional self-dual tensor, *Phys. Lett.* B **400**, 255.
29. Bandos, I. , Lechner, K. , Nurmagambetov, A. , Pasti, P. , Sorokin, D. and Tonin, M. (1997) Covariant action for the super-five-brane of M-theory, *Phys. Rev. Lett.* **78**, 4332.
Bandos, I. , Lechner, K. , Nurmagambetov, A. , Pasti, P. , Sorokin, D. and Tonin, M. (1997) On the equivalence of different formulations of the M theory five-brane, *Phys. Lett.* B **408**, 135.
30. Perry, M. and Schwarz, J. H. (1997) *Nucl. Phys.* B **489**, 47.
Schwarz, J. H. (1997) Coupling a self-dual tensor to gravity in six dimensions, *Phys. Lett.* B **395**, 191.
Aganagic, M., Park, J. , Popescu, C. and Schwarz, J. H. (1997) World-volume

action of the M-theory five-brane, *Nucl. Phys.* B **496**, 191.

31. Sorokin, D. and Townsend, P. K. (1997) M-theory superalgebra from the M-5-brane, *Phys. Lett.* B **412**, 265.
32. Bergshoeff, E., Sorokin, D. and Townsend, P. K. (1998) The M5-brane Hamiltonian, *Nucl. Phys.* B **533**, 303.
33. Dolan, L. and Nappi, C. R. (1998) A modular invariant partition function for the fivebrane, *Nucl. Phys.* B **530**, 683.
34. Bandos, I. , Berkovits, N. and Sorokin, D. (1998) Duality-symmetric eleven-dimensional supergravity and its coupling to M-branes, Nucl. Phys. B **522**, 214.
    Sorokin D. (1998) Coupling of M-branes in M-theory, hep-th/9806175.
35. Lechner K. and Marchetti, P. (2001) Dirac branes, characteristic currents and anomaly cancellations in 5-branes, Nucl. Phys. Proc. Suppl. **102**, 94.
36. Claus, P. Kallosh, R. and Van Proeyen, A. (1998) M 5-brane and superconformal (0,2) tensor multiplet in 6 dimensions, *Nucl. Phys.* B **518**, 117.
    Claus,P. (1999) Super M-brane actions in AdS(4) x S(7) and AdS(7) x S(4), *Phys. Rev.* D **59**, 066003.
    Grojean, C. and Mourad, J. (2000) Super fivebranes near the boundary of AdS(7) x S(4), *Nucl. Phys.* B **567**, 133.
37. Lechner, K. (1999) Self-dual tensors and gravitational anomalies in 4n+2 dimensions, *Nucl. Phys.* B **537**, 361.
    Lechner, K. , Marchetti, P. A. and Tonin, M. (2001) Anomaly free effective action for the elementary M5-brane, hep-th/0107061.
38. Cherkis, S. and Schwarz, J. H. (1997) Wrapping the M theory five-brane on K3, *Phys. Lett.* B **403**, 225. Berman, D. (1998) M5 on a torus and the three brane, *Nucl. Phys.* B **533**, 317.
    De Castro, A. and Restuccia, A. (1999) On the interacting chiral gauge field theory in D = 6 and the off-shell equivalence of dual Born-Infeld-like actions, *Class. Quant. Grav.* **16**, 2729.
    Sato, T. (1999) Various supersymmetric brane configurations from superalgebras in many types of M-brane backgrounds, *Nucl. Phys.* B **548**, 231.
    Kimura, T. and Oda, I. (2001) Duality of super D-brane actions in general type II supergravity background, *Int. J. Mod. Phys.* A **16**, 503.
    Minasian, R. , Moore, G. and Tsimpis, D. (2000) Calabi-Yau black holes and (0,4) sigma models, *Commun. Math. Phys.* **209**, 325.
    Ovrut, B. A. (1999) N = 1 supersymmetric vacua in heterotic M-theory, hep-th/9905115.
    Gomis, J. , Ramallo, A. V. , Simon, J. and Townsend, P. K. (1999) Supersymmetric baryonic branes, *JHEP* **9911**, 019.
    Marino, M. , Minasian, R.,Moore, G. and Strominger, A. (2000) Nonlinear instantons from supersymmetric p-branes, *JHEP* **0001**, 005.
    Bandos, I. , Nurmagambetov, A. and Sorokin, D. (2000) The type IIA NS5-brane, *Nucl. Phys.* B **586**, 315.
    Derendinger, J. and Sauser, R. (2001) A five-brane modulus in the effective N = 1 supergravity of M-theory, *Nucl. Phys.* B **598**, 87.
    Lima, E. , Ovrut, B. and Park, J. (2001) Five-brane superpotentials in heterotic M-theory, hep-th/0102046.
    Moore, G. , Peradze, G. and Saulina, N. (2001) Instabilities in heterotic M-theory induced by open membrane instantons, *Nucl. Phys.* B **607**, 117.
    De Castro, A. and Restuccia, A. (2001) Master canonical action and BRST charge of the M theory bosonic five brane, hep-th/0103123.
    Bonelli, G. (2001) On the supersymmetric index of the M-theory 5-brane and little string theory, hep-th/0107051.
39. Dall'Agata, G. and Lechner, K. (1998) N = 1, D = 6 supergravity: Duality and non-minimal couplings, *Nucl. Phys.* B **511**, 326.
    Dall'Agata, G. and Lechner, K. and Tonin, M. (1998) Large covariant actions for

N = 1, D = 6 supergravity theories with chiral bosons, *Nucl. Phys.* B **512**, 179.
Riccioni, F. and Sagnotti, A. (1998) Some properties of tensor multiplets in six-dimensional supergravity, *Nucl. Phys. Proc. Suppl.* **67**, 68.
Riccioni, F. and Sagnotti, A. (1998) Self-dual tensors in six-dimensional supergravity, hep-th/9812042.
Riccioni, F. (1998) Tensor multiplets in six-dimensional (2,0) supergravity, *Phys. Lett.* B **422**, 126. Riccioni, F. (2000) Abelian vector multiplets in six-dimensional supergravity, *Phys. Lett.* B **474**, 79.
Van Hoof, K. (2000) An action for the (2,0) self-dual tensor multiplet in a conformal supergravity background, *Class. Quant. Grav.* **17**, 2093.

40. Dall'Agata, G. Lechner, K. and Sorokin, D. (1997) Covariant actions for the bosonic sector of D = 10 IIB supergravity, *Class. Quant. Grav.* **14**, L195. Dall'Agata, G. Lechner, K. and Tonin, M. (1998) D = 10, N = IIB supergravity: Lorentz-invariant actions and duality, *JHEP* **9807**, 017.
41. Oda, I. (1998) SL(2,Z) self-duality of super D3-brane action on AdS(5) x S(5), *JHEP* **9810**, 015.
Lee, S. , Minwalla, S. , Rangamani M. and Seiberg, N. (1998) Three-point functions of chiral operators in D = 4, N = 4 SYM at large N, *Adv. Theor. Math. Phys.* **2**, 697.
Arutyunov, G. E. and Frolov, S. A. (1999) On the origin of supergravity boundary terms in the AdS/CFT correspondence, *Nucl. Phys.* B **544**, 576 [hep-th/9806216].
Arutyunov, G. E. and Frolov, S. A. (1999) Quadratic action for type IIB supergravity on AdS(5) x S(5), *JHEP* **9908**, 024.
Green, M. B. (1999) Interconnections Between Type Ii Superstrings, M Theory And N = 4 Yang-Mills, hep-th/9903124.
Arutyunov, G. E. and Frolov, S. A. (2000) Some cubic couplings in type IIB supergravity on AdS(5) x S(5) and three-point functions in SYM(4) at large N, *Phys. Rev.* D **61**, 064009.
Bastianelli, F. and Zucchini, R. (2000) Three point functions for a class of chiral operators in maximally supersymmetric CFT at large N, *Nucl. Phys.* B **574**, 107.
42. Bastianelli, F. , Frolov, S. and Tseytlin, A. A. (2000) Conformal anomaly of (2,0) tensor multiplet in six dimensions and AdS/CFT correspondence, *JHEP* **0002**, 013.

THE COSMOLOGICAL INFLUENCE ON MATTER (COSMIC OBJECTS, DARK MATTER AND PARTICLES) PREDICTED BY THE 5-DIMENSIONAL PROJECTIVE UNIFIED FIELD THEORY

E. SCHMUTZER
University of Jena, Germany

Abstract. The last version of the Projective Unified Field Theory of the author (non-Einstein-like 5-dimensional field equations, but original projection formalism) overcomes the physical difficulties of the former variants (equivalence principle, equation of motion, spinor theory etc.). This version has been applied to cosmology and cosmogony in a series of publications. Some main results are: 1. Closed homogeneous isotropic cosmological model without big bang, exhibiting satisfying values for the age of the cosmos [19 billion years], Hubble factor H [67.8 km/(s Mpc)], relative temporal change of the effective gravitational constant [$1.58 \cdot 10^{-11}$/year] etc. 2. Dominating repulsion interaction between the cosmic particles in the first stage after the "big start" (instead of big bang). 3. Decreasing radius and decreasing excentricity of circle-like orbiting bodies moving about a center etc. 4. Heat production ("bremsheat") within a moving body, induced by the cosmological expansion and being possibly interesting with respect to seismic and volcanic activities of celestial bodies. 5. Bremsheat induced expansion of celestial bodies with realistic expansion velocities (Earth: $6.8 \cdot 10^{-2}$ cm/year; Sun: 0.39 cm/year). 6. Accretion theory of dark matter around a center with a possible explanation of the "Pioneer effect" of satellites. 7. Theory of the "rotation curves" of orbiting stars in galaxies. 8. The new concept of mass partly fulfills Mach's principle.

1. Retrospect

My education in differential geometry was based on the compact vector calculus. Therefore in my axiomatics of the geometry of the 5-dimensional projective space I took the basis vectors (spanning the space) as the genuine carriers of information on curvature, torsion etc. The further historical research on differential geometry went another way by introducing the concept of the "external differential calculus" into geometry and particularly into geometrized field theories. It

P.G. Bergmann and V. de Sabbata (eds.),
Advances in the Interplay Between Quantum and Gravity Physics, 387–433.

was A. Gorbatsievich [1] who recently could prove that my results can fully be reproduced within the framework of this contemporary geometrical language.

The Projective Unified Field Theory (PUFT) has been developed in 3 stages: first versions 1957 and 1980 [2] and final version 1995 [3], which differ by different field equations and projection formalisms for the transition from the 5-dimensional projective space to the 4-dimensional space-time. The last version (non-Einstein-like 5-dimensional field equations, but original projection formalism) overcomes the physical difficulties of the former variants (equivalence principle, equation of motion, spinor theory etc.). This version has been applied to cosmology and cosmogony in a series of publications.

Since my lectures at several Erice Courses on Gravitation and Cosmology were published in the proceedings of these courses [4], I can restrict myself here to the presentation of the last version of the field equations and conservation laws. Therefore the main part of this paper will treat the application of PUFT to cosmology, cosmogony and astrophysics. Work on a detailed monograph started recently, in which the historical development, the mathematical and physical basis of PUFT, the physico-philosophical interpretation and various applications will be presented [5].

Further I would like to mention that in my treatises on PUFT I had to use a special terminology in order to cope with the physically new situation within this theory.

2. Five-dimensional Hamilton principle, field equation and conservation law

Following conventions are used: The geometrical basis of the 5-dimensional projective space (X^μ 5-dimensional homogeneous coordinates) as well as of the 4-dimensional space-time (x^i 4-dimensional curvilinear coordinates) is the Riemannian geometry. Greek indices run from 1 to 5, Latin indices from 1 to 4. Comma and semicolon mean the partial derivative and the covariant derivative, respectively. Further the Gauss system of units (as mostly in astrophysics) is used.

The squared 5-dimensional line element is defined by ($g_{\mu\nu}$ 5-dimensional metric with the signature {+,+,+,+,-})

$$\left(\mathrm{d}\overset{5}{s}\right)^2 = g_{\mu\nu}\mathrm{d}X^\mu \, \mathrm{d}X^\nu \, . \tag{1}$$

In our theory the scalar field S means the length of the 5-dimensional radius vector existing in our geometrical concept. This quantity is in a specific way related to the scalar field of the various Kaluza-Klein theories and the Jordan theory. As it is well known, Jordan was the first who dropped the norm condition (constant scalar field) in the older projective relativity theories [6].

The definition of S results from

$$S^2 = g_{\mu\nu} X^\mu \, X^\nu \, . \tag{2}$$

In our theory it is more appropriate to use the quantity σ which we call for certain definite reasons scalaric field:

$$\sigma = \ln \frac{S}{S_0} \tag{3}$$

(Here S_0 is a basic constant of the physical dimension of length. It does not occur in our papers on cosmology, cosmogony and astrophysics, but it plays a certain role in the theory of spinors).

Starting point of our theory is the 5-dimensional action integral [3]

$$\text{a)}\quad W = \frac{1}{c}\int\limits_{V_5} \overset{5}{L} d^{(5)} f = \frac{1}{c}\int\limits_{V_5} \overset{5}{\mathcal{L}} d^{(5)} X \quad \text{with} \quad \text{b)}\ \overset{5}{\mathcal{L}} = \overset{5}{L}\sqrt{\overset{5}{g}} \tag{4}$$

($\overset{5}{L}$ Lagrange density, $\overset{5}{\mathcal{L}}$ Lagrange function, $\overset{5}{g}$ = - $\det(g_{\mu\nu})$ metric determinant, $d^{(5)} f$ 5-dimensional volume element, c velocity of light in vacuo).

The Hamilton principle

$$\begin{aligned}
&\text{a)}\quad \delta \int\limits_{V_5} \overset{5}{\mathcal{L}} d^{(5)} X = 0 \quad \text{with} \\
&\text{b)}\quad \delta g_{\mu\nu} \Big|_{(V_5)} = 0\,,\ \delta g_{\mu\nu,\tau} \Big|_{(V_5)} = 0 \quad \text{(Hamilton boundary conditions)}
\end{aligned} \tag{5}$$

leads to the Lagrange equation

$$\frac{\delta \overset{5}{\mathcal{L}}}{\delta g_{\mu\nu}} = \frac{\partial \overset{5}{\mathcal{L}}}{\partial g_{\mu\nu}} - \left(\frac{\partial \overset{5}{\mathcal{L}}}{\partial g_{\mu\nu,\tau}}\right)_{,\tau} + \left(\frac{\partial \overset{5}{\mathcal{L}}}{\partial g_{\mu\nu,\tau,\sigma}}\right)_{,\tau,\sigma} = 0\,. \tag{6}$$

According to our concept we split the Lagrange density into the geometrizable part (fields $g_{\mu\nu}$ and σ) and the up to now non-geometrizable part named substrate:

$$\overset{5}{L} = \overset{5}{L}_{geom} + \overset{5}{L}_{sub}. \tag{7}$$

The quantities being at our disposal for constructing the geometrized Lagrange density with appropriate transformation properties are the metric $g_{\mu\nu}$ and the scalar field S as well as the derivatives of both quantities. For the following we refer to a second order theory. Nevertheless the variety of construction is enormously large. Let us here repeat our ideas which led us in our paper [3] to a basic selection principle which was used as the fundament of PUFT (version 1995).

We remembered the fascinating property of Dirac's field theory of the electron/positron: The Lagrange density has the well-known form constructed of the bispinor field. With the help of the Hamilton principle the field equation of the actual motion of the bispinor field is derived. Inserting of the field equation into the Lagrange density leads to the surprising result that the Lagrange density of the actual motion is zero.

We considered this special field-theoretical case as an indication of a deeper founded phenomenon in nature and tried to test tentatively this hypothesis in our unified field theory. Following this idea we started with the ansatz

$$\overset{5}{L}_{geom} = f(S)\overset{5}{R} + g(S)S^{,\alpha}{}_{;\alpha} + h(S)S^{,\alpha}S_{,\alpha} + k(S). \tag{8}$$

In this context it is advisable to introduce the energy projector of the substrate ($\varkappa_0$ Einstein's gravitational constant as a true constant of nature)

$$\Theta^{\mu\nu} = \frac{1}{\varkappa_0 f\sqrt{\overset{5}{g}}}\frac{\delta\overset{5}{\mathcal{L}}_{sub}}{\delta g_{\mu\nu}}\,. \tag{9}$$

If one uses the ansatz (8) for deriving the field equations, one receives a certain result. Inserting this in the Lagrange density leads to the structure of the Lagrange density of the actual motion of the fields

$$\overset{5}{L}_{act} = \overset{5}{L}_{metric,act} + \overset{5}{L}_{sub,act} - \frac{2\varkappa_0 f^2}{3f + Sf'}\Theta^{\mu}_{\mu}\,. \tag{10}$$

Following the line described above, for determining the four free functions f, g, h, k and by further considerations with respect to positive definiteness of the scalaric field energy we demand

$$\overset{5}{L}_{metric,act} = 0\,. \tag{11}$$

Now two possibilities for further restriction appear:

Substrate null principle:

$$\text{a)}\quad \overset{5}{L}_{sub,act} = 0 \quad \text{with} \quad \text{b)}\ \overset{5}{L}_{act} = -\frac{2\varkappa_0 f^2}{3f + Sf'}\Theta^{\mu}_{\mu}\,; \tag{12}$$

Strong null principle:

$$\text{a)}\overset{5}{L}_{sub,act} = \frac{2\varkappa_0 f^2}{3f + Sf'}\Theta^{\mu}_{\mu} \quad \text{with} \quad \text{b)}\ \overset{5}{L}_{act} = 0\,. \tag{13}$$

Our further calculations are based on the “substrate null principle” (“next to nothing principle”). Here it is not possible to go into details.

The 5–dimensional field equation reads:

$$\begin{aligned} & R_{\mu\nu} - \frac{1}{2}\overset{5}{R} g_{\mu\nu} - \frac{1}{S} S_{,\mu;\nu} - \frac{2}{S^2} S_{,\mu} S_{,\nu} \\ & - \frac{1}{S} s_\mu s_\nu \left(4S^{,\tau}{}_{;\tau} - \frac{6}{S} S_{,\tau} S^{,\tau} + \frac{3\lambda_0}{S} - \frac{1}{2}\overset{5}{R} S \right) \\ & + \frac{1}{S} g_{\mu\nu} \left(S^{,\tau}{}_{;\tau} + \frac{\lambda_0}{S} \right) = \varkappa_0 \Theta_{\mu\nu} \end{aligned} \tag{14}$$

($R_{\mu\nu}$ Ricci tensor, $\overset{5}{R}$ 5-dimensional curvature invariant, $s_\mu = X_\mu / S$ radial unite vector, λ_0 quasi–cosmological constant similar to the cosmological constant of the Einstein theory). From (14) follows the conservation law

$$\left(\frac{1}{S} \Theta_\mu{}^\nu \right)_{;\nu} = 0 \,. \tag{15}$$

Further we mention the cyclic relation

$$s_{<\rho\sigma,\tau>} = 0 \tag{16}$$

being an identity within the framework of the theory of projectors.

Later it becomes obvious that the Einstein theory is the special case $\sigma = const$ of our theory.

3. Four-dimensional field equations and balance equations (conservation laws)

For our further investigations we drop the quasi-cosmological term: $\lambda_0 = 0$. The projection from the 5–dimensional projective space onto the 4–dimensional space–time leads to following sets of equations.

3.1. GENERALIZED GRAVITATIONAL FIELD EQUATIONS

$$R^{ij} - \frac{1}{2} R g^{ij} = \varkappa_0 T^{ij} \tag{17}$$

with the usual meaning of the symbols. The energy tensor

$$T^{ij} = E^{ij} + S^{ij} + \Theta^{ij} \tag{18}$$

consists of the following three terms:

$$\begin{aligned} \text{a)} \quad E^{ij} &= \frac{1}{4\pi} \left(B^{ik} H_k{}^j + \frac{1}{4} g^{ij} B_{kl} H^{kl} \right) , \\ \text{b)} \quad S^{ij} &= \frac{2}{\varkappa_0} \left(\sigma^{,i} \sigma^{,j} - \frac{1}{2} g^{ij} \sigma_{,k} \sigma^{,k} \right) \end{aligned} \tag{19}$$

and the substrate energy tensor Θ^{ij} describing the substrate (B_{ij} electromagnetic field strength tensor, H_{ij} electromagnetic induction tensor).

3.2. GENERALIZED ELECTROMAGNETIC FIELD EQUATIONS

$$\text{a)}\quad H^{ij}{}_{;j} = \frac{4\pi}{c} j^i\,, \quad \text{b)}\ B_{\langle ij,k \rangle} = 0\,, \quad \text{c)}\ H_{ij} = e^{2\sigma} B_{ij} \tag{20}$$

(j^i electric current density).

3.3. SCALARIC FIELD EQUATION

$$\sigma^{,j}{}_{;j} = -\frac{\varkappa_0}{2}\left(\vartheta + \frac{1}{8\pi} B_{kl} H^{kl}\right), \tag{21}$$

where

$$\vartheta = \Theta_\mu{}^\mu - \Theta_k{}^k \tag{22}$$

is the scalaric substrate energy density (short: scalerg density). The corresponding scalmass density is defined by

$$\mu^{(\sigma)} = \frac{1}{c^2}\vartheta\,. \tag{23}$$

From the above field equations the following balance equations (conservation laws) result:

$$\text{a)}\quad \Theta_i{}^k{}_{;k} = -\frac{1}{c} B_{ik} j^k + \vartheta \sigma_{,i}\,, \quad \text{b)}\quad j^k{}_{;k} = 0\,. \tag{24}$$

If we consider an electrically charged mechanical continuum (without conductive effects), the substrate energy tensor and the electric current density take the form

$$\text{a)}\quad \Theta_{ij} = -\left(\mu + \frac{p}{c^2}\right) u_i u_j - p g_{ij}\,, \quad \text{b)}\quad j^k = \rho_0 u^k\,, \tag{25}$$

respectively (μ mass density of the substrate, p mechanical pressure, ρ_0 electrical rest charge density, $u^i = \mathrm{d}\,x^i/\mathrm{d}\,\tau$ 4-velocity with $u_i u^i = -c^2$, τ proper time). Then from (25) result the mechanical equation of motion

$$\left(\mu + \frac{p}{c^2}\right) u^k{}_{;l} u^l = \frac{1}{c}\rho_0 B^k{}_l u^l - \left(p^{,k} + \frac{1}{c^2}\frac{\mathrm{d}p}{\mathrm{d}\tau} u^k\right) - \vartheta\left(\sigma^{,k} + \frac{1}{c^2}\frac{\mathrm{d}\sigma}{\mathrm{d}\tau} u^k\right) \tag{26}$$

and the mechanical balance equation

$$\left[\left(\mu + \frac{p}{c^2}\right) u^k\right]_{;k} = \frac{1}{c^2}\left(\frac{\mathrm{d}p}{\mathrm{d}\tau} + \vartheta \frac{\mathrm{d}\sigma}{\mathrm{d}\tau}\right). \tag{27}$$

In context with the equivalence principle we postulate

$$\mu = \mu^{(\sigma)}\sigma\,. \tag{28}$$

Then from (27) we obtain

$$\left[\left(\mu^{(\sigma)} + \frac{p}{\sigma c^2}\right) u^k\right]_{;k} = \frac{1}{c^2}\frac{\mathrm{d}}{\mathrm{d}\tau}\left(\frac{p}{\sigma}\right)\,. \tag{29}$$

The transition from (26) to a point-like test body by spatial integration leads to the equation of motion

$$M u^k{}_{;l} u^l = \frac{Q}{c} B^k{}_l u^l - D\left(\sigma^{,k} + \frac{1}{c^2}\frac{\mathrm{d}\sigma}{\mathrm{d}\tau} u^k\right)\,. \tag{30}$$

Here mean: M inertial mass (including pressure), Q electric charge, D scalerg of the test body corresponding to the scalerg density (22). Furthermore, the scalmass $\mathcal{M}$ of the body corresponding to the scalmass density (23) is given by

$$\mathcal{M} = \frac{D}{c^2}\,, \tag{31}$$

where

$$M = \mathcal{M}\sigma \tag{32}$$

holds.

Illuminating the step to (28) resp. (32), we add the following basic annotation: As usual in mechanics resulting from unified field theories, fulfilling of the equivalence principle is a difficult challenge. We had to cope with this question up to 1995. In the paper [3] by investigations on canonical mechanics corresponding to our Projective Unified Field Theory we came to the finding that the scheme of canonical mechanics (apart from the usual dependence of the mass on velocity and metric) leads to the following dependence of the mass on the scalaric field:

$$M = M^{(N)} + \int \mathcal{M}(\sigma) d\sigma \qquad (M^{(N)} = \mathrm{const})\,. \tag{33}$$

If the scalmass is independent of σ, it follows

$$M = M^{(N)} + \mathcal{M}\sigma\,. \tag{34}$$

Obviously, according to Newton's concept of mass, the constant mass $M^{(N)}$ in (33) has to be interpreted as the Newtonian mass of the body considered. If this quantity is unequal to zero, then our scalaric force term in the equation of motion (30)

$$F^k_{(S)} = -D\left(\sigma^{,k} + \frac{1}{c^2}\frac{\mathrm{d}\sigma}{\mathrm{d}\tau} u^k\right) \tag{35}$$

prevents fulfilling of the equivalence principle. Therefore starting with the papers [3] we studied the consequences of a tentative new concept of mass on the basis $M^{(N)} = 0$. This attempted idea is the basis of the formulas (28) to (32).Furthermore, in this way an important philosophical aspect of Mach's principle is fulfilled: The inertial mass M depends on the cosmological constellation ("far masses") expressed by the scalaric function σ.

4. Closed homogeneous isotropic cosmological model

4.1. COSMOLOGICAL SCENARIO AND BASIC COSMOLOGICAL EQUATIONS

The metric of such a world model reads

$$ds^2 = K(\xi)\left[d\vartheta^2 + \sin^2\vartheta\left(d\theta^2 + \sin^2\theta d\varphi^2\right)\right] - d\xi^2\,, \tag{36}$$

where $K(\xi)$ with $\xi = x^4 = ct$ is the time-dependent world radius, and $\vartheta, \theta, \varphi$ are the polar angles.By arguments arising from the equation of motion (explained later) we were led to the following cosmological scenario [7] :

1. **Repulsion era (antigravitational era):** This era characterized by the repulsion between the existing particles starts at the time $t = 0$ by a (soft) big start (Urstart) with the value $\sigma(t = 0) = 0$ of the scalaric world function σ and ends at the time $t = t_0$ with the value $\sigma(t = t_0) = 1$. According to the present ideas in this period a gas mixture of quantum particles (quarks, leptons, gluons, protons, neutrons etc.) of an extremely high temperature existed.
2. **Attraction era (gravitative era):** This era characterized by the gravitative attraction between the particles and bodies begins at the time t_0 with the value $\sigma = 1$ of the scalaric world function σ and lasts till now. The gravitative force consists of the dominating attractive gravitational (metric) force and the repulsive scalaric force. In this era the nucleosynthesis and the forming of galaxies and stars happens.

Now the assumptions $\rho_0 = 0$ and $B_{kl}H^{kl} = 0$ (incoherent electromagnetic background radiation) are chosen. As it is well known, for the electromagnetic field the following relations hold:

$$\text{a)}\quad p_{(r)} = \frac{1}{3}w_{(r)}\,, \qquad \text{b)}\ \mu_{(r)} = \frac{1}{c^2}w_{(r)} \tag{37}$$

($p_{(r)}$ radiation pressure, $w_{(r)}$ radiation energy density, $\mu_{(r)}$ radiation mass density).

Further it is convenient to introduce the following quantities for the closed universe:

$$
\begin{aligned}
&\text{a)}\quad V_{\cos} = 2\pi^2 K^3 && \text{(volume of the cosmos)},\\
&\text{b)}\quad M_{\cos} = 2\pi^2 \mu K^3 && \text{(mass of the cosmos)},\\
&\text{c)}\quad \mathcal{M}_{\cos} = 2\pi^2 \mu^{(\sigma)} K^3 && \text{(scalmass of the cosmos)},\\
&\text{d)}\quad D_{\cos} = 2\pi^2 \vartheta K^3 && \text{(scalerg of the cosmos)}\,.
\end{aligned}
\tag{38}
$$

Under the above assumptions from the general field equations (17) and (21) follow — partially after integration — the specialized equations (prime means derivative with respect to ξ)

$$
\begin{aligned}
&\text{a)}\quad \frac{K''}{K} - \frac{2}{3}\sigma'^2 + \frac{\varkappa_0 \Lambda_0}{K^4} + \frac{\varkappa_0 \sigma}{6}\left(\vartheta + \frac{3p}{\sigma}\right) = 0\,,\\
&\text{b)}\quad \frac{1}{K^2}\left(K'^2 + 1\right) + \frac{1}{3}\sigma'^2 - \frac{\varkappa_0 \Lambda_0}{K^4} - \frac{\varkappa_0 \sigma}{3}\vartheta = 0\,,\\
&\text{c)}\quad \sigma'' + \frac{3}{K}K'\sigma' - \frac{\varkappa_0 \vartheta}{2} = 0\,,\\
&\text{d)}\quad \Lambda_0 = \frac{1}{3} w_{(r)} K^4
\end{aligned}
\tag{39}
$$

(Λ_0 electromagnetic radiation constant of the cosmos). Further from the continuum-mechanical equations (26) and (29) we obtain

$$
\vartheta' + \frac{3}{K}\left(\vartheta + \frac{p}{\sigma}\right) K' = 0\,. \tag{40}
$$

The equation (39a) can be derived by differentiation of (39b) and subsequent eliminations, i.e. the system to treat consists of the equations (39a), (39c) and (40).

In our first papers on cosmology on the basis of the field equations (1995) we followed the usual arguments to consider the cosmological matter as a dust-like incoherent quasi-perfect gas (stars, galaxies, intergalactic matter etc. as particles) with pressure $p = 0$. Later we generalized this simple model, already leading to a cosmology without big bang, to the following more realistic cosmological model with pressure:

The closed homogeneous isotropic cosmos mainly consists of a proper cosmological gas of dark matter particles (dm-particles) which strongly dominates the visual (normal) matter of stars, galaxies etc. (percentage perhaps less than 1%) which "swim" in this "cosmological sea of the dm-particle gas" ("dm-gas"). For simplicity we restricted our analytical calculations to a one-sort gas obeying the Boltzmann-Maxwell statistics. Based on a relativistic treatment of the dm-particles whose velocity may be related to the cosmological expansion process we were led to the equation of state for a perfect gas

$$
p = n\mathrm{k}T \tag{41}
$$

(n particle number density, T kinetic temperature of the gas, k Boltzmann constant), but with the following specific relationship of the kinetic temperature to the cosmological fields K and σ:

$$T = \frac{m_0^{(\sigma)} c^2 B_0 \sqrt{-g_{44}}}{3\mathrm{k}K\sqrt{1+\frac{\sigma^2 K^2}{B_0^2}}}, \tag{42}$$

where $m_0^{(\sigma)}$ is the scalmass of the dm-particles, B_0 is a constant of integration determined later.

For practical reasons it is convenient to perform the following cosmological rescaling

$$\text{a)} \quad K = A_0 L\,, \quad \xi = A_0 \eta \quad \text{with} \quad \text{b)} \quad \eta = \frac{ct}{A_0} \tag{43}$$

(L rescaled world radius, η rescaled time-like coordinate, $A_0 = 10^{27}$ cm is a rescaling factor). Now factorizing $B_0 = A_0/\sqrt{C_0}$ is comfortable. Then the equation of state takes the shape

$$p = \frac{\vartheta \sigma \sqrt{-g_{44}}}{3(1 + C_0 L^2 \sigma^2)}\,. \tag{44}$$

This expression has to be inserted into the system of differential equations (39a,c) and (40). After rescaling results

$$\begin{aligned}
&\text{a)} \quad \ddot{L} - \frac{2}{3} L \dot{\sigma}^2 + \frac{\Delta}{L^3} + \widehat{\varkappa} \sigma L \vartheta \left(1 + \frac{1}{1 + C_0 L^2 \sigma^2}\right) = 0\,,\\
&\text{b)} \quad \ddot{\sigma} + \frac{3}{L} \dot{L} \dot{\sigma} - 3 \widehat{\varkappa} \vartheta = 0\,,\\
&\text{c)} \quad \dot{\vartheta} + \frac{1}{L} \vartheta \dot{L} \left(3 + \frac{1}{1 + C_0 L^2 \sigma^2}\right) = 0
\end{aligned} \tag{45}$$

(dot means derivative with respect to η), where the abbreviations

$$\text{a)} \quad \Delta = \frac{\varkappa_0 \Lambda_0}{A_0^2}\,, \quad \text{b)} \quad \widehat{\varkappa} = \frac{\varkappa_0 A_0^2}{6} = 3.4617 \times 10^5 \ \mathrm{g}^{-1} \mathrm{cm\ s}^2 \tag{46}$$

are used.

4.2. NUMERICAL TREATMENT OF THE SYSTEM OF DIFFERENTIAL EQUATIONS

The subsequent problem consists in the choice of a minimum of free parameters with the aim to describe a maximum of empirical facts. For this purpose we remember the present empirical value (index p means present time)

$$T_{ph,p} = 2.728 \ \mathrm{K} \tag{47}$$

of the temperature of the electromagnetic background radiation (photon temperature). Finally we arrived via (39d) at the choice

$$\Delta = 2.22415 \times 10^{-7} \tag{48}$$

for one of the free parameter in the differential equation (45a).

Further the initial conditions had to be chosen such that a reasonable outcome, compared with the empirical astrophysical facts, could be reached. We were satisfied with the results by choosing the value $C_0 = 5{\cdot}10^{-4}$ and following initial conditions:

$$\begin{aligned} &\text{a)} \quad L(\eta = 0) = 10^{-4}\,, \dot{L}(\eta = 0) = 4.6088486771411\,, \\ &\text{b)} \quad \sigma(\eta = 0) = 0\,, \dot{\sigma}(\eta = 0) = 0\,, \\ &\text{c)} \quad \vartheta(\eta = 0) = 2.1 \times 10^{8} \text{ g cm}^{-1}\text{s}^{-2}\,, \end{aligned} \tag{49}$$

where the second (numerical) initial condition resulted by exploiting of (39b). The subsequent numerical calculations were performed with the help of Maple and Mathematica for control (partly with an accuracy of digits = 27).

Our main result is a cosmological model without big bang. We called the basic temporal begin of the cosmos received “big start” (Urstart).

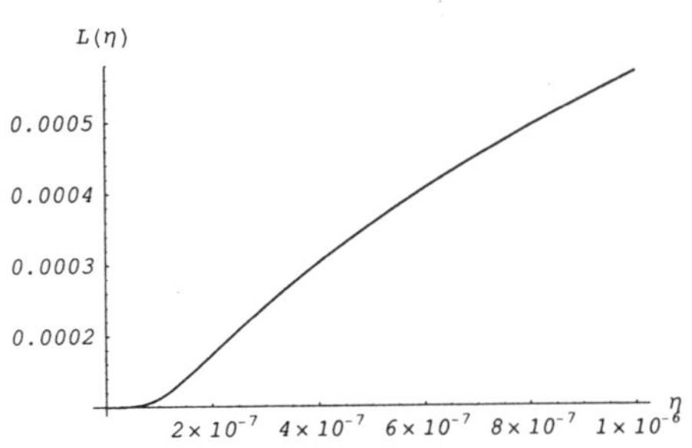

Figure 1. Temporal course of the rescaled world radius immediately after the big start

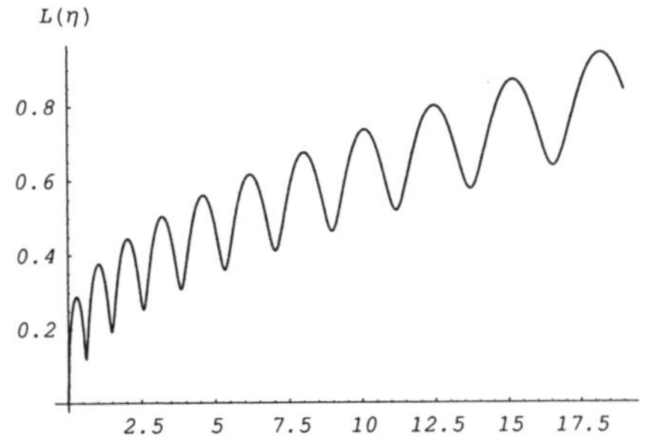

Figure 2. Temporal course of the rescaled world radius for the whole time of existence of the cosmos

Fig. 1 shows the temporal course of the rescaled world radius L (ordinate) as a function of the rescaled time η (abscissa) immediately after the big start. The unit $\Delta\eta = 1$ approximately corresponds to 1 billion y (y= year), i.e. the range on the abscissa means about 1000 y. Fig. 2 shows the temporal course of the rescaled world radius for the whole time scale between the big start and about 19 billion y (with a period of about 2.5 billion y, possibly being interesting with respect to the heat balance of the planets, too). Fig. 3 exhibits the temporal course of the scalaric world function σ between the big start and about 19 billion y.

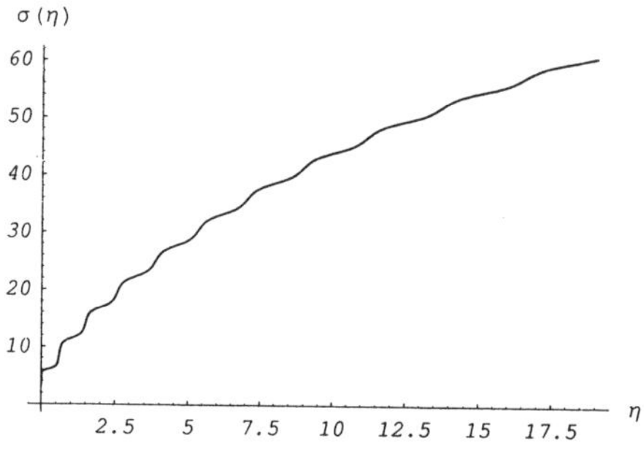

Figure 3. Temporal course of the scalaric world function for the whole time of existence of the cosmos

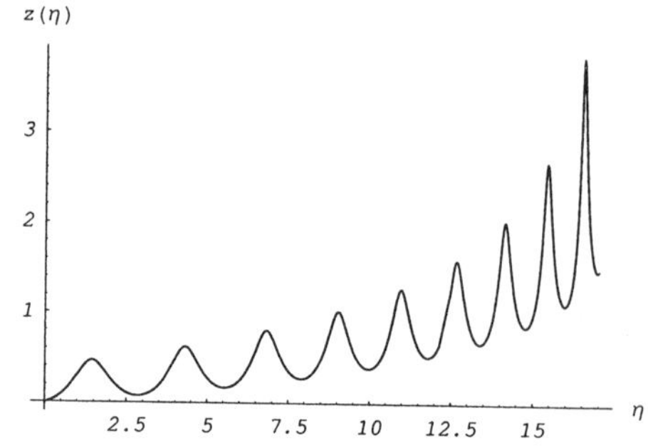

Figure 4. Temporal course of the relative frequency shift backwards from the present time (unit $\Delta\eta = 1$ corresponds to the distance of circa 1 billion light years)

The results for some cosmological quantities at the big start are:

$$\text{a)}\quad T_{ph}(\eta = 0) = 2.55 \cdot 10^4 \text{ K}, \quad \text{b)}\, T(\eta = 0) = 9.72 \cdot 10^{41} \cdot m_0^{(\sigma)} \text{ g}^{-1} \text{ K}\,. \tag{50}$$

The duration of the repulsion era is given by $\eta_{RP} = 1.0257 \cdot 10^{-7}$, i.e $t_{RP} =$ 108.5 y.

For the astrophysical evaluation of this cosmological theory it is necessary to present some main numerical values for the present time:

$$\begin{aligned} &\text{a)}\quad \eta_p = 18\,, \quad \text{b)}\, t_p = 19 \times 10^9 \text{ y} && \text{(age of the cosmos)},\\ &\text{c)}\quad K_p = 9.357.. \times 10^{26} \text{ cm} && \text{(radius of the cosmos)},\\ &\text{d)}\quad L_p = 0.9357...\,, && \text{e)}\quad \dot{L}_p = 0.0684...\,,\\ &\text{f)}\quad \sigma_{cp} = 59.7731\,, && \text{g)}\quad \dot{\sigma}_{cp} = 1.0018,\\ &\text{h)}\quad \mu_p = 2.4 \times 10^{-27} \text{ g cm}^{-3}, && \text{i)}\quad H_p = 67.8 \text{ km/(sMpc)}\,. \end{aligned} \tag{51}$$

In Fig. 4 we present the relative frequency shift (redshift) $z(\eta)$ as a function of the distance (backwards from the present time). Astrophysically interesting in this figure is the waviness (non-monotony) of the curve (equal redshift values for different distances !).

5. Solution for a point-like body

5.1. EXACT SOLUTION

Here we refer to our publication [8]. Using the well–known metric

$$ds^2 = e^{\alpha(r)} dr^2 + r^2 \left(\sin^2\theta d\varphi^2 + d\theta^2\right) - e^{\beta(r)} (dx^4)^2 \tag{52}$$

we obtain from the gravitational field equation (17)

$$\begin{aligned} &\text{a)}\quad \beta' - \alpha' + \frac{2}{r}(1 - e^{\alpha}) = 0\,, \quad \text{b)}\quad \alpha' + \beta' + 2r\sigma'^2 = 0\,, \\ &\text{c)}\quad \beta'' + \frac{1}{2}\beta'(\beta' - \alpha') + \frac{1}{r}(3\beta' - \alpha') + \frac{2}{r^2}(1 - e^{\alpha}) = 0 \end{aligned} \tag{53}$$

and from the scalaric field equation (21)

$$\left(r^2\sigma' e^{\frac{1}{2}(\beta-\alpha)}\right)' = 0 \tag{54}$$

(prime means derivative with respect to r). Rather lengthy calculations lead to the exact solution

$$r = \widehat{r}\frac{\sqrt{1-\tau_1-\Lambda^2}}{1-\Lambda}\left(\frac{\sqrt{1-\tau_1}-\Lambda}{\sqrt{1-\tau_1}+\Lambda}\right)^{\frac{1}{2\sqrt{1-\tau_1}}} \tag{55}$$

with

$$\Lambda = \sqrt{1 + \tau_1(e^{\alpha} - 1)} \tag{56}$$

and

$$e^{\beta} = \left(\frac{\tau_1\alpha_1}{2r}\right)^2 \frac{1 + \frac{1}{\tau_1}(\Lambda^2 - 1)}{(\Lambda - 1)^2}\,, \tag{57}$$

$$\sigma = \bar{\sigma} - \frac{1}{2}\sqrt{\tau_1}\beta \tag{58}$$

($\widehat{r}$, τ_1, α_1 and $\bar{\sigma}$ are constants of integration). Let us mention that first investigations apparently lead to the conclusion that our above exact solution for a point-like mass does not exhibit a horizon (as in the Einstein theory), i.e. black holes in the usual sense do not exist in PUFT.

The Schwarzschild solution is the special case $\tau_1 = 0, \sigma = \bar{\sigma}$. For physical applications we need explicite solutions up to the second order in $1/r$. By Taylor expansion, in the first steps needed up to the third order, we arrive at ($\alpha_1 = \widehat{r} = r_g$ gravitational radius)

$$\begin{aligned} &\text{a)}\quad e^{\alpha} = 1 + \frac{r_g}{r} + \frac{r_g^2\left(1 + \frac{\tau_1}{4}\right)}{r^2}\,, \\ &\text{b)}\quad e^{\beta} = 1 - \frac{r_g}{r}\quad, \quad \beta = -\frac{r_g}{r}\left(1 + \frac{r_g}{2r}\right), \\ &\text{c)}\quad \sigma = \bar{\sigma} + \frac{1}{2r}r_g\sqrt{\tau_1}\left(1 + \frac{r_g}{2r}\right). \end{aligned} \tag{59}$$

We are left with the 3 parameters r_g, τ_1 and $\bar{\sigma}$ which now have to be determined further. For this reason we present the non-relativistic results for the fields of a static sphere (r_0 = radius of the sphere) .

5.2. APPROXIMATE SOLUTION

In the following we list the fields up to the first order in the gravitational constant.

5.2.1. *External fields:*

$$\text{a)} \quad \phi_e = -\frac{\gamma_N \mathcal{M}_c \sigma_c}{r} + \phi_{e0}\,, \qquad \text{b)}\ \sigma_e = \frac{\gamma_N \mathcal{M}_c}{rc^2} + \sigma_c \tag{60}$$

(ϕ_{e0}, σ_c constants of integration), where ϕ is the Newtonian gravitational potential related to g_{44} by

$$g_{44} = -\left(1 + \frac{2\phi}{c^2}\right). \tag{61}$$

Furthermore, γ_N is the Newtonian gravitational constant related to the Einsteinian gravitational constant $\varkappa_0$ by

$$\varkappa_0 = \frac{8\pi\gamma_N}{c^4}\,. \tag{62}$$

$\mathcal{M}_c$ is the scalmass of the central body, and $\sigma_c = \sigma_c(t)$ is (treated as a constant of integration) the scalaric world function with its slight temporal (adiabatic) variation, compared to the temporal variations by the motion of the bodies considered later.

5.2.2. *Internal fields:*

$$\begin{aligned} \text{a)} \quad \phi_i &= \frac{\gamma_N \mathcal{M}_c \sigma_c}{2r_0}\left(\frac{r^2}{r_0^2} - 3\right) + \phi_{e0}\,, \\ \text{b)} \quad \sigma_i &= -\frac{\gamma_N \mathcal{M}_c}{2c^2 r_0}\left(\frac{r^2}{r_0^2} - 3\right) + \sigma_c\,. \end{aligned} \tag{63}$$

Let us mention that we fulfilled the continuity conditions for ϕ and σ as well as for the first derivatives of these functions.

6. Mechanical motion of a test body

6.1. RELATIVISTIC MOTION AND EINSTEIN EFFECTS

As mentioned above, our concept of mass and scalmass has been founded such that the equivalence principle is fulfilled completely. Starting point for treating the relativistic motion of a test body is the equation of motion (30) for the case of vanishing electric charge of the body: $Q = 0$. We obtain the specialized equation of motion

$$u^k{}_{;l}u^l = -c^2\left((\ln\sigma)^{,k} + \frac{1}{c^2}\frac{\mathrm{d}\ln\sigma}{\mathrm{d}\tau}u^k\right). \tag{64}$$

Comparing the results (59) and (60) we obtain the following identification of the parameters

$$\text{a)} \quad r_g = \frac{2\gamma_N \mathcal{M}_c \sigma_c}{c^2}\,, \quad \text{b)} \quad \sqrt{\tau_1} = \frac{1}{\sigma_c}\,, \quad \text{c)} \quad \sigma_c = \bar{\sigma}\,. \tag{65}$$

Applying equation (64) to the Einstein effects and comparing the results with the Newtonian mechanics leads to the relationship between the gravitational radius r_g in our theory (65a) and the gravitational radius $r_g^{(N)}$ in the Einstein theory referring to the Newton theory:

$$r_g^{(N)} = r_g \left(1 - \frac{1}{\sigma_c^2}\right) \tag{66}$$

where the definition

$$r_g^{(N)} = \frac{2\gamma_N M_c^{(N)}}{c^2} \tag{67}$$

holds ($\sigma_c = \sigma_p$ present value of the scalaric world function, $M_c^{(N)}$ Newtonian gravitational mass of the center).

In the following we list our results for the Einstein effects.

6.1.1. *Advance of the periastron:*

$$\frac{\Delta\varphi}{(\Delta\varphi)_{\text{Einstein}}} = \frac{1 - \frac{2}{3\sigma_c^2} - \frac{1}{6\sigma_c^4}}{\left(1 - \frac{1}{\sigma_c^2}\right)^2} \,. \tag{68}$$

Using for the scalaric world function the present value (51) $\sigma_c = 59.7731$, we obtain the universal correction

$$\left(\frac{\Delta\varphi}{(\Delta\varphi)_{\text{Einstein}}}\right)_p = 1.0004 \tag{69}$$

induced by the cosmological expansion.

6.1.2. *Deflection of electromagnetic waves:*

$$\text{a)} \quad \frac{\Delta\chi}{(\Delta\chi)_{\text{Einstein}}} = \frac{1}{1 - \frac{1}{\sigma_c^2}} \,, \qquad \text{b)} \quad \left(\frac{\Delta\chi}{(\Delta\chi)_{\text{Einstein}}}\right)_p = 1.0003 \,.$$

6.1.3. *Frequency shift of electromagnetic waves:*

$$\text{a)} \quad \frac{\Delta\omega}{(\Delta\omega)_{\text{Einstein}}} = \frac{1}{1 - \frac{1}{\sigma_c^2}} \,, \qquad \text{b)} \quad \left(\frac{\Delta\omega}{(\Delta\omega)_{\text{Einstein}}}\right)_p = 1.0003 \,.$$

Referring to the empirically oriented literature, we find that the results of the measuring values do not contradict the above theoretical values.

6.2. NON-RELATIVISTIC MOTION

Applying the approximate result (59) to the equation of motion (30) leads to the following non–relativistic equation of motion of a test body:

$$M\left(\frac{\mathrm{d}\boldsymbol{v}}{\mathrm{d}t} + \mathrm{grad}\,\Psi + \boldsymbol{v}\frac{\mathrm{d}\ln\sigma}{\mathrm{d}t}\right) = Q\left(\boldsymbol{E} + \frac{1}{c}\boldsymbol{v}\times\boldsymbol{B}\right) \tag{70}$$

with the following meaning of the symbols used (M inertial mass and $\boldsymbol{v}$ velocity of the moving body, respectively):

$$\Psi = -\frac{\gamma_N M_c^{(S)}}{r} \tag{71}$$

with ($\mathcal{M}_c$ scalmass of the central body)

$$M_c^{(S)} = \mathcal{M}_c\sigma^{(S)} \tag{72}$$

and

$$\sigma^{(S)} = \frac{\sigma^2 - 1}{\sigma} \text{ (scalaric correction factor)}\,. \tag{73}$$

As shown in detail in [8], from (70) it becomes obvious that the acting "gravitative force" on a moving test body

$$\boldsymbol{F}_{\text{gravitative}} = \boldsymbol{F}_{\text{gravitational}} + \boldsymbol{F}_{\text{scalaric}} = -M\,\mathrm{grad}\Psi = -\frac{\gamma_N M M_c^{(S)}}{r^3}\boldsymbol{r} \tag{74}$$

consists of two parts, namely the attracting gravitational force

$$\boldsymbol{F}_{\text{gravitational}} = -M\,\mathrm{grad}\phi = -\frac{\gamma_N M M_c}{r^3}\boldsymbol{r} \tag{75}$$

and the repulsing scalaric force

$$\boldsymbol{F}_{\text{scalaric}} = -M\,\mathrm{grad}\Sigma = \frac{\gamma_N M \mathcal{M}_c}{\sigma r^3}\boldsymbol{r}\,, \tag{76}$$

where

$$\Psi = \phi + \Sigma \tag{77}$$

with

$$\text{a)}\quad \phi = -\frac{\gamma_N M_c}{r}\,,\qquad \text{b)}\ \Sigma = \frac{\gamma_N \mathcal{M}_c}{\sigma r} \tag{78}$$

holds. This finding was the reason for our dividing the evolution of the cosmos into the repulsion era ($0 \leq \sigma < 1$) and the attraction era ($\sigma \geq 1$).

These considerations open the possibility of defining the "effective gravitational factor" being time-dependent (we don't use the term "gravitational constant") by

$$G = \gamma_N \sigma^{(S)}\,, \tag{79}$$

since we believe that this definition is appropriate for the experiments (based on the equation of motion of a test body but not on the definition of the gravitative potential energy, as we did earlier). In this context one should realize that this definition of G, being different from the usual one, is based on our scalmass concept of time-independent masses. Detailed calculations lead to the following expression for the relative time dependence (as a universal cosmological effect) of the effective empirical gravitational factor:

$$\frac{1}{G}\frac{\mathrm{d}G}{\mathrm{d}t} = \frac{1+1/\sigma^2}{1-1/\sigma^2}\frac{\mathrm{d}\ln\sigma}{\mathrm{d}t}\,. \tag{80}$$

Our numerical choice of the parameters obtained above leads to the value

$$\left(\frac{1}{G}\frac{\mathrm{d}G}{\mathrm{d}t}\right)_p = 1.58 \times 10^{-11}\ \mathrm{y}^{-1}. \tag{81}$$

As it is well known, since a rather long time intensive international affords were undertaken in order to measure this quantity (see the corresponding papers of G. T. Gillies, V. Melnikov, A. J. Sanders etc. in this volume).

Let us finally mention that on the basis of this mass concept the gravitative potential energy between two bodies with the (inertial) masses $\mathcal{M}_1$ and $\mathcal{M}_2$ (r distance between both bodies) reads

$$V = -\frac{\gamma_N \mathcal{M}_1 \mathcal{M}_2 (\sigma^2 - 1)}{r}\,. \tag{82}$$

7. Adiabatic scalaric approximation of the equation of motion

As shown recently [8], the equation of motion (70) of an electrically uncharged test body orbiting around a central body because of the conservation of the angular momentum of the body (R radial coordinate)

$$\text{a)}\quad L_0 = \mathcal{M}F_0 \quad \text{with} \quad \text{b)}\quad F_0 = \sigma R v_\phi \tag{83}$$

takes the form (in this section and later the dot means derivative with respect to the time t)

$$\ddot{R} - \frac{F_0^2}{\sigma^2 R^3} + \frac{\bar{K}}{R^2} + \frac{1}{\sigma}\dot{R}\dot{\sigma} = 0\,, \tag{84}$$

where

$$\text{a)}\quad \bar{K} = \gamma_N \mathcal{M}_c \sigma^{(S)}\,, \quad \text{b)}\quad v_\phi = R\dot{\phi} \tag{85}$$

holds ($\mathcal{M}_c$ scalaric mass of the central body, v_ϕ azimuthal velocity of the moving body). The transition $R(t)$ to $R(\phi)$ performed as usual and the transformation $R = 1/u$ lead to

$$\frac{\mathrm{d}^2u}{\mathrm{d}\phi^2} + u = \frac{\bar{K}\sigma^2}{F_0^2}\,. \tag{86}$$

In Newtonian mechanics the inhomogeneity on the right hand side of this equation is a constant, i.e. we have to solve a linear inhomogeneous differential equation. In our theory the inhomogeneity contains a slight cosmological time dependence, i.e. we are confronted with a rather complicated mathematical problem. But since the motion of the body considered may be rather fast compared to the cosmological temporal variation, we can restrict our further investigation to this case. We call the procedure applied "adiabatic scalaric approximation". Then the result of integrating (86) is

$$u = A_S\cos(\phi - \phi_0) + \frac{\gamma_N\mathcal{M}_c\sigma(\sigma^2-1)}{F_0^2} \tag{87}$$

(A_S scalaric orbital coefficient as a constant of integration, ϕ_0 constant of integration). Using the scalaric orbital radius R_S and the scalaric excentricity ϵ_S both defined by

$$\text{a)}\quad R_S = \frac{F_0^2}{\gamma_N\mathcal{M}_c\sigma(\sigma^2-1)}\,,\qquad \text{b)}\quad \epsilon_S = A_S R_S\,, \tag{88}$$

respectively, then equation (87) takes the form

$$R = \frac{R_S}{1+\epsilon_S\cos(\phi-\phi_0)}\,. \tag{89}$$

By differentiation of (88a) we find the result

$$\dot{R}_S = -\frac{F_0^2(3\sigma^2-1)}{\gamma_N\mathcal{M}_c\sigma^2(\sigma^2-1)^2}\dot{\sigma}. \tag{90}$$

Let us now define and list some quantities further used:

$$\text{a)}\quad \omega_S = \dot{\phi} = \frac{\gamma_N^2\mathcal{M}_c^2\sigma(\sigma^2-1)^2}{F_0^3}\,,$$

$$\text{b)}\quad \dot{\omega}_S = \ddot{\phi} = \frac{(\gamma_N\mathcal{M}_c)^2(\sigma^2-1)(5\sigma^2-1)}{F_0^3}\dot{\sigma}\,,$$

$$\text{c)}\quad v_{\phi S} = \frac{F_0}{\sigma R_S} = \frac{\gamma_N\mathcal{M}_c(\sigma^2-1)}{F_0}\,,\qquad \text{d)}\quad a_S = \dot{v}_{\phi S} = \frac{2\gamma_N\mathcal{M}_c\sigma}{F_0}\dot{\sigma}\,,$$

$$\text{e)}\quad \tau_S = \frac{2\pi}{\omega_S} = \frac{2\pi F_0^3}{\gamma_N^2\mathcal{M}_c^2\sigma(\sigma^2-1)^2}\,,\quad \text{f)}\ \dot{\tau}_S = -\frac{2\pi}{\omega_S^2}\dot{\omega}_S\,,\quad \text{g)}\ \dot{\epsilon}_S = A_S\dot{R}_S\,. \tag{91}$$

Physically remarkable are the signs of these cosmological universal effects: positive value of the angular acceleration (secular acceleration); negative values of the time derivatives of the scalaric orbital radius (decrease), revolution period (decrease), excentricity (transition from elliptic to cyclic orbits).

Using some values of (51) we find following numerical results:

$$\begin{aligned} &\text{a)} \quad \left(\frac{\dot{R}_S}{R_S}\right)_p = \left(\frac{\dot{\epsilon}_S}{\epsilon_S}\right)_p = -4.74 \times 10^{-11}\mathrm{y}^{-1} , \\ &\text{b)} \quad \left(\frac{\dot{\tau}_S}{\tau_S}\right)_p = -\left(\frac{\dot{\omega}_S}{\omega_S}\right)_p = -7.9 \times 10^{-11}\ \mathrm{y}^{-1} , \\ &\text{c)} \quad \left(\frac{\dot{v}_{\phi S}}{v_{\phi S}}\right)_p = 3.16 \times 10^{-11}\ \mathrm{y}^{-1} . \end{aligned} \tag{92}$$

Finally in this section we give numeric examples for the orbital motion of some objects (departures of our theory from Newtonian mechanics):

Sun around the center of the galaxy:

$$\text{a)}\ (\dot{R}_S)_p = -3.94 \times 10^4\ \mathrm{cm\ s}^{-1} , \qquad \text{b)}\ (\ddot{\phi})_p = 4.31'' \times 10^{-9}/(100\ \mathrm{y})^2 . \tag{93}$$

Earth around the sun:

$$\text{a)}\ (\dot{R}_S)_p = -2.24 \times 10^{-5}\ \mathrm{cm\ s}^{-1} , \qquad \text{b)}\ (\ddot{\phi})_p = 1.03''/(100\ \mathrm{y})^2 . \tag{94}$$

Moon around the earth:

$$\text{a)}\ (\dot{R}_S)_p = -5.77 \times 10^{-8}\ \mathrm{cm\ s}^{-1} , \qquad \text{b)}\ (\ddot{\phi})_p = 13.52''/(100\ \mathrm{y})^2 . \tag{95}$$

Satellite around the earth (height over the earth = 200 km):

$$\text{a)}\ (\dot{R}_S)_p = -9.86 \times 10^{-10}\ \mathrm{cm\ s}^{-1} , \qquad \text{b)}\ (\ddot{\phi})_p = 6.13'' \times 10^3/(100\ \mathrm{y})^2 . \tag{96}$$

In order to avoid misunderstandings, we would like to emphasize that these effects treated are (pure) universal cosmological effects. Of course, the orbiting bodies are subjected to a lot of additional influences as friction etc. As an example we mention the tidal effects acting on the moon with the empirical result of its drifting apart from the earth with the radial velocity of $v_R \approx 3.8$ cm/s.

8. Heat production in a moving body

8.1. TEST BODY, RIGID BODY AND BINARY SYSTEM

Considering the equation of motion (70) one immediately recognizes the force term

$$\boldsymbol{F}_B = -\mathcal{M}\frac{\mathrm{d}\sigma}{\mathrm{d}t}\boldsymbol{v} \tag{97}$$

being linear in the velocity and acting as a friction force term for the expansion era of the cosmos. We understand that this cosmological deceleration force (short: bremsforce) leads to the cosmologically induced internal production of deceleration heat (short: bremsheat):

$$P_B \equiv -\frac{\mathrm{d}Q_B}{\mathrm{d}t} = -\boldsymbol{F}_B\boldsymbol{v} = \mathcal{M}\boldsymbol{v}^2\frac{\mathrm{d}\sigma}{\mathrm{d}t} \; . \tag{98}$$

The corresponding expressions for a continuum read

$$\text{a)}\quad \boldsymbol{F}_B = -\mu^{(\sigma)}\frac{\mathrm{d}\sigma}{\mathrm{d}t}\boldsymbol{v}\,, \quad \text{b)}\quad \frac{\mathrm{d}q_B}{\mathrm{d}t} = -\mu^{(\sigma)}\boldsymbol{v}^2\frac{\mathrm{d}\sigma}{\mathrm{d}t}\,. \tag{99}$$

In order to apply this result to a rigid body with translational and rotational motion we have generalized the well-known Newtonian mechanics of a rigid body to our mechanics (the difference lies in the time dependence of the factor σ). The main results are:

Kinetic energy:

$$T = T^{\text{transl}} + T^{\text{rot}} \tag{100}$$

with

$$\text{a)}\quad T^{\text{transl}} = \frac{1}{2}M\dot{\boldsymbol{r}}_c^2\,, \quad \text{b)}\quad T^{\text{rot}} = \frac{1}{2}\sum_{a,b=1}^{3} I_{ab}\omega_a\omega_b\,, \tag{101}$$

where

$$I_{11} = \sum_{\Omega} M_{\Omega}(\eta_{\Omega}^2 + \zeta_{\Omega}^2)\,, \quad I_{12} = -\sum_{\Omega} M_{\Omega}\xi_{\Omega}\eta_{\Omega}\,, \tag{102}$$

is the inertial tensor ($M = \mathcal{M}\sigma$ mass of the rigid body, $\boldsymbol{r}_c$ radius vector to the center of mass, ω_a components of the angular velocity vector).

Angular momentum:

$$\boldsymbol{L} = \boldsymbol{L}^{\text{transl}} + \boldsymbol{L}^{\text{rot}} \tag{103}$$

with

$$\text{a)}\quad \boldsymbol{L}^{\text{transl}} = M\boldsymbol{r}_c \times \dot{\boldsymbol{r}}_c\,, \quad \text{b)}\quad \boldsymbol{L}^{\text{rot}} = \sum_{\Omega} M_{\Omega}\bar{\boldsymbol{r}}_{\Omega} \times (\boldsymbol{\omega} \times \bar{\boldsymbol{r}}_{\Omega})\,, \tag{104}$$

where

$$L_a^{\text{rot}} = \sum_{b=1}^{3} I_{ab}\omega_b\,. \tag{105}$$

Force moment:

$$\boldsymbol{M} = \boldsymbol{M}^{\text{transl}} + \boldsymbol{M}^{\text{rot}} \tag{106}$$

with

$$\text{a)}\quad \boldsymbol{M}^{\text{transl}} = \boldsymbol{r}_c \times \boldsymbol{F}\,, \quad \text{b)}\quad \boldsymbol{M}^{\text{rot}} = \sum_{\Omega} \bar{\boldsymbol{r}}_{\Omega} \times \boldsymbol{F}_{\Omega}\,, \tag{107}$$

where

$$\boldsymbol{F} = \sum_{\Omega} \boldsymbol{F}_{\Omega} = \boldsymbol{F}^{\text{ext}} + \boldsymbol{F}^{\text{int}} \tag{108}$$

($\mathbf{F}_{\Omega}$ force on the particle Ω, $\bar{\boldsymbol{r}}_{\Omega}$ radius vector to the particle Ω from the center of mass).

The equation of motion of a rigid body

$$\frac{\mathrm{d}\boldsymbol{L}}{\mathrm{d}t} = \boldsymbol{M} \tag{109}$$

splits into the two equations

$$\text{a)} \quad \frac{\mathrm{d}\boldsymbol{L}^{\text{transl}}}{\mathrm{d}t} = \boldsymbol{M}^{\text{transl}}, \quad \text{b)} \ \frac{\mathrm{d}\boldsymbol{L}^{\text{rot}}}{\mathrm{d}t} = \boldsymbol{M}^{\text{rot}}. \tag{110}$$

The first equation is identical with the translational equation of motion

$$\text{a)} \quad \frac{\mathrm{d}(M\dot{\boldsymbol{r}}_c)}{\mathrm{d}t} = \boldsymbol{F} \quad \text{with} \quad \text{b)}\ \boldsymbol{p}_c = M\dot{\boldsymbol{r}}_c \quad \text{(momentum)}. \tag{111}$$

If no force is acting, then follows

$$\dot{\boldsymbol{r}}_c = \frac{\boldsymbol{v}_0}{\sigma} \ \text{(constant of integration)}.$$

From this result we learn that the velocity of a force-free body decreases by the expansion of the cosmos (effect of homogenization).

If we refer to a rigid body with compensating internal forces, then in the case of gravitative forces the force reads

$$\text{a)} \quad \boldsymbol{F} = \boldsymbol{F}^{\text{ext}} \quad \text{with} \quad \text{b)}\ \boldsymbol{F}^{\text{ext}} \approx -\frac{\gamma_N M M_c^{(S)}}{r_c^3}\boldsymbol{r}_c \tag{112}$$

(M mass of the rigid body, $M_c^{(S)}$ gravitative mass of the external body, $r_c = |\boldsymbol{r}_c|$).

Evaluating the above result (98) and applying the findings here obtained, we get for the bremsheat production of a rigid body with translational and rotational motion

$$P_B = \left(\frac{\gamma_N M M_c^{(S)}}{R_S} + \sum_{a,b=1}^{3} I_{ab}\omega_a\omega_b \right) \frac{\mathrm{d}\ln\sigma}{\mathrm{d}t} \tag{113}$$

and hence for a sphere (r_0 radius of the sphere)

$$P_B = M \left(\frac{\gamma_N M_c^{(S)}}{R_S} + \frac{2}{5} r_0^2 \omega^2 \right) \frac{\mathrm{d}\ln\sigma}{\mathrm{d}t} \tag{114}$$

with

$$I = \frac{2}{5} M r_0^2 \tag{115}$$

It is convenient to define the ratio of the orbital to the rotational effect:

$$S_B = \frac{5\gamma_N M_c^{(S)}}{2R_S r_0^2 \omega^2} \, . \tag{116}$$

For a binary system the bremsheat production is given by ($\boldsymbol{r} = \boldsymbol{r}_1 - \boldsymbol{r}_2$ relative radius vector)

$$\text{a)} \quad P_B = M^{\text{red}} \dot{\boldsymbol{r}}^2 \frac{d \ln \sigma}{dt} \quad \text{with} \quad \text{b)} \quad M^{\text{red}} = \frac{M_1 M_2}{M_1 + M_2} \text{ (reduced mass)} \, . \tag{117}$$

8.2. NUMERICAL EVALUATION

Here we evaluate the formulas (114) and (116) for the present time. Remembering numerical results of (51) and referring to Newton's mass interpretation in this context

$$\text{a)} \quad M^{(N)} = \mathcal{M}\sigma_p \, , \quad \text{b)} \; M_c^{(N)} = \mathcal{M}_c \sigma_p^{(S)} , \tag{118}$$

we obtain

$$\text{a)} \quad P_B = 5.03 \cdot 10^{-19} \, M^{(N)} \left(\frac{\gamma_N M_c^{(N)}}{R_S} + 0.4 \, r_0^2 \omega^2 \right) \text{s}^{-1} \, , \tag{119}$$

$$\text{b)} \quad S_B = 1.67 \cdot 10^{-7} \frac{M_c^{(N)}}{R_S r_0^2 \omega^2} \text{g}^{-1} \, \text{cm}^3 \, \text{s}^{-2} .$$

There is not space enough here to list the used astronomical values for masses, distances, radii, angular velocities etc. For details one should consult special literature [9], [10] and [11]. Therefore, in order to give a numerical impression we only list some final results with respect to considerations on volcanicity and seismicity induced by the cosmological expansion.

Earth (as center):

Moon:

$P_B = 3.8 \times 10^{17}$erg s^{-1}, $S_B = 1.2 \times 10^5$ (orbital effect prevailing).
Empirical value: $P_{\text{emp}} \approx 10^{19}$erg s^{-1}.

Satellite (mass: 10^6 g, height over the earth: 200 km):

$P_B = 0.3$ erg s^{-1}.

Jupiter (as center):

Moon Io:

$P_B = 1.4 \times 10^{20}$ erg s^{-1}.

Moon Kallisto:

$P_B = 3.6 \times 10^{19}$ erg s^{-1}.

Sun (as center):

Earth:
$P_B = 2.7 \times 10^{22}$ erg s^{-1}, $S_B = 10^4$ (orbital effect prevailing).
Empirical value: $P_{\text{emp}} = 2.5 \times 10^{20}$ erg s^{-1}.

Jupiter:
$P_B = 2.3 \times 10^{24}$ erg s^{-1}, $S_B = 2.4$ (orbital effect slightly prevailing).
Empirical value: $P_{\text{emp}} = 6.4 \times 10^{24}$erg s^{-1}.

Saturn:
$P_B = 3.8 \times 10^{23}$ erg s^{-1}, $S_B = 2.4$ (orbital effect slightly prevailing).
Empirical value: $P_{\text{emp}} = 4.6 \times 10^{23}$ erg s^{-1}.

Center of the galaxy:

Sun:
$P_B = 4.9 \times 10^{29}$ erg s^{-1}, $S_B = 3 \times 10^4$ (orbital effect prevailing).
Empirical value (solar constant): $P_{\text{emp}} = L_\odot = 3.847 \times 10^{33}$ erg s^{-1} (measured effect explained by nuclear fusion).

White dwarves (mean values):
$P_B = 2.7 \times 10^{29}$ erg s^{-1} (orbital effect).
Empirical value: $P_{\text{emp}} \approx 10^{-4} L_\odot = 3.847 \times 10^{29}$ erg s^{-1}.

Sagittarius *:
If we take the empirical values [11]: $M^{(N)} = 5.19 \times 10^{39}$ (mass), $L = 10^6 L_\odot$ (luminosity), we arrive at ($L \to P_B$)
$P_B = \frac{1}{R_S} \times 3.1 \times 10^{58}$ g cm^3 s^{-3} (orbital effect). Comparing both last formulas gives the radius of the orbit $R_S = 8.1 \times 10^{18}$ cm .

Sagittarius * (as center):

Object S1:
Tentatively using for the mass the value of the mass of the sun and further from the references quoted above for the radius of the orbit $R_S = 1.8 \times 10^{16}$ cm, we find $P_B = 1.9 \times 10^{31}$ erg s^{-1} $\approx 5 \times 10^{-3} L_\odot$ (orbital effect).

Rotating body:

Let us finally treat the separate rotational effect contained in (119a). If we take into account the stability at the equator of the rotating body ($M_c^{(N)} \to M^{(N)}$),

$$(r_0 \omega)^2 \approx \frac{\gamma_N M^{(N)}}{r_0}, \tag{120}$$

we arrive at

$$P_B = 1.34 \cdot 10^{-26} \frac{(M^{(N)})^2}{r_0} \text{ g}^{-1} \text{ cm}^3 \text{ s}^{-3}. \tag{121}$$

For illustration numerical values of the above quantities were recently calculated for specific data of pulsars and quasars. For the special pulsar values $M^{(N)} =$

$1.4 \cdot M_\odot$ (mass) and $r_0 = 30$ km (radius) we obtained $P_B = 9L_\odot$ (bremsheat production) and $\nu = 3.5 \cdot 10^2\ \mathrm{s}^{-1}$ (rotational frequency). For the special quasar values $M^{(N)} = 10^8 \cdot M_\odot$ (mass) and $r_0 = 10^{16}$ cm (radius) follows $P_B = 1.38 \cdot 10^7 L_\odot$ and $\nu = 0.58\ \mathrm{y}^{-1}$.

In order to avoid misunderstandings we emphasize: By presenting the numerical evaluation of some theoretical conclusions of our theory we do not claim that the mechanisms of astrophysical effects considered here explain the full empirical experience. For illumination of this situation a further detailed study of the subject is essential. Nevertheless our predicted bremsheat production could be interesting for the energy balance of astrophysical bodies, e.g.with respect to volcanicity, plate tectonics and seismicity.

9. Thermal expansion of a sphere-like cosmic object

9.1. GENERAL THEORY

The idea of this section is to study the thermal heat expansion of a sphere-like cosmic object induced by the bremsheat production investigated in the previous section [12].

As it is well known, the differential formula for the linear thermal dilation of a rod is given by

$$d\ell = \alpha_\ell \ell dT \tag{122}$$

(ℓ length of the rod, T temperature, α_ℓ linear thermal coefficient of dilation). Using the formula (V volume of the sphere with the radius r)

$$\text{a)}\quad V = \frac{4\pi}{3} r^3 \quad \text{with} \quad \text{b)}\ \mathrm{dV} = 4\pi \mathrm{r}^2 \mathrm{dr}\ , \tag{123}$$

one finds for the differential dilations

$$\text{a)}\ \mathrm{d}r = \alpha_\ell\, r\, \mathrm{d}T, \quad \text{b)}\ \mathrm{d}V = \alpha_c V \mathrm{d}T, \tag{124}$$

where $\alpha_c = 3\alpha_\ell$ is the cubic thermal coefficient of dilation.

For later application we need the relationship between the heat content Q and the temperature T of a body of the mass M. The well-known formula reads (c_Q specific heat)

$$\frac{dQ}{dt} = c_Q M \frac{dT}{dt}\ . \tag{125}$$

Let us now perform a rough global treatment of the problem posed in order the get some estimates on the expansion process of the body considered.

Our basic idea is such that the scalaric cosmological bremsheat production leads to two main effects: release of thermal radiation and increase of the temperature of the body, i.e. expansion of the body. For this reason we physically identify up to a free individual factor f_S the quantities (125) and (98):

$$\frac{dQ}{dt} = -f_S \frac{dQ_B}{dt}. \tag{126}$$

The "heat consumption factor" f_S describes the portion of the scalaric bremsheat which is used for the dilation of the volume of the body. Then from (125) we get

$$\frac{\mathrm{d}T}{\mathrm{d}t} = -\frac{f_S}{Mc_Q}\frac{\mathrm{d}Q_B}{\mathrm{d}t}. \tag{127}$$

Eliminating by means of (124a) leads to the temporal change of the radius of the sphere:

$$\frac{\mathrm{d}r}{\mathrm{d}t} = -\frac{f_S\,\alpha_\ell\, r}{Mc_Q}\frac{\mathrm{d}Q_B}{\mathrm{d}t}. \tag{128}$$

For later application it is convenient to introduce the "scalaric material factor" of the body

$$\Sigma_S = \frac{f_S\,\alpha_\ell}{c_Q}. \tag{129}$$

Then by means of (98)and (101) we get from (128)

$$\frac{\mathrm{d}r}{\mathrm{d}t} = \frac{\Sigma_S\, r}{M}\left[M\dot{\boldsymbol{r}}_c^2 + \sum_{a,b=1}^{3} I_{ab}\,\omega_a\omega_b\right]\frac{\mathrm{d}\ln\sigma}{\mathrm{d}t}. \tag{130}$$

If the translational effect considerably dominates the rotational one, which for cosmic bodies within our planetary system is the case, the latter formula reduces to

$$\frac{\mathrm{d}r}{\mathrm{d}t} = \Sigma_S r\dot{\boldsymbol{r}}_c^2\frac{\mathrm{d}\ln\sigma}{\mathrm{d}t}. \tag{131}$$

From (130) or (131) we learn that the relative scalaric thermal expansion effect of a body is independent of the mass of the body, i.e. this effect is a universal cosmological effect.

9.2. THERMAL EXPANSION OF A BODY ORBITING ABOUT A CENTRAL MASS

Now we apply the result (131) to a body moving approximately on a circular orbit with the radius R_S. According to our theory in this case the formula

$$\dot{\boldsymbol{r}}_c^2 = \frac{\gamma_N M_c^{(S)}}{R_S} \tag{132}$$

is valid, where $M_c^{(S)}$ (72) is the gravitative mass of the central body. Eliminating in (131) gives the general relation

$$\frac{\mathrm{d}r}{\mathrm{d}t} = \frac{\gamma_N M_c^{(S)}\Sigma_S\, r}{R_S}\frac{\mathrm{d}\ln\sigma}{\mathrm{d}t}. \tag{133}$$

One should remember that for the present time $M_{c|p}^{(S)} = M_c^{(N)}$ (usual Newtonian mass) is valid.

To investigate the long-time course of the radius of the spherical cosmic object we have to integrate (133). For this reason we first rearrange it by means of (72):

$$\frac{\mathrm{d}\ln r}{\mathrm{d}t} = \Sigma_S \left(\frac{\gamma_N \mathcal{M}_c}{F_0}\right)^2 \frac{(\sigma^2-1)^2}{\sigma} \frac{\mathrm{d}\sigma}{\mathrm{d}t}. \tag{134}$$

Referring to the case of validity of the inequality $\sigma^2 \gg 1$ being fulfilled for the time of about 1000 y after the big start and later, i.e. valid at the time of birth of cosmic objects. Therefore we are allowed to go over from (134) to

$$\frac{\mathrm{d}\ln r}{\mathrm{d}t} = \frac{1}{4}\,\Sigma_S \left(\frac{\gamma_N \mathcal{M}_c}{F_0}\right)^2 \frac{\mathrm{d}\sigma^4}{\mathrm{d}t}. \tag{135}$$

Now it is convenient to introduce the "scalaric cosmic expansion factor"

$$E_S = \frac{1}{4}\,\Sigma_S \left(\frac{\gamma_N \mathcal{M}_c}{F_0}\right)^2 . \tag{136}$$

Then (135) reads

$$\frac{\mathrm{d}\ln r}{\mathrm{d}t} = E_S \frac{\mathrm{d}\sigma^4}{\mathrm{d}t}\,. \tag{137}$$

Let us further assume that this factor may be a constant: E_S=const.

Now we are able to integrate with the result

$$r = \bar{r}\exp\left(E_S\,\sigma^4\right) \tag{138}$$

($\bar{r}$ constant of integration). For practical use it is comfortable to write this equation in the form

$$r = r_b \exp\left[E_S\left(\sigma^4 - \sigma_b^4\right)\right], \tag{139}$$

where

$$\text{a)}\quad r_b = r(t=t_b) \quad \text{and} \quad \text{b)}\quad \sigma_b = \sigma(t=t_b) \tag{140}$$

mean the values of the radius of the sphere and of the scalaric world function at the time of birth of the cosmic object considered (index b refers to birth). By rearranging (137) we find the corresponding formula for the expansion velocity of the sphere:

$$\frac{\mathrm{d}r}{\mathrm{d}t} = 4\,E_S\, r\, \sigma^3 \frac{\mathrm{d}\sigma}{\mathrm{d}t}\,. \tag{141}$$

9.3. SOME GLOBAL NUMERICAL RESULTS FOR AN EARTH-LIKE MODEL

Let us start with some historical annotations: For a long time geophysicists have discussed the history of the evolution of the earth, particularly also the problem of contraction or expansion. In the first period of research the opinion of a contracting earth strongly dominated. In context with A. Wegener's hypothesis of continental drift (1915) the idea of an expanding earth won followers as O. C. Hilgenberg (1933), V. B. Nejman (1962), S. W. Carey (1976), J. Koziar (1980), N. J. Osipisin and V. F. Blinov (1987), G. O. W. Kremp (1992) and many others. Nevertheless in spite of all scientific activities the supporters of this concept form a small minority.

With respect to a theoretical foundation of the concept of an expanding earth we remind of P. Jordan (1955) who followed P. A. M. Dirac's idea (1938) of a decreasing "gravitational constant" in the course of time. The theories of Jordan and some other theoreticians came in contradiction particularly to the Einstein effects within our planetary system. The reason for our present treatise is to show that our PUFT, which is in agreement with the Einstein effects, leads to some predictions which could be interesting for astrophysicists and geophysicists. Therefore we hope that our new astrophysical and geophysical effects (if they exist at all) will be taken into account among the large variety of effects being under investigation.

In comparing our results with the geophysical experience we get some help by the empirical exponential Koziar formula (1994) on the expansion of the earth [13]:

$$r = r_1 + r_2\,\mathrm{e}^{\varrho t}, \tag{142}$$

where $r_1 = 2800$ km, $r_2 = 3570$ km, $\varrho = 7.25 \cdot 10^{-9}$ y^{-1}. From this formula the present velocity of expansion of the earth (surface)

$$\left(\frac{\mathrm{d}r}{\mathrm{d}t}\right)_p = 2.6 \text{ cm y}^{-1} \tag{143}$$

results. But we mention that other geophysicists (e.g. L. Egyed) take into consideration the considerably smaller value

$$\left(\frac{\mathrm{d}r}{\mathrm{d}t}\right)_p \approx 5\,(10^{-2} \text{ to } 10^{-1}) \text{ cm y}^{-1}\,. \tag{144}$$

Let us in the sense of a rough estimate now exploit the formulas (139) and (141). First we are able to calculate the quantity $(\gamma_N \mathcal{M}_c/F_0)^2$, since we know the data

$$\text{a)}\quad M_c^{(N)} = 1.989 \cdot 10^{33} \text{ g} \quad \text{(mass of the sun)}\,, \tag{145}$$

$$\text{b)}\ |\,\boldsymbol{r} \times \dot{\boldsymbol{r}}\,| = 4.46 \cdot 10^{19} \text{ cm}^2 \text{ s}^{-1} \quad \text{(areal velocity of the earth)}\,.$$

Remembering the values $\sigma_p = 59.7731$ and $(\frac{\mathrm{d}\sigma}{\mathrm{d}t})_p = 9.48 \cdot 10^{-10}$ y^{-1} (51), with the help of (145a) and (72) we find

$$\text{a)}\quad \mathcal{M}_c = 3.3 \cdot 10^{31} \text{ g}\,, \quad \text{b)}\quad \mathrm{F}_0 = 2.66 \cdot 10^{21} \text{ cm}^2 \text{ s}^{-1}. \tag{146}$$

Thus we arrive at

$$\left(\frac{\gamma_N \mathcal{M}_c}{F_0}\right)^2 = 6.96 \cdot 10^5 \text{ cm}^2 \text{ s}^{-2}. \tag{147}$$

Applying this result to (136) we get

$$E_S = 1.74 \cdot \Sigma_S \cdot 10^5 \text{ cm}^2 \text{ s}^{-2}. \tag{148}$$

Decisive for the following is now some knowledge about the parameter Σ_S (129) in which the linear thermal coefficient of dilation α_ℓ and the specific heat c_Q occur.

In new geophysical literature on the earth one finds the average values [14]

$$\text{a)} \quad \alpha_\ell = 2.28 \cdot 10^{-5} \text{ K}^{-1}, \quad \text{b)} \quad c_Q = 1.5 \cdot 10^7 \text{ cm}^2\text{s}^{-2}\text{K}^{-1}. \tag{149}$$

If we use for the heat consumption factor the rough value $f_S = 0.5$, we obtain from (129)

$$\Sigma_S = 7.6 \cdot 10^{-13} \text{ cm}^{-2}\text{s}^2 \tag{150}$$

and further from (148)

$$E_S = 1.32 \cdot 10^{-7} . \tag{151}$$

If we take for the age of the earth (with an open value of the radius at birth) the value $(\triangle t)_e = 4.6 \cdot 10^9$ y, then the birth of the earth took place in $t_b = 14.4 \cdot 10^9$ y after the big start (Urstart). For the scalaric world function at this time the value $\sigma_b = 51.64$ and ${\sigma_b}^4 = 7.11 \cdot 10^6$ (numerical results) holds. Applying these results to (139) we arrive at

$$\text{a)} \quad \frac{r}{r_b} = \exp\left[1.32 \cdot 10^{-7}\left(\sigma^4 - 7.11 \cdot 10^6\right)\right], \quad \text{b)} \quad \frac{r_p}{r_b} = 2.12. \tag{152}$$

Let us next apply (141) to the present time:

$$\left(\frac{dr}{dt}\right)_p = 4E_S\, r_p\, \sigma_p^3 \left(\frac{d\sigma}{dt}\right)_p . \tag{153}$$

Inserting the values $r_p = 6371$ km (present radius of the earth) and further the result (151), we get for the expansion velocity of the earth (surface)

$$\left(\frac{dr}{dt}\right)_p = 6.8 \cdot 10^{-2} \text{ cm y}^{-1}. \tag{154}$$

Thanks to the immense progress in superconductivity and laser physics nowadays such an extreme precision in length measuring technique was reached that it is possible to measure an expansion of the earth of about 1 cm per year and better. We hope that this technical development can help to decide soon the question after the expansion of the earth.

Similar calculations have been performed for moon, Mars and Jupiter.

For plotting of the temporal course of the radius and the radial velocity we annotate: The abscissa of the figures presented in the following refers to the rescaled time $\eta = 0.945 \cdot 10^{-9} t\, \text{y}^{-1}$. As mentioned above this practically means that the unit $\Delta\eta = 1$ roughly corresponds to 10^9 y. Referring to the figures for the radius,

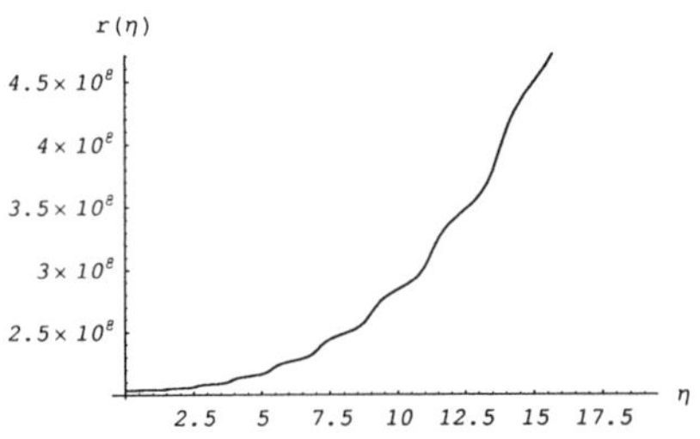

Figure 5. Temporal course of the radius of an earth-like cosmic object

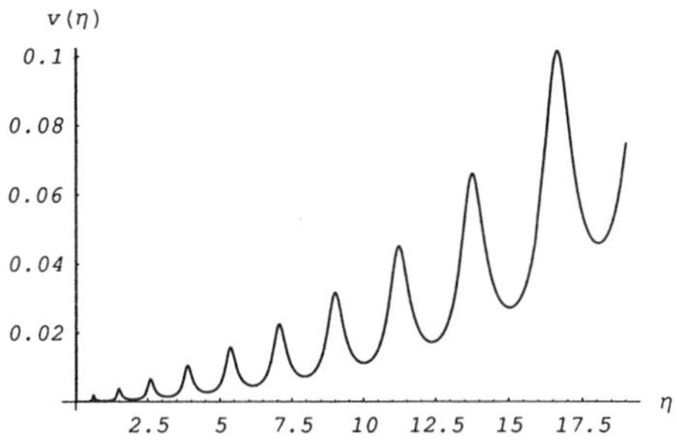

Figure 6. Temporal course of the radial velocity of an earth-like cosmic object

the unit of the radius at the ordinate is given in cm; referring to the figures for the velocity, the unit of the velocity at the ordinate is $\mathrm{cm/y}$.

The temporal course of the radius of the spherical body is presented in Fig. 5 for the whole time scale since the big start. The temporal course of the radial velocity of the surface of the spherical body is presented in Fig. 6 which shows the course of the velocity for the whole time scale since the big start.

10. Some global numerical results for a star-like cosmic object (Sun as a model)

10.1. PRELIMINARY CONSIDERATIONS

These days it is generally recognized in science that, as the Sun burns through its hydrogen on the main sequence, it steadily grows hotter and therefore more luminous. It is also accepted that through this fact the Sun continuously grows larger. Recently an interesting paper on this subject has been distributed by Korycansky et al. [15].

For rough application of the theory presented above we refer to approximate numerical values of the Sun as a numerical model (further short: sun) orbiting about the center of the Galaxy):

$$\begin{array}{llll} \text{a)} & R & = & 2.6 \cdot 10^{22}\,\mathrm{cm} & \text{(radius of the orbit),} \\ \text{b)} & V & = & 2.2 \cdot 10^{7}\,\mathrm{cm\,s^{-1}} & \text{(orbital velocity of the sun),} \\ \text{c)} & M_c & = & 1.8 \cdot 10^{44}\,\mathrm{g} & \text{(acting mass of the Galaxy),} \\ \text{d)} & r_p & = & 6.96 \cdot 10^{10}\,\mathrm{cm} & \text{(present radius of the sun),} \\ \text{e)} & t_{sun} & = & 4.66 \cdot 10^{9}\,\mathrm{y} & \text{(age of the sun).} \end{array} \tag{155}$$

Let us mention that the value of the central mass acting at the place of the sun resulted from our theory of the rotation curve [16]. Obviously this value is different from the usually accepted value of the mass of the whole Galaxy which is somewhat greater.

By means of the above values we find for the sun

$$F_0 = 3.46 \cdot 10^{31}\,\mathrm{cm}^2\,\mathrm{s}^{-1}\,. \tag{156}$$

The main problem arising now is to find a way to get information on the scalaric expansion factor E_S of the sun. For comparing it is appropriate to remember the results obtained for planet-like cosmic bodies (151).

The situation with respect to the sun is insofar qualitatively different from the situation just mentioned, since the interior of the sun consists of gas (mainly H, partly atomic burning to He) with a very high temperature. For simplicity we approximately treat this problem by referring to a perfect gas for which the relation

$$\overline{\varepsilon} = \frac{m_0}{2}\,\overline{u^2} = \frac{3\mathrm{k}T}{2} \tag{157}$$

is valid (m_0 mass of a gas particle, u velocity of the gas particles subjected to averaging, k Boltzmann constant, T kinetic temperature). As it is well known, for a perfect gas the formulas

$$\text{a)}\quad \alpha_c \;=\; \frac{1}{T} = \frac{3\mathrm{k}}{m_0\overline{u^2}}\,,\qquad \text{b)}\quad c_Q \;=\; \frac{3\mathrm{k}}{2m_0} \tag{158}$$

hold. Hence the relations (129) and (136) take the form

$$\Sigma_S = \frac{2f_S m_0}{3\mathrm{k}T} \tag{159}$$

and

$$E_S = \frac{f_S m_0}{6\mathrm{k}T}\left(\frac{\gamma_N M_c}{\sigma F_0}\right)^2 . \tag{160}$$

Using the numerical values (155) and (156) for the sun, we find

$$\text{a)}\quad \left(\frac{\gamma_N M_c}{\sigma F_0}\right)^2 = 3.38 \cdot 10^7\,\mathrm{cm}^2\,\mathrm{s}^{-2} \quad \text{and} \quad \text{b)}\quad \Sigma_S = 8.12 \cdot 10^{-16}\,\mathrm{cm}^{-2}\,\mathrm{s}^2. \tag{161}$$

As mentioned above, it is our aim to treat cosmic bodies globally. Therefore it seems to be acceptable to take the value $T = 5 \cdot 10^6$ K as an appropriate mean value of the kinetic temperature of the sun. Let us further remember the value of the Boltzmann constant $\mathrm{k} = 1.38 \cdot 10^{-16}\,\mathrm{g\,cm^2 s^{-2}\,K^{-1}}$ and the value of the proton mass $m_p = 1.68 \cdot 10^{-24}$ g (being used as the mass of the H-atom constituent of the gas considered). Hence we obtain from (160) the numerical result ($f_S = 1$)

$$E_S = 6.865 \cdot 10^{-9}\,. \tag{162}$$

10.2. NUMERICAL EVALUATION AND PLOTTING FOR THE SUN (AS A MODEL)

10.2.1. *Numerical formulas for radius and expansion velocity*

Inserting the values (162), (155d) and (51f, g) into formula (153), we find for the present expansion radial velocity of the surface of the sun

$$\left(\frac{\mathrm{d}r}{\mathrm{d}t}\right)_p = 0.387\,\mathrm{cm\,y}^{-1}\,. \tag{163}$$

Further we get information from formula (139) on the temporal behavior of the radius of the sun. Inserting (162) into this equation we arrive at

$$r_p = r_b \exp[6.865 \cdot 10^{-9}(\sigma_p^4 - \sigma_b^4)]\,. \tag{164}$$

Now from (155d) we know the quantity r_p and from (51f) the quantity σ_p. As mentioned above, there is a good consensus among astrophysicists that the age of the sun is about 4.66 billion years. Therefore we need the value of σ at the time of birth of the sun. Since the ages of the sun and the earth approximately coincide, we can take over the values applied above for the earth: $\sigma_b = \sigma(t = t_b) = 51.64$ and $\sigma_b^4 = 7.11 \cdot 10^6$.

Now we are prepared to calculate the radius of the sun at the birth with the result

$$r_b = 6.695 \cdot 10^{10}\mathrm{cm}\,. \tag{165}$$

10.2.2. *Annotation to the mass of the Galaxy*

In investigating the orbital motion of the sun about the central body of the Galaxy, for physical reasons it is necessary to use the gravitational force just at the place of the moving sun. Because of the matter (particularly dark matter) outside the sun it is obvious that the mass causing the gravitational force mentioned is not the total mass of the Galaxy. We recently treated the problem of the radial mass distribution in the Galaxy within the framework of our theory of the rotation curve of stars [16]. In that paper we came to the numerical value (155c) of the gravitationally acting mass at the place of the sun. All numerical calculations performed above are based on this value.

From formula (160) we learn that the scalaric cosmic expansion factor E_S contains M_c quadratically. That means under the circumstances of the exponential character of the expansion law (139) an extraordinary sensitivity of the radius on the mass M_c.

Let us conclude these considerations with the general remark that the curves show a quasi-continuous course through the whole time scale from near the big start to the presence, i.e. there is no hint at the time of birth of the cosmic object considered. All in all it seems that this theory corresponds to a evolution of the star-like cosmic objects from "small to large" and not according to the accretion concept from "large to small".

11. Theory of dark matter accretion around a center

Empirical astrophysical estimates of the visible mass density of the normal (mostly baryonic) matter come to the maximum value $\approx 10^{-30}$ $\mathrm{g\,cm^{-3}}$. We interpreted our cosmological result shown above with respect to this difference in the mass densities as a hint at the existence of a dominating dark matter part with a mass density more than two orders of magnitude greater than that of the normal matter. Our theory of dark matter accretion around a gravitating central body is presented in paper [17].

Our hypothetical model of the cosmos investigated, roughly corresponds to the following picture of the present cosmos: Our cosmos consists of a gas of dark matter particles (dm-particles) penetrating all matter of the cosmos, particularly also the existing compact objects (stars, nuclei of galaxies etc.) which look like buoys in the dark matter sea. Assuming a homogeneous gas of one sort of particles, by some hypothetical arguments we recently were led to the following hypothetical numerical values:

$$\begin{aligned} \text{a) } m_p &= 2\cdot 10^{-36}\mathrm{g} = 1.13\cdot 10^{-3}\mathrm{eV/c^2} \quad \text{(rest mass of the dm-particle)}, \\ \text{b) } T_p &= 1.7\,\mathrm{K} \quad \text{(kinetic temperature of the dm-gas)}. \end{aligned} \tag{166}$$

As a result of the attractive interaction of the dm-particles with the compact object considered we expect a statistical distribution of the dm-particles within and around the compact object being embedded in this dm-gas. Let us introduce the notion "accretion cloud" (Umwolke) for this part of the dm-gas which by accretion exceeds the cosmological average distribution of the dm-gas. For simplicity we model the compact object by a homogeneous mass sphere. Further for simplicity we apply the Boltzmann-Maxwell statistics describing spinless particles without degeneracy.

Let us here mention that in a different context and on a different theoretical basis H. Dehnen et al. [18] used the Bose-Einstein statistics (preferred by physical arguments) and the Fermi-Dirac statistics (dropped) in treating the dark matter in galaxies in order to find a theoretical concept in understanding the problem of the "flat rotation curves" of stars in galaxies. In interesting papers detailed quantum-statistical calculations were performed. We think that for a rough understanding of our approach it is legitimate to simplify the situation by referring to the Boltzmann-Maxwell statistics.

11.1. BASIC EQUATIONS

In the subsection 5.2 we treated the non-relativistic weak field approximation of the PUFT field equations for a point-like central body. In the following these field equations are specialized to the non-relativistic and weak field case for a continuous matter distribution, particularly around a sphere-like body. Then the gravitational field equation is given by

$$\Delta\Phi = 4\pi\gamma_N \mu \tag{167}$$

($\Phi(\boldsymbol{r}, t)$ local Newtonian gravitational field, $\mu(\boldsymbol{r}, t)$ mass density).

In this approximation the scalaric field may consist of two parts according to

$$\text{a)} \quad \sigma = \sigma_c + s \quad \text{with} \quad \text{b)}\ |s| \ll |\sigma_c| \,, \tag{168}$$

where $\sigma_c(t)$ is the global scalaric cosmological field and $s(\boldsymbol{r}, t)$ the local scalaric field of the central body. Then the scalaric field equation reads

$$\Delta\sigma = \Delta s = -\frac{4\pi\gamma_N \mu}{\sigma_c c^2}. \tag{169}$$

In the general case of a test body moving in the external fields Φ and s as well as in eventually existing external electromagnetic fields (neglecting in all fields the back reaction) the equation of motion reads:

$$M\left(\frac{\mathrm{d}\boldsymbol{v}}{\mathrm{d}t} + \operatorname{grad}\Phi + \frac{c^2}{\sigma_c}\operatorname{grad} s + \boldsymbol{v}\frac{\mathrm{d}\ln\sigma}{\mathrm{d}t}\right) = Q\left(\boldsymbol{E} + \frac{\boldsymbol{v}\times\boldsymbol{B}}{c}\right). \tag{170}$$

The special case of a point-like central body has been presented in equation (70).

In the following we refer to a test body without electric charge, i.e. $Q = 0$. Because of the same mathematical structure of both the equations (169) and (167), in this approximation it is possible to introduce the combined field function

$$\Psi = \Phi\left(1 - \frac{1}{\sigma_c^2}\right) + c^2 \ln\sigma_c + \frac{1}{\sigma_c^2}\Phi_c \tag{171}$$

($\Phi_c(t)$ cosmological value of the Newtonian potential). Hence we obtain the common field equation

$$\Delta\Psi = 4\pi\gamma_N\left(1 - \frac{1}{\sigma_c^2}\right)\mu \tag{172}$$

and the equation of motion

$$\frac{\mathrm{d}\boldsymbol{v}}{\mathrm{d}t} + \operatorname{grad}\Psi + \boldsymbol{v}\frac{\mathrm{d}\ln\sigma}{\mathrm{d}t} = 0. \tag{173}$$

Let us now use following notation:

$$\mu = \mu_n + \mu_{dm} \tag{174}$$

(μ_n mass density of normal matter, μ_{dm} mass density of dark matter). Application of this latter formula to the interior (i) and to the exterior (e) of the spherical central body leads to the equations

$$\text{a)} \quad \mu_i = \mu_{ni} + \mu_{dmi} \quad \text{and} \quad \text{b)} \quad \mu_e = \mu_{ne} + \mu_{dme}\,. \tag{175}$$

Next we split the mass densities into both the parts coming from the global cosmological matter (normal matter and dm-matter) and from the local matter (central body and accretion cloud):

$$\text{a)}\quad \mu_{ni} = \mu_{ng} + \mu_{ns}, \qquad \text{b)}\quad \mu_{dmi} = \mu_{dmg} + \hat{\mu}_{dmi} \tag{176}$$

and

$$\text{a)}\quad \mu_{ne} = \mu_{ng}\,, \qquad \text{b)}\quad \mu_{dme} = \mu_{dmg} + \hat{\mu}_{dme}\,. \tag{177}$$

Here the indices refer to: $g \to$ global (cosmological contribution) and $s \to$ sphere (central body contribution). The roof denotes the dm- contribution induced by the central body, i.e. surplus dark matter to be treated statistically.

For our further calculations it is convenient to introduce the abbreviations

$$\begin{aligned} &\text{a)}\quad \widetilde{\mu}_e = \mu_G + \hat{\mu}_{dme} \quad \text{with} \quad \text{b)}\quad \mu_G = \mu_{ng} + \mu_{dmg}, \quad \text{i.e.} \\ &\text{c)}\quad \mu_i = \mu_G + \mu_{ns} + \hat{\mu}_{dmi}\,. \end{aligned} \tag{178}$$

Let us remind that our cosmological model led us to following numerical values of the scalaric cosmological world function:

$$\begin{aligned} &\text{a)}\quad \sigma_c(t \approx 1000\ \text{y}) \approx 3.3\,, \quad \text{b)}\quad \sigma_{cp} = 59.7731\,, \quad \text{i.e.} \\ &\text{c)}\quad \sigma_c^2 \gg 1 \quad \text{(for the time scale under consideration).} \end{aligned} \tag{179}$$

Then the field equation (172) for the interior and exterior reads:

$$\text{a)}\quad \Delta\Psi_i = 4\pi\gamma_N\mu_i, \quad \text{b)}\quad \Delta\Psi_e = 4\pi\gamma_N\widetilde{\mu}_e. \tag{180}$$

Further from (171) results

$$\Psi = \Phi + c^2 \ln \sigma_c\,. \tag{181}$$

Next, analogously to the above formulas for the mass density of the dm-particles we now write down the corresponding formulas for the particle number density n_{dm} being related to the mass density as follows (m mass of a dm-particle):

$$\text{a)}\quad \mu_{dm} = m n_{dm}\,, \quad \text{b)}\quad \mu_{dmg} = m n_{dmg}\,, \quad \text{c)}\quad \hat{\mu}_{dm} = m\hat{n}_{dm}\,. \tag{182}$$

We find

$$n_{dm} = n_{dmg} + \hat{n}_{dm}\,. \tag{183}$$

In (182) and (183) we omitted the indices i and e.

Since the time dependence of the cosmological quantities (e.g. of $\sigma_c(t)$) is extremely slow (adiabatic time dependence), compared with the time dependence of the cosmogonic processes of the cosmic objects, we omit explicit writing of the time t in such cosmological quantities. Particularly constants of spatial integration implicitly involve this adiabatic time dependence.

11.2. STATISTICAL DM-PARTICLE DISTRIBUTION IN THE ACCRETION CLOUD

According to our accretion concept the statistical dm-particle number density for the case of spherical symmetry reads:

$$n \equiv \hat{n}_{dm} = \bar{n}\left[\exp\left(-\frac{m\chi(r)}{\mathrm{k}T}\right) - 1\right], \qquad (184)$$

where $\bar{n}$ denotes the dm-particle number density far away from the cental body:

$$\text{a)}\quad \bar{n} = n(r=\infty) \quad \text{with} \quad \text{b)}\quad \chi_\infty \equiv \chi_e(r=\infty) = 0\,. \qquad (185)$$

The potential function χ used is related to Ψ by

$$\text{a)}\quad \Psi = \chi + \Psi_c\,, \quad \text{where} \quad \text{b)}\quad \Psi_c = \Phi_c + c^2 \ln \sigma_c\,. \qquad (186)$$

Using (184) we can write the equations (180) as

$$\begin{aligned} \text{a)}\quad \Delta\chi_i &= 4\pi\gamma_N\left[\mu_0 + m\bar{n}\left\{\exp(-\tfrac{m\chi_i}{\mathrm{k}T}) - 1\right\}\right], \\ \text{b)}\quad \Delta\chi_e &= 4\pi\gamma_N\left[\mu_G + m\bar{n}\left\{\exp(-\tfrac{m\chi_e}{\mathrm{k}T}) - 1\right\}\right], \end{aligned} \qquad (187)$$

where the quantity

$$\mu_0 = \mu_G + \mu_{ns} = \text{const} \qquad (188)$$

(referring to a homogeneous sphere with radius r_0) means the mass density of the sphere.

Let us for the following use the approximation assumption

$$\left|\frac{m\chi}{\mathrm{k}T}\right| \ll 1 \qquad (189)$$

being well fulfilled in our further applications. Hence series expansion of (184) leads to the linearity

$$n = -\frac{\bar{n}m\chi}{\mathrm{k}T} \qquad (190)$$

between n and χ. This simplification means that we are able to perform the further calculations analytically. Then the equations (187) commonly treated (omitting the indices i and e) read

$$\Delta\chi + \varkappa^2\chi = 4\pi\gamma_N\hat{\mu}\,, \qquad (191)$$

where $\hat{\mu}$ means μ_0 for the interior and μ_G for the exterior, and $\varkappa^2$ is defined by

$$\varkappa^2 = \frac{4\pi\gamma_N m^2\bar{n}}{\mathrm{k}T}\,. \qquad (192)$$

Let us here mention that the author's dissertation had referred to the theory of strong electrolytes, where according to the Debye-Milner theory the statistical treatment of a (negative) ion cloud around a fixed (positive) ion (and vice versa) plays an important role. This knowledge induced the author's idea to treat the dark matter gas analogously. But in the first moment he was surprised that, though attractive forces act between positive and negative electric charges as well as between masses, in his basic equation (191) the second term on the left hand side appeared with a positive sign, in contrast to the theory of strong electrolytes. The consequence of this fact is that in the Debye-Milner theory a Yukawa-like term with exponential decrease occurs, whereas in this theory an interesting phenomenon of periodicity is met.

In solving the equation (191) we first go over to spherical polar coordinates:

$$\frac{\mathrm{d}}{\mathrm{d}r}\left(r^2\frac{\mathrm{d}\chi}{\mathrm{d}r}\right) + \varkappa^2 r^2 \chi = 4\pi\gamma_N \hat{\mu} r^2 . \qquad (193)$$

In performing the further calculations being presented in detail in [17] we realized that following basic functions play an important role:

$$\text{a)}\quad U(z) = z\sin z + \cos z \quad \text{and} \quad \text{b)}\ V(z) = \sin z - z\cos z . \qquad (194)$$

Finally we arrived at following results:

$$\text{a)}\quad \chi_i = B_0\left\{1 - \frac{1}{\varkappa r}U(\varkappa r_0)\sin(\varkappa r)\right\},$$
$$\text{b)}\quad \frac{\mathrm{d}\chi_i}{\mathrm{d}r} = \frac{B_0}{\varkappa r^2}U(\varkappa r_0)V(\varkappa r); \qquad (195)$$

$$\text{a)}\quad \chi_e = -\frac{B_0}{\varkappa r}V(\varkappa r_0)\cos(\varkappa r),$$
$$\text{b)}\quad \frac{\mathrm{d}\chi_e}{\mathrm{d}r} = \frac{B_0}{\varkappa r^2}V(\varkappa r_0)U(\varkappa r), \qquad (196)$$

where

$$\text{a)}\quad B_0 = \frac{\mu_0 kT}{m^2\bar{n}} = \frac{3\gamma_N M_c}{\varkappa^2 r_0^3} \quad \text{with} \quad \text{b)}\ M_c = \frac{4\pi\mu_0 r_0^3}{3} \qquad (197)$$

(M_c mass of the sphere) holds.

11.3. RADIAL DRIFT OF A TEST BODY ("PIONEER EFFECT")

Recent evaluation of radio metric data from Pioneer 10/11, Galileo and Ulysses spacecraft by Anderson et al. [19] indicates an anomalous, constant acceleration acting on the spacecraft, being about 50 AU far from the Sun, with a magnitude $\approx 8.5\cdot 10^{-8}\mathrm{cm/s^2}$, directed towards the Sun. Apparently the explanation of this new effect seems to be related to a correction of Newtonian mechanics by additional terms. In the following we try to find an explanation of this radial drift effect by applying our theory.

The equation of motion of a test body (173) with the help of (186a) takes the form

$$\frac{d\boldsymbol{v}}{dt} + \operatorname{grad}\chi + \boldsymbol{v}\frac{d\ln\sigma_c}{dt} = 0\,. \tag{198}$$

From this equation we learn that for a spherically symmetric potential, as investigated above, the appearing acceleration A of the body considered can be split into the Newtonian part A_N and the scalaric part A_S according to

$$A = \frac{d\chi}{dr} = A_N + A_S\,. \tag{199}$$

Referring to the interior and exterior of the central body we identify

$$\text{a)}\quad A_{Ni} = \frac{\gamma_N M_c r}{r_0^3} \quad \text{and} \quad \text{b)}\quad A_{Ne} = \frac{\gamma_N M_c}{r^2} \tag{200}$$

(interior and exterior Newtonian accelerations). With the help of (195b) and (196b) the additional scalaric parts take the shape

$$A_{Si} = \frac{\gamma_N M_c \varkappa^2}{(\varkappa r_0)^3}\left[\frac{3U(\varkappa r_0)V(\varkappa r)}{(\varkappa r)^2} - \varkappa r\right] \approx \frac{\gamma_N M_c \varkappa^2 r}{2r_0}\left[1 - \frac{1}{5}\left(\frac{r}{r_0}\right)^2\right],$$

$$A_{Se} = \frac{\gamma_N M_c}{r^2}\left[\frac{3V(\varkappa r_0)U(\varkappa r)}{(\varkappa r_0)^3} - 1\right] \approx \frac{\gamma_N M_c \varkappa^2}{2}\left[1 - \frac{1}{4}(\varkappa r)^2\right]. \tag{201}$$

Taking into account that the series expansion applied is valid for $(\varkappa r)^2 \ll 1$, we compare the latter formula with the empirical facts mentioned above:

$$\frac{\gamma_N M_c \varkappa^2}{2}\left[1 - \frac{1}{4}(\varkappa r)^2\right] = 8.5 \cdot 10^{-8}\,\mathrm{cm\,s^{-2}}\,. \tag{202}$$

Inserting the well-known values of the Newtonian gravitational constant, the mean distance of the spacecraft considered from the Sun ($r = 50$ AU), and the mass of the Sun ($M_c = 1.989 \cdot 10^{33}$ g), we arrive at the value of $\varkappa$ for the Sun (1 AU $= 1.496 \cdot 10^{13}$ cm)

$$\varkappa = 3.58 \cdot 10^{-17}\,\mathrm{cm}^{-1} = 5.36 \cdot 10^{-4}(\mathrm{AU})^{-1}\,. \tag{203}$$

Though our theory predicts a rather constant value of the radial acceleration far away from the Sun, nevertheless this acceleration value decreases weakly at greater distance.

12. Rotation curves of stars in galaxies

In a recent paper we applied our dark matter accretion theory presented in the previous section to the orbital motion of the stars around the center of the Galaxy with the aim to find an explanation of the approximately flat rotation curves of the orbiting stars [16].

12.1. MOTION OF A TEST BODY IN POLAR COORDINATES

Starting point for our further studies is the equation of motion (198). Now we use the polar coordinates $\{R, \Phi\}$. Then after integration this equation reads (dot means derivative with respect to the time t)

$$\begin{aligned}
&\text{a)} \quad \ddot{R} - R\dot{\Phi}^2 + \frac{\mathrm{d}\chi}{\mathrm{d}R} + \dot{R}\Sigma = 0, \\
&\text{b)} \quad R^2\dot{\Phi}\sigma_c = \mid \boldsymbol{R} \times \dot{\boldsymbol{R}} \mid \sigma_c = F_0 \quad \text{with} \\
&\text{c)} \quad \Sigma = \frac{\mathrm{d}\ln\sigma_c}{\mathrm{d}t}.
\end{aligned} \tag{204}$$

As it is well known, the azimuthal velocity is defined by

$$v_\Phi = R\dot{\Phi} = \frac{F_0}{R\sigma_c}. \tag{205}$$

With the aid of this expression formula (204a) leads to

$$\ddot{R} - \frac{v_\Phi^2}{R} + \frac{\mathrm{d}\chi}{\mathrm{d}R} + \dot{R}\Sigma = 0\,. \tag{206}$$

Hence the equation

$$v_\Phi = \sqrt{R\ddot{R} + R\dot{R}\Sigma + R\frac{\mathrm{d}\chi}{\mathrm{d}R}} \tag{207}$$

follows. Now we introduce the abbreviation

$$A = \ddot{R} + \dot{R}\Sigma\,. \tag{208}$$

In the usual theory of the rotation curves on the basis of Newtonian mechanics approximately the motion of the body on a circle (R =const) is considered. We think this approximation is not good enough. Therefore, induced by the empirical situation for small values of the radial coordinate R we try the basic ansatz

$$A = \alpha R = 10^{-30} a R\,, \tag{209}$$

where α or a (the factor 10^{-30} is an adaption factor with respect to the large numerical astronomical values) are quantities to be specified by further concrete physical assumptions. Substituting this expression into (207) gives the formula for the so-called rotation curve $V = v_\Phi(R)$:

$$V = \sqrt{R\left(A + \frac{\mathrm{d}\chi}{\mathrm{d}R}\right)}. \tag{210}$$

Finishing this subsection we would like to emphasize that in our theory the situation is basically different to the usual theory of the rotation curves based on the assumption R =const, since here because of the periodicity occurring in the potential the expression $\frac{\mathrm{d}\chi}{\mathrm{d}R}$ is indefinite, i.e. that for $A = 0$ the radicand exhibits negative-valued regions.

12.2. ROTATION CURVE FOR THE EXTERIOR OF THE CENTRAL BODY IN GENERAL (AZIMUTHAL MOTION)

In our paper on rotation curves mentioned above [16] we treated the posed task in detail for the interior and exterior of the central body. Here we have to restrict to a short report for the exterior (index e). In this case with the help of (196b) and (209) equation (210) takes the shape

$$V_e = R\sqrt{\alpha_e + \frac{3\gamma_N M_c}{(\varkappa r_0)^3 R^3} V(\varkappa r_0) U(\varkappa R)}\,. \tag{211}$$

Analogously follows here the angular velocity

$$\omega_e = \frac{V_e}{R} = \sqrt{\alpha_e + \frac{3\gamma_N M_c}{(\varkappa r_0)^3 R^3} V(\varkappa r_0) U(\varkappa R)}\,. \tag{212}$$

12.3. PHYSICAL PROCESSES IN THE INTERIOR OF THE CENTRAL BODY (RADIAL MOTION)

Our next task is to get some information about the quantity A (208) by investigating physical main processes in the interior of the central body. We approach this problem tentatively by considering two opposite processes: 1. Expansion process of the central body by scalaric cosmological bremsheat production according to our hypothesis. 2. Scalaric-adiabatic cosmological contraction process according to our approximate treatment of the orbital motion of a test body considered.

12.3.1. *Expansion process (radial motion)*

Now it is convenient to introduce the scalaric-thermal velocity (bremsheat induced radial velocity). For this quantity we receive from (130) by means of (127)

$$v_{rQ} \equiv \frac{\mathrm{d}r_Q}{\mathrm{d}t} = \alpha_Q r\,, \tag{213}$$

where

a) $\alpha_Q = \dfrac{f_S J \alpha_c \Sigma}{3W}$ (scalaric-thermal velocity coefficient), (214)

b) $W = M_c c_Q$ (heat capacity of the central body with mass M_c) ,

c) $J = M_c \dot{r}_c^2 + \dfrac{2M_c}{5}(r_0\omega)^2$

is valid. Here one should remember that the coefficient α_Q as well as other cosmological quantities exhibit a slight scalaric-adiabatic cosmological time dependence.

12.3.2. *Scalaric-adiabatic contraction process (radial motion)*
Starting point is here the formulas (88a) and (90). As pointed out above, since about 1000 years after the big start the inequality $\sigma_c^2 \gg 1$ is well justified, i.e. both the equations mentioned read much simpler ($R_S \to r_S$):

$$\text{a)} \quad r_S = \frac{F_0^2}{\gamma_N \mathcal{M}_c {\sigma_c}^3}\,, \qquad \text{b)} \quad \dot{r}_S = -\frac{3F_0^2}{\gamma_N \mathcal{M}_c {\sigma_c}^4}\dot{\sigma}_c\,. \tag{215}$$

By differentiation we find the expression

$$\ddot{r}_S = -\frac{3F_0^2}{\gamma_N \mathcal{M}_c {\sigma_c}^4}\left(\ddot{\sigma}_c - \frac{4}{\sigma}\dot{\sigma}_c^2\right)\,. \tag{216}$$

Substituting into the relation (208), being applied to this case, yields for the rearranged formula (209) the result

$$\alpha_S = \frac{A_S}{r_S} = -\frac{3}{\sigma_c}\left(\ddot{\sigma}_c - \frac{3}{\sigma_c}\dot{\sigma}_c^2\right)\,. \tag{217}$$

For eliminating the second order derivative of the scalaric field function σ_c we have to go back to the corresponding field equation contained in our publications mentioned above. After lengthy calculations one obtains

$$\mathrm{d}r_S = -3r\mathrm{d}\ln\sigma_c\,. \tag{218}$$

12.3.3. *Superposition of both radial effects*
In both preceding subsections we treated in detail the expansion effect and the contraction effect. By adding both effects we arrive at the total effect for the interior

$$\mathrm{d}r_i = \mathrm{d}r_Q + \mathrm{d}r_S\,. \tag{219}$$

Introducing the abbreviation

$$\tilde{\alpha} = \alpha_Q - 3\Sigma \tag{220}$$

leads to the relationship between the radius r and the total radial velocity

$$\dot{r}_i = \tilde{\alpha} r_i\,. \tag{221}$$

For further calculations it is convenient to use the scalaric bremsheat parameter

$$\beta_Q = \frac{Jf_S\alpha_c}{3W}\,. \tag{222}$$

Then (214a) and (220) take the form

$$\text{a)}\quad \alpha_Q = \beta_Q\Sigma\,, \qquad \text{b)}\quad \tilde{\alpha} = (\beta_Q - 3)\Sigma\,. \tag{223}$$

For the following we specialize our considerations to the case of a central body with a resting center of mass ($\dot{\boldsymbol{r}}_c = 0$). Then formula (214c) reduces to

$$J = \frac{2M_c}{5}(r_0\omega)^2\,. \tag{224}$$

Neglecting the scalaric-adiabatic time dependence of the material constitution quantity β_Q, then by differentiation of (221) we arrive at the relation

$$\ddot{r}_i = (\beta_Q - 3)\left[\dot{\Sigma} + (\beta_Q - 3)\Sigma^2\right] r_i\,. \tag{225}$$

Inserting the expressions (221) and (225) into the formula (208), applied to the interior ($R \to r_i$), leads to

$$A_i = \alpha_i r_i \tag{226}$$

with

$$\alpha_i = (\beta_Q - 3)\left[\dot{\Sigma} + (\beta_Q - 2)\,\Sigma^2\right]\,. \tag{227}$$

12.4. PHYSICAL PROCESSES IN THE EXTERIOR OF THE CENTRAL BODY (RADIAL MOTION)

The difference between the exterior and the interior consists in the fact that in the exterior no bremsheat effect takes place. This means that a moving body firstly, when leaving by jet the rotating central body, starts with the constant initial velocity

$$\dot{r}_{jet} = \dot{r}_Q(r = r_0) = \alpha_Q\, r_0\,, \tag{228}$$

and secondly underlies the scalaric-adiabatic cosmological contraction (218). Superposition of both effects gives

$$v_e \equiv \dot{r}_e = \alpha_Q r_0 - 3r\Sigma\,. \tag{229}$$

In this case from (208) follows

$$A_e = \ddot{r}_e + \dot{r}_e\Sigma = \alpha_e r\,, \tag{230}$$

where

$$\alpha_e = \Lambda - 3\Sigma^2 + \beta_Q\Sigma^2\frac{r_0}{r} \tag{231}$$

with

$$\Lambda = -3\left[\dot{\Sigma} - 3\Sigma^2\right] \tag{232}$$

holds.

12.5. ORBITAL MOTION OF A TEST BODY IN THE EXTERIOR OF THE CENTRAL BODY

12.5.1. *Rearrangement of the velocity equations in usual astronomical units*

In presenting the plots of the rotation curve and the spiral motion, in astrophysics the azimuthal velocity of the moving body (in astrophysics the notion "rotation velocity" is used) is indicated in $\frac{\text{km}}{\text{s}}$, whereas its distance from the central body is given in kpc. In adapting to this practice we have to rearrange the equations (211) and (229), using the relationship $1\text{kpc} = 3.086 \cdot 10^{16}$ km.The result of this recalculation is the following:

$$V_e = 30.86 \cdot \text{km} \cdot R \cdot (\text{kpc})^{-1} \sqrt{a_e + \frac{3\gamma_N M_c \cdot 10^{30}}{(\varkappa r_0)^3 R^3} V(\varkappa r_0) U(\varkappa R)} \quad (233)$$

and

$$v_e = 30.86 \cdot 10^{15} \cdot \text{km} \cdot [\alpha_Q r_0 - 3R\Sigma] \cdot (\text{kpc})^{-1} . \quad (234)$$

Both equations are the basis for our numerical investigations, particularly for plotting the rotation curve and the spiral, that will be done later.

12.5.2. *Spirality of the motion of the test body in the exterior*

The spiral form of the orbiting test body results by superposition of the radial velocity (234) and the azimuthal velocity (233). Let us introduce the notion "spirality" by the definition

$$Sp = \frac{V_e}{v_e} = \tan \chi , \quad (235)$$

where χ will be called spirality angle. Then the final formula for the spirality reads

$$Sp = \tan \chi = \frac{\sqrt{a_e + \frac{3\gamma_N M_c \cdot 10^{30}}{(\varkappa r_0)^3 R^3} V(\varkappa r_0) U(\varkappa R)}}{10^{15} \left(\alpha_Q \frac{r_0}{R} - 3\Sigma \right)} . \quad (236)$$

From this expression we learn that the reversion point of the radial motion is determined by the quantity

$$R_{rev} = \frac{\alpha_Q r_0}{3\Sigma} . \quad (237)$$

Remembering the relation

$$\text{a)} \quad v_e = \dot{R} \quad \text{and} \quad \text{b)} \quad V_e = R\dot{\Phi} \quad (238)$$

in polar coordinates, we arrive at the following relationship between the spiral angle χ and the polar angle Φ in form of the differential equation

$$\tan\chi = R\frac{\mathrm{d}\Phi}{\mathrm{d}R}, \tag{239}$$

where $\tan\chi$ is given by (236).

For our further calculations it is convenient to introduce the quantity $u = \varkappa R$. Then we arrive at the differential equation

$$\begin{aligned}
\text{a)}\ \frac{\mathrm{d}\Phi}{\mathrm{d}u} &= \frac{1}{P}\sqrt{\alpha_S + \frac{A_Q\Sigma}{u} + \frac{Q_1 U(u)}{u^3}} \quad \text{with} \\
\text{b)}\ Q_1 &= \frac{3\gamma_N M_c V(\varkappa r_0)}{r_0^3}, \ \text{c)}\ P = A_Q - 3u\Sigma, \ \text{d)}\ A_Q = \alpha_Q \varkappa r_0 .
\end{aligned} \tag{240}$$

Integration leads to the implicit form $\Phi = \Phi(R)$ of the spiral equation (Φ_0 constant of integration, $u_0 = \varkappa r_0$) :

$$\Phi = \Phi_0 + \int\limits_{\xi=u_0}^{u} \frac{\sqrt{\alpha_S + \frac{A_Q\Sigma}{\xi} + \frac{Q_1 U(\xi)}{\xi^3}}}{A_Q - 3\xi\Sigma}\mathrm{d}\xi . \tag{241}$$

The intervals of the variables are given by

$$\text{a)}\ r_0 \leqq R \leqq R_{rev}, \quad \text{b)}\ u_0 \leqq u \leqq \varkappa R_{rev} . \tag{242}$$

For a better astrophysical understanding of the spiral function (241) and particularly for plotting we are forced to approximate this integral representation by series expansion about the expansion center u_0 according to $\xi = u_0 + \eta$. Expanding up to the second order in η, after a rather lengthy calculation we arrive at the result ($\frac{\eta}{u_0} \ll 1$)

$$\Phi - \Phi_0 = A_0\left[u - u_0 + \frac{N}{2}(u - u_0)^2\right], \tag{243}$$

where A_0 and N are defined by

$$\begin{aligned}
\text{a)}\ A_0 &= \frac{1}{P_0}\sqrt{\alpha_S + \frac{A_Q\Sigma}{u_0} + \frac{Q_1 U(u_0)}{u_0^3}}, \\
\text{b)}\ N &= \frac{3\Sigma}{P_0} + \frac{-A_Q\Sigma u_0^2 + Q_1 u_0^2 \cos u_0 - 3Q_1 U(u_0)}{2u_0[\alpha_S u_0^3 + A_Q\Sigma u_0^2 + Q_1 U(u_0)]}, \\
\text{c)}\ P_0 &= A_Q - 3u_0\Sigma .
\end{aligned} \tag{244}$$

Solving the quadratic equation (243) we arrive at

$$R = \rho \pm \sqrt{\rho^2 - C_0}\,\sqrt{1 + D(\Phi - \Phi_0)} \qquad (245)$$

with

$$\text{a)}\ C_0 = \frac{(Nu_0 - 2)u_0}{N\varkappa^2}\,, \quad \text{b)}\ D = \frac{2}{A_0 N\varkappa^2(\rho^2 - C_0)}\,, \quad \text{c)}\ \rho = \frac{Nu_0 - 1}{N\varkappa}\,. \qquad (246)$$

This is the explicit form of the spiral equation received by approximation up to the second order in Φ.

By further power series expansion we get in first order (linearized form) the Archimedes spiral

$$\text{a)}\quad R = r_0 \pm S_0(\Phi - \Phi_0) \quad \text{with} \quad \text{b)}\quad S_0 = \tfrac{1}{2}D\sqrt{\rho^2 - C_0}\,. \qquad (247)$$

Hence we find for the distance (at a fixed angle) between two spiral arms the formula

$$\Delta R = 2\pi|S_0|\,. \qquad (248)$$

13. Numerical evaluation of the theory (model Milky Way)

Our next aim is to evaluate numerically for an orbiting test body the general formulas for its radial velocity and azimuthal velocity (rotation curve) as well as the formula for its spiral motion (superposition of the radial and azimuthal velocities).

The formulas presented above show that we need two different categories of parameters, namely first the "present" cosmological parameters as adiabatically fixed parameters determined by our cosmological model which we treated in our previous papers, and second the individual parameters which partly are rather well known by empirical observations. Partly we have to guess two parameters (dark matter parameter $\varkappa$, scalaric bremsheat parameter β_Q) in order to be in rough accordance with the observations from the motion of the orbiting stars (empirical rotation curves). Of course, our simplified model can only reflect some basic features of the astrophysics of galaxies. Specifically our further application of the above theory refers to a rough model of the Milky Way (Galaxy).

13.1. COSMOLOGICAL AND INDIVIDUAL PARAMETERS

Numerical cosmological values for the "present time": Here we use our cosmological results to derive from (51):

$$\begin{aligned} \text{a)}\quad \Sigma &= 5.03 \cdot 10^{-19}\,\mathrm{s}\,, \quad \text{b)}\quad \dot{\Sigma} = -2.81 \cdot 10^{-36}\,\mathrm{s}^{-2}\,, \\ \text{c)}\quad \alpha_S &= 1.05 \cdot 10^{-35}\,\mathrm{s}^{-2}\,. \end{aligned} \qquad (249)$$

Numerical values for rough modelling of the Galaxy:

$$\begin{aligned} &\text{a)} \quad M_c &&= 1.8 \cdot 10^{44}\,\text{g} &&\text{(central mass acting on the sun)}\,, \\ &\text{b)} \quad r_0 &&= 0.8\ \text{kpc} &&\text{(radius of the inner central body)}\,, \\ &\text{c)} \quad \varkappa &&= 0.04\ (\text{kpc})^{-1} &&\text{(dark matter fitting parameter)}\,. \end{aligned} \tag{250}$$

Numerical value of the scalaric bremsheat parameter (analog to sun matter, being explained later):

$$\beta_Q = 5.32 \cdot 10^3\,. \tag{251}$$

Hence the parameter (227) and the quantity (231) take the form

$$\begin{aligned} &\text{a)} \quad \alpha_i &&= 7.16 \cdot 10^{-30}\,\text{s}^{-2}, \\ &\text{b)} \quad \alpha_e &&= \left(1.05 \cdot 10^{-5} + \tfrac{1.1 \cdot 10^{-3}}{R}\ \text{kpc}\right) \cdot 10^{-30}\,\text{s}^{-2}\,. \end{aligned} \tag{252}$$

Concluding this subsection let us add some annotations to the scalaric bremsheat

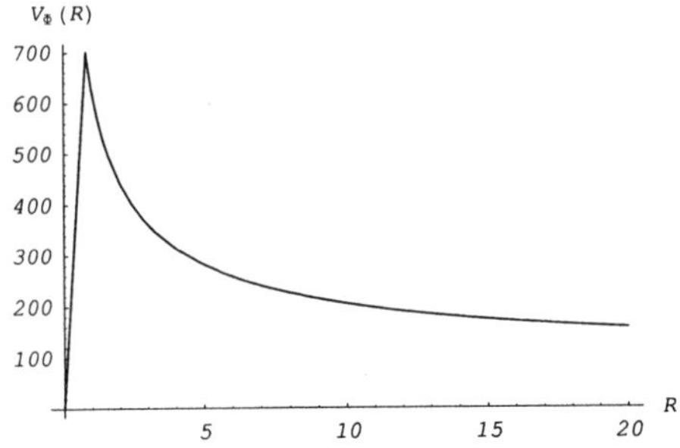

Figure 7. Rotation curve (radial course of the azimuthal velocity) of stars about the center of a galaxy (model: sun)

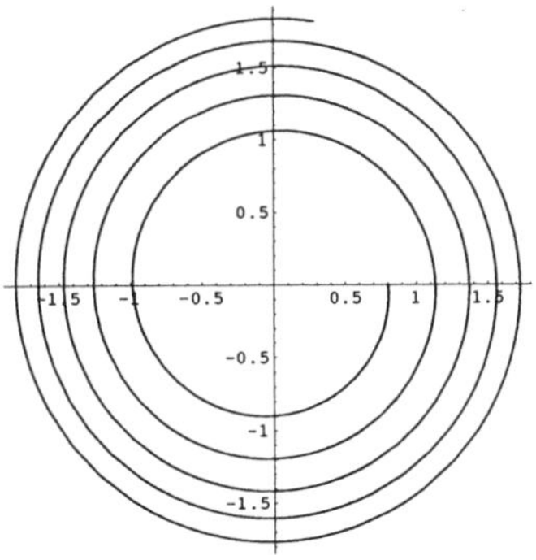

Figure 8. Spiral motion of a cosmic object (star) ejected by the rotation of the central body of a galaxy

parameter. According to its definition (222) it mainly depends on following values of the inner central body of the Galaxy: quantity J (214c), cubic heat expansion coefficient α_c and heat capacity W (214b) which contains the specific heat c_Q of the central body. The heat consumption factor f_S seems to be of the order of magnitude 1. Let us further mention that we know the radius r_0 of the central body and roughly its angular velocity ω. Further we realize that the mass of the central body cancels. Therefore we are mainly left with the parameters α_c and c_Q . Let us for numerical estimates tentatively use the values for the sun, already applied above (158):

$$\text{a)} \quad \alpha_{c\text{sun}} \approx 3.66 \cdot 10^{-3}\text{K}^{-1}, \quad \text{b)} \quad c_{Q\text{sun}} \approx 1.24 \cdot 10^8\,\text{cm}^2\,\text{s}^{-2}\,\text{K}^{-1}\,. \tag{253}$$

Hence we obtain for the sun the value $\beta_Q(251)$.

Fig. 7 shows the radial course of the azimuthal velocity V_Φ (abscissa R in kpc, ordinate V_Φ in km s^{-1}). This is the "rotation curve" following from our theory.

Plotting of the spiral of the orbiting body is based on the formula (245) exhibiting the parameters ρ, C_0 and D, which after some substitutions are determined by the equations (246). A rather lengthy numerical calculation leads to the values

$$\text{a)} \quad \rho = 1.3 \text{ kpc}, \quad \text{b)}\ C_0 = 1.5 \text{ (kpc)}^2, \quad \text{c)}\ D = -0.29\,. \tag{254}$$

Fig. 8 shows the spiral, where the radial distance is given in kpc.

I would like to express my thanks to the E. Majorana Center Erice for hospitality as well as to the professors Venzo de Sabbata and A. Gorbatsievich (University of Minsk) for scientific discussions and help.

14. References

1. Gorbatsievich, A. (2001) *Gen. Rel. Grav.* **33**, 965
2. Schmutzer, E. (1957) *Zeitschr.Phys.* **149**, 329; (1968) *Relativistische Physik*, Teubner, Leipzig; (1982) in E. Schmutzer (ed.), *Proceedings of GR 9*, DVW Berlin and CUP Cambridge
3. Schmutzer, E. (1995) *Fortschritte der Physik* **43**, 613; (1995) *Ann. Physik (Leipzig)* **4**, 251
4. Schmutzer, E. (1983): in V. de Sabbata and E. Schmutzer (eds.), *Unified Field Theories of more than 4 Dimensions including Exact Solutions*, World Scientific Singapore
5. Schmutzer, E. (2002) *Projektive Einheitliche Feldtheorie mit Anwendungen in Kosmologie und Astrophysik*, Harri Deutsch Verlag Frankfurt (in preparation)
6. Jordan, P. (1955) *Schwerkraft und Weltall*, Vieweg Verlag Braunschweig
7. Schmutzer, E. (2000) *Astron. Nachr.* **321**, 209
8. Schmutzer, E. (2000) *Astron. Nachr.* **321**, 137; (2001) *Gen. Rel. Grav.* **33**, 843
9. Landolt-Börnstein (1996), H.H. Voigt (ed.) *Astronomy and Astrophysics — Stars and Clusters 3B* (ed. H.H. Voigt), Springer Verlag Berlin etc.
10. Zimmermann, H. and Weigert, A. (1995) *ABC - Lexikon Astronomie* (8th edition), Spektrum Akad. Verlag Heidelberg etc.
11. Genzel, R., Eckart, A., Ott, T. and Eisenhauer, F. (1997) *Mon. Not. R. Astr. Soc.* **291**, 219
12. Schmutzer, E. (2000) *Astron. Nachr.* **321**, 227; (2001) *Astr. Nachr.* **321** (for stars in press)
13. Ciechanowicz, S. and Koziar, J. (1994) in M. Barone and F. Selleri (eds.), *Frontiers of Fundamental Physics*, Plenum Press New York
14. Walzer U., Hendel, R. and Baumgardner, J.R. (2001) *Physics of the Earth and Planetary Interiors* (in press)

15. Korycansky, D. G., Laughlin, G. and Adams, F. C. (2001) *Astrophysics and Space Science* (in press)
16. Schmutzer, E. (2001) *Astron. Nachr.* **322**, 97
17. Schmutzer, E. (2001) *Astron. Nachr.* **322**, 87
18. Dehnen, H., Rose, B. and Amer, K. (1995) *Astrophysics and Space Science* **234**, 69
19. Anderson, J. D., Laing, Ph. A., Lau, E. L., Liu, A.S., Nieto, M.M. and Turyshev, S. G. (1998) *Phys. Rev. Lett.* **81**, 2558

Ernst Schmutzer
Cospedaer Grund 57
D-07743 Jena
Germany

e-mail: eschmu@aol.com

ETTORE MAJORANA AND MATVEI BRONSTEIN (1906-1938): MEN AND SCIENTISTS

YU.P. STEPANOVSKY
National Scientific Centre "Kharkov Institute of Physics and Technology"
1 Akademicheskaja Street, Kharkov — 61108, Ukraine

There are many categories of scientists in the world; people of second and third rank who do their best but do not go very far. There are people of the first class who make great discoveries, fundamental to the development of science. And then there are the geniuses like Galileo and Newton. Well, Ettore Majorana was one of these.

E. Fermi [1]

Men like Matvei Bronstein are born to beautify mankind and to highlight some outlines of the Universe.

V.Ya. Frenkel, G.E. Gorelik [2]

P.G. Bergmann and V. de Sabbata (eds.),
Advances in the Interplay Between Quantum and Gravity Physics, 435–458.

1. Introduction

The first meeting of two prominent physicists — Ettore Majorana and Matvei Bronstein — happened in Erice in 1991, more than 50 years after they died [2]. In July 1991 Ettore Majorana Centre for Scientific Culture organized 29th Course: *Physics at the Highest Energy and Luminosity: To Understand the Origin of Mass* at the International School of Subnuclear Physics. A well-known Russian physicist L.B. Okun' in his lecture discussed classification of physical theories based on fundamental constants c, G and $\hbar$. This classification was first introduced by the Soviet physicist Matvei Bronstein in 1933. L.B. Okun' told about the short life and tragic death of Matvei Petrovich Bronstein. Matvei Bronstein and Ettore Majorana had similar lives in a particular sense. They both were born in 1906 and died in 1938. The scientific works of the both were connected with each other and are of interest up to nowadays. They both possessed great personalities and were great scientists. And in this article they are meeting again.

2. The Life and Disappearance of Ettore Majorana

It is said that for over two-thousand years history only two great men of exact science were brought to mankind by wonderful land of Sicily. The first was Archimedes born in Syracuse in 290/280 BC and killed by the Romans in autumn of 212 or spring of 211 BC in the sack of the city. The second one was Ettore Majorana born in Catania, which lies on the east coast of Sicily 53 km to the north of Syracuse.

Ettore Majorana was born on August 5, 1906, of a well-known family of Catania.[1] The uncle of Ettore, Quirino Majorana, was a respected professor of experimental physics at the University of Bologna. Ettore was a fourth child in the family. He had two brothers, Salvatore and Luciano, and two sisters, Rosina and Maria. Ettore had begun to show signs of unusual ability for numerical calculation when he was just four. His favourite game was to multiply in his head in few seconds two three-figure numbers given to him by relatives or their friends. His elementary education Ettore obtained in the school directed by Jesuit fathers.

In 1921 the Ettore family moved to Rome, where in 1923 Ettore became a student of the School of Engineering at the University of Rome. At the end of 1927 Emilio Segrè introduced E. Majorana to the new circle of physicists, which had grown up around Enrico Fermi at the Institute on the via Panisperna. E. Majorana took up physics at the beginning of 1928 after a conversation with E. Fermi.

[1]Detailed account of life and the works of Ettore Majorana see in the biographical sketch by E. Amaldi [3], in the book and paper of E. Recami [1, 4] and in the book of L. Sciascia [5].

His first 6 papers (**M1** — **M6**) all deal with problems of atomic and molecular physics. In the paper **M6** the problem of non-adiabatic spin flip in a beam of polarized atoms was considered. The papers **M7** — **M9** were devoted to the theory of nuclear forces and elementary particles. And the posthumous paper **M10** concerned relations between physics and social sciences. In the papers **M6** — **M9** E. Majorana introduced such important notions as the sphere of Majorana, infinite component Majorana equation, Majorana exchange forces, Majorana representation of Dirac matrices and Majorana neutrino.

E. Majorana was very clever and penetrating man. He much sooner than others in the circle of E. Fermi understood that mysterious beryllium radiation consists of the neutrons, that atomic nuclei are built from protons and neutrons, and that nuclear forces have the exchange character. However, E. Majorana had published neither of these ideas. E. Fermi tried to persuade him to publish at least something, but all his efforts were futile. Nevertheless, E. Fermi succeeded in persuading E. Majorana to go abroad to Leipzig and Copenhagen, where E. Majorana meet W. Heisenberg and N. Bohr. In Leipzig E. Majorana made friends with W. Heisenberg. Before leaving Rome for Leipzig E. Majorana published the paper **M7** on the relativistic theory of particles with arbitrary intrinsic angular momentum. It was namely W. Heisenberg, who persuaded E. Majorana to publish his well-known paper on nuclear theory **M8**.

When E. Majorana returned to Rome in autumn of 1933 he was not in the good health. He rarely visited Institute on the via Panisperna and after some months no longer came at all. The attempts of E. Majorana friends to bring him back to living a normal life were unsuccessful. One of his friends E. Amaldi wrote [3]: "None of us succeeded in finding out whether he was still doing theoretical physics research; I believe he was, but I have no proof".

At the beginning of 1937 the University of Palermo advertised the competition for three professor chairs in theoretical physics. The first three leading contenders were Gian Carlo Wick, Giulio Racah, and Giovanni Gentile Jr. Suddenly, E. Majorana also decided to take part in the competition. He had not published any scientific paper for some years. To enter the competition E. Majorana sent for publication in the Nuovo Cimento his paper on the symmetrical theory of the electrons and positrons, **M9**. The decision of E. Majorana to enter the competition was a scandal because the winners were predetermined definitely: E. Majorana, G.C. Wick, and G. Racah. The friend of E. Majorana G. Gentile Jr. would not have passed the competition. But G. Gentile Jr. was the son of G. Gentile Sr., the latter was a famous philosopher and fascist politician.[2] The solution of the problem was found: E. Majorana *in recognition of his special merit* without competition was appointed as ordinary professor of theoretical physics in the Royal University of Naples. On Thursday, January 13, 1938, E. Majorana had given his first lecture.

On March 26, 1938, he left Naples for Palermo, had arrived to Palermo, then, most probably, on March 27 returned to Naples and disappeared forever. The prominent Italian writer Leonardo Sciascia in his book "The disappearance of Majorana" [5] supported the point of view that E. Majorana decided to escape in religious life and entered an Italian monastery. L. Sciascia held a romantic view that E. Majorana disappearance was attributed to unwillingness to collaborate in the releasing of the destructive power of the atom. The famous researcher of E. Majorana life and works, the well-known physicist-theoretician Erasmo Recami published recently a new edition of his book "The Case of Majorana" [1]. In his another work [4] E. Recami outlines human and scientific personality of E. Majorana (on the basis of letters, documents, testimonies, collected by him in about twenty years). However, even Professor E. Recami does not know what happened to E. Majorana, and we have to agree with E. Fermi [3], who observed: "With his intelligence, once as he decided to disappear, Majorana would certainly have succeeded".

3. The Life and Death of Matvei Bronstein

Let us digress on Kharkov physics in the 1930s, which is directly relevant to our subject.

From 1918 to 1934 Kharkov was the capital of the Soviet Ukraine and in 1930 the Ukrainian Institute of Physics and Technology was founded there. It was known under the acronym 'UFTI' (now — NSC "KIPT", National Scientific Centre "Kharkov Institute of Physics and Technology"). One could say that Kharkov in 1930s played nowaday's role of Erice. Paul Ehrenfest, Niels Bohr, Yakov Frenkel, Igor Tamm, Boris Podolsky, Vladimir Fock, Wolfgang Pauli, Yurii Rumer, Paul Dirac, Pascual Jordan, Friedrich Houtermans, Walter Heitler, George Gamov, Dmitrii Ivanenko, Leon Rosenfeld, George Plachek, Rudolf Peierls, Laszlo Tisza, Lev Landau, Victor Weisskopf, [3] and many others lived and worked for some time in the city or took part in conferences on nuclear or theoretical physics that took place in Kharkov. One of the organizers of these conferences was Matvei Bronstein,

[2]Giovanni Gentile Sr. was the founder of Domus Galileana in Piza, where unpublished manuscripts of E. Majorana are kept. In the April of 1944 anti-fascist communists came across Giovanni Gentile Sr. going back to his home. "Are you Professor Giovanni Gentile?" — "Yes, I am." Giovanni Gentile Sr. was shot immediately... He was guilty of being the president of the Academy of Italy in the Fascist Social Republic established by the Germans.

[3]The names are ordered by the date of birth

a remarkable Leningrad physicist-theoretician, who had wide interests in astrophysics, cosmology, quantum gravitation, nuclear and semiconductor physics.

On October 1932 a group of Kharkov UFTI physicists repeated the Cambridge experiment of J.D. Cockroft and E.T.S. Walton, who in the April of the same year just broke-up the nucleus of ^7Li by bombarding it with the protons.

$$p + {^7Li} \rightarrow {^4He} + {^4He}.$$

The report about this achievement of Kharkov scientists was published in a telegram send to Moscow that looked approximately in the following way:

```
                    MOSCOW, KREMLIN,
               TO COMRADE JOSEPH STALIN

THE NUCLEUS OF LITHIUM
WAS DISINTEGRATED.
THE WORK IS CONTINUING.
```

Working at Kharkov UFTI at the time Lev Landau mocked this message to J. Stalin and narrated that he and his students send to Moskow another one:

```
                    MOSCOW, KREMLIN,
               TO COMRADE JOSEPH STALIN

THE SINE WAS DIFFERENTIATED,
THE COSINE WAS OBTAINED.
THE WORK IS CONTINUING.
```

L. Landau was wrong to consider the disintegration of ^7Li as unimportant experiment. Cambridge and Kharkov physicists for the first time experimentally demonstrated the validity of the famous Einstein formula $E = mc^2$. J.D. Cockroft and E.T.S. Walton received the 1951 Nobel Prize *for their pioneer work on the transmutation of atomic nuclei by artificially accelerated atomic particles.*[4]

And afterwards the time came that is known as the "Year 1937". In 1937 — 1938 Kharkov UFTI was smashed up [6, 7, 8]. Many noted Soviet physicists were arrested and executed. Among them — the head of cryogenic

[4]One should keep in mind that at the time Stalin and his assistants organized in the Ukraine genocidal artificial famine and about seven million people perished. People were dying straight in the streets of big cities. When P. Ehrenfest was invited during his short visit to hold a full-time position, live and work at Kharkov, he pointed at dead bodies lying at the streets and said: "Shall I really come here?" P. Ehrenfest left Kharkov on January 11, 1933. On September 25, 1933 he committed suicide.

laboratory of Kharkov UFTI and prominent scientist Lev Shubnikov, who was shot dead for "economic sabotage in the low temperature physics". Lev Landau, who moved to work at Moscow by then, was arrested and spent a year in prison. Under tortures he confessed to "counter-soviet activity in the theoretical physics" and only by a miracle was saved by Pyotr Kapitsa. Similarly, Matvei Bronstein was arrested and shot for fictitious "terrorist activity".

Matwei Bronstein was born on December 2, 1906 in Vinnitsa, Ukraine.[5] Matvei Bronstein was the second child in the family. He had the elder sister Mikhalina and the younger (on 10 minutes) twin-brother Isidor. The father of Matvej Bronstein was a physician, the family was poor, only the daughter Mikhalina attended the school, the brothers have had no such a possibility, they never were given a formal education. They were self-taught persons and both were highly educated men. Matvei Bronstein was a man-encyclopedia and all that he knew he got to know without any teacher only from numerous books he read. In 1915 the Matvei family moved to Kiev. The Bronstein's family was of Jewish origin but not religious. When brothers were twelve they decided to test whether the God exists. They carried out such a dangerous experiment — they both cried loudly: "The God is a fool!" This terrible crime was not followed any punishment. So it was tested that the God does not exist! In 1923 the brothers became students of a professional electro-technical school but in 1924 they abandoned it because of their family poverty and started working at a factory. In autumn 1925 M. Bronstein begun to visit "Section of Physics" of undergraduates study group at Kiev University. Later on he with sincere gratitude remembered the founder and head of that undergraduates study group Pyotr Tartakovsky. In 1926 M. Bronstein went to Leningrad and became a student of the Leningrad University. Previously to his enrolment the student M. Bronstein had published three scientific papers **B1** — **B3** on quantum mechanics (in 1925), two of them in Zeitschrift für Physik.

In spring of 1927 M. Bronstein made acquaintance with Lev Landau, George Gamov, and Dmitrii Ivanenko and made friends with them. They formed brilliant "quartet" that lived full life. They all loved physics devotedly and this love was mutual [9]. There was a somewhat curious story how M. Bronstein came to know Landau and Gamov [2]. This was because of Landau and Gamov close friend Evgeniya Kanegisser, later the lady Peierls. Lady Peierls told about her first meeting with M. Bronstein:

> There were pools, sparrows twittered, a warm wind blew... I have turned to short young person, wearing big spectacles, and told him: "Again gentle wind made drunk my heart...". Then he had immediately recited

[5]Detailed account on the life and works of Matvei Bronstein see in the book [2]

> the introduction to this poem of Gumilyov.[6] I joyfully screamed, and on the way to the university we were reciting our favourite poetry.

She was greatly impressed and immediately introduced this a great admirer and connoisseur of poetry to Landau and Gamov circle.

From 1930 to his tragical death M. Bronstein was the research assistant of headed by Yakov Frenkel Theoretical Department of the Leningrad Institute of Physics and Technology. In 1930 together with Ya. Frenkel M. Bronstein published a paper on "Quantization of Free Electrons in Magnetic Field" (**B4**). The Bronstein's book "The Structure of Matter" (1935, **B6**) contains the Part "The Relationship Between the Physical Theories and the Problem of Relativistic Quantum Theory", which he regarded as very important to his philosophical view of the natural world. In 1936 there was published the most significant M. Bronstein work "Quantization of Gravitational Waves" (**B7**) and in 1937 — the last scientific work of M. Bronstein "On Possibility of Spontaneous Splitting of the Photons" (**B9**). His scientific activity was impressive: though M. Bronstein's life was very brief, he published 35 scientific papers, 32 popular scientific articles and books, 6 articles in scientific encyclopedias; he translated and edited scientific books, delivered lectures to school pupils and undergraduates, postgraduates and research fellows.

M. Bronstein was the son-in-law of the famous Russian kids writer Kornej Chukovsky and a remarkable writer for children himself. He became a writer with the help of his wife Lidiya Chukovskaya (1907-1996), well-known as an eminent writer and fighter for human rights. M. Bronstein published three beautiful scientific popular books for children: "Solar Matter" (1934), "X-rays" (1936) and "Radiotelegraph Inventors" (1936); all of which were reprinted in 1990 (**B10**). M. Bronstein started to write one more book for children — about Galileo Galilei [10]. But alas! He was forced to leave this World without finishing the book about the great Italian scientist.

M. Bronstein was arrested on August 6, 1937 and was shot on February 18, 1938 when he was only thirty-one. Matvej Bronstein was guilty not only of fictitious "terrorist activity" but mainly of having the same surname as the original surname of Lev Trotskii. M. Bronstein was often been heard saying for fun, "...if Trotskii comes to power, I will call him my nephew..." [11]. But it was the old Russian tradition that a joke as much as one's life was worth.

[6]Nikolay Gumilyov was an eminent Russian poet. In August 1921 he was arrested and shot for fictitious counter-revolutionary activities.

4. The Works of Majorana and Bronstein and Related Topics

4.1. QUATERNION FORM OF MAXWELL EQUATIONS, DIRAC AND WEYL EQUATIONS AND UNPUBLISHED MAJORANA'S WORK ON PHOTONS

Let us consider Maxwell equations describing electromagnetic field in vacuum (with units adopted $c = G = \hbar = 1$).

$$\begin{aligned} \frac{\partial \vec{H}}{\partial t} &= -\mathrm{rot}\vec{E}, \quad \mathrm{div}\vec{H} = 0, \\ \frac{\partial \vec{E}}{\partial t} &= \mathrm{rot}\vec{H}, \quad \mathrm{div}\vec{E} = 0, \end{aligned} \tag{1}$$

or equivalently

$$\begin{aligned} \frac{\partial}{\partial t} H_i &= -\varepsilon_{ikl}\frac{\partial}{\partial x_k} E_l, \quad \frac{\partial}{\partial x_i} H_i = 0, \\ \frac{\partial}{\partial t} E_i &= \varepsilon_{ikl}\frac{\partial}{\partial x_k} H_l, \quad \frac{\partial}{\partial x_i} E_i = 0, \end{aligned} \tag{2}$$

were ε_{ikl} — Levi-Civita tensor, $\varepsilon_{123} = 1$. After introducing complex quantities

$$\begin{aligned} \vec{\psi}_R &= \vec{E} + i\vec{H}, \\ \vec{\psi}_L &= \vec{E} - i\vec{H}, \end{aligned} \tag{3}$$

we can rewrite Maxwell equations in the form:

$$\begin{aligned} \frac{\partial \vec{\psi}_R}{\partial t} &= \mathrm{rot}\ \vec{\psi}_R, \quad \mathrm{div}\ \vec{\psi}_R = 0, \\ \frac{\partial \vec{\psi}_L}{\partial t} &= -\mathrm{rot}\ \vec{\psi}_L, \quad \mathrm{div}\ \vec{\psi}_L = 0, \end{aligned} \tag{4}$$

Let us make use of the Pauli matrices σ_i and their usual multiplication rules

$$\sigma_i \sigma_k = \delta_{ik} I + i\varepsilon_{ikl}\sigma_l. \tag{5}$$

Now we can rewrite the Maxwell equations (4) in the form:

$$\begin{aligned} \left(\frac{\partial}{\partial t} + \vec{\sigma}\frac{\partial}{\partial \vec{x}}\right)\vec{\sigma}\vec{\psi}_R &= 0, \\ \left(\frac{\partial}{\partial t} - \vec{\sigma}\frac{\partial}{\partial \vec{x}}\right)\vec{\sigma}\vec{\psi}_L &= 0. \end{aligned} \tag{6}$$

So, we have obtained the spinor form of the Maxwell equations that is equivalent to the quaternion form, which was originally considered by Maxwell [12].

In 1928 P.A.M. Dirac discovered his famous equation [13]. The first edition of the well-known H. Weyl's book [14] was published in the same year. In his book H. Weyl represented 4-component Dirac equation as a pair of two 2-component equations,

$$
\begin{aligned}
&(\tfrac{\partial}{\partial t} + \vec{\sigma}\tfrac{\partial}{\partial \vec{x}})\psi_R + im\psi_L = 0,\\
&(\tfrac{\partial}{\partial t} - \vec{\sigma}\tfrac{\partial}{\partial \vec{x}})\psi_L + im\psi_R = 0,
\end{aligned}
\tag{7}
$$

where ψ_R and ψ_L are 2-component spinors. In the limit $m \to 0$, the equations (7) transform into the Weyl equations [15] that are very similar to the Maxwell ones (6),

$$
\begin{aligned}
&(\tfrac{\partial}{\partial t} + \vec{\sigma}\tfrac{\partial}{\partial \vec{x}})\psi_R = 0,\\
&(\tfrac{\partial}{\partial t} - \vec{\sigma}\tfrac{\partial}{\partial \vec{x}})\psi_L = 0.
\end{aligned}
\tag{8}
$$

It was E. Majorana who for the first time in his unpublished manuscript [16] made clear the near resemblance of the Maxwell and Weyl equations. The manuscript was written between 1928 and 1932 and is kept in Domus Galileana in Piza. E. Majorana used the transformation law of vectors $\vec{\psi}_R$ and $\vec{\psi}_L$ under a rotation on infinitesimal angle $\vec{\varphi}$,

$$\vec{\psi}'_{R,L} = \vec{\psi}_{R,L} - [\vec{\varphi}, \vec{\psi}_{R,L}]. \tag{9}$$

Omitting labels R and L, we have

$$\psi'_k = \psi_k - \varepsilon_{kil}\varphi_i\psi_l. \tag{10}$$

Let us introduce a notation

$$(S_i)_{kl} = -i\varepsilon_{ikl}. \tag{11}$$

From (10) we obtain

$$\psi'_k = \psi_k + i(S_i)_{kl}\varphi_i\psi_l. \tag{12}$$

Therefore $(S_i)_{kl}$ are spin angular momentum operators, and by using them we can represent the Maxwell equations in the same form as the Weyl equations:

$$
\begin{aligned}
&(\tfrac{\partial}{\partial t} + \vec{S}\tfrac{\partial}{\partial \vec{x}})\psi_R = 0, &\quad \mathrm{div}\vec{\psi}_R = 0,\\
&(\tfrac{\partial}{\partial t} - \vec{S}\tfrac{\partial}{\partial \vec{x}})\psi_L = 0, &\quad \mathrm{div}\vec{\psi}_L = 0.
\end{aligned}
\tag{13}
$$

Thus we see that E. Majorana was also the first who understood a quantum mechanical nature of the Maxwell equations. He discovered that Maxwell equations may be represented as a Schrödinger equation with a very simple Hamiltonian. In the general case the wave equations for a massless field of arbitrary spin also may be represented in the Majorana form (of course those equations must be supplemented by some relations of the type $\mathrm{div}\vec{\psi} = 0$).

4.2. BRONSTEIN'S WORK ON WEAK GRAVITATIONAL WAVES

Matwei Bronstein's work "The Quantization of Gravitational Waves" (**B7**) was the first deep investigation of quantum gravity. This is the summary of the article written by M. Bronstein

> A consequent quantum theory of weak gravitational field is constructed. The gravitational field in empty space is treated as a quantum-mechanical system, and relativistic-invariant commutation rules are introduced. The gravitational interaction of material bodies is established through the intermediate agency of gravitational quanta. Two physical applications are considered: 1) the loss of energy of material systems by means of the radiation of gravitational waves, and 2) the derivation of Newton's Law of Gravitation.

The Bronstein's derivation of Newton's Law of Gravitation was based on the work of Fock and Podolsky [17], who derived the Coulomb Law from Dirac's theory. M. Bronstein represented metric tensor for weak gravitational field in the form

$$g_{\mu\nu} = \Delta_{\mu\nu} + 2h_{\mu\nu}, \tag{14}$$

where $\Delta_{\mu\nu}$ — metric tensor with nonzero diagonal components (1, -1, -1, -1). We introduced factor 2 to the Bronstein's expression (14) for more similarity with the case of electromagnetic field. M. Bronstein obtained the following commutation relation for the components of $h_{\mu\nu}$:

$$[h_{\mu\nu}(x), h_{\rho\sigma}(x)] = -i(\Delta_{\mu\rho}\Delta_{\nu\sigma} + \Delta_{\mu\sigma}\Delta_{\nu\rho} - \Delta_{\mu\nu}\Delta_{\rho\sigma})D_0(x - x'), \tag{15}$$

where D_0 is the Green function,

$$D_0(x) = \frac{1}{(2\pi)^3}\int e^{ikx}\mathrm{sgn}k_0\delta(k^2)d^4k. \tag{16}$$

The corresponding commutation relation for electromagnetic 4-potentials are

$$[A_\mu(x), A_\nu(x')] = i\Delta_{\mu\nu}D_0(x - x'). \tag{17}$$

It follows from (15) and (17) that

$$[h_{12}(x), h_{12}(x')] = -iD_0(x - x'), \tag{18}$$

and

$$[A_1(x), A_1(x')] = -iD_0(x - x'). \tag{19}$$

Let us put

$$h_{00} = \Phi, \quad A_0 = \phi,$$

where Φ and ϕ are the gravitational and electric potentials. From (15) and (17) we obtain

$$[h_{00}(x), h_{00}(x')] = [\Phi(x), \Phi(x')] = -iD_0(x - x'), \tag{20}$$

$$[A_0(x), A_0(x')] = [\phi(x), \phi(x')] = iD_0(x - x'). \tag{21}$$

As was shown by M. Bronstein, just this difference of commutators signs leads to the attraction in gravitation and repulsion in electricity.

The Bronstein's paper contains also the quantum-mechanical derivation of the Einstein's formula for the loss of energy by radiating of gravitational waves.

$$I = \frac{1}{80\pi}(\frac{d^3}{dt^3}Q_{ik})^2, \tag{22}$$

where Q is quadruple momentum of the system. M. Bronstein was right when he predicted that for first time the Einstein's formula will be confirmed by an observation of slowing rotation of binary stars [2]. The 1993 Nobel Prize was awarded jointly to: Russell A. Hulse and Joseph H. Taylor Jr. "for the discovery of a new type of pulsar, a discovery that has opened up new possibilities for the study of gravitation". R.A. Hulse and J.H. Teylor discovered binary pulsar PSR 1913 + 16 of very strange behaviour and investigated their characteristics with wonderful accuracy [18]. They determined the mass of visible component of pulsar

$$m_1 = 1,4411\ m_{\odot},$$

and the mass of nonvisible component of pulsar

$$m_2 = 1,3874\ m_{\odot},$$

the pulsar period

$$\tau = 0,059029997929613\ s,$$

the orbital motion period

$$T = 27906,980895\ s,$$

the rate of increasing of the pulsar period

$$\Delta\tau = 0,272246\ s/year$$

and the rate of decreasing the orbital motion period

$$\Delta\ T = 76,0(3)\ \mu s/year.$$

The last value was found to be in good agreement with Einstein theory that gives for this quantity the following value,

$$\Delta\, T\ (Einstein\ theory) = 75,8\ \mu s/year.$$

In his paper M. Bronstein represented the linearized Einstein equations in the following form:

$$\begin{aligned} \frac{1}{c}\frac{\partial}{\partial t}H_{ij} &= -\varepsilon_{ikl}\frac{\partial}{\partial x_k}E_{lj}, \qquad \frac{\partial}{\partial x_i}H_{ij} = 0, \\ \frac{1}{c}\frac{\partial}{\partial t}E_{ij} &= \quad \varepsilon_{ikl}\frac{\partial}{\partial x_k}H_{lj}, \qquad \frac{\partial}{\partial x_i}E_{ij} = 0, \end{aligned} \tag{23}$$

where E_{ij} and H_{ij} are symmetrical and traceless tensors,

$$\begin{aligned} E_{ij} &= R_{4j4i} = \tfrac{1}{4}\varepsilon_{ikl}\varepsilon_{jmn}R_{klmn}, \\ H_{ij} &= \tfrac{i}{2}\varepsilon_{imn}R_{4jmn} \\ &= \tfrac{i}{2}\varepsilon_{imn}R_{mn4j}, \end{aligned} \tag{24}$$

$R_{\mu\nu\rho\sigma}$ is the curvature tensor.

The equations (23) are very similar to the Maxwell ones. We will make sure that they are very similar to the Weyl equations also. Let us introduce the complex fields

$$\begin{aligned} \psi_{Rij} &= E_{ij} + iH_{ij}, \\ \psi_{Lij} &= E_{ij} - iH_{ij}. \end{aligned} \tag{25}$$

Since rotation operators for the fields ψ_{ij} are

$$(S_k)_{ijlm} = -i\varepsilon_{kil}\delta_{jm} - i\varepsilon_{kjm}\delta_{il}, \tag{26}$$

we can rewrite the equations (23) in the form

$$\begin{aligned} (2\tfrac{\partial}{\partial t} + \vec{S}\tfrac{\partial}{\partial \vec{x}})\psi_R = 0, &\qquad \partial_i\psi_{Rij} = 0, \\ (2\tfrac{\partial}{\partial t} - \vec{S}\tfrac{\partial}{\partial \vec{x}})\psi_L = 0, &\qquad \partial_i\psi_{Lij} = 0. \end{aligned} \tag{27}$$

There is direct evidence of near resemblance of Bronstein (23), Maxwell (13) and Weyl (8) equations.

A very simple demonstration of impossibility to explain the cosmological red shift by splitting of photons was given by M. Bronstein in his last paper **B7**. It was shown by him that the splitting of photons leads to the dependence of the red shift on the wavelength, which is in contradiction with the experiment.

4.3. EXTENDED LITTLE LORENTZ GROUP AND THE WAVE EQUATIONS OF MASSLESS FIELDS

The great similarity of Bronstein, Maxwell and Weyl equations is not accidental because the arbitrary spin wave equations for the massless fields may be written in the following form [19, 20]:

$$(iS_{\mu\nu} + S\delta_{\mu\nu})\frac{\partial}{\partial x_\nu}\Psi(x) = 0, \tag{28}$$

where $S_{\mu\nu}$ are infinitesimal operators of the (s_1, s_2) — representation of the proper Lorentz group, $S = s_1 + s_2$. The well-known Weinberg theorem [21] states that the helicity of the massless field $\lambda = s_1 - s_2$.

The wave equations (28) follow from the existence of the *Extended Little Lorentz Group* [20, 22], i.e. *Little Lorentz Group* [23] (Lorentz transformations leaving invariant the 4-momentum p_μ) *enlarged by dilatations*, and a very simple assumption that *the state of an elementary particle with arbitrary spin may be completely described by its 4-momentum* [19].

For particular values of (s_1, s_2) one has the following correspondence:

(1,0) and (0,1)	J.C. Maxwell, 1864 (photons) [24],
(1/2,0) and (0,1/2)	H. Weyl, 1929 (neutrinos) [15],
(2,0) and (0,2)	M.P. Bronstein, 1936 (gravitons) [**B7**],
(1/2,1/2)	V.I. Ogievetsky, I.V. Polubarinov, 1966 (notophs) [25],
(-1/4,-1/4)	P.A.M. Dirac, 1971 [26],
(-1/4,-3/4) and (-3/4,-1/4)	L.P. Staunton, 1974 [27].

It is necessary to note that infinite component (-1/4,-1/4), (-1/4,-3/4) and (-3/4,-1/4) representations of the proper Lorentz group were for the first time introduced in the theory of relativistic wave equations by E. Majorana in his paper **M7** in 1932.

4.4. MAJORANA-PENROSE REPRESENTATION OF SPIN-TENSORS

In 1960 Roger Penrose [28] discovered the following remarkable property of symmetrical spin-tensor $\Psi_{\alpha_1\alpha_2...\alpha_n}$, $\alpha_i = 1, 2$,

$$\Psi_{\alpha_1\alpha_2...\alpha_n} = A\psi^1_{(\alpha_1}\psi^2_{\alpha_2}...\psi^n_{\alpha_n)}, \tag{29}$$

where A is some constant, ψ^1, ψ^2, ... ψ^n are 2-component spinors, $_{(...)}$ denotes symmetrization. The existence and uniqueness of (29) follows from the fundamental theorem of algebra. To get the representation (29), let us multiply the spin-tensor $\Psi_{\alpha_1\alpha_2...\alpha_n}$ on spin-tensor $\chi^{\alpha_1}\chi^{\alpha_2}...\chi^{\alpha_n}$, where χ^α is the spinor with the components $\chi^1 = z$, $\chi^2 = 1$,

$$\Psi_{\alpha_1\alpha_2...\alpha_n}\chi^{\alpha_1}\chi^{\alpha_2}...\chi^{\alpha_n} = P_n(z), \tag{30}$$

where

$$P_n(z) = \Psi_{111...1}z^n + n\Psi_{211...1} + \frac{n(n-1)}{2}\Psi_{221...1} + ... + \Psi_{222...2} \qquad (31)$$

is some polynomial of complex variable z. The factorization of polynomial $P_n(z)$ gives

$$P_n(z) = \Psi_{111...1}(z - a_1)(z - a_2)...(z - a_n). \qquad (32)$$

Let us inroduce the spinors ψ^i with components $\psi_1^i = 1, \quad \psi_2^i = -a_i$. Then we have

$$\Psi_{\alpha_1\alpha_2...\alpha_n}\chi^{\alpha_1}\chi^{\alpha_2}...\chi^{\alpha_n} = A(\psi^1_{\alpha_1}\chi^{\alpha^1})(\psi^2_{\alpha_2}\chi^{\alpha^2})...(\psi^n_{\alpha_n}\chi^{\alpha^n}), \qquad (33)$$

where $A = \Psi_{111...1}$. The relation (33) expresses the general binary multilinear form as a product of linear factors. These factors are essentially unique. Omitting the spinors χ's in (33) we obtain finally (29). Thus the validity of Penrose representation (29) is proved.

The spin-tensor of rank $2S$ describes the paricle with spin S. So, according to Penrose, a state of a particle with arbitrary spin S can be represented as a set of $2S$ spin 1/2 particles states:

$$\Psi_S = A\Big\{\psi^{(1)} \times \psi^{(2)} \times \psi^{(3)} \times \ldots \times \psi^{(2S)}\Big\}, \qquad (34)$$

For the above mentioned, we can attach to each normalized spinor some unit vector $\vec{n} = (\psi^\star\vec{\sigma}\psi)/(\psi^\star\psi)$ and the following visual picture arises for description of an arbitrary state of spin S particle (see Figure 1).

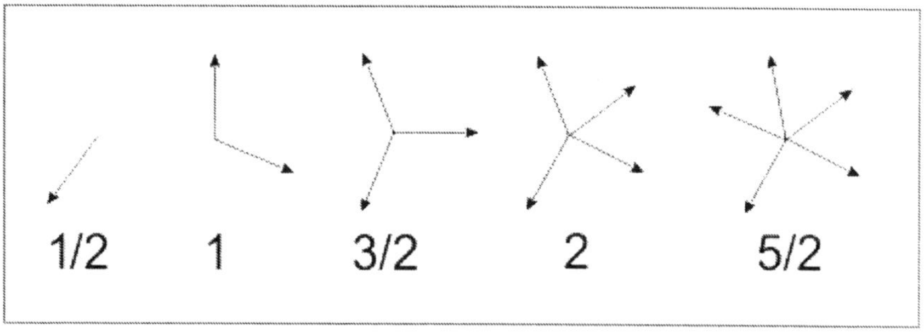

Figure 1. Majorana-Penrose visual description of spin S particle

Thus the state of spin S particle can be represented as a set of $2S$ unit vectors. R. Penrose did not know that he rediscovered the visual description of spin S particle that E. Majorana found in 1932 (**M6**) and that known as sphere of Majorana. Later on R. Penrose payed tribute to work of E. Majorana [29]. Majorana sphere is the unit sphere on which the ends of unit vectors $\vec{n}_i$ are lying. E. Majorana obtained the formulae (31) and (32), so his approach to the problem was identical to Penrose.

R. Penrose used the Majorana representation for elementary description of the gravity field classification by Petrov [28]. Thus, according to Penrose, the following pictures correspond to Petrov's types I, II, III, D, and N (Figure 2)

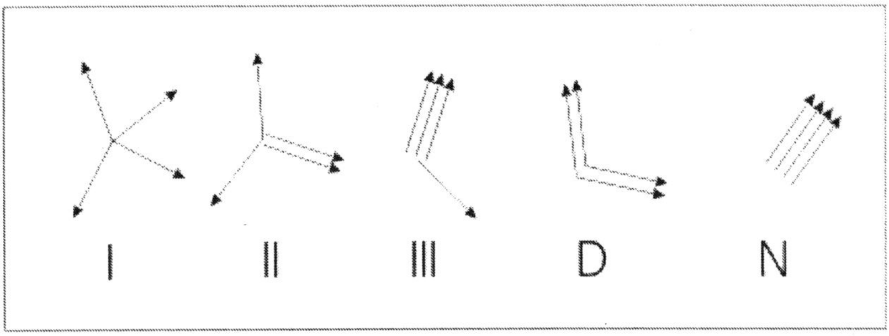

Figure 2. Gravity fields classification by Petrov according R. Penrose

Majorana-Penrose representation may be used also for description and calculating of the Maxwell axes in electrodynamics [30] and in the earth magnetism theory [31] and for simple treatment of conical refraction in crystal optics [32].

4.5. MAJORANA AND DIRAC INFINITE COMPONENT RELATIVISTIC WAVE EQUATIONS WITHOUT NEGATIVE ENERGY SOLUTIONS

Let us consider a bispinor A_α

$$A_\alpha == \begin{pmatrix} a_1 \\ a_2 \\ -a_2^+ \\ a_1^+ \end{pmatrix}, \tag{35}$$

where a_1^+, a_2^+ and a_1, a_2 are the usual creation and annihilation operators. The commutator of A_α's is

$$A_\alpha A_\beta - A_\beta A_\alpha = C_{\alpha\beta}, \tag{36}$$

where C is the charge conjugation matrix. If we define

$$S_{\mu\nu} = \frac{1}{8i}\left(\overline{A}(\gamma_\mu\gamma_\nu - \gamma_\nu\gamma_\mu)A\right) \tag{37}$$

as the infinitesimal operators of the corresponding representation of the proper Lorentz group, then the quantity

$$\Phi_\alpha = \langle\psi|A_\alpha|\psi\rangle \tag{38}$$

is the bispinor and $|\psi\rangle$ is in a particular sense the square root of bispinor Φ_α.

Now we can write the following equations:

$$(\gamma_\mu \frac{\partial}{\partial x_\mu} - m)^\beta_\alpha A_\beta |\psi(x)\rangle = 0. \tag{39}$$

These equations, discovered by Dirac in 1971 [26], have only the positive energy plane wave solutions

$$|\psi(p,x)\rangle = exp(ipx)|u(p)\rangle. \tag{40}$$

Dirac didn't know, that the same representation of the Lorentz group was successfully used by E. Majorana in 1932 (**M7**) for the same purpose, namely, to obtain the equations with no negative energy solutions.

Unfortunately, no one can introduce interaction in the positive energy Dirac equations. We have four conditions on one function

$$K_\alpha |\psi(x)\rangle = 0 \tag{41}$$

and ten conditions of compatibility

$$(K_\alpha K_\beta - K_\beta K_\alpha)|\psi(x)\rangle = 0. \tag{42}$$

For free fields these ten conditions give

$$(\Box - m^2)C_{\alpha\beta}|\psi(x)\rangle = 0. \tag{43}$$

Unluckily, if we introduced an interaction those conditions would contradict one another.

The Majorana equation discovered in 1932 can be obtained from the positive energy Dirac equation (39) by multiplication on $A^\alpha = C^{\alpha\beta} A_\beta$,

$$A^\alpha (\gamma_\mu \frac{\partial}{\partial x_\mu} - m)^\beta_\alpha A_\beta |\psi(x)\rangle = 0. \tag{44}$$

Whereas the Dirac-1971 equation describes the particle with zero spin, the Majorana-1932 equation describes the particle with arbitrary spin S.

Positive energy Majorana-Dirac equation may be used for the treatment of squeezed states in quantum optics. Let Q_α be a spinor of the $SO(2,1)$ group,

$$Q_\alpha = \begin{pmatrix} a \\ a^+ \end{pmatrix}. \tag{45}$$

It is evident that

$$Q_\alpha Q_\beta - Q_\beta Q_\alpha = \varepsilon_{\alpha\beta}. \tag{46}$$

Let us consider the plane wave solution of the Majorana-Dirac equation

$$(\sigma_i \frac{\partial}{\partial x_i} - m)^{\beta}_{\alpha} Q_\beta |\psi\rangle = 0, \tag{47}$$

$$|\psi\rangle = e^{ipx}|u\rangle. \tag{48}$$

From (47) and (48) we have

$$(ip_i\sigma_i - m)^{\beta}_{\alpha} Q_\beta |u\rangle = 0. \tag{49}$$

For the $p_i = (0, p, i\varepsilon)$ we obtain

$$[(\varepsilon + m)a - pa^+]|u\rangle = 0, \tag{50}$$

i.e. the equation determining the squeezed vacuum state. If $p = 0$, we derive

$$a|u_0\rangle = 0, \tag{51}$$

so

$$|u_0\rangle = |0\rangle. \tag{52}$$

The plain wave solution for $p_i = (0, p, i\varepsilon)$ may be obtained by the Lorentz transformation

$$|u_p\rangle = \exp J_1 \mathrm{Arth}\frac{p}{\varepsilon}|0\rangle. \tag{53}$$

Let us introduce the Lorentz group generators [33]

$$J_i = \frac{1}{4}(\sigma_i)^{\alpha\beta} Q_\alpha Q_\beta, \qquad J_\pm = J_1 \pm iJ_2. \tag{54}$$

On using the relation

$$\begin{aligned} \exp(J_1 \mathrm{Arth}\tfrac{p}{\varepsilon}) &= \exp \tfrac{\varepsilon - m}{p} J_+ \times \\ \exp(-\ln \tfrac{\varepsilon + m}{2m}) J_3 &\times \exp \tfrac{\varepsilon - m}{p} J_-, \end{aligned} \tag{55}$$

we get

$$|u_p\rangle = \sqrt{\frac{2m}{\varepsilon + m}} \exp(-\frac{p}{\varepsilon + m})(a^+)^2|0\rangle. \tag{56}$$

As shown by S. Hawking in his classical paper, the black hole radiation is characterized by the Plank's spectrum with some temperature, and the state of radiation is a squeezed vacuum state. However, we can see from (56) that allowed photon numbers n for the state of squeezed vacuum are 0, 2, 4, 6, ..., which is in a sharp contradiction with photon numbers for the thermal radiation state 0, 1, 2, 3, ... So Hawking radiation is not thermal radiation [34].

4.6. ON MAJORANA REPRESENTATION OF DIRAC MATRICES AND $SO(3,3) \approx SL(4R)/Z_2$ ISOMORPHISM

May be, the last work of E. Majorana "Symmetric Theory of Electrons and Positrons", published in 1937, is the most popular one. In this work were introduced the new conceptions now well known as Majorana representation of Dirac matrices, Majorana neutrino, Majorana particles. E. Majorana discovered that irreducible two-valued representation of proper Lorentz Group may be realized by real 4 × 4-matrices. Then the wave function of antiparticle $\psi(antiparticle)$ is simply complex conjugate of the negative energy solution ψ_- of Dirac equation,

$$\psi(antiparticle) = \psi^\star_-, \tag{57}$$

and operator $\gamma_\mu \frac{\partial}{\partial x_\mu} + m$ that is a part of Dirac equation

$$(\gamma_\mu \frac{\partial}{\partial x_\mu} + m)\psi = 0 \tag{58}$$

is real operator. We will construct some simple generalization of Majorana representation. Let us consider arbitrary real antisymmetric matrix with components

$$R = \begin{pmatrix} 0 & x_3 + t_3 & -x_2 - t_2 & x_1 - t_1 \\ -x_3 - t_3 & 0 & x_1 + t_1 & x_2 - t_2 \\ x_2 + t_2 & -x_1 - t_1 & 0 & x_3 - t_3 \\ -x_1 + t_1 & -x_2 - t_2 & -x_3 + t_3 & 0 \end{pmatrix}. \tag{59}$$

The linear transformation

$$R_{\alpha\beta} \to R'_{\alpha'\beta'} = \Lambda^\alpha_{\alpha'} \Lambda^\beta_{\beta'} R_{\alpha\beta} \tag{60}$$

gives new matrix R' with components

$$R' = \begin{pmatrix} 0 & x'_3 + t'_3 & -x'_2 - t'_2 & x'_1 - t'_1 \\ -x'_3 - t'_3 & 0 & x'_1 + t'_1 & x'_2 - t'_2 \\ x'_2 + t'_2 & -x'_1 - t'_1 & 0 & x'_3 - t'_3 \\ -x'_1 + t'_1 & -x'_2 - t'_2 & -x'_3 + t'_3 & 0 \end{pmatrix}. \tag{61}$$

If the matrices Λ are real unimodular matrices, $\Lambda \in SL(4R)$, then the quantity

$$\varepsilon^{\alpha\beta\gamma\delta} R_{\alpha\beta} R_{\gamma\delta} = x_1^2 + x_2^2 + x_3^2 - t_1^2 - t_2^2 - t_3^2, \tag{62}$$

where $\varepsilon^{\alpha\beta\gamma\delta}$ is Levi-Civita symbol, $\varepsilon^{1234} = 1$, is invariant. Thus we see that real unimodular 4×4-matrices correspond to Lorentz transformations $SO(3,3)$ of six-dimensional space that has three space and three time coordinates. The six-dimensional structure of 4×4 matrices was in details considered in the papers [35, 36, 37]

4.7. $CG\hbar$-CLASSIFICATION OF PHYSICAL THEORIES BY M. BRONSTEIN, QUANTUM HALL EFFECT AND SOME PROPERTIES OF THE NUMBER 137

M. Bronstein had great interest to fundamental constants of physics. In his popular scientific books he discussed the following $cG\hbar$-scheme (see Figure 3). For example this scheme may be found in the book **B6** in the Part "The Relationship Between the Physical Theories and the Problem of Relativistic Quantum Theory". The scheme on Figure 3 is given in some modification [2].

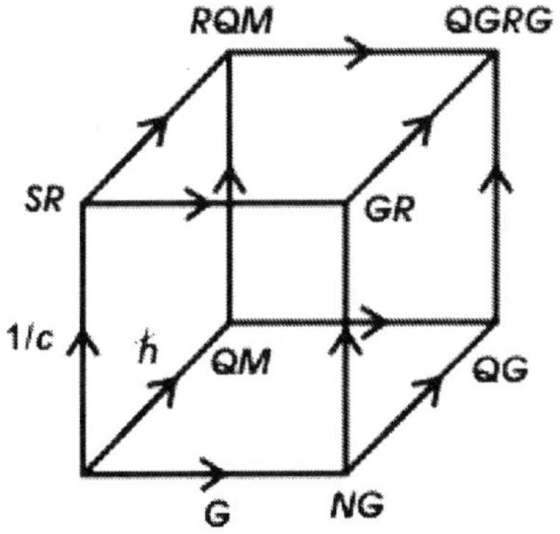

Figure 3. cGh-classification of physical theories by M. Bronstein

There are following abbreviations on the Figure 3:

NG	**Newtonian Gravity**
QM	**Quantum Mechanics**
SR	**Special Relativity**
GR	**General Relativity**
QG	**Quantum Gravity**
RQM	**Relativistic Quantum Mechanics**
QGRG	**Quantum General Relativistic Gravity**

From 35 articles of M. Bronstein only two were written with coauthors. The first coauthor was Ya. Frenkel (**B4**) and the second one was L. Landau (**B5**). In their paper Ya. Frenkel and M. Bronstein considered the quantum mechanical motion of an electron in magnetic field. It was the actual problem at the time. In the same 1930 L. Landau published his famous paper on this subject [38]. However Landau, Frenkel, and Bronstein were not the firsts who solved the problem. The firsts who solved in 1928 the problem on quantum mechanical motion of electron in magnetic field were V.Fock [39] and I.Rabi [40], one of the founders of Ettore Majorana Centre for Scientific Culture. V. Fock at the first time considered in his paper the object that now is well known as the "Quantum Dot". I. Rabi solved the problem on the quantum motion of a relativistic electron in magnetic field using recently discovered Dirac equation. With connection with all these works let us briefly consider the remarkable phenomenon Quantum Hall Effect.

In 1980 Klaus von Klitzing [41] (the 1985 Nobel Prize) discovered experimentally that the Hall resistance of a thin semi-conductor film is quantized with great accuracy according to the formula

$$R_H = \frac{h}{e^2}\frac{1}{N}, \qquad N = 1, 2, 3, ..., \tag{63}$$

In the paper [41] was proposed new method of very precise establishment of the value of the fine structure constant. To make clear the connection of the Hall resistance and of the fine structure constant

$$\frac{e^2}{\hbar c} = \frac{1}{137,0359895(61)}$$

we have to make more precise the well-known connection of light velocity and resistance unit, $1/c = 30$ *ohm*. Indeed,

$$ohm = \frac{volt}{ampere} = \frac{1}{9 \cdot 10^9} \cdot \frac{\frac{coulomb}{metr}}{\frac{coulomb}{second}} = \frac{1}{30 \cdot 3 \cdot 10^8 \frac{metr}{second}}.$$

To obtain more precise connection we have to make the substitution $9 \to (2,99792458)^2$ that finally gives

$$\frac{1}{c} = 29,9792458 \; ohm.$$

Now we can express h/e^2 through *ohm*:

$$\frac{h}{e^2} = 2\pi\cdot\frac{\hbar c}{e^2}\cdot\frac{1}{c} = 2\pi\cdot 137,0359895(61)\cdot 29,9792458 \; ohm = 25812,8056(11) \; ohm.$$

The resistance 25812,806 *ohm* is measured with grate accuracy. Thus we see that measurement of "resistance quantum" $\frac{h}{e^2}$ leads to precise establishment of the fine structure constant value.

Let us talk over now two curious properties of fine structure constant or more precisely of the number 137, which is approximately equal to the inverse value of this remarkable constant. The first property was announced by V. Weisskopf [42]. Once V. Weisskopf has talk about CABALAH[7] with one of great connoisseur in this matter Gershom Shalom. Gershom Shalom was explaining to V. Weisskopf how to put in correspondence the numbers to the words in Hebrew and suddenly asked: "Say me, please, have you some numbers of interest in physics?" V. Weisskopf answered: "Of course, we have, the number 137, for example!" Gershom Shalom was shocked. "O, — he exclaimed — but CABALAH is just 137!"

קבלה.

Figure 4. The number that corresponds to the world CABALAH is 137

Indeed, (see Figure 4) there are four letters that make up the world CABALAH: 'kof'=100, 'bet'=2, 'lamed'=30, 'hay'=5, 100+2+30+5=137.

And another one fanny property of the number 137 (it is a modest observation of the author of these lines): if everyone will write twice the number of the year he (or she) was born, then obtained eight-figure number will be divided on 137 without remainder. For example, Ettore Majorana and Matwei Bronstein were born in 1906. It is easily to verify that 19061906:137=139138.

5. Acknowledgements

Ettore Majorana and Matvei Bronstein are heros of mine for over than 40 years since I have read about E. Majorana in Laura Fermi's book "Atoms in the Family" and for the first time have gotten to know about the remarkable works and tragic lot of M. Bronstein. I am very grateful to Professor Venzo De Sabbata for the opportunity to visit the Ettore Majorana Centre for Scientific Culture and for giving me the possibility to deliver the lectures about E. Majorana and M. Bronstein. I appreciate deeply the people at the Ettore Majorana Centre for their great hospitality. I am very thankful to Alexander Zheltukhin, who introduced me to Professor De Sabbata; to Dmitrii Sorokin for communicating me two inaccessible in Ukraine Majorana's manuscripts; and to Kostyantin Ilyenko for essential help in writing this text.

[7]CABALAH (Hebrew: "Tradition") is the secret mystical knowledge (divine revelation) that was communicated by God to Moses and Adam and that provides a means of approaching God directly.

The works of Ettore Majorana

M1 (1928) Sullo sdoppiamento dei termini Roentgen ottici a causa dell'elettrone rotante e sulla intensità delle righe del Cesio *Rendiconti Accademia Lincei*, **8**, 229-233, (with Giovanni Gentile Jr.).
M2 (1931) Sulla formazione dello ione molecolare di He, *Nuovo Cimento*, **8**, pp.22-28.
M3 (1931) I presunti termini anomali dell'Elio, *Nuovo Cimento*, **8**, 78-83.
M4 (1931) Reazione pseudopolare fra atomi di Idrogeno, *Rendiconti Accademia Lincei*, **13**, 58-61.
M5 (1931) Teoria dei tripletti P' incompleti, *Nuovo Cimento*, **8**, 107-113.
M6 (1932) Atomi orientati in campo magnetico variabile, *Nuovo Cimento*, **9**, 43-50.
M7 (1932) Teoria relativistica di particelle con momento intrinseco arbitrario, *Nuovo Cimento*, **9**, 335-344.
M8 (1933) Über die Kerntheorie *Zeitschrift für Physik*, **82**, 137-145.
M8' (1933) Sulla teoria dei nuclei, *La Ricerca Scientifica*, **4**, 559-565.
M9 (1937) Teoria simmetrica dell'elettrone e del positrone, *Nuovo Cimento*, **14**, 171-184.
M10 (1942) Il valore delle leggi statistiche nella fisica e nelle scienze sociali, *Scientia*, **36**, 55-66.

Some works of Matvei Bronstein

B1 (1925) On one conclusion from the light quanta hypothesis, *Zhurnal Russkogo Fisiko-khimicheskogo obshhestva*, **57**, 321-325.
B2 (1925) Zur Theorie des kontinuierlischen Röntgenspectrums, *Zeitschrift für Physik*, **32**, 881-885.
B3 (1925) Bemerkung zur Quantentheorie des Laue-Effektes, *Zeitschrift für Physik*, **32**, 886-893.
B4 (1930) Quantization of free electrons in magnetic field, *Zhurnal Russkogo Fisiko-khimicheskogo obshhestva*, **62**, 485-494 (with Ya.I. Frenkel).
B5 (1933) The second law of thermodynamics and Universe, *Physikalische Zeitschrift der Sowjetunion*, **4**, 114-118 (with L.Landau).
B6 (1935) *The Structure of Matter*, ONTI, Leningrad, Moscow.
B7 (1936) Quantization of Gravitational Waves, *Zhurnal Eksperimentalnoj i Teoreticheskoj Fisiki*, **6**, 195-236.
B8 (1936) Quantum theory of weak gravitational field, *Physikalische Zeitschrift der Sowjetunion*, **9**, 140-157.
B9 (1937) On the possibility of spontaneous splitting of the photons, *Zhurnal Eksperimentalnoj i Teoreticheskoj Fisiki*, **7**, 335-338.
B10 (1990) *Solar Matter*, Nauka, Moscow.

References

1. Recami, E. (2001) *Il caso Majorana: Epistolario, Documenti, Testimonianze*, Di Renco Editore, Roma.
2. Frenkel V.Ya. and Gorelik G.E. (1994) *Matvei Petrovich Bronstein and Soviet Theoretical Physics in the Thirties*, Birkhauser Verlag, Basel-Boston-Berlin.
3. Amaldi E. (1966) Ettore Majorana: Man and Scientist, in A.Zichichi (ed.), *Strong and Weak Interactions*, Academic Press Inc., New York, pp. 9-77.
4. Recami E. (1998)*Ricordo di Ettore Majorana a sessant'anni dalla sua scomparsa*, physics/9810023.
5. Sciascia L. (1975) *La scomparsa di Majorana*, Einaudi, Torino.
6. Weisberg A. (1951) *The Accused*, Simon and Schuster, New York.
7. Pavlenko V.Yu., Raniuk Yu.N., Khramov Yu.A. (1998) *Delo UFTI*, Feniks, Kiev.
8. Bakai O. and Raniuk Yu. (1993) *History of physics research in Ukraine*, http://meltingpot.fortunecity.com/pakistan/83/physics/bakai01.html
9. Frenkel V.Ya. (1994) George Gamow: World line 1904-1933 (On the ninetieth anniversary of G.A. Gamow's birth) *Physics-Uspekhi* **37**, 767-789.
10. Gorelik G.E. (1999) *Gloria Mundi*, http://sites.netscape.net/gegorelik/Gloria_W.htm
11. Josephson P. R. (1999) *Physics and Politics in Revolutionary Russia*, University of California Press, Berkeley, Los Angeles, Oxford.
12. Maxwell J.C. (1873) *A Treatise on Electricity and Magnetism*, V. 2, Oxford University Press, Oxford.
13. Dirac P.A.M. (1928) The quantum theory of the electron, *Proc.Roy. Soc.*, **A117**, 610-624.
14. Weyl H. (1928) *Gruppentheorie und Quantummechanik*, Verlag von S.Hirzel, Leipzig.
15. Weyl H. (1929) Elektron und Gravitation, *Z. Phys.*, **56**, 330-352.
16. Mignani R., Recami E., and Baldo M. (1974) About a Dirac-like equation for the photon according to Ettore Majorana, *Lett. al Nuovo Cimento*,**11**, 568-572.
17. Fock V. and Podolsky B. (1932) On the quantization of electromagnetic waves and the interaction of charges on Dirac's theory, *Sow. Phys.* **1**, 801-817.
18. Will C.M. (1993) *Theory and Experiment in Gravitational Physics*, Cambridge University Press, Cambridge.
19. Stepanovsky Yu.P. (1964) Little Lorentz group and wave equations of free massless fields with arbitrary spin, *Ukrainskij fizicheskij zhurnal*, **9**, 1165-1957.
20. Stepanovsky Yu.P. (1981) On wave equations of massless fields, *Teoreticheskaja i matematicheskaja fizika*, **47**, 343-351.
21. Weinberg S. (1964) Feynman rules for any spin. II. Massless particles, *Phys. Rev.*, **134B**, 882-896.
22. Kim Y.S. and Wigner E. (1987) Cylindrical group and massless particles, *J. Math. Phys.*, **28**, 1175-1179.
23. Wigner E. (1939) On unitary representations of the inhomogeneous Lorentz group, *Ann. Math.*, **40**, 149-204.
24. Maxwell J.C. (1864) A dynamical theory of the electromagnetic field, *Roy. Soc. Trans.*, **155** 459-530.
25. Ogievecky V.I. and Polubarinov I.V. (1966) Notophs and their interactions, *Jadernaja fizika*, **4**, 216-224.
26. Dirac P.A.M. (1971) A positive-energy relativistic wave equation, *Proc. Roy. Soc.*, **A322**, 435-445.
27. Staunton L.P. (1974) Spin 1/2 positive-energy wave equation, *Phys. Rev.*, **D10**, 1760-1766.
28. Penrose R. (1960) A spinor approach to general relativity, *Ann. of Phys.*, **10**, 171-201.
29. Penrose R. (1995) *Shadows of the Mind*, Vintage, London.

30. Stepanovsky Yu.P. (1981) Maxwell poles and spinor theory, *Ukrainskij fizicheskij zhurnal*, **26**, 1768-1773.
31. Stepanovsky Yu.P. (1982) On the equation determining geomagnetic multipole axes, *Geomagnetism i aeronomija*, **22**.
32. Stepanovsky Yu.P. (1982) Conical refraction of spin 1/2 particles, *Jadernaja fizika*, **35**, 336-339.
33. Sannikov S.S. (1965) On noncompact symmetry group of oscillator, *Zhurnal chksperimental'noj i teoreticheskoj fiziki*, **49**, 1913-1922.
34. Stepanovsky Yu.P. (1998) Sonoluminescence and black holes as sources of squeezed light, in J. Wess, V.P. Akulov (Eds.), *Supersymmetry and Quantum Field Theory*, Springer, Berlin, pp.
35. Stepanovsky Yu.P. (1966) Six-dimensional form of Dirac matrix algebra, *Ukrainskij fizicheskij zhurnal*, **11**, 813-824.
36. Stepanovsky Yu.P. (1966) Complete set of Fierz's relations in six-dimensional form, *Ukrainskij fizicheskij zhurnal*, **11**, 1191-1197.
37. Buchdahl H.A. (1968) On the calculus of four-spinors, *Proc. Roy. Soc.*, **A303**, 355-379.
38. Landau L. (1930) Diamagnetismus der Metalle, *Zs. Phys.*, **64**, 629-638.
39. Fock V. (1928) Bemerkung zur Quantelung des harmonischen Oszillators im Magnetfeld, *Zs. Phys.*, **47**, 446-448.
40. Rabi I. (1928) Der freie Elektron in homogen Magnetfeld, *Zs. Phys.*, **49**, 507-511.
41. Klitzing K., von, Dorda G., Pepper M. (1980) New method of very precise establishment of the fine structure constant value based on measurement of quantized Hall resistance, *Phys. Rev. Lett.*, **45**, 494-499.
42. Weisskopf V.F. (1982) Growing up with field theory, *Uspekhi fizicheskikh nauk*, **138**, 455-475.

DUALISM BETWEEN SPACE-TIME AND MATTER ANALYZED FROM EPISTEMOLOGICAL POINT OF VIEW

RENATE WAHSNER
Max Planck Institute for the History of Science
Wilhelmstr. 44, D-10117 Berlin
wahsner@mpiwg-berlin.mpg.de

Abstract

The forming of concepts and theories in physics is essentially determined by measurement requirements. Thus it is not possible to discuss about problems of the forming of physical concepts and theories or about physical truth by abstracting from measurement and its requirements. The fact that measurement-theoretical determinations must be both presupposed and attached to dynamics is called here "dualism".

The basic dualism of physics is that of space-time and dynamics. Classical mechanics determined this dualism for whole physics as far as it occupies a specific position in physics, a position that, again and again, makes it necessary to refer to the basic concepts developed by it. However, this dualism must be given a new definition by the respective physical theory.*

1. Introduction

The specific principle of physics to think motion, in its fundamental epistemological structure, was established by classical atomism and perfected in the following centuries for the purpose of physics. The philosophical system of Greek atomism developed for the first time the principle of a physical representation of nature [57]. This principle bases on the atomistic main idea to reduce the whole world, all existing, to the atom and the void, to the full and the empty, to being and non-being, to body and space. By this reduction it was established the conception of the reality of entities being opposed to each other: The being was presupposed as real as the non-being, discontinuity as real as continuity, rest as real as motion. Both momenta are to exist really but they must have existences that are separated from each other and that exist as two determined beings that are differentiated from each other and belong together at the same time. This distribution of conceptual momenta really existing only in unity, the distribution of opposite determinations and their reformulation into relatively independent entities made it possible to grasp the motion in such a manner that motion could be thought free from logical contradiction – motion that according to Hegel is the existing contradiction. Hegel explains: "Something moves only, not when it is in this Now here and in another Now there, but when it is in one and the same Now here and not here, when it is and is not in this Here at the same time. We have to admit that the old dialecticians were right in indicating contradictions in the motion; however, it does not follow from it that the motion is not but rather that the motion is the existing contradiction itself."[1]

P.G. Bergmann and V. de Sabbata (eds.),
Advances in the Interplay Between Quantum and Gravity Physics, 459–475.

The ancient thinker Lucretius Carus composed poetry:

"All nature, then, as self-sustained, consists
Of twain of things: of bodies and of void
In which they're set, and where they're moved around.
...
And thus,
Beside the inane and bodies, is no third" [23, book 1, verse 419-438].

Their philosophical system was the atomists' reaction to the recognition, arrived at by predecessors, the Eleatics and the Heracliteans, that motion cannot be conceived without contradiction and therefore, as they concluded, cannot be conceived at all. This recognition took form in two opposing viewpoints. The Eleatics considered that in truth the world was motionless and comprehensible; it was conceivable, just because of its motionlessness. In the eyes of the Heracliteans, on the other hand, the world is in permanent motion, but precisely for that reason it cannot be known or conceived but can at most be adequately reflected in sensual perception. Motion seemed unthinkable both in rationalist-Eleatic and in the empirist-Heraclitean conception, because not-being was seen as not thinkable. For what was – like not-being – not an object, what thus has no specificable features, cannot be conceived in thought. Consequently the idea that something is and is not, that something by being in motion is at one place and simultaneously is not there, is likewise taken to be an unthinkable contradiction. The atomists solved the problem of motion, of conception and of reality which their predecessors had posed from opposing viewpoints – a problem which sprang from the insight that the thinking (always in some way discontinuous) and real existence (in the last analysis always continuos) cannot directly correspond. Thus the atomists solved a problem springing from the fundamental philosophical question which here was explicitly posed for the first time.

The ancient atomists found a highly ingenious escape from this dilemma: by means of their reduction of everything that exists to atoms and void they exhibited the possibility of conceiving not only what is but also what is not as object, so that being and not-being can be thought of as coexisting in the world. Atoms represent being, the void represents not-being, and one is as real as the other. The reality of what is not splits up what is, so that the being may now have different and contrary descriptions. What is is continuos and discontinuous, indivisible and divisible, constant and variable, infinite and finite. (And it is no longer impossible to think that something, by being in motion, both is and not is.) The first term in each pair concerns the single atom, the atom by itself, the second concerns compositions of atoms, that is to say macroscopic bodies and more in general atoms in relation among each other. That it is the same being to which these contrary descriptions apply is guaranteed by the fact that all atoms are qualitatively equal: they all have only size and form, the differences among them are merely quantitative.

This atomistic distribution of momenta for the first time in the history of human thought permitted discriminating between matter and space. It made it possible to grasp motion conceptually, in such a fashion that the contradiction constituting motion can be thought physically, i.e., in such a way that motion later could be made measurable and calculable. Without the concept of space the simultaneous reality of being and not-being could not have been maintained. But now space appeared as the condition for making motion possible and therefore also as the condition for the possibility of comprehending, of thinking about the world – such that logical contradictions do not arise.

To recognize that thought and reality are not identical and to recognize the contradictoriness of motion thus formed the condition for developing the physical thought-principle. However, the atomists' solution for making motion thinkable and bringing thought and reality into coincidence was only one possible solution of this problem of motion, thought and reality. It was that particular solution that took nature as its undoubted presupposition. It conceived the entire universe as nature; consciousness, the subject of cognition, functions merely as an external observer.[2] This is just the point of view of science. The other possible (and indeed necessary) solution of the problem is the philosophical one, embodied in Platon's dialectics.

The atomist conception made motion physically thinkable; but it could not establish a necessary connection between this thought-principle and sensual perception. In other words, while ancient atomism developed the principle of physical thinking, it could not establish a principle of physical sensibility. For this purpose atomism had to be modified.

In the original atomist view the atom is in essence postulated as a single, unrelated individual. But making the particle's individuality absolute in this way cannot provide a sufficient gnoseological basis for physics, at least not in a way that would link theory and measurement. The limit of the atomist view appears in its attempt to explain the diversity of the world, the genesis and decline of nature, by means of properties that may be ascribed to a single body, to an atom as a single. The isolation of the physical body, conceived as an atom, later was eliminated by the physical (not the mechanistic) concept of force, by the laws of physical interactions. The physical concept of force comprises the idea that natural bodies act on each other, that in referring or better said relating to each other they exert a force, and therefore are not something strictly isolated and essentially singular.

The atomistic approach, of course, brings about metaphysical mechanicism if reality is assumed to be so *per se*. If it is taken as a method, however, which does not operate with real entities but with thought-entities (physical quantities, for example), this approach founds the principle of physical thinking.

The characteristic of this above-mentioned principle implies that dualisms occur in physics necessarily: the dualism of the space-time structure and physical interactions, that is, of geochronometry and dynamics, of particles and field, of passive and active principles, of topology and metrics etc.[3] In these cases, the two sides of a dualism within the framework of a respective physical theory are not on the same footing but the first side bears the character of a precondition related to the basic equations of the respective theory. This different weighting of the two sides was not yet given by classical atomism because classical atomism made the physical motion thinkable but – as indicated above – could not establish a necessary connection between the principle of thinking as developed by it and the sensual perception (because it could not say how to *produce* thought-entities, measuring quantities). The different weighting of the two poles of dualism as appearing in (developed) physics is conditioned by the requirements of measurement and calculation, by the physical method to link sensuality and understanding.

The way towards physics as an empirical mathematized natural science can also be characterized as endeavor to find out what has to be determined as the presupposed side of dualism and what as the side that has to be explained physically. Galilei was the first to go the physically constructive step in this direction when he presupposed the rectilinear uniform motion as the basic property of bodies. On the basis of making a difference between rectilinear uniform motion and motions deviating from them Newton was able to develop the classical mechanics where, by means of physical dynamics, only motions that deviate from rectilinear uniform motions are explained.

As physics conceives the contradiction constituting the motion by the above-mentioned distribution of momenta, by the distribution of opposite conceptual determinations it must make a certain divide between the two sides of the dualisms. The history of the development of physics has shown that it is of importance how this divide between the existing-in-unity momenta of real motion is made.

For instance, the duality of natural and forced motions – the conception developed by Aristotle and having its effects until the time of Renaissance – proved to be inadequate for being used as the conceptual basis of physics. As the concept "natural motion" was determined on the basis of a geocentric point of view and was deduced from the material nature of bodies, it was not possible to determine it in a uniform manner for all motions in nature: a difference had been made between supralunear and sublunear (and in the latter between falling and rising) natural motions. Thus physics as a science of the whole nature could not be established, even not a physical dynamics of the terrestrial nature.[4]

But even a heliocentrically oriented distribution of momenta abstracting from the material nature of bodies and avoiding a breaking of the world into two fields is physically not capable of bearing if, like in the conception of Kepler, for all motions existing in the world the rest is determined as the moment which must be presupposed.[5] Kepler said that if there was only one body in the world, it would be in rest because rest was the basic property of matter; if a second body was brought into the world, however, these two bodies bring each other into motion. As a conclusion for the system of planets he wrote that as all matter tends to rest in a place where it is (unless attracted by a magnetic power of a neighboring body), it happens that the moving power of the sun gets into conflict with this inertia of matter, like two weights struggle with each other on the balance, and that from the ratio of the two forces it results a higher or lower speed of the planets.[6] Later one succeeded in describing the motion of celestial bodies as a conflict between inertia and gravitation but one succeeded in doing so only by another determination of inertia.[7] The principle of inertia being a fundamental of classical mechanics postulates the rectilinear uniform motion equivalent to rest as the basic property of physical bodies and thus the gap between rest and motion is excluded right from the beginning.[8] In this sense, d'Alembert writes: "The motion and its general properties are the first and most essential subject of mechanics; this science presupposes the existence of the motion."[9] Thus mechanics makes the necessary divide not between rest and motion but between rectilinear uniform motion and accelerated motion, divide in the sense that the first is supposed to be a motion etalon and that only motions deviating from it are explained by physical dynamics. – Of course, this determination of a standard motion by Newton is opposite to the basic principle of the general reciprocity of motion as formulated by Galilei and Huygens, supported by Leibniz and discussed by Kant. But the approach of Newton was the only one which made it possible to measure the motion.[10]

Seen from this viewpoint, the prehistory of physics presents itself as endeavor to find how the opposite conceptual momenta really existing in a unity must be distributed in order to make the motion not only thinkable (physically thinkable) but also measurable and calculable. In the following history of physics, one aim was to find out whether the divide can be made in another way (for example, in such a manner that less "*a priori*" determinations were necessary) [40; 45; 54; 60; 62] or whether it can be avoided generally (what is attended by unified theories).

2. The Galilean-Newtonian determination of the dualism of space-time and dynamics

Thus classical mechanics as the first natural science in its proper sense defines the presupposed "natural" motion by the principle of inertia. In his *Principia* Newton presented this principle as the first law in the following form: "Every body remains in its state of rest or in rectilinear uniform motion unless forced to change its state by acting forces."[11]

The term "natural motion" used for historical reason is not to provoke the idea that this is a motion that can be seen from every-day experience. It is very unlikely to find a rectilinear and uniform motion in nature. To realize it, one has to create ideal conditions artificially.[12] The equivalence of such a motion with rest is also not sensually perceptible but a theoretical conclusion that was only possible after finding out that the rectilinear uniform motion can be defined uniquely by the quantity *velocity*.[13] Compared with this statement, the Aristotelian "natural motion" was far closer to every-day experience. But this was just the reason why it was only possible to establish a phenomenological natural philosophy on this basis, and why these conceptual determinations could not be used for developing a physical theory [64, pp. 242-256].

The formulation of the principle of inertia proves that physics does not think the separation of body and motion even under the aspect of force-free motion and does not presuppose an unmoved body or an unbodily (*ungegenständliche*) motion. That assumed separation presupposes that it is reasonable or even necessary in physics to make a difference between a body in rest and a body in motion in such a manner that the motion, as an additional property, is attached to the body definable *per se*. But the principle of inertia excludes a qualitative difference between rest and motion. This does not imply, of course, that no difference could be made between rest and motion in physics. A science that is to measure the motion can and must be able to state whether a velocity is equal to zero or different from it. The fact that rest and motion are equated qualitatively and do not represent two different qualities is shown in that they are given by different values of one and the same quantity, that is, of the velocity.

To avoid misunderstandings let us to say that the body is not liquidated in kinematics which studies the motion (the accelerated, too) of physical bodies and systems without taking the moving forces into consideration. However, the body can be characterized more precisely only by the subsequent dynamics for which kinematics is constructed as a presupposition. Seen from a purely kinematic point of view, it occurs only as a moved "something". If one wants to interpret kinematics as a natural theory – as Galilei had done because of the lack of physical dynamics – this will be possible only to the extent of succeeding in replacing the properties of bodies, following from the dynamics in case of an elaborated physics, by additional assumptions. When Newton elaborated the classical mechanics, he presumed the moving "something" as mass – thus anticipating certain results of the dynamics – that can be measured by weight of the respective body as he had found out by his "very precise pendulum tests". Newton designated the resistance against a change in the state of motion as the basic property of this mass. Definition 3 of his *Principles* reads: "Matter is able to resist; therefore, each body, as far as itself it is concerned, remains in a state of rest or in uniform rectilinear motion."[14] This ability proportional to the body or the mass "is different from inertia only as to the approach", Newton said [27, p. 3].

As mentioned above, it is necessary to create ideal conditions for producing a uniform rectilinear motion; this means that the respective body has to be isolated from its physical environment so that it cannot be effected by any forces. Since we can suppose that it is possible to realize a uniform rectilinear motion, one has to assume the possibility of isolating physical bodies or systems. Reversibly, the principle of inertia shows what one considers as an isolated body or system in physics.

However, the principle of inertia does not only contain the thought of the ability to isolate but also the necessary reference to a second object. Without such an object, motion cannot be noticed generally. But talking about physics, the object of reference has to satisfy demands: There must not be any interaction between the body in rectilinear uniform motion and the object of reference (as it has to be a force-free moved body), and the object of reference be of such a nature that a rectilinear uniform motion can be noticed in reference to it, that an Euclidean straight line can be defined by it. Such an object of reference is just the homogeneous, infinitely expanded Euclidean space. This space can be considered as a specifically prepared body which makes it possible to define the rectilinear uniform motion for measurement. This necessity was pointed out by d'Alembert: "In order to get a clear idea about motion one is not able to dispense with two space ideas at least in silence. ... One considers the space as impermeable and as the thing what is called the body; the other considers it simply as a spatial extension without investigating the fact whether it is permeable or not, that is as the measure of the distance between one body and the other whose parts conceived solidly and may serve for evaluating rest and motion of bodies."[15]

The fact that, for reasons mentioned above, the principle of inertia contains the Euclidean space as a necessary presupposition clearly shows that this physics can think motion only as a correlation of bodies even in the case where the mutual effects of bodies must be excluded; thus the concept of an unbodily motion does not occur in physics even not in connection with the purely kinematic aspect.[16] The object of reference is *quasi* taken as a body without behavior but it is determined as body in such a way that it contains the reference to other bodies implicitly.

Thus, the "natural" motion presupposed by classical mechanics is determined via the principle of inertia; as etalon of motion, it is used to measure accelerated motions deviating from it and being produced by forces. Basically, the sense of physics consists in the lawful representation of these motions by means of dynamics; on the other hand, only dynamics clearly shows the genuine importance of the principle of inertia as described so far. To consider the problem how the above-cited contradiction is grasped through the physical motion, it is indispensable to show how in dynamics in connection with its kinematic presuppositions the relationship between physical body and physical motion is described.

The core of the Newtonian dynamics is the second axiom saying that the change of the velocity of a body, that is, the acceleration, is proportional to the acting force.[17] To come to a law describing a specific physical interaction, a force term is to be inserted for the force that is limited by axiomatics but not determined by it. For example, when the force term of gravitational force is used in the second axiom of Newton, one gets the law of gravitation. This law grasps the gravitational interaction of n bodies in such a way that the acceleration of each of these bodies can be determined through its gravitational interaction with the other $n-1$ bodies. The second Newtonian axiom just wins a physical sense from the fact that it is clarified in relation to which reference systems the forces and accelerations contained in this axiom are defined; these are those determined by motions of n bodies if the forces acting between them were excluded (that is, if the bodies were isolated in the sense of the first axiom), and by all motions being rectilinear and uniform with respect to them. The fact that, in case of a given term of force $\mathfrak{F}$, the equation of motion determined by the second axiom of Newton are valid with respect to systems of reference (inertial systems) defined in this way is expressed in the Galilean principle of relativity [38, pp. 15-23, 65]. The class of inertial systems defines the physical space (within the framework of classical mechanics) on which, as background, the (accelerated) motion of the physical matter (example is the above-mentioned gravitational interaction of n bodies) is described.[18] Thus the absolute space of Newton is not a transcendental phantom but a product of idealization that is to determine the standard motion. This function of the absolute space was not recognized for a long time [51]. Even at the end of the 19th century, it had been necessary to clarify this point [20; 24; 26].

The distribution of momenta of the contradiction that appeared in kinematics as the dualism of space-time and velocity is changed into the dualism of space-time and acceleration by passing over to dynamics so that the kinematic space-time is different from the dynamic space-time due to the different distribution. The new, dynamic dualism makes it possible and requires the introduction of new quantities (mass, charge) that cannot be created by the combination of space and time quantities alone. As to the conceptual basis of physics, the dualism of space-time and matter proved to be decisive, and all the other dualisms or conceptual distribution had to be subordinated to it.

3. Dualisms in modern physics

The dualism of space-time and matter, being a basic feature of classical physics as described above, must be given a new definition by respective physical theory. Classical mechanics determined this dualism for whole physics as far as it occupies a specific position in it, a position that makes it necessary to refer to the basic concepts developed by it again and again [cf., e.g., 5; 6; 14; 31; cf. also 32, pp. 21, 50, 55, 71; 22; 39; 46; 52]. But this does not mean that the border between the two sides of dualism could not be shifted by other physical theories.

General relativity theory took the gravitational interaction from the side of dynamics to the side of space-time by the geometrization of gravitation. Thus the dualism was defined basically newly: the gravitational ether – a term introduced by Einstein and Weyl – replaced the class of inertial systems forming the absolute space; the gravitational ether unifies gravitation and inertia to form a new background on which all non-gravitational interactions are described and measured. The former rigid Newtonian space-time background is changed in a what Einstein called reference mollusc (*Bezugsmolluske*). Otherwise, because rigid measurement rods are necessary for measurement, the theory of general relativity is also forced to refer to a *quasi*-Newtonian situation if it is to establish a connection to the experiment, if it is to establish the possibility of physical interpretation. As long as relatively weak gravitational fields are concerned, this is possible rather easily. However, in case of strong gravitational fields, the modeling of the reference mollusc by means of *quasi*-rigid Newtonian space-time is not possible, even not approximately. The resulting difficulties linked with the revealing of the physical content of general relativity theory are one of the most important topics in discussions dealing with modern gravitational research today. What matters is to establish a connection with classical mechanics for measurement-theoretical reasons and, at the same time, to elaborate a theory with which it is possible to grasp physical situations that differ widely from situations still described by classical mechanics.

Since the geometrization of gravitation also means a physicalization of geometry, as far as space-time structures were identified with the physical field of gravitation, not all space-time determinations are presupposed now but are dynamically determined partly. Thus the *a priori* measurement-theoretical part of general relativity theory was reduced as compared to that in classical mechanics [64, pp. 242-259].

In quantum theory, the necessity to determine the dualism of space-time and dynamics in a new manner is especially evident [48; 49; 54]: The basic dynamic law of that theory, the Schrödinger equation, can only be formulated in a many-dimensional configuration space which is basically different from the space determined measurement-theoretically. Therefore, Heisenberg designated this dualism of space-time and dynamics even as a cut. He said: "The situation of any experiment in nuclear physics is as follows: We use more or less complicated apparata for putting questions to nature, questions that are always directed to the definition of any objective process in space and time ... . Thus it is an automatic result that in dealing with the process mathematically, we make a cut between the apparata we use as auxiliary means for putting our questions, on the one hand, ... and the physical systems about which we want to learn something, on the other. The latter, we represent mathematically by a wave function and for this wave function a differential equation exists according to the quantum theory which determines the future state of the wave function on the basis of the present one. As to the apparata, however, we content ourselves with laws formulated in classical concepts which provided us with the right to use them as apparata of measurement. The cut between the system to be observed and the measurement apparata is given by our questions naturally but it does not show a discontinuity of the physical process evidently."[19]

To make at all a linkage to experience, that is, to establish a connection between dynamics and measurement determinations, axioms relating to the measurement process are added to the Schrödinger equation. They indicate how to come to the three- or four-dimensional space-time of measurement via the projection of the multi-dimensional configuration space. By this, the gap is closed between the dynamic and the measurement spaces in such a manner that space-time of measurement and the Schrödinger equation do not exist out of context with each other but both of them become part of one and the same physical theory.

Thus, we are dealt here with the case that space in which basic physical equations are formulated does not coincide with the three- or four-dimensional space of experience where measurement apparata are acting. In classical mechanics, the specific case occurs that dynamics is related to this measurement space (the absolute Newtonian space). As long, as one only knew classical physical theories, the opinion was prevailing that both kinds of space were identical. One knew, of course, that mathematically abstract spaces can be introduced that allow a more elegant formulation of dynamics under specific conditions. However, it was always possible to formulate dynamics relating to the measurement space, too. This is different as to quantum mechanics.

The dualism of space-time and dynamics leads "from the determination of Schrödinger's wave equation in the configuration space to probabilistics in the physical space".[20] The probabilistic aspects of quantum mechanics result from a linkage of the Schrödinger equation (or its relativistic generalization like Klein-Gordon equation, Dirac equation or Fock equation) to measurement. In a lecture given in 1932, Max Planck expressed this situation. He said: "Generally, the laws of material waves are completely different from those inherent in the classical mechanics of material points. What matters, however, is the fact that the function characteristic for material waves, the wave function or probability function (it does not matter which name is used), is completely determined by the initial and boundary conditions for all places and times, according to certain rules of calculation, irrespective of the fact whether the *Schrödinger* operators are used or the *Heisenberg* matrixes or the *Dirac* q-numbers. – We see: determinism governs in the world picture of classical physics, it is only the symbols and stipulations of calculation that are different. According to this, in quantum physics, like in classical physics before, the uncertainty of predictions relating to events in the sensual world is reduced to the uncertainty of the connection between world picture and sensual world, that is, to the uncertainty of translation of world-picture symbols to sensual world, and vice versa."[21]

Thus it is clearly shown that it is a misinterpretation saying that the so-called statistical causality brings a subjective element into the description of phenomena. This epistemological error – the founders of the quantum mechanics did not make it – results from the insufficient analysis of the conditions under which phenomena in physics are studied [9, p. 593].

These considerations invalidate arguments saying that quantum mechanics disproves the concept of reality. According to these arguments, the reality concept was said to have been falsified because quantum mechanics had proved the impossibility to describe nature in a causal manner. The scheme of such an argumentation starts from the difference between quantum mechanics and classical mechanics and concludes that, as quantum mechanics would be modern physics at all, the concept of a causality finds no place in modern description of nature and thus the strict lawfulness would be disproved. This approach, however, is an absolutization of classical mechanics by designating its laws alone as adequate to a complete, lawful description of nature. However, this is not the standpoint of physics but of philosophical mechanicism. On the other hand, one underestimates the importance of classical mechanics by ignoring the necessary linkage of quantum mechanics, expressed through the Bohr principle of complementary and the principle of correspondence (Ehrenfest theorems), to classical mechanics; but only this linkage makes the quantum mechanics physics.

Thus, Einstein's discontentedness with quantum mechanics did not result from the fact that he considered this physical theory as being indeterministic (what he did not do). He rejected the attempts of Bohm and de Broglie as too simple attempts to describe the quantum mechanics deterministically [8, p. 258] and he denied having used the question "Is the theory strictly deterministic?" as a criterion for an admissible theory [28, p. 293]. As he said, his concern was the so-called real description. Einstein regarded the quantum mechanics as an incomplete and indirect description of reality, which would be replaced by a complete and direct one later [7, p. 234]. Rather he hoped that the development of physics would result in a theory in which measurement space and dynamic space would coincide. Einstein's effort aiming at achieving a real description did not arise from a naive philosophical conception, a mechanistic one, for example, but from a profound physical program. His aim was to link modern physics to classical physics without any sharp cuts by formulating the dynamics again with an immediately reference to the experience space.[22]

All discussions about quantum mechanics are concentrated on the topic to solve the problem of the cut between systems to be studied and the measurement apparatus, a problem occurred in quantum mechanics *explicitly* for the first time. As this cut is due to different spaces (dynamic space and experience space), the attempt to unsharpen it results in a revision of the respective space concepts. (Discontinuities occurring in the projection of dynamic space to the measurement space, for example, that is, discontinuities occurring through the reduction of the wave function, would be different or would be avoided if the space to which the projection is made had an other structure than the Galilean-Newtonian space.) By founding the special and general relativity theories, Einstein just had made an extremely successful revision of space-time and thus of the physical experience space, and he hoped to achieve a success when he continued his work in this direction. For this aim he developed his program of a unified geometric field theory.

But to what an extent a modification of the experience space is actually possible, can only be judged by means of real measurement possibilities. For what matters is not the space of everyday experience but the space of scientific (measuring) experience of physics. Thus it is not enough to designate the abstract mathematical configuration space as the physical experience space, but this new space had to be linked with an expansion of the presupposed measurement basis of physics. – This is an opposite result compared with that one of relativity theory.

Contrary to Einstein's program, many authors start from the credo "All nature is quantum", or to say it in Weinberg's words: "... quantum mechanics governs eveything in the universe, not only individual electrons and atoms and molecules but also the experimental apparatus and the physicists who use it" [66, p. 64]. As a consequence of this standpoint, those authors try to eliminate the cut by describing the measurement apparatus as well as the systems observed and the "conscious observer" by wave function with certainty evolving to a final wave function, in which the probabilities of the individual measurements are what are prescribed in the Copenhagen interpretation. So the cut in quantum mechanics is eliminated by an extension of the representation-space dynamics.

While the dualism between space-time and matter was weakened in general relativity it was made tenser in quantum physics. This contrary situation becomes relevant seeking to unificate these both theories or, respectively, to quantize gravity.

To solve this difficult problem, from a standpoint outside of physics one often argues that all existing consists of particles and therefore gravitational field should consist of gravitons. But this view overlooks the fact that gravitational field is a thought-entity not a sensuous-concrete object. Of course, from the standpoint of physics, such an argumentation is not suggested. From this standpoint one is aware of the specific nature of physical objects.

The difficulty one meets in unifying general relativity and quantum physics becomes especially obvious when one casts a short look at the attempt to do this in analogy to the foundation of quantum electrodynamics. The arguments used here are based on the paper "Zur Frage der Meßbarkeit elektromagnetischer Feldgrößen" presented by Bohr and Rosenfeld in 1933 [2].

Bohr and Rosenfeld analyzed here the connection between classical physics and quantum theory and used their results as the starting point for the quantum theory of fields. They showed that classical physics is important for the quantum field theory in a protophysical sense because measuring bodies are always macrophysical bodies and thus objects of classical mechanics as to essential properties. The dualism occurring in the physical determination of motion appears here in a certain manner as the connection between classical mechanics and quantum field dynamics.

Bohr and Rosenfeld demonstrated that there are two aspects under which the measuring bodies have to be considered as classical ones. First, the quantum structure of the measuring body has to be ignored, the measurement body has to be taken as classical as to its structure. Bohr and Rosenfeld write: "It is true that the measurement of fields requires the use of material charged test bodies [i.e., bodies with a quantum structure], but their unambiguous application as measuring instruments depends exactly on the extent to which we can treat their response to the fields as well as their influence as field sources on the basis of classical electrodynamics" [2, p. 126]. Second, measuring bodies must be uniformly charged, finely extended, rigid bodies in the sense of classical mechanics. The argument that the existence of rigid bodies contradicts the finite propagation velocity of all physical actions, that is, the theory of special relativity, has to be rejected by the argument that one must not think of rigid bodies *per se* but of an artificial product à la Galilei. This body can be an extremely complicated physical system which may differ from a rigid body in a sense of mechanics in many aspects. It has only to be constructed in such a manner that, relating to field measurements mentioned above, it closely approximates the behavior of a rigid body.

Bohr and Rosenfeld showed that the measuring bodies must have the two classical properties mentioned above in order to avoid contradictions between the measurement determinations and the quantum field theory. Such contradictions must, however, be avoided because their occurrence would mean that the dynamic field theory – since not resulting in measurable consequences – would have no physical meaning. They stated: "All measurements of physical quantities, by definition, must be a matter of the application of classical concepts; and that, therefore, in field measurements any consideration of limitations on the strict applicability of classical electrodynamics would be in contradiction with the measurement concept itself" [2, p. 153].

The precedeness of classical physics in the sense of the necessity to presume a classical structure of measuring bodies, however, does not mean – as repeatedly mentioned above – that one has solved for all times the question which aspects of classical physics are important or not for a new theory or where the "cut" has to be made between the measurement-theoretical part and dynamics. Here again, we come to the conclusion also resulting from the discussion about the ambiguous position of the second Newtonian axiom and saying that the etalon of motion cannot be gained independently of dynamics.

It is quite evident that the clarification of the measurement-theoretical basis of the quantum field theory is more than a subsequent justification of functioning physics but it can help develop physical theories. The discussion about the principles of measuring electromagnetic fields shows how the quantum theory of fields has to be developed. It must be such a theory that fundamental relations like the uncertainty relation arising from the measurement determinations also result from the formalism of the quantum field theory. Thus it is explained, so to speak, what it means to transfer quantum mechanics to fields.

The great value of these investigations became evident, above all, in the foundation of the quantum theory of gravitational field [42; 43; 44]. Indeed, the arguments of Bohr and Rosenfeld were used for the gravitational field more than four decades ago in order to get guide-lines for quantisizing the gravitational field [4; 29; 33; 34; 37; 67; 68]. In particular, the demands of Bohr and Rosenfeld on a physically reasonable measurement together with the principle of strong equivalence lead to the existence of a smallest length as a limitation of quantum field theory.

Since, due to the principle of equivalence, the given relations impose stronger limitations on the measuring of the gravitational field than on the measuring of electromagnetic field, one obtains a quantum theory of gravitation which is strongly different from the quantum theory of the electromagnetic field. The principle of equivalence makes it possible to geometrize the gravitation field by identifying the potential functions describing the gravitational field with the metrical functions $g_{\mu\nu}$ characterizing the geometry of a Riemannian space-time. This so-called geometrization of gravitation can also be read reversibly, namely as the physicalization of geometric quantities in the sense of their dynamization. This can be done because in general relativity theory the geometrical quantities $g_{\mu\nu}$ become physical field quantities determined by basic dynamical equations, namely the Einstein gravitational equations. There were given arguments showing that, as a consequence of the principle of equivalence in connection with the quantum principle, the quantization of the gravitational field leads to limitations on the measurability of this field given by the Rosenfeld conditions. They say that the regions of the dimension of the Planck length are not accessible to measurement. Therefore, on the one hand, the principle of equivalence provides the possibility to geometrisize gravitational fields and, on the other, it gives limits, together with the quantum principle, under which the developed quantum field theory loses its physical meaning, under which gravitation cannot be interpreted as a usual field. This principle makes the geometrization of gravitation (or the physical dynamization of geometry) possible, i.e., it provides the possibility to liquidate partly the dualism of space-time and dynamics. And at the same time, it watches over that this geometrization, i.e., the liquidation of the dualism, is done only to such extent that space-time can continue to play its role as protophysically (measurement-theoretically) founded background. All this shows clearly that the dualism of measurement determinations and dynamics as developed by classical mechanics and substantiated, in different ways, by theory of general relativity and quantum theory must not be abandoned in principle unless one takes the risk to rob physics of its content.

Our considerations demonstrate the close connection between measurement-theoretical presuppositions and physical dynamics. On the one hand, just the analysis of measurement determinations shows the ways towards a quantum field theory and its limits. Otherwise, this example shows also that the measurement theory does not precede dynamics absolutely but dynamics also decides upon the fact what can be measured and how the measuring body has to be. In particular, the demand for the existence of a field theory the quantum effects of which can be measured, leads to presume a measuring body whose mechanism of displacement and momentum transfer are subject to the laws of quantum mechanics. For the measurement of a classical (not quantized) field theory, however, a measuring body would have been necessary which moves according to the laws of classical mechanics.

4. Conclusion

Physics takes the dialectical contradiction constituting motion as dualism; this results from the fact that it has to determine and presuppose etalons in order to become a measuring science. Naturally, the etalons must not act on physical bodies (or: they may only act in *one* well-defined manner, under *one* well-defined aspect) and they must not be effected by them (or: they are only influenced by this *one* action). On the other hand, physical bodies are considered as mutually effecting. Of course, such a situation does not exist in nature and thus it has to be constructed in a experimentally and theoretically "artificial" way.[23] Therefore, one is not dealing with concrete bodies but with abstract ones to which, however, one specific property or behavior is ascribed. Thus a measuring science will never be able to conceive the world in its totality; it can only be valid to the extent to which the abstraction made reflects an essential connection in a good approximation. It does not describe real but possible modes of behavior of real objects.

In discussing typical features of classical mechanics, of the theory of relativity, and of quantum mechanics one can see that measurement-theoretical requirements find their expression in the appearance of different dualisms. By them, physics was founded as discipline that is able to measure and to calculate motion. As a consequence, the dualisms are characteristics of physics. They can have different forms in the respective theories but it seems that they can never be avoided.

The fundamental physical dualism is that one of space-time and matter or, respectively, of geometry and dynamics, which was for the first time successfully determined by classical mechanics. This theory determined the dualism under consideration for whole physics as far as it occupies a specific position in physics, a position that, again and again, makes it necessary to refer to the basic concepts developed by it. However, this dualism must be given a new definition by the respective physical theory. In particular, concerning the new determination of this dualism, general relativity theory and quantum theory took completely different ways. From this result the problems one meets in the unification of both these theories.

Moreover, the necessity of the dualism imposes limitations on physical conceptions like geometrodynamics, cosmology, quantum theory without any cut, and the quantization of gravity.

5. Notes

*This contribution is based on a joint paper with H.-H. v. Borzeszkowski [55, parts 3, 4.0, 4.2].

1 "Es bewegt sich etwas nur, nicht indem es in diesem Jetzt hier ist und in einem anderen Jetzt dort, sondern indem es in einem und demselben Jetzt hier und nicht hier, indem es in diesem Hier zugleich ist und nicht ist. Man muß den alten Dialektikern die Widersprüche zugeben, die sie in der Bewegung aufzeigen; aber daraus folgt nicht, daß darum die Bewegung nicht ist, sondern vielmehr, daß die Bewegung der daseiende Widerspruch selbst ist" [11, p. 76; see also 12, p. 314].

2 See Schrödinger's principle of objectivization [36;54;59].

3 This point of view was founded in [47]; it was elaborated in [50].

4 As to the problems of the conception of peripatetic motion and its reception, see [69; cf. also 56, pp. 21-31].

5 As to the contribution of Kepler to the development of the theory of physical motion, see [41; 52; 58].

6 "Da jegliche Materie zur Ruhe neigt, an dem Ort, an dem sie sich befindet (wenn nicht ein benachbarter Körper sie durch eine magnetische Kraft an sich zieht), geschieht es, daß die bewegende Kraft der Sonne in Widerstreit gerät mit dieser Trägheit der Materie, wie an der Waage zwei Gewichte miteinander streiten, und daß aus dem Verhältnis der beiden Kräfte schließlich eine größere oder geringere Geschwindigkeit der Planeten herauskommt" [18, pp. 105 f.].

7 Koyré also underlined this point of view strongly when he said: "We see that Kepler, the great Kepler, the founder of modern astronomy, the same man who proclaimed the unity of matter in the whole universe and stated that *ubi materia, ibi geometria*, failed to established the basis of modern physical science for one and only one reason: he still believed that motion is ontologically on a higher level of being than rest" [19, p. 12].

8 In his Introduction to the *Mathematical Principles*, Newton writes: "... motion and rest alone as they are seen normally, differ from each other by the kind of relation, and those bodies are not always in rest, that are considered to be in rest normally". ("... sed motus et quies, uti vulgo concipiuntur, respectu solo distinuuntur ab invicem; neque semper vere quiescunt quae vulgo tanquam quiescentia spectantur" [27, p. 4].)

9 "Le Mouvement et ses propiétés générales, sont le premier et le principal objectde la Méchanique; cette Science suppose l'exostence du Mouvement ... " [3, p. V].

10 This statement does not deny the importance of the contributions of other scientists before and after Newton for the development of what one calls today Newtonian mechanics.

11 "*Lex I.* Corpus omne perseverare in statu suo, quiescendi vel movendi uniformiter in directum nisi quâtenus a viribus impressis cogitur statum illum mutare" [27, p. 15].

12 In recognizing this necessity and the importance of this necessity for the formulation of physical concepts, the constructive philosophy of science in its protophysical form makes a difference between two processes of forming concepts – the *abstraction* and the *ideation*. According to this conception of philosophy of science, concept to be used in the theory of measurement can only be formulated by the latter. Paul Lorenzen wrote: "In contrast to the Aristotelian abstraction ... the ideation also contains standards of action, for example for the production of planes and orthogonalities" [21, p. 63].

13 See also P. Jaeglé, who writes: "The velocity can become an essential feature and one of the dimensions of the motion only when it is presented *as constantl, as invariable in the course of time and as preserved in the course of motion*. Place and moment are changing but they remain identical as to their intervals irrespective of the form of the latter and how small they are thought. Outside this identity which changes an interval of time in such one of space, there is no exact concept of velocity and there is no mechanics!" ("La vitesse ne prend corps comme caractéristique essentielle du mouvement – comme l'une de ses dimensions – que pour autant qu'elle puisse être conçue comme *constante, invariable au cours du temps, conservée au cours du mouvement*. Le lieu et l'instant changent mais restent identiques dans le rapport de leurs intervalles, quels que soient ces derniers, aussi petits qu'on puisse les imaginer. Hors de cette identité qui transforme un intervalle de temps en un autre d'espace, point de notion précise de vitesse, point de méchanique!" [16, p. 36].)

14 "Materiae vis insita est potentia resistendi, qua corpus unumquodque, quantum in se est, perseverat in statu suo vel quiescendi vel movendi uniformiter in directum" [27, p. 3].

15 "... nuos contenderous de remarquer, que pour avoir une idee claire du Mouvement, on ne peut se dispenser de distinguer au moins par l'esprit deux sortes d'étendue: l'une, qui soit regardée comme impénétrable, et qui constitue ce qu'on appelle proprement les corps; l'autre, qui étant considérée simplement comme

étendue, sans examiner si elle est pénétrable ou non, soit la mesure de la disrance d'un corps à un autre, et dont les parties envisagées comme fixes et immobiles, puissent servir à juger du repos ou du mouvement des corps" [3, p. VI]. (Here d'Alembert used the term "two space ideas" instead of terms "moved body" and "reference body" used above.)

16 In his paper "Galilei and the Scientific Revolution of the Seventeenth Century," A. Koyré comprehensively examined the physical and epistemological presupposition implied by the principle of inertia. However, he misunderstands its measurement-theoretical function. Moreover, he does not see the bodily (*gegenständlich*) character of measurement. Thus he understands the Galilean teaching of motion as a victory of Plato over Aristotle [19, pp. 1-15].

17 Sometimes one meets with the opinion that Newton's second law "is not to be confused with the one expressed in differential terms that we know today as 'Newton's law'" [1, p. 56]. Of course, in Newton's *Principia*, one does not find this law in an analytic form but in a corresponding geometric formulation. Now and again this fact leads, for example, to the curious misunderstanding that *time* does not occur in this theory. Such a view ignores that Newton defines a *change of motion* as change per time-interval. Possibly, the misunderstanding arises because Newton does not mentioned this explicitely, and for people who do not follow his deductions this might be difficult to see. (For the relation between the geometric and analytic formulation in Newton, see, e.g., [35].)

18 On the relationship between Newton's three laws see: the editors' annotations and epilogue to [25, pp. 534-536, 600-604; 50, in particular pp. 31-39].

19 "Die Situation irgendeines Experiments in der Atomphysik ist die folgende: Wir stellen mit Hilfe von mehr oder weniger komplizierten Apparaten Fragen an die Natur, die stets auf die Feststellung irgendeines objektiven Ablaufs in Raum und Zeit gerichtet sind ... . Aus dieser Sachlage ergibt sich automatisch, daß wir bei der mathematischen Behand lung des Vorganges einen Schnitt ziehen einerseits zwischen den Apparaten, die wir als Hilfsmittel zum Stellen der Frage betrachten, ... und an dererseits den physikalischen Systemen, über die wir etwas wissen wollen. Die letzteren stellen wir mathematisch durch eine Wellenfunktion dar, wobei für diese Wellenfunktion nach der Quantentheorie eine Differentialgeichung besteht, die den zukünftigen Zustand der Wellenfunktion aus dem gegenwärtigen bestimmt. Für die Apparate dagegen begnügen wir uns mit den in den klassischen Begriffen formulierbaren Gesetzen, die uns das Recht geben, sie eben als Meßapparate zu verwenden. Der Schnitt zwischen dem zu beobachtenden System und den Meßapparaten ist durch unsere Fragestellung naturgemäß gegeben, bezeichnet aber offenbar keine Diskontinuität des physikalischen Geschehens" [14, pp. 46-47].

20 "Die Reduzierung der Freiheitsgrade – vom Konfigurationsraum wird in den dreidimensionalen Raum projiziert – führt von der Determiniertheit der Schrödingerschen Wellengleichung im Konfigurationsraum auf die Probabilistik im physikalischen Raum" [32, p. 69].

21 "Im allgemeinen sind die Gesetze der Materiewellen grundverschieden von denen der klassischen Mechanik materieller Punkte. Wesentlich ist aber der Umstand, daß die für die Materiewellen charakteristische Funktion: die Wellenfunktion oder Wahrscheinlichkeitsfunktion – der Name tut hier nichts zur Sache – durch die Anfangsbedingungen und die Randbedingungen für alle Orte und Zeiten vollständig determiniert ist, nach ganz bestimmten Rechenregeln, sei es, daß man sich dabei der *Schrödinger*schen Operatoren oder der *Heisenberg*schen Matrizen oder der *Dirac*schen q-Zahlen bedient. – Wir sehen: In dem Weltbild der Quantenphysik herrscht der Determinismus ebenso wie in der klassischen Physik, nur sind die benutzten Symbole andere, und es wird mit anderen Rechnungsvorschriften operiert. Dementsprechend wird in der Quantenphysik, ebenso wie früher in der klassischen Physik, die Unsicherheit in der Voraussage von Ereignissen der Sinnenwelt reduziert auf die Unsicherheit des Zusammenhangs zwischen Weltbild und Sinnenwelt, d.h. auf die Unsicherheit der Übertragung der Symbole des Weltbildes auf die Sinnenwelt, und umgekehrt" [30, pp. 13 f.]. (The fact has to be taken into account that Planck uses the term "sensous world" instead of the terms "space of physical experience" or "space-time of measurement".)

22 Compared with the discussion between Einstein and Bohr, Boltzmann had used quite analoguous arguments against the mechanics elaborated by Heinrich Hertz [32, pp. 25-29].

23 Thus it is an error to think like Hegel that only philosophy – unlike other sciences – has not the advantage to admit their subjects immediately from the idea and to presuppose the method of recognizing for the beginning and continuation as already given. Hegel writes: "Die Philosophie entbehrt des Vorteils, der den anderen Wissenschaften zugute kommt, ihre *Gegenstände* als unmittelbar von der Vorstellung zugegeben, sowie die *Methode* des Erkennens für Anfang und Fortgang als bereits angenommen *voraussetzen* zu können" [10, § 1]. For a discussion of this view of Hegel see [41; 61; 63; 64].

6. References

1. Blay, M. (1998) *Reasoning with the Infinite. From the Closed World to the Mathematical Universe*, University of Chicago Press, Chicago–London.
2. Bohr, N., and Rosenfeld, L. (1933) Zur Frage der Meßbarkeit der elektromagnetischen Feldgrößen, *Kgl. Danske Vidensk. Selskab. Mat-fyz. Meddelelser* **12**, No. 8. [English translation in *Niels Bohr, Collected Works*, vol. 7, J. Kalckar (ed.), North- Holland, Amsterdam et al., 1996, pp. 123-166.]

3. d'Alembert, J.B. (1976) *Traité de dynamique*, Chez Fuchs Libraire, Paris.
4. DeWitt, B. (1962) The quantization of Geometry, in L. Witten (ed.), *Gravitation*, John Wiley and Sons, NewYork–London.
5. Einstein, A. (1934) The Mechanics of Newton and their influence on the development of theoretical physics, in A. Einstein, *The World as I See It*, translat. A. Harris, Covici Friede Inc., NewYork.
6. Einstein, A. (1950) Isaac Newton, in A. Einstein, *Out of My Later Years*, Philosophical Library, NewYork.
7. Einstein, A. (1969) Letter to M. Born, April 5, 1948, in *Albert Einstein – Hedwig und Max Born. Briefwechsel 1916-1955, commented by M. Born*, Ullstein, Frankfurt a.M. et al.
8. Einstein, A. (1969) Letter to M. Born, May 12, 1952, in *Albert Einstein – Hedwig und Max Born. Briefwechsel 1916-1955, commented by M. Born*, Ullstein, Frankfurt a.M. et al.
9. George, C., Prigogine, I., and Rosenfeld, L. (1979) The macroscopic level of quantum mechanics, in R. S. Cohen and J. Stachel (eds.), *Selected Papers of Léon Rosenfeld*, Reidel, Dordrecht et al.
10. Hegel, G.W.F. (1986) *Enzyklopädie der philosophische Wissenschaften im Grundrisse. (1830). Erster Teil. Die Wissenschaft der Logik, mit den mündlichen Zusätzen*, in E. Moldenhauer and K.M. Michel (eds.),*Works in 20 vols.*, Suhrkamp, Frankfurt a.M., vol. 8.
11. Hegel, G.W.F. (1986) *Wissenschaft der Logik. Zweiter Teil*, in E. Moldenhauer and K.M. Michel (eds.),*Works in 20 vols.*, Suhrkamp, Frankfurt a.M., vol. 6.
12. Hegel, G.W.F. (1986) *Vorlesungen über die Geschichte der Philosophie I* in E. Moldenhauer and K.M. Michel (eds.),*Works in 20 vols.*, Suhrkamp, Frankfurt a.M., vol. 18.
13. Heisenberg, W. (1930) *The Physical Principles of the Quantum Theory*, University of Chicago Press, Chicago.
14. Heisenberg, W. (1942) Prinzipielle Fragen der modernen Physik, in W. Heisenberg, *Wandlungen in den Grundlagen der Naturwissenschaft*, Hirzel, Leipzig.
15. Heisenberg, W. (1971) *Physics and Beyond. Encounters and Conversations*, Harper and Row, NewYork et al.
16. Jaeglé, P. (1976) *Essai sur l'espace et la temps*, Editions sociales, Paris.
17. Kant, I. (1968) *Kritik der reinen Vernunft*, in W. Weischedel (ed.), *Immanuel Kant, Works*, Suhrkamp, Frankfurt a.M., vols. III-IV.
18. Kepler, J. (1923) *Mysterium Cosmographicum – Das Weltgeheimnis*, B. Filser, Augsburg [German translation].
19. Koyré, A. (1968) Galilei and the scientific revolution of the seventeenth century, in A. Koyré, *Metaphysics and Measurement*, Chapman and Hall, London.
20. Lange, L. (1886) *Die geschichtliche Entwickelung des Bewegungsbegriffs und ihr voraussichtliches Endergebnis. Ein Beitrag zur historischen Kritik der mechanischen Prinzipien*, W. Engelmann, Leipzig.
21. Lorenzen, P. (1976) Theorie des technischen Wissens, in G. Böhme (ed.), *Protophysik. Für und wider eine konstruktive Wissenschaftstheorie*, Suhrkamp, Frankfurt a.M.
22. Lorenzen, P., v. Borzeszkowski, H.-H., and Wahsner, R. (1995) *Messung als Begründung oder Vermittlung? Ein Briefwechsel mit Paul Lorenzen über Protophysik und ein paar andere Dinge*, H.-H. v. Borzeszkowski and R. Wahsner (eds.), Academia, Sankt Augustin.
23. Lucretius Carus (1949) *Of the Nature of Things*, transl. W.E. Leonard, J.M.Dent & Sons Ltd., London et al.
24. Mach, E. (1883) *Die Mechanik in ihrer Entwickelung – historisch-kritisch dargestellt*, F.A. Brockhaus, Leipzig.
25. Mach, E. (1988) *Die Mechanik in ihrer Entwicklung – historisch-kritisch dargestellt*, supplemented by an appendix, ed. R. Wahsner and H.-H. von Borzeszkowski, Akademie Verlag, Berlin.
26. Neumann, C. (1870) *Über die Principien der Galilei-Newtonschen Theorie*, Teubner, Leipzig.
27. Newton, I. (1822) *Philosophiae Naturalis Principia Mathematica*, Typis Andreae et Johannis M. Ducan, Glasgue.
28. Pauli, W. (1969) Letter to M. Born, March 31, 1954, in *Albert Einstein – Hedwig und Max Born. Briefwechsel 1916-1955, commented by M. Born*, Ullstein, Frankfurt a.M. et al.
29. Peres, A., and Rosen, N. (1960) Quantum limitations on the measurement of gravitational fields, *Physical Review* **118**, 335.
30. Planck, M. (1958) *Der Kausalbegriff in der Physik*, Joh. A. Barth, Leipzig.
31. Planck, M. (1958) *Das Weltbild der neuen Physik*, Joh. A. Barth, Leipzig.
32. Rompe, R., and Treder, H.-J. (1979) *Über Physik. Studien zu ihrer Stellung in Wissenschaft und Gesellschaft*, Akademie Verlag, Berlin.
33. Rosenfeld, L. (1957) in *Report of the Conference on the Role of Gravitation in Physics*, Chapel Hill, North Carolina.
34. Rosenfeld, L. (1966) Quantentheorie und Gravitation, in H.-J. Treder (ed.), *Entstehung, Entwicklung und Perspektiven der Einsteinschen Gravitationstheorie*, Akademie Verlag, Berlin.
35. Schöpf, H.-G. (1991) Newton zwischen Geometrie und Erfahrung, in H. Hecht (ed.), *Gottfried Wilhelm Leibniz im philosophischen Diskurs über Geometrie und Erfahrung*, Akademie Verlag, Berlin.
36. Schrödinger, E. (1996) *Nature and the Greeks*, in *Nature and the Greeks/ Science and Humanism*, Cambridge U. P., Cambridge.
37. Treder, H.-J. (1963) Gravitonen, *Fortschritte der Physik* **11**, 81.

38. Treder, H.-J. (1974) *Über Prinzipien der Dynamik von Einstein, Hertz, Mach und Poincaré*, Akademie Verlag, Berlin.
39. Treder, H.-J. (1976) Das Korrespondenzprinzip und die Bedeutung von Quanten- und Relativitätstheorie, *Spektrum* **7**, 37.
40. v. Borzeszkowski, H.-H. (1992) Kantscher Raumbegriff und Einsteins Theorie, *Dt. Zs. Philosophie* **40**, 36.
41. v. Borzeszkowski, H.-H. (1993) Hegel's interpretation of classical mechanics, in M.J. Petry (ed.), *Hegel and Newtonianism*, International Archives of the History of Ideas, vol. 136, Kluwer Academic Publishers, Dordrecht.
42. v. Borzeszkowski, H.-H., and Treder, H.-J. (1982) Quantum theory and Einstein's General Relativity, *Found. Phys.* **12**, 1113.
43. v. Borzeszkowski, H.-H., and Treder, H.-J. (1988) *The Meaning of Quantum Gravity*, Reidel, Dordrecht.
44. v. Borzeszkowski, H.-H., and Treder, H.-J. (1993) The physical meaning of canonical quantization in terms of reference systems, *General Relativity and Gravitation* **25**, 291.
45. v. Borzeszkowski, H.-H., and Wahsner, R. (1979) Erkenntnistheoretischer Apriorismus und Einsteins Theorie. Einstein in seiner Beziehung zu Newton und Kant, *Dt. Zs. f. Philosophie* **27**, 213.
46. v. Borzeszkowski, H.-H., and Wahsner, R. (1980) *Newton und Voltaire. Zur Begründung und Interpretation der klassischen Mechanik*, Akademie Verlag, Berlin.
47. v. Borzeszkowski, H.-H., and Wahsner, R. (1980) Über die Notwendigkeit der Philosophie für die Naturwissenschaft, in B. Heidtmann (ed.), *Dialektik 1. Beiträge zu Philosophie und Wissenschaften. Orientierungen der Philosophie*, Pahl-Rugenstein, Cologne.
48. v. Borzeszkowski, H.-H., and Wahsner, R. (1988) Quantum mechanics and the physical reality concept, *Found. Phys.* **18**, 669.
49. v. Borzeszkowski, H.-H., and Wahsner, R. (1988) Heisenberg's cut and the limitations of its movability, *Ann. Phys. (Leipzig)* **45**, 522.
50. v. Borzeszkowski, H.-H., and Wahsner, R. (1989) *Physikalischer Dualismus und dialektischer Widerspruch. Studien zum physikalischen Bewegungsbegriff*, Wiss. Buch-gesellschaft, Darmstadt.
51. v. Borzeszkowski, H.-H., and Wahsner, R. (1995) Mach's criticism to Newton and Einstein's reading of Mach: The stimulating role of two misunderstandings, in J.B. Barbour and H. Pfister (eds.), *Mach's Principle: From Newton's Bucket to Quantum Gravity,* Einstein Studies, vol. 6, Birkhäuser, Boston et al.
52. v. Borzeszkowski, H.-H., and Wahsner, R. (1998) Die Natur technisch denken? Zur Synthese von *τέχνη* und *φύσις* in der Newtonschen Mechanik oder das Verhältnis von praktischer und theoretischer Mechanik in Newtons Physik, in M. Weingarten (ed.), *Zur Kultur der Moral. Praktische Philosophie in der posttraditionalen Gesellschaft*, Philo, Berlin (in print). [See also: *Preprint 87 of the Max Planck Institute for the History of Science*, Berlin.]
53. v. Borzeszkowski, H.-H. and Wahsner, R. (1999) Not even classical mechanics is mechanistic, in D. Aerts, J. van der Veken, and H. van Belle (eds.),*Worldviews and the Problems of Synthesis. Vol 4: The yellow book of "Einstein Meets Magritte"*, Kluwer Academic Publishers, Dordrecht–NewYork.
54. v. Borzeszkowski, H.-H. and Wahsner, R. (2000) Non-locality versus locality: The epistemological background of the Einstein-Bohr debate, in P.G. Bergmann, V. de Sabbata, J. N. Goldberg (eds.), *Classical and QuantumNonlocality*, World Scientific, Singapore et al.
55. v. Borzeszkowski, H.- H., and Wahsner, R. (2000) *Action and Reaction. Studies on Motion and Contradiction in Physics* (submitted for publication).
56. Wahsner, R. (1978) *Mensch und Kosmos. Die copernicanische Wende*, Akademie Verlag, Berlin.
57. Wahsner, R. (1981) *Das Aktive und das Passive. Zur erkenntnistheoretischen Begründung der Physik durch den Atomismus – dargestellt an Newton und Kant*, Akademie Verlag, Berlin.
58. Wahsner, R. (1981) Weltharmonie und Naturgesetz. Zur wissenschaftstheoretischen und wissenschaftshistorischen Bedeutung der Keplerschen Harmonienlehre, *Dt. Zs. f. Philosophie* **29**, 531.
59. Wahsner, R. (1989) Schrödinger's reception of Greek atomism, in E.I. Bitsakis and C.A. Nicolaides (eds.), *The Concept of Probability*, Kluwer Academic Publishers, Dordrecht.
60. Wahsner, R. (1992) Erkenntnistheoretischer Apriorismus und die neuzeitliche Physik, *Dt. Zs. Philosophie* **40**, 24.
61. Wahsner, R. (1993) The philosophical background to Hegel's criticism of Newton, in M.J. Petry (ed.), *Hegel and Newtonianism*, International Archives of the History of Ideas, vol. 136, Kluwer Academic Publishers, Dordrecht.
62. Wahsner, R. (1994) Apriorische Funktion und aposteriorische Herkunft. Hermann von Helmholtz' Untersuchungen zum Erfahrungsstatus der Geometrie," in L. Krüger (ed.), *Universalgenie Helmholtz. Rückblick nach 100 Jahren*, Akademie Verlag, Berlin.
63. Wahsner, R. (1996) *Zur Kritik der Hegelschen Naturphilosophie. Über ihren Sinn im Lichte der heutigen Naturerkenntnis*, Hegeliana. Studien und Quellen zu Hegel und zum Hegelianismus, Bd. 7, Peter Lang, Frankfurt a.M.et al.
64. Wahsner, R., and v. Borzeszkowski, H.-H. (1992) *Die Wirklichkeit der Physik. Studien zu Idealität und Realität in einer messenden Wissenschaft*, Peter Lang, Frankfurt a.M. et al.
65. Wahsner, R., and v. Borzeszkowski, H.-H. (1997) Vorwort, Einleitung, Anmerkungen, Anhang, in Voltaire, *Elemente der Philosophie Newtons/ Verteidigung des Newtonianismus/ Die Metaphysik des Neuton*, ed. R. Wahsner and H.-H. v. Borzeszkowski, de Gruyter, Berlin.

66. Weinberg, S. (1993) *Dreams of a Final Theory. The Search for the Fundamental Laws of Nature*, Vintage, London.
67. Wheeler, J.A. (1957) On the nature of quantum geometrodynamics, *Annals of Physics* **2**, 604.
68. Wheeler, J. A. (1964) Geometrodynamics and the issue of the final state, in C. DeWitt and B. DeWitt (eds.), *Relativity, Groups and Topology*, Blackie and Sons, London–Glasgow.
69. Wolff, M. (1978) *Geschichte der Impetustheorie. Untersuchungen zum Ursprung der klassischen Mechanik*, Suhrkamp, Frankfurt a.M.

The Thermodynamics of Black Holes*

Robert M. Wald
Enrico Fermi Institute and Department of Physics
University of Chicago
5640 S. Ellis Avenue
Chicago, Illinois 60637-1433
email:rmwa@midway.uchicago.edu

Abstract

We review the present status of black hole thermodynamics. Our review includes discussion of classical black hole thermodynamics, Hawking radiation from black holes, the generalized second law, and the issue of entropy bounds. A brief survey also is given of approaches to the calculation of black hole entropy. We conclude with a discussion of some unresolved open issues.

*This article is based upon an article of the same title published in Living Reviews in Relativity, http://www.livingreviews.org

P.G. Bergmann and V. de Sabbata (eds.),
Advances in the Interplay Between Quantum and Gravity Physics, 477–522.

1 Introduction

During the past 30 years, research in the theory of black holes in general relativity has brought to light strong hints of a very deep and fundamental relationship between gravitation, thermodynamics, and quantum theory. The cornerstone of this relationship is black hole thermodynamics, where it appears that certain laws of black hole mechanics are, in fact, simply the ordinary laws of thermodynamics applied to a system containing a black hole. Indeed, the discovery of the thermodynamic behavior of black holes – achieved primarily by classical and semiclassical analyses – has given rise to most of our present physical insights into the nature of quantum phenomena occurring in strong gravitational fields.

The purpose of this article is to provide a review of the following aspects of black hole thermodynamics:

- At the purely classical level, black holes in general relativity (as well as in other diffeomorphism covariant theories of gravity) obey certain laws which bear a remarkable mathematical resemblance to the ordinary laws of thermodynamics. The derivation of these laws of classical black hole mechanics is reviewed in section 2.

- Classically, black holes are perfect absorbers but do not emit anything; their physical temperature is absolute zero. However, in quantum theory black holes emit Hawking radiation with a perfect thermal spectrum. This allows a consistent interpretation of the laws of black hole mechanics as physically corresponding to the ordinary laws of thermodynamics. The status of the derivation of Hawking radiation is reviewed in section 3.

- The *generalized second law* (GSL) directly links the laws of black hole mechanics to the ordinary laws of thermodynamics. The arguments in favor of the GSL are reviewed in section 4. A discussion of entropy bounds is also included in this section.

- The classical laws of black hole mechanics together with the formula for the temperature of Hawking radiation allow one to identify a quantity associated with black holes – namely $A/4$ in general relativity – as playing the mathematical role of entropy. The apparent validity of the

GSL provides strong evidence that this quantity truly is the physical entropy of a black hole. A major goal of research in quantum gravity is to provide an explanation for – and direct derivation of – the formula for the entropy of a black hole. A brief survey of work along these lines is provided in section 5.

- Although much progress has been made in our understanding of black hole thermodynamics, many important issues remain unresolved. Primary among these are the "black hole information paradox" and issues related to the degrees of freedom responsible for the entropy of a black hole. These unresolved issues are briefly discussed in section 6.

Throughout this article, we shall set $G = \hbar = c = k = 1$, and we shall follow the sign and notational conventions of [1]. Although I have attempted to make this review be reasonably comprehensive and balanced, it should be understood that my choices of topics and emphasis naturally reflect my own personal viewpoints, expertise, and biases.

2 Classical Black Hole Thermodynamics

In this section, I will give a brief review of the laws of classical black hole mechanics.

In physical terms, a black hole is a region where gravity is so strong that nothing can escape. In order to make this notion precise, one must have in mind a region of spacetime to which one can contemplate escaping. For an asymptotically flat spacetime (M, g_{ab}) (representing an isolated system), the asymptotic portion of the spacetime "near infinity" is such a region. The *black hole* region, $\mathcal{B}$, of an asymptotically flat spacetime, (M, g_{ab}), is defined as

$$\mathcal{B} \equiv M - I^-(\mathcal{I}^+), \tag{1}$$

where $\mathcal{I}^+$ denotes future null infinity and I^- denotes the chronological past. Similar definitions of a black hole can be given in other contexts (such as asymptotically anti-deSitter spacetimes) where there is a well defined asymptotic region.

The *event horizon*, $\mathcal{H}$, of a black hole is defined to be the boundary of $\mathcal{B}$. Thus, $\mathcal{H}$ is the boundary of the past of $\mathcal{I}^+$. Consequently, $\mathcal{H}$ automatically satisfies all of the properties possessed by past boundaries (see, e.g., [2] or [1] for further discussion). In particular, $\mathcal{H}$ is a null hypersurface which is composed of future inextendible null geodesics without caustics, i.e., the expansion, θ, of the null geodesics comprising the horizon cannot become negatively infinite. Note that the entire future history of the spacetime must be known before the location of $\mathcal{H}$ can be determined, i.e., $\mathcal{H}$ possesses no distinguished local significance.

If Einstein's equation holds with matter satisfying the null energy condition (i.e., if $T_{ab}k^a k^b \geq 0$ for all null k^a), then it follows immediately from the Raychauduri equation (see, e.g., [1]) that if the expansion, θ, of any null geodesic congruence ever became negative, then θ would become infinite within a finite affine parameter, provided, of course, that the geodesic can be extended that far. If the black hole is *strongly asymptotically predictable* – i.e., if there is a globally hyperbolic region containing $I^-(\mathcal{I}^+) \cup \mathcal{H}$ – it can be shown that this implies that $\theta \geq 0$ everywhere on $\mathcal{H}$ (see, e.g., [2, 1]). It then follows that the surface area, A, of the event horizon of a black hole can never decrease with time, as discovered by Haw'-ing [4].

It is worth remarking that since $\mathcal{H}$ is a past boundary, it automatically must be a C^0 embedded submanifold (see, e.g., [1]), but it need not be C^1. However, essentially all discussions and analyses of black hole event horizons implicitly assume C^1 or higher order differentiability of $\mathcal{H}$. Recently, this higher order differentiability assumption has been eliminated for the proof of the area theorem [3].

The area increase law bears a resemblance to the second law of thermodynamics in that both laws assert that a certain quantity has the property of never decreasing with time. It might seem that this resemblance is a very superficial one, since the area law is a theorem in differential geometry whereas the second law of thermodynamics is understood to have a statistical origin. Nevertheless, this resemblance together with the idea that information is irretrievably lost when a body falls into a black hole led Bekenstein to propose [5, 6] that a suitable multiple of the area of the event horizon of a black hole should be interpreted as its entropy, and that a *generalized second law* (GSL) should hold: The sum of the ordinary entropy of matter outside of a black hole plus a suitable multiple of the area of a black hole never decreases. We will discuss this law in detail in section 4.

The remaining laws of thermodynamics deal with equilibrium and quasi-equilibrium processes. At nearly the same time as Bekenstein proposed a relationship between the area theorem and the second law of thermodynamics, Bardeen, Carter, and Hawking [7] provided a general proof of certain laws of "black hole mechanics" which are direct mathematical analogs of the zeroth and first laws of thermodynamics. These laws of black hole mechanics apply to stationary black holes (although a formulation of these laws in terms of isolated horizons will be briefly described at the end of this section).

In order to discuss the zeroth and first laws of black hole mechanics, we must introduce the notions of stationary, static, and axisymmetric black holes as well as the notion of a Killing horizon. If an asymptotically flat spacetime (M, g_{ab}) contains a black hole, $\mathcal{B}$, then $\mathcal{B}$ is said to be *stationary* if there exists a one-parameter group of isometries on (M, g_{ab}) generated by a Killing field t^a which is unit timelike at infinity. The black hole is said to be *static* if it is stationary and if, in addition, t^a is hypersurface orthogonal. The black hole is said to be *axisymmetric* if there exists a one parameter group of isometries which correspond to rotations at infinity. A stationary, axisymmetric black hole is said to possess the "t–ϕ orthogonality property" if the 2-planes spanned by t^a and the rotational Killing field ϕ^a are orthogonal to a family of 2-dimensional surfaces. The t–ϕ orthogonality property holds for all stationary-axisymmetric black hole solutions to the vacuum Einstein or Einstein-Maxwell equations (see, e.g., [8]).

A null surface, $\mathcal{K}$, whose null generators coincide with the orbits of a one-parameter group of isometries (so that there is a Killing field ξ^a normal to $\mathcal{K}$) is called a *Killing horizon*. There are two independent results (usually referred to as "rigidity theorems") that show that in a wide variety of cases of interest, the event horizon, $\mathcal{H}$, of a stationary black hole must be a Killing horizon. The first, due to Carter [9], states that for a static black hole, the static Killing field t^a must be normal to the horizon, whereas for a stationary-axisymmetric black hole with the t–ϕ orthogonality property there exists a Killing field ξ^a of the form

$$\xi^a = t^a + \Omega\phi^a \tag{2}$$

which is normal to the event horizon. The constant Ω defined by Eq. (2) is called the *angular velocity of the horizon*. Carter's result does not rely on any field equations, but leaves open the possibility that there could exist stationary black holes without the above symmetries whose event horizons are not Killing horizons. The second result, due to Hawking [2] (see also [10]), directly proves that in vacuum or electrovac general relativity, the event horizon of any stationary black hole must be a Killing horizon. Consequently, if t^a fails to be normal to the horizon, then there must exist an additional Killing field, ξ^a, which is normal to the horizon, i.e., a stationary black hole must be nonrotating (from which staticity follows [11, 12, 13]) or axisymmetric (though not necessarily with the t–ϕ orthogonality property). Note that Hawking's theorem makes no assumptions of symmetries beyond stationarity, but it does rely on the properties of the field equations of general relativity.

Now, let $\mathcal{K}$ be any Killing horizon (not necessarily required to be the event horizon, $\mathcal{H}$, of a black hole), with normal Killing field ξ^a. Since $\nabla^a(\xi^b\xi_b)$ also is normal to $\mathcal{K}$, these vectors must be proportional at every point on $\mathcal{K}$. Hence, there exists a function, κ, on $\mathcal{K}$, known as the *surface gravity* of $\mathcal{K}$, which is defined by the equation

$$\nabla^a(\xi^b\xi_b) = -2\kappa\xi^a. \tag{3}$$

It follows immediately that κ must be constant along each null geodesic generator of $\mathcal{K}$, but, in general, κ can vary from generator to generator. It is not difficult to show (see, e.g., [1]) that

$$\kappa = \lim(Va), \tag{4}$$

where a is the magnitude of the acceleration of the orbits of ξ^a in the region off of $\mathcal{K}$ where they are timelike, $V \equiv (-\xi^a \xi_a)^{1/2}$ is the "redshift factor" of ξ^a, and the limit as one approaches $\mathcal{K}$ is taken. Equation (4) motivates the terminology "surface gravity". Note that the surface gravity of a black hole is defined only when it is "in equilibrium", i.e., stationary, so that its event horizon is a Killing horizon. There is no notion of the surface gravity of a general, non-stationary black hole, although the definition of surface gravity can be extended to isolated horizons (see below).

In parallel with the two independent "rigidity theorems" mentioned above, there are two independent versions of the zeroth law of black hole mechanics. The first, due to Carter [9] (see also [14]), states that for any black hole which is static or is stationary-axisymmetric with the t–ϕ orthogonality property, the surface gravity κ, must be constant over its event horizon $\mathcal{H}$. This result is purely geometrical, i.e., it involves no use of any field equations. The second, due to Bardeen, Carter, and Hawking [7] states that if Einstein's equation holds with the matter stress-energy tensor satisfying the dominant energy condition, then κ must be constant on any Killing horizon. Thus, in the second version of the zeroth law, the hypothesis that the t–ϕ orthogonality property holds is eliminated, but use is made of the field equations of general relativity.

A *bifurcate Killing horizon* is a pair of null surfaces, $\mathcal{K}_A$ and $\mathcal{K}_B$, which intersect on a spacelike 2-surface, $\mathcal{C}$ (called the "bifurcation surface"), such that $\mathcal{K}_A$ and $\mathcal{K}_B$ are each Killing horizons with respect to the same Killing field ξ^a. It follows that ξ^a must vanish on $\mathcal{C}$; conversely, if a Killing field, ξ^a, vanishes on a two-dimensional spacelike surface, $\mathcal{C}$, then $\mathcal{C}$ will be the bifurcation surface of a bifurcate Killing horizon associated with ξ^a (see [15] for further discussion). An important consequence of the zeroth law is that if $\kappa \neq 0$, then in the "maximally extended" spacetime representing a stationary black hole, the event horizon, $\mathcal{H}$, comprises a branch of a bifurcate Killing horizon [14]. This result is purely geometrical – involving no use of any field equations. As a consequence, the study of stationary black holes which satisfy the zeroth law divides into two cases: "extremal" black holes (for which, by definition, $\kappa = 0$), and black holes with bifurcate horizons.

The first law of black hole mechanics is simply an identity relating the changes in mass, M, angular momentum, J, and horizon area, A, of a stationary black hole when it is perturbed. To first order, the variations of these quantities in the vacuum case always satisfy

$$\delta M = \frac{1}{8\pi}\kappa\delta A + \Omega\delta J. \tag{5}$$

In the original derivation of this law [7], it was required that the perturbation be stationary. Furthermore, the original derivation made use of the detailed form of Einstein's equation. Subsequently, the derivation has been generalized to hold for non-stationary perturbations [11, 16], provided that the change in area is evaluated at the bifurcation surface, $\mathcal{C}$, of the unperturbed black hole (see, however, [17] for a derivation of the first law for non-stationary perturbations that does not require evaluation at the bifurcation surface). More significantly, it has been shown [16] that the validity of this law depends only on very general properties of the field equations. Specifically, a version of this law holds for any field equations derived from a diffeomorphism covariant Lagrangian, L. Such a Lagrangian can always be written in the form

$$L = L\left(g_{ab}; R_{abcd}, \nabla_a R_{bcde}, ...; \psi, \nabla_a \psi, ...\right), \tag{6}$$

where ∇_a denotes the derivative operator associated with g_{ab}, R_{abcd} denotes the Riemann curvature tensor of g_{ab}, and ψ denotes the collection of all matter fields of the theory (with indices suppressed). An arbitrary (but finite) number of derivatives of R_{abcd} and ψ are permitted to appear in L. In this more general context, the first law of black hole mechanics is seen to be a direct consequence of an identity holding for the variation of the Noether current. The general form of the first law takes the form

$$\delta M = \frac{\kappa}{2\pi}\delta S_{\text{bh}} + \Omega\delta J + ..., \tag{7}$$

where the "..." denote possible additional contributions from long range matter fields, and where

$$S_{\text{bh}} \equiv -2\pi \int_{\mathcal{C}} \frac{\delta L}{\delta R_{abcd}} n_{ab} n_{cd}. \tag{8}$$

Here n_{ab} is the binormal to the bifurcation surface $\mathcal{C}$ (normalized so that $n_{ab}n^{ab} = -2$), and the functional derivative is taken by formally viewing the Riemann tensor as a field which is independent of the metric in Eq. (6). For the case of vacuum general relativity, where $L = R\sqrt{-g}$, a simple calculation yields

$$S_{\text{bh}} = A/4, \tag{9}$$

and Eq. (7) reduces to Eq. (5).

The close mathematical analogy of the zeroth, first, and second laws of thermodynamics to corresponding laws of classical black hole mechanics is broken by the Planck-Nernst form of the third law of thermodynamics, which states that $S \to 0$ (or a "universal constant") as $T \to 0$. The analog of this law fails in black hole mechanics – although analogs of alternative formulations of the third law do appear to hold for black holes [18] – since there exist extremal black holes (i.e., black holes with $\kappa = 0$) with finite A. However, there is good reason to believe that the "Planck-Nernst theorem" should not be viewed as a fundamental law of thermodynamics [19] but rather as a property of the density of states near the ground state in the thermodynamic limit, which happens to be valid for commonly studied materials. Indeed, examples can be given of ordinary quantum systems that violate the Planck-Nernst form of the third law in a manner very similar to the violations of the analog of this law that occur for black holes [20].

As discussed above, the zeroth and first laws of black hole mechanics have been formulated in the mathematical setting of stationary black holes whose event horizons are Killing horizons. The requirement of stationarity applies to the entire spacetime and, indeed, for the first law, stationarity of the entire spacetime is essential in order to relate variations of quantities defined at the horizon (like A) to variations of quantities defined at infinity (like M and J). However, it would seem reasonable to expect that the equilibrium thermodynamic behavior of a black hole would require only a form of local stationarity at the event horizon. For the formulation of the first law of black hole mechanics, one would also then need local definitions of quantities like M and J at the horizon. Such an approach toward the formulation of the laws of black hole mechanics has recently been taken via the notion of an *isolated horizon*, defined as a null hypersurface with vanishing shear and expansion satisfying the additional properties stated in [21]. (This definition supersedes the more restrictive definitions given, e.g., in [22, 23, 24].) The presence of an isolated horizon does not require the entire spacetime to be stationary [25]. A direct analog of the zeroth law for stationary event horizons can be shown to hold for isolated horizons [26]. In the Einstein-Maxwell case, one can demand (via a choice of scaling of the normal to the isolated horizon as well as a choice of gauge for the Maxwell field) that the surface gravity and electrostatic potential of the isolated horizon be functions of only its area and charge. The requirement that time evolution be symplectic then leads to a version of the first law of black hole mechanics as well as a (in general, non-unique) local notion of the energy of the isolated horizon [26]. These results also have been generalized to allow dilaton couplings [24] and Yang-Mills fields [27, 26].

In comparing the laws of black hole mechanics in classical general relativity with the laws of thermodynamics, it should first be noted that the black hole uniqueness theorems (see, e.g., [8]) establish that stationary black holes – i.e., black holes "in equilibrium" – are characterized by a small number of parameters, analogous to the "state parameters" of ordinary thermodynamics. In the corresponding laws, the role of energy, E, is played by the mass, M, of the black hole; the role of temperature, T, is played by a constant times the surface gravity, κ, of the black hole; and the role of entropy, S, is played by a constant times the area, A, of the black hole. The fact that E and M represent the same physical quantity provides a strong hint that the mathematical analogy between the laws of black hole mechanics and the laws of thermodynamics might be of physical significance. However, as argued in [7], this cannot be the case in classical general relativity. The physical temperature of a black hole is absolute zero (see subsection 4.1 below), so there can be no physical relationship between T and κ. Consequently, it also would be inconsistent to assume a physical relationship between S and A. As we shall now see, this situation changes dramatically when quantum effects are taken into account.

3 Hawking Radiation

In 1974, Hawking [28] made the startling discovery that the physical temperature of a black hole is not absolute zero: As a result of quantum particle creation effects, a black hole radiates to infinity all species of particles with a perfect black body spectrum, at temperature (in units with $G = c = \hbar = k = 1$)

$$T = \frac{\kappa}{2\pi}. \tag{10}$$

Thus, $\kappa/2\pi$ truly is the *physical* temperature of a black hole, not merely a quantity playing a role mathematically analogous to temperature in the laws of black hole mechanics. In this section, we review the status of the derivation of the Hawking effect and also discuss the closely related Unruh effect.

The original derivation of the Hawking effect [28] made direct use of the formalism for calculating particle creation in a curved spacetime that had been developed by Parker [29] and others. Hawking considered a classical spacetime (M, g_{ab}) describing gravitational collapse to a Schwarzschild black hole. He then considered a free (i.e., linear) quantum field propagating in this background spacetime, which is initially in its vacuum state prior to the collapse, and he computed the particle content of the field at infinity at late times. This calculation involves taking the positive frequency mode function corresponding to a particle state at late times, propagating it backwards in time, and determining its positive and negative frequency parts in the asymptotic past. His calculation revealed that at late times, the expected number of particles at infinity corresponds to emission from a perfect black body (of finite size) at the Hawking temperature (Eq. (10)). It should be noted that this result relies only on the analysis of quantum fields in the region exterior to the black hole, and it does not make use of any gravitational field equations.

The original Hawking calculation can be straightforwardly generalized and extended in the following ways. First, one may consider a spacetime representing an arbitrary gravitational collapse to a black hole such that the black hole "settles down" to a stationary final state satisfying the zeroth law of black hole mechanics (so that the surface gravity, κ, of the black hole final state is constant over its event horizon). The initial state of the quantum field may be taken to be any nonsingular state (i.e., any Hadamard state

– see, e.g., [15]) rather than the initial vacuum state. Finally, it can be shown [30] that all aspects of the final state at late times (i.e., not merely the expected number of particles in each mode) correspond to black body[1] thermal radiation emanating from the black hole at temperature (Eq. (10)).

It should be noted that no infinities arise in the calculation of the Hawking effect for a free field, so the results are mathematically well defined, without any need for regularization or renormalization. The original derivations [28, 30] made use of notions of "particles propagating into the black hole", but the results for what an observer sees at infinity were shown to be independent of the ambiguities inherent in such notions and, indeed, a derivation of the Hawking effect has been given [31] which entirely avoids the introduction of any notion of "particles". However, there remains one significant difficultly with the Hawking derivation: In the calculation of the backward-in-time propagation of a mode, it is found that the mode undergoes a large blueshift as it propagates near the event horizon, but there is no correspondingly large redshift as the mode propagates back through the collapsing matter into the asymptotic past. Indeed, the net blueshift factor of the mode is proportional to $\exp(\kappa t)$, where t is the time that the mode would reach an observer at infinity. Thus, within a time of order $1/\kappa$ of the formation of a black hole (i.e., $\sim 10^{-5}$ seconds for a one solar mass Schwarzschild black hole), the Hawking derivation involves (in its intermediate steps) the propagation of modes of frequency much higher than the Planck frequency. In this regime, it is difficult to believe in the accuracy of free field theory – or any other theory known to mankind.

An approach to investigating this issue was first suggested by Unruh [32], who noted that a close analog of the Hawking effect occurs for quantized sound waves in a fluid undergoing supersonic flow. A similar blueshifting of the modes quickly brings one into a regime well outside the domain of validity of the continuum fluid equations. Unruh suggested replacing the continuum fluid equations with a more realistic model at high frequencies to see if the fluid analog of the Hawking effect would still occur. More recently, Unruh investigated models where the dispersion relation is altered at ultra-high frequencies, and he found no deviation from the Hawking prediction [33]. A variety of alternative models have been considered by other researchers [34,

[1]If the black hole is rotating, the the spectrum seen by an observer at infinity corresponds to what would emerge from a "rotating black body".

35, 36, 37, 38, 39, 40]. Again, agreement with the Hawking effect prediction was found in all cases, despite significant modifications of the theory at high frequencies.

The robustness of the Hawking effect with respect to modifications of the theory at ultra-high frequency probably can be understood on the following grounds. One may view the backward-in-time propagation of modes as consisting of two stages: a first stage where the blueshifting of the mode brings it into a WKB regime but the frequencies remain well below the Planck scale, and a second stage where the continued blueshifting takes one to the Planck scale and beyond. In the first stage, the usual field theory calculations should be reliable. On the other hand, after the mode has entered a WKB regime, it seems plausible that the kinds of modifications to its propagation laws considered in [33, 34, 35, 36, 37, 38, 39, 40] should not affect its essential properties, in particular the magnitude of its negative frequency part.

Indeed, an issue closely related to the validity of the original Hawking derivation arises if one asks how a uniformly accelerating observer in Minkowski spacetime perceives the ordinary (inertial) vacuum state (see below). The outgoing modes of a given frequency ω as seen by the accelerating observer at proper time τ along his worldline correspond to modes of frequency $\sim \omega \exp(a\tau)$ in a fixed inertial frame. Therefore, at time $\tau \gg 1/a$ one might worry about field-theoretic derivations of what the accelerating observer would see. However, in this case one can appeal to Lorentz invariance to argue that what the accelerating observer sees cannot change with time. It seems likely that one could similarly argue that the Hawking effect cannot be altered by modifications of the theory at ultra-high frequencies, provided that these modifications preserve an appropriate "local Lorentz invariance" of the theory. Thus, there appears to be strong reasons for believing in the validity of the Hawking effect despite the occurrence of ultra-high-frequency modes in the derivation.

There is a second, logically independent result – namely, the Unruh effect [41] and its generalization to curved spacetime – which also gives rise to the formula (10). Although the Unruh effect is mathematically very closely related to the Hawking effect, it is important to distinguish clearly between them. In its most general form, the Unruh effect may be stated as follows (see [42, 15] for further discussion): Consider a classical spacetime (M, g_{ab}) that contains a bifurcate Killing horizon, $\mathcal{K} = \mathcal{K}_A \cup \mathcal{K}_B$, so that there is a one-parameter group of isometries whose associated Killing field, ξ^a, is normal to $\mathcal{K}$. Consider a free quantum field on this spacetime. Then there exists at most one globally nonsingular state of the field which is invariant under the isometries. Furthermore, in the "wedges" of the spacetime where the isometries have timelike orbits, this state (if it exists) is a KMS (i.e., thermal equilibrium) state at temperature (10) with respect to the isometries.

Note that in Minkowski spacetime, any one-parameter group of Lorentz boosts has an associated bifurcate Killing horizon, comprised by two intersecting null planes. The unique, globally nonsingular state which is invariant under these isometries is simply the usual ("inertial") vacuum state, $|0\rangle$. In the "right and left wedges" of Minkowski spacetime defined by the Killing horizon, the orbits of the Lorentz boost isometries are timelike, and, indeed, these orbits correspond to worldlines of uniformly accelerating observers. If we normalize the boost Killing field, b^a, so that Killing time equals proper time on an orbit with acceleration a, then the surface gravity of the Killing horizon is $\kappa = a$. An observer following this orbit would naturally use b^a to define a notion of "time translation symmetry". Consequently, by the above general result, when the field is in the inertial vacuum state, a uniformly accelerating observer would describe the field as being in a thermal equilibrium state at temperature

$$T = \frac{a}{2\pi} \tag{11}$$

as originally discovered by Unruh [41]. A mathematically rigorous proof of the Unruh effect in Minkowski spacetime was given by Bisognano and Wichmann [43] in work motivated by entirely different considerations (and done independently of and nearly simultaneously with the work of Unruh). Furthermore, the Bisognano-Wichmann theorem is formulated in the general context of axiomatic quantum field theory, thus establishing that the Unruh effect is not limited to free field theory.

Although there is a close mathematical relationship between the Unruh effect and the Hawking effect, it should be emphasized that these results refer to *different* states of the quantum field. We can divide the late time modes of the quantum field in the following manner, according to the properties that they would have in the analytically continued spacetime [14] representing the asymptotic final stationary state of the black hole: We refer to modes that would have emanated from the white hole region of the analytically continued spacetime as "UP modes" and those that would have originated from infinity as "IN modes". In the Hawking effect, the asymptotic final state of the quantum field is a state in which the UP modes of the quantum field are thermally populated at temperature (10), but the IN modes are unpopulated. This state (usually referred to as the "Unruh vacuum") would be singular on the white hole horizon in the analytically continued spacetime. On the other hand, in the Unruh effect and its generalization to curved spacetimes, the state in question (usually referred to as the "Hartle-Hawking vacuum" [44]) is globally nonsingular, and *all* modes of the quantum field in the "left and right wedges" are thermally populated.[2]

The differences between the Unruh and Hawking effects can be seen dramatically in the case of a Kerr black hole. For the Kerr black hole, it can be shown [42] that there does not exist any globally nonsingular state of the field which is invariant under the isometries associated with the Killing horizon, i.e., there does not exist a "Hartle-Hawking vacuum state" on Kerr spacetime. However, there is no difficultly with the derivation of the Hawking effect for Kerr black holes, i.e., the "Unruh vacuum state" does exist.

It should be emphasized that in the Hawking effect, the temperature (10) represents the temperature as measured by an observer near infinity. For any observer following an orbit of the Killing field, ξ^a, normal to the horizon, the locally measured temperature of the UP modes is given by

$$T = \frac{\kappa}{2\pi V}, \tag{12}$$

where $V = (-\xi^a \xi_a)^{1/2}$. In other words, the locally measured temperature of the Hawking radiation follows the Tolman law. Now, as one approaches the horizon of the black hole, the UP modes dominate over the IN modes. Taking Eq. (4) into account, we see that $T \to a/2\pi$ as the black hole horizon, $\mathcal{H}$, is approached, i.e., in this limit Eq. (12) corresponds to the flat spacetime Unruh effect.

Equation (12) shows that when quantum effects are taken into account, a black hole is surrounded by a "thermal atmosphere" whose local temperature as measured by observers following orbits of ξ^a becomes divergent as one approaches the horizon. As we shall see in the next section, this thermal atmosphere produces important physical effects on quasi-stationary bodies near the black hole. On the other hand, it should be emphasized that for a

[2]The state in which none of the modes in the region exterior to the black hole are populated is usually referred to as the "Boulware vacuum". The Boulware vacuum is singular on both the black hole and white hole horizons.

macroscopic black hole, observers who freely fall into the black hole would not notice any important quantum effects as they approach and cross the horizon.

4 The Generalized Second Law (GSL)

In this section, we shall review some arguments for the validity of the generalized second law (GSL). We also shall review the status of several proposed entropy bounds on matter that have played a role in discussions and analyses of the GSL.

4.1 Arguments for the validity of the GSL

Even in classical general relativity, there is a serious difficulty with the ordinary second law of thermodynamics when a black hole is present, as originally emphasized by J.A. Wheeler: One can simply take some ordinary matter and drop it into a black hole, where, according to classical general relativity, it will disappear into a spacetime singularity. In this process, one loses the entropy initially present in the matter, and no compensating gain of ordinary entropy occurs, so the total entropy, S, of matter in the universe decreases. One could attempt to salvage the ordinary second law by invoking the bookkeeping rule that one must continue to count the entropy of matter dropped into a black hole as still contributing to the total entropy of the universe. However, the second law would then have the status of being observationally unverifiable.

As already mentioned in section 2, after the area theorem was proven, Bekenstein [5, 6] proposed a way out of this difficulty: Assign an entropy, S_{bh}, to a black hole given by a numerical factor of order unity times the area, A, of the black hole in Planck units. Define the *generalized entropy*, S', to be the sum of the ordinary entropy, S, of matter outside of a black hole plus the black hole entropy

$$S' \equiv S + S_{\text{bh}}. \tag{13}$$

Finally, replace the ordinary second law of thermodynamics by the *generalized second law* (GSL): The total generalized entropy of the universe never decreases with time,

$$\Delta S' \geq 0. \tag{14}$$

Although the ordinary second law will fail when matter is dropped into a black hole, such a process will tend to increase the area of the black hole, so there is a possibility that the GSL will hold.

Bekenstein's proposal of the GSL was made prior to the discovery of Hawking radiation. When Hawking radiation is taken into account, a serious problem also arises with the second law of black hole mechanics (i.e., the area theorem): Conservation of energy requires that an isolated black hole must lose mass in order to compensate for the energy radiated to infinity by the Hawking process. Indeed, if one equates the rate of mass loss of the black hole to the energy flux at infinity due to particle creation, one arrives at the startling conclusion that an isolated black hole will radiate away all of its mass within a finite time. During this process of black hole "evaporation", A will decrease. Such an area decrease can occur because the expected stress-energy tensor of quantum matter does not satisfy the null energy condition – even for matter for which this condition holds classically – in violation of a key hypothesis of the area theorem.

However, although the second law of black hole mechanics fails during the black hole evaporation process, if we adjust the numerical factor in the definition of $S_{\rm bh}$ to correspond to the identification of $\kappa/2\pi$ as temperature in the first law of black hole mechanics – so that, as in Eq. (9) above, we have $S_{\rm bh} = A/4$ in Planck units – then the GSL continues to hold: Although A decreases, there is at least as much ordinary entropy generated outside the black hole by the Hawking process. Thus, although the ordinary second law fails in the presence of black holes and the second law of black hole mechanics fails when quantum effects are taken into account, there is a possibility that the GSL may always hold. If the GSL does hold, it seems clear that we must interpret $S_{\rm bh}$ as representing the *physical* entropy of a black hole, and that the laws of black hole mechanics must truly represent the ordinary laws of thermodynamics as applied to black holes. Thus, a central issue in black hole thermodynamics is whether the GSL holds in all processes.

It was immediately recognized by Bekenstein [5] (see also [7]) that there is a serious difficulty with the GSL if one considers a process wherein one carefully lowers a box containing matter with entropy S and energy E very close to the horizon of a black hole before dropping it in. Classically, if one could lower the box arbitrarily close to the horizon before dropping it in, one would recover all of the energy originally in the box as "work" at infinity. No energy would be delivered to the black hole, so by the first law of black hole mechanics, Eq. (7), the black hole area, A, would not increase. However, one would still get rid of all of the entropy, S, originally in the box, in violation of the GSL.

Indeed, this process makes manifest the fact that in classical general relativity, the physical temperature of a black hole is absolute zero: The above process is, in effect, a Carnot cycle which converts "heat" into "work" with 100% efficiency [45]. The difficulty with the GSL in the above process can be viewed as stemming from an inconsistency of this fact with the mathematical assignment of a finite (non-zero) temperature to the black hole required by the first law of black hole mechanics if one assigns a finite (non-infinite) entropy to the black hole.

Bekenstein proposed a resolution of the above difficulty with the GSL in a quasi-static lowering process by arguing [5, 6] that it would not be possible to lower a box containing physically reasonable matter close enough to the horizon of the black hole to violate the GSL. As will be discussed further in the next sub-section, this proposed resolution was later refined by postulating a universal bound on the entropy of systems with a given energy and size [46]. However, an alternate resolution was proposed in [47], based upon the idea that, when quantum effects are taken into account, the physical temperature of a black hole is no longer absolute zero, but rather is the Hawking temperature, $\kappa/2\pi$. Since the Hawking temperature goes to zero in the limit of a large black hole, it might appear that quantum effects could not be of much relevance in this case. However, despite the fact that Hawking radiation at infinity is indeed negligible for large black holes, the effects of the quantum "thermal atmosphere" surrounding the black hole are not negligible on bodies that are quasi-statically lowered toward the black hole. The temperature gradient in the thermal atmosphere (see Eq. (12)) implies that there is a pressure gradient and, consequently, a buoyancy force on the box. This buoyancy force becomes infinitely large in the limit as the box is lowered to the horizon. As a result of this buoyancy force, the optimal place to drop the box into the black hole is no longer the horizon but rather the "floating point" of the box, where its weight is equal to the weight of the displaced thermal atmosphere. The minimum area increase given to the black hole in the process is no longer zero, but rather turns out to be an amount just sufficient to prevent any violation of the GSL from occurring in this process [47].

The analysis of [47] considered only a particular class of gedankenexperiments for violating the GSL involving the quasi-static lowering of a box near a black hole. Of course, since one does not have a general proof of the ordinary second law of thermodynamics – and, indeed, for finite systems, there should always be a nonvanishing probability of violating the ordinary second law – it would not be reasonable to expect to obtain a completely general proof of the GSL. However, general arguments within the semiclassical approximation for the validity of the GSL for arbitrary infinitesimal quasi-static processes have been given in [48, 49, 15]. These arguments crucially rely on the presence of the thermal atmosphere surrounding the black hole. Related arguments for the validity of the GSL have been given in [50, 51]. In [50], it is assumed that the incoming state is a product state of radiation originating from infinity (i.e., IN modes) and radiation that would appear to emanate from the white hole region of the analytically continued spacetime (i.e., UP modes), and it is argued that the generalized entropy must increase under unitary evolution. In [51], it is argued on quite general grounds that the (generalized) entropy of the state of the region exterior to the black hole must increase under the assumption that it undergoes autonomous evolution.

Indeed, it should be noted that if one could violate the GSL for an infinitesimal quasi-static process in a regime where the black hole can be treated semi-classically, then it also should be possible to violate the ordinary second law for a corresponding process involving a self-gravitating body. Namely, suppose that the GSL could be violated for an infinitesimal quasi-static process involving, say, a Schwarzschild black hole of mass M (with M much larger than the Planck mass). This process might involve lowering matter towards the black hole and possibly dropping the matter into it. However, an observer doing this lowering or dropping can "probe" only the region outside of the black hole, so there will be some $r_0 > 2M$ such that the detailed structure of the black hole will directly enter the analysis of the process only for $r > r_0$. Now replace the black hole by a shell of matter of mass M and radius r_0, and surround this shell with a "*real*" atmosphere of radiation in thermal equilibrium at the Hawking temperature (10) as measured by an observer at infinity. Then the ordinary second law should be violated when one performs the same process to the shell surrounded by the ("real") thermal atmosphere as one performs to violate the GSL when the black hole is present. Indeed, the arguments of [48, 49, 15] do not distinguish between infinitesimal quasi-static processes involving a black hole as compared with a shell surrounded by a ("real") thermal atmosphere at the Hawking temperature.

In summary, there appear to be strong grounds for believing in the validity of the GSL.

4.2 Entropy bounds

As discussed in the previous subsection, for a classical black hole the GSL would be violated if one could lower a box containing matter sufficiently close to the black hole before dropping it in. Indeed, for a Schwarzschild black hole, a simple calculation reveals that if the size of the box can be neglected, then the GSL would be violated if one lowered a box containing energy E and entropy S to within a proper distance D of the bifurcation surface of the event horizon before dropping it in, where

$$D < \frac{S}{(2\pi E)}. \tag{15}$$

(This formula holds independently of the mass, M, of the black hole.) However, it is far from clear that the finite size of the box can be neglected if one lowers a box containing physically reasonable matter this close to the black hole. If it cannot be neglected, then this proposed counterexample to the GSL would be invalidated.

As already mentioned in the previous subsection, these considerations led Bekenstein [46] to propose a universal bound on the entropy-to-energy ratio of bounded matter, given by

$$S/E \leq 2\pi R, \tag{16}$$

where R denotes the "circumscribing radius" of the body. Here "E" is normally interpreted as the energy above the ground state; otherwise, Eq. (16) would be trivially violated in cases where the Casimir energy is negative [52] – although in such cases in may still be possible to rescue Eq. (16) by postulating a suitable minimum energy of the box walls [53].

Two key questions one can ask about this bound are: (1) Does it hold in nature? (2) Is it needed for the validity of the GSL? With regard to question (1), even in Minkowski spacetime, there exist many model systems that are physically reasonable (in the sense of positive energies, causal equations of state, etc.) for which Eq. (16) fails. (For a recent discussion of such counterexamples to Eq. (16), see [54, 55, 52]; for counter-arguments to these references, see [53].) In particular it is easily seen that for a system consisting of N non-interacting species of particles with identical properties, Eq. (16) must fail when N becomes sufficiently large. However, for a system of N species of free, massless bosons or fermions, one must take N to be enormously large [56] to violate Eq. (16), so it does not appear that nature has chosen to take advantage of this possible means of violating (16). Equation (16) also is violated at sufficiently low temperatures if one defines the entropy, S, of the system via the canonical ensemble, i.e., $S(T) = -\mathrm{tr}[\rho \ln \rho]$, where ρ denotes the canonical ensemble density matrix,

$$\rho = \exp(-H/T)\mathrm{tr}[\exp(-H/T)], \tag{17}$$

where H is the Hamiltonian. However, a study of a variety of model systems [56] indicates that (16) holds at low temperatures when S is defined via the microcanonical ensemble, i.e., $S(E) = \ln n$ where n is the density of quantum states with energy E. More generally, Eq. (16) has been shown to hold for a wide variety of systems in flat spacetime [56, 57].

The status of Eq. (16) in curved spacetime is unclear; indeed, while there is some ambiguity in how "E" and "R" are defined in Minkowski spacetime [52], it is very unclear what these quantities would mean in a general, non-spherically-symmetric spacetime. (These same difficulties also plague attempts to give a mathematically rigorous formulation of the "hoop conjecture" [58].) With regard to "E", it has long been recognized that there is no meaningful local notion of gravitational energy density in general relativity. Although numerous proposals have been made to define a notion of "quasi-local mass" associated with a closed 2-surface (see, e.g., [59, 60]), none appear to have fully satisfactory properties. Although the difficulties with defining a localized notion of energy are well known, it does not seem to be as widely recognized that there also are serious difficulties in defining "R": Given any spacelike 2-surface, $\mathcal{C}$, in a 4-dimensional spacetime and given any open neighborhood, $\mathcal{O}$, of $\mathcal{C}$, there exists a spacelike 2-surface, $\mathcal{C}'$ (composed of nearly null portions) contained within $\mathcal{O}$ with arbitrarily small area and circumscribing radius. Thus, if one is given a system confined to a world tube in spacetime, it is far from clear how to define any notion of the "externally measured size" of the region unless, say, one is given a preferred slicing by spacelike hypersurfaces. Nevertheless, the fact that Eq. (16) holds for the known black hole solutions (and, indeed, is saturated by the Schwarzschild black hole) and also plausibly holds for a self-gravitating spherically symmetric body [61] provides an indication that some version of (16) may hold in curved spacetime.

With regard to question (2), in the previous section we reviewed arguments for the validity of the GSL that did not require the invocation of any entropy bounds. Thus, the answer to question (2) is "no" unless there are deficiencies in the arguments of the previous section that invalidate their conclusions. A number of such potential deficiencies have been pointed out by Bekenstein. Specifically, the analysis and conclusions of [47] have been criticized by Bekenstein on the grounds that:

i. A "thin box" approximation was made [62].

ii. It is possible to have a box whose *contents* have a greater entropy than unconfined thermal radiation of the same energy and volume [62].

iii. Under certain assumptions concerning the size/shape of the box, the nature of the thermal atmosphere, and the location of the floating point, the buoyancy force of the thermal atmosphere can be shown to be negligible and thus cannot play a role in enforcing the GSL [63].

iv. Under certain other assumptions, the box size at the floating point will be smaller than the typical wavelengths in the ambient thermal atmosphere, thus likely decreasing the magnitude of the buoyancy force [64].

Responses to criticism (i) were given in [65] and [66]; a response to criticism (ii) was given in [65]; and a response to (iii) was given in [66]. As far as I am a aware, no response to (iv) has yet been given in the literature except to note [67] that the arguments of [64] should pose similar difficulties for the ordinary second law for gedankenexperiments involving a self-gravitating body (see the end of subsection 4.1 above). Thus, my own view is that Eq. (16) is not necessary for the validity of the GSL[3]. However, this conclusion remains controversial; see [68] for a recent discussion.

More recently, an alternative entropy bound has been proposed: It has been suggested that the entropy contained within a region whose boundary

[3]It is worth noting that if the buoyancy effects of the thermal atmosphere were negligible, the bound (16) also would not be sufficient to ensure the validity of the GSL for non-spherical bodies: The bound (16) is formulated in terms of the "circumscribing radius", i.e., the *largest* linear dimension, whereas if buoyancy effects were negligible, then to enforce the GSL one would need a bound of the form (16) with R being the *smallest* linear dimension.

has area A must satisfy [69, 70, 71]

$$S \leq A/4. \tag{18}$$

This proposal is closely related to the "holographic principle", which, roughly speaking, states that the physics in any spatial region can be fully described in terms of the degrees of freedom associated with the boundary of that region. (The literature on the holographic principle is far too extensive and rapidly developing to attempt to give any review of it here.) The bound (18) would follow from (16) under the additional assumption of small self-gravitation (so that $E \lesssim R$). Thus, many of the arguments in favor of (16) are also applicable to (18). Similarly, the counterexample to (16) obtained by taking the number, N, of particle species sufficiently large also provides a counterexample to (18), so it appears that (18) can, in principle, be violated by physically reasonable systems (although not necessarily by any systems actually occurring in nature).

Unlike Eq. (16), the bound (18) explicitly involves the gravitational constant G (although we have set $G = 1$ in all of our formulas), so there is no flat spacetime version of (18) applicable when gravity is "turned off". Also unlike (16), the bound (18) does not make reference to the energy, E, contained within the region, so the difficulty in defining E in curved spacetime does not affect the formulation of (18). However, the above difficulty in defining the "bounding area", A, of a world tube in a general, curved spacetime remains present (but see below).

The following argument has been given that the bound (18) is necessary for the validity of the GSL [71]: Suppose we had a spherically symmetric system that was not a black hole (so $R > 2E$) and which violated the bound (18), so that $S > A/4 = \pi R^2$. Now collapse a spherical shell of mass $M = R/2 - E$ onto the system. A Schwarzschild black hole of radius R should result. But the entropy of such a black hole is $A/4$, so the generalized entropy will decrease in this process.

I am not aware of any counter-argument in the literature to the argument given in the previous paragraph, so I will take the opportunity to give one here. If there were a system which violated the bound (18), then the above argument shows that it would be (generalized) entropically unfavorable to collapse that system to a black hole. I believe that the conclusion one should draw from this is that, in this circumstance, it should not be possible to form a black hole. In other words, the bound (18) should be necessary in order for black holes to be stable or metastable states, but should not be needed for the validity of the GSL.

This viewpoint is supported by a simple model calculation. Consider a massless gas composed of N species of (boson or fermion) particles confined by a spherical box of radius R. Then (neglecting self-gravitational effects and any corrections due to discreteness of modes) we have

$$S \sim N^{1/4} R^{3/4} E^{3/4}. \tag{19}$$

We wish to consider a configuration that is not already a black hole, so we need $E < R/2$. To violate (18) – and thereby threaten to violate the GSL by collapsing a shell upon the system – we need to have $S > \pi R^2$. This means that we need to consider a model with $N \gtrsim R^2$. For such a model, start with a region R containing matter with $S > \pi R^2$ but with $E < R/2$. If we try to collapse a shell upon the system to form a black hole of radius R, the collapse time will be $\gtrsim R$. But the Hawking evaporation timescale in this model is $t_{\rm H} \sim R^3/N$, since the flux of Hawking radiation is proportional to N. Since $N \gtrsim R^2$, we have $t_{\rm H} \lesssim R$, so the Hawking evaporation time is shorter than the collapse time! Consequently, the black hole will never actually form. Rather, at best it will merely act as a catalyst for converting the original high entropy confined state into an even higher entropy state of unconfined Hawking radiation.

As mentioned above, the proposed bound (18) is ill defined in a general (non-spherically-symmetric) curved spacetime. There also are other difficulties with (18): In a closed universe, it is not obvious what constitutes the "inside" versus the "outside" of the bounding area. In addition, (18) can be violated near cosmological and other singularities, where the entropy of suitably chosen comoving volumes remains bounded away from zero but the area of the boundary of the region goes to zero. However, a reformulation of (18) which is well defined in a general curved spacetime and which avoids these difficulties has been given by Bousso [72, 73, 74]. Bousso's reformulation can be stated as follows: Let $\mathcal{L}$ be a null hypersurface such that the expansion, θ, of $\mathcal{L}$ is everywhere non-positive, $\theta \leq 0$ (or, alternatively, is everywhere non-negative, $\theta \geq 0$). In particular, $\mathcal{L}$ is not allowed to contain caustics, where θ changes sign from $-\infty$ to $+\infty$. Let B be a spacelike cross-section of $\mathcal{L}$. Bousso's reformulation conjectures that

$$S_{\mathcal{L}} \leq A_B/4, \tag{20}$$

where A_B denotes the area of B and $S_{\mathcal{L}}$ denotes the entropy flux through $\mathcal{L}$ to the future (or, respectively, the past) of B.

In [67] it was argued that the bound (21) should be valid in certain "classical regimes" (see [67]) wherein the local entropy density of matter is bounded in a suitable manner by the energy density of matter. Furthermore, the following generalization of Bousso's bound was proposed: Let $\mathcal{L}$ be a null hypersurface which starts at a cross-section, B, and terminates at a cross-section B'. Suppose further that $\mathcal{L}$ is such that its expansion, θ, is either everywhere non-negative or everywhere non-positive. Then

$$S_{\mathcal{L}} \leq |A_B - A_{B'}|/4. \tag{21}$$

Although we have argued above that the validity of the GSL should not depend upon the validity of the entropy bounds (16) or (18), there is a close relationship between the GSL and the generalized Bousso bound (21). Namely, as discussed in section 2 above, classically, the event horizon of a black hole is a null hypersurface satisfying $\theta \geq 0$. Thus, in a classical regime, the GSL itself would correspond to a special case of the generalized Bousso bound (21). This suggests the intriguing possibility that, in quantum gravity, there might be a more general formulation of the GSL – perhaps applicable to an arbitrary horizon as defined on p. 134 of [15], not merely to an event horizon of a black hole – which would reduce to (21) in a suitable classical limit.

5 Calculations of Black Hole Entropy

The considerations of the previous sections make a compelling case for the merger of the laws of black hole mechanics with the laws of thermodynamics. In particular, they strongly suggest that $S_{\rm bh}$ ($= A/4$ in general relativity – see Eqs.(8) and (9) above) truly represents the physical entropy of a black hole. Now, the entropy of ordinary matter is understood to arise from the number of quantum states accessible to the matter at given values of the energy and other state parameters. One would like to obtain a similar understanding of why $A/4$ represents the entropy of a black hole in general relativity by identifying (and counting) the quantum dynamical degrees of freedom of a black hole. In order to do so, it clearly will be necessary to go beyond the classical and semiclassical considerations of the previous sections and consider black holes within a fully quantum theory of gravity. In this section, we will briefly summarize some of the main approaches that have been taken to the direct calculation of the entropy of a black hole.

The first direct quantum calculation of black hole entropy was given by Gibbons and Hawking [75] in the context of Euclidean quantum gravity. They started with a formal, functional integral expression for the canonical ensemble partition function in Euclidean quantum gravity and evaluated it for a black hole in the "zero loop" (i.e, classical) approximation. As shown in [77], the mathematical steps in this procedure are in direct correspondence with the purely classical determination of the entropy from the form of the first law of black hole mechanics. A number of other entropy calculations that have been given within the formal framework of Euclidean quantum gravity also can be shown to be equivalent to the classical derivation (see [78] for further discussion). Thus, although the derivation of [75] and other related derivations give some intriguing glimpses into possible deep relationships between black hole thermodynamics and Euclidean quantum gravity, they do not appear to provide any more insight than the classical derivation into accounting for the quantum degrees of freedom that are responsible for black hole entropy.

It should be noted that there is actually an inconsistency in the use of the canonical ensemble to derive a formula for black hole entropy, since the entropy of a black hole grows too rapidly with energy for the canonical ensemble to be defined. (Equivalently, the heat capacity of a Schwarzschild black hole is negative, so it cannot come to equilibrium with an infinite heat bath.) A derivation of black hole entropy using the microcanonical ensemble has been given in [76].

Another approach to the calculation of black hole entropy has been to attribute it to the "entanglement entropy" resulting from quantum field correlations between the exterior and interior of the black hole [79, 80, 81]. As a result of these correlations across the event horizon, the state of a quantum field when restricted to the exterior of the black hole is mixed. Indeed, in the absence of a short distance cutoff, the von Neumann entropy, $-\mathrm{tr}[\rho \ln \rho]$, of any physically reasonable state would diverge. If one now inserts a short distance cutoff of the order of the Planck scale, one obtains a von Neumann entropy of the order of the horizon area, A. Thus, this approach provides a natural way of accounting for why the entropy of a black hole is proportional to its surface area. However, the constant of proportionality depends upon a cutoff and is not (presently) calculable within this approach. (Indeed, one might argue that in this approach, the constant of proportionality between S_{bh} and A should depend upon the number, N, of species of particles, and thus could not equal $1/4$ (independently of N). However, it is possible that the N-dependence in the number of states is compensated by an N-dependent renormalization of G [82] and, hence, of the Planck scale cutoff.) More generally, it is far from clear why the black hole horizon should be singled out for a such special treatment of the quantum degrees of freedom in its vicinity, since similar quantum field correlations will exist across any other null surface. It is particularly puzzling why the local degrees of freedom associated with the horizon should be singled out since, as already noted in section 2 above, the black hole horizon at a given time is defined in terms of the entire future history of the spacetime and thus has no distinguished local significance. Finally, since the gravitational action and field equations play no role in the above derivation, it is difficult to see how this approach could give rise to a black hole entropy proportional to Eq. (8) (rather than proportional to A) in a more general theory of gravity. Similar remarks apply to approaches which attribute the relevant degrees of freedom to the "shape" of the horizon [83] or to causal links crossing the horizon [84].

A closely related idea has been to attribute the entropy of the black hole to the ordinary entropy of its thermal atmosphere [85]). If we assume that the thermal atmosphere behaves like a free, massless (boson or fermion) gas, its entropy density will be (roughly) proportional to T^3. However, since T diverges near the horizon in the manner specified by Eq. (12), we find that the total entropy of the thermal atmosphere near the horizon diverges. This is, in effect, a new type of ultraviolet catastrophe. It arises because, on account of arbitrarily large redshifts, there now are infinitely many modes – of arbitrarily high locally measured frequency – that contribute a bounded energy as measured at infinity. To cure this divergence, it is necessary to impose a cutoff on the locally measured frequency of the modes. If we impose a cutoff of the order of the Planck scale, then the thermal atmosphere contributes an entropy of order the horizon area, A, just as in the entanglement entropy analysis. Indeed, this calculation is really the same as the entanglement entropy calculation, since the state of a quantum field outside of the black hole is thermal, so its von Neumann entropy is equal to its thermodynamic entropy (see also [86]). Note that the bulk of the entropy of the thermal atmosphere is highly localized in a "skin" surrounding the horizon, whose thickness is of order of the Planck length.

Since the attribution of black hole entropy to its thermal atmosphere is essentially equivalent to the entanglement entropy proposal, this approach has essentially the same strengths and weaknesses as the entanglement entropy approach. On one hand, it naturally accounts for a black hole entropy proportional to A. On the other hand, this result depends in an essential way on an uncalculable cutoff, and it is difficult to see how the analysis could give rise to Eq. (8) in a more general theory of gravity. The preferred status of the event horizon and the localization of the degrees of freedom responsible for black hole entropy to a "Planck length skin" surrounding the horizon also remain puzzling in this approach. To see this more graphically, consider the collapse of a massive spherical shell of matter. Then, as the shell crosses its Schwarzschild radius, the spacetime curvature outside of the shell is still negligibly small. Nevertheless, within a time of order the Planck time after the crossing of the Schwarzschild radius, the "skin" of thermal atmosphere surrounding the newly formed black hole will come to equilibrium with respect to the notion of time translation symmetry for the static Schwarzschild exterior. Thus, if an entropy is to be assigned to the thermal atmosphere in the manner suggested by this proposal, then the degrees of freedom of the thermal atmosphere – which previously were viewed as irrelevant vacuum fluctuations making no contribution to entropy – suddenly become "activated" by the passage of the shell for the purpose of counting their entropy. A momentous change in the entropy of matter in the universe has occurred, even though observers riding on or near the shell see nothing of significance occurring.

Another approach that is closely related to the entanglement entropy and thermal atmosphere approaches – and which also contains elements closely related to the Euclidean approach and the classical derivation of Eq. (8) – attempts to account for black hole entropy in the context of Sakharov's theory of induced gravity [87, 88]. In Sakharov's proposal, the dynamical aspects of gravity arise from the collective excitations of massive fields. Constraints are then placed on these massive fields to cancel divergences and ensure that the effective cosmological constant vanishes. Sakharov's proposal is not expected to provide a fundamental description of quantum gravity, but at scales below the Planck scale it may possess features in common with other more fundamental descriptions. In common with the entanglement entropy and thermal atmosphere approaches, black hole entropy is explained as arising from the quantum field degrees of freedom outside the black hole. However, in this case the formula for black hole entropy involves a subtraction of the (divergent) mode counting expression and an (equally divergent) expression for the Noether charge operator, so that, in effect, only the massive fields contribute to black hole entropy. The result of this subtraction yields Eq. (9).

More recently, another approach to the calculation of black hole entropy has been developed in the framework of quantum geometry [89, 90]. In this approach, if one considers a spacetime containing an isolated horizon (see section 2 above), the classical symplectic form and classical Hamiltonian each acquire an additional boundary term arising from the isolated horizon [26]. (It should be noted that the phase space [91] considered here incorporates the isolated horizon boundary conditions, i.e., only field variations that preserve the isolated horizon structure are admitted.) These additional terms are identical in form to that of a Chern-Simons theory defined on the isolated horizon. Classically, the fields on the isolated horizon are determined by continuity from the fields in the "bulk" and do not represent additional degrees of freedom. However, in the quantum theory – where distributional fields are allowed – these fields are interpreted as providing additional, independent degrees of freedom associated with the isolated horizon. One then counts the "surface states" of these fields on the isolated horizon subject to a boundary condition relating the surface states to "volume states" and subject to the condition that the area of the isolated horizon (as determined by the volume state) lies within a squared Planck length of the value A. This state counting yields an entropy proportional to A for black holes much larger than the Planck scale. Unlike the entanglement entropy and thermal atmosphere calculations, the state counting here yields finite results and no cutoff need be introduced. However, the formula for entropy contains a free parameter (the "Immirzi parameter"), which arises from an ambiguity in the loop quantization procedure, so the constant of proportionality between S and A is not calculable.

The most quantitatively successful calculations of black hole entropy to date are ones arising from string theory. It is believed that at "low energies", string theory should reduce to a 10-dimensional supergravity theory (see [92] for considerable further discussion of the relationship between string theory and 10-dimensional and 11-dimensional supergravity). If one treats this supergravity theory as a classical theory involving a spacetime metric, g_{ab}, and other classical fields, one can find solutions describing black holes. On the other hand, one also can consider a "weak coupling" limit of string theory, wherein the states are treated perturbatively. In the weak coupling limit, there is no literal notion of a black hole, just as there is no notion of a black hole in linearized general relativity. Nevertheless, certain weak coupling states can be identified with certain black hole solutions of the low energy limit of the theory by a correspondence of their energy and charges. (Here, it is necessary to introduce "D-branes" into string perturbation theory in order to obtain weak coupling states with the desired charges.) Now, the weak coupling states are, in essence, ordinary quantum dynamical degrees of freedom, so their entropy can be computed by the usual methods of statistical physics. Remarkably, for certain classes of extremal and nearly extremal black holes, the ordinary entropy of the weak coupling states agrees exactly with the expression for $A/4$ for the corresponding classical black hole states; see [93] and [94] for reviews of these results. Recently, it also has been shown [95] that for certain black holes, subleading corrections to the state counting formula for entropy correspond to higher order string corrections to the effective gravitational action, in precise agreement with Eq. (8).

Since the formula for entropy has a nontrivial functional dependence on energy and charges, it is hard to imagine that the agreement between the ordinary entropy of the weak coupling states and black hole entropy could be the result of a random coincidence. Furthermore, for low energy scattering, the absorption/emission coefficients ("gray body factors") of the corresponding weak coupling states and black holes also agree [96]. This suggests that there may be a close physical association between the weak coupling states and black holes, and that the dynamical degrees of freedom of the weak coupling states are likely to at least be closely related to the dynamical degrees of freedom responsible for black hole entropy. However, it remains a challenge to understand in what sense the weak coupling states could be giving an accurate picture of the local physics occurring near (and within) the region classically described as a black hole.

The relevant degrees of freedom responsible for entropy in the weak coupling string theory models are associated with conformal field theories. Recently Carlip [97, 98] has attempted to obtain a direct relationship between the string theory state counting results for black hole entropy and the classical Poisson bracket algebra of general relativity. After imposing certain boundary conditions corresponding to the presence of a local Killing horizon, Carlip chooses a particular subgroup of spacetime diffeomorphisms, generated by vector fields ξ^a. The transformations on the phase space of classical general relativity corresponding to these diffeomorphisms are generated by Hamiltonians H_ξ. However, the Poisson bracket algebra of these Hamiltonians is not isomorphic to the Lie bracket algebra of the vector fields ξ^a but rather corresponds to a central extension of this algebra. A Virasoro algebra is thereby obtained. Now, it is known that the asymptotic density of states in a conformal field theory based upon a Virasoro algebra is given by a universal expression (the "Cardy formula") that depends only on the Virasoro algebra. For the Virasoro algebra obtained by Carlip, the Cardy formula yields an entropy in agreement with Eq. (9). Since the Hamiltonians, H_ξ, are closely related to the corresponding Noether currents and charges occurring in the derivation of Eqs. (8) and (9), Carlip's approach holds out the possibility of providing a direct, general explanation of the remarkable agreement between the string theory state counting results and the classical formula for the entropy of a black hole.

6 Open Issues

The results described in the previous sections provide a remarkably compelling case that stationary black holes are localized thermal equilibrium states of the quantum gravitational field, and that the laws of black hole mechanics are simply the ordinary laws of thermodynamics applied to a system containing a black hole. Although no results on black hole thermodynamics have been subject to any experimental or observational tests, the theoretical foundation of black hole thermodynamics appears to be sufficiently firm as to provide a solid basis for further research and speculation on the nature of quantum gravitational phenomena. In this section, I will briefly discuss two key unresolved issues in black hole thermodynamics which may shed considerable further light upon quantum gravitational physics.

6.1 Does a pure quantum state evolve to a mixed state in the process of black hole formation and evaporation?

In classical general relativity, the matter responsible for the formation of a black hole propagates into a singularity lying within the deep interior of the black hole. Suppose that the matter which forms a black hole possesses quantum correlations with matter that remains far outside of the black hole. Then it is hard to imagine how these correlations could be restored during the process of black hole evaporation unless gross violations of causality occur. In fact, the semiclassical analyses of the Hawking process show that, on the contrary, correlations between the exterior and interior of the black hole are continually built up as it evaporates (see [15] for further discussion). Indeed, these correlations play an essential role in giving the Hawking radiation an exactly thermal character [30].

As already mentioned in subsection 4.1 above, an isolated black hole will "evaporate" completely via the Hawking process within a finite time. If the correlations between the inside and outside of the black hole are not restored during the evaporation process, then by the time that the black hole has evaporated completely, an initial pure state will have evolved to a mixed state, i.e., "information" will have been lost. In a semiclassical analysis of the evaporation process, such information loss does occur and is ascribable to the propagation of the quantum correlations into the singularity within the black hole. A key unresolved issue in black hole thermodynamics is whether this conclusion continues to hold in a complete quantum theory of gravity. On one hand, arguments can be given [15] that alternatives to information loss – such as the formation of a high entropy "remnant" or the gradual restoration of correlations during the late stages of the evaporation process – seem highly implausible. On the other hand, it is commonly asserted that the evolution of an initial pure state to a final mixed state is in conflict with quantum mechanics. For this reason, the issue of whether a pure state can evolve to a mixed state in the process of black hole formation and evaporation is usually referred to as the "*black hole information paradox*".

There appear to be two logically independent grounds for the claim that the evolution of an initial pure state to a final mixed state is in conflict with quantum mechanics:

i. Such evolution is asserted to be incompatible with the fundamental principles of quantum theory, which postulates a unitary time evolution of a state vector in a Hilbert space.

ii. Such evolution necessarily gives rise to violations of causality and/or energy-momentum conservation and, if it occurred in the black hole formation and evaporation process, there would be large violations of causality and/or energy-momentum (via processes involving "virtual black holes") in ordinary laboratory physics.

With regard to (1), within the semiclassical framework, the evolution of an initial pure state to a final mixed state in the process of black hole formation and evaporation can be attributed to the fact that the final time slice fails to be a Cauchy surface for the spacetime [15]. No violation of any of the local laws of quantum field theory occurs. In fact, a closely analogous evolution of an initial pure state to a final mixed state occurs for a free, massless field in Minkowski spacetime if one chooses the final "time" to be a hyperboloid rather than a hyperplane [15]. (Here, the "information loss" occurring during the time evolution results from radiation to infinity rather than into a black hole.) Indeed, the evolution of an initial pure state to a final mixed state is naturally accommodated within the framework of the algebraic approach to quantum theory [15] as well as in the framework of generalized quantum theory [99].

The main arguments for (2) were given in [100] (see also [101]). However, these arguments assume that the effective evolution law governing laboratory physics has a "Markovian" character, so that it is purely local in time. As pointed out in [102], one would expect a black hole to retain a "memory" (stored in its external gravitational field) of its energy-momentum, so it is far from clear that an effective evolution law modeling the process of black hole formation and evaporation should be Markovian in nature. Furthermore, even within the Markovian context, it is not difficult to construct models where rapid information loss occurs at the Planck scale, but negligible deviations from ordinary dynamics occur at laboratory scales [102].

For the above reasons, I do not feel that the issue of whether a pure state evolves to a mixed state in the process of black hole formation and evaporation should be referred to as a "paradox". Nevertheless, the resolution of this issue is of great importance: If pure states remain pure, then our basic understanding of black holes in classical and semiclassical gravity will have to undergo significant revision in quantum gravity. On the other hand, if pure states evolve to mixed states in a fully quantum treatment of the gravitational field, then at least the aspect of the classical singularity as a place where "information can get lost" must continue to remain present in quantum gravity. In that case, rather than "smooth out" the singularities of classical general relativity, one might expect singularities to play a fundamental role in the formulation of quantum gravity [103]. Thus, the resolution of this issue would tell us a great deal about both the nature of black holes and the existence of singularities in quantum gravity.

6.2 What (and where) are the degrees of freedom responsible for black hole entropy?

The calculations described in section 5 yield a seemingly contradictory picture of the degrees of freedom responsible for black hole entropy. In the entanglement entropy and thermal atmosphere approaches, the relevant degrees of freedom are those associated with the ordinary degrees of freedom of quantum fields outside of the black hole. However, the dominant contribution to these degrees of freedom comes from (nearly) Planck scale modes localized to (nearly) a Planck length of the black hole, so, effectively, the relevant degrees of freedom are associated with the horizon. In the quantum geometry approach, the relevant degrees of freedom are also associated with the horizon but appear to have a different character in that they reside directly on the horizon (although they are constrained by the exterior state). Finally the string theory calculations involve weak coupling states, so it is not clear what the degrees of freedom of these weak coupling states would correspond to in a low energy limit where these states may admit a black hole interpretation. However, there is no indication in the calculations that these degrees of freedom should be viewed as being localized near the black hole horizon.

The above calculations are not necessarily in conflict with each other, since it is possible that they each could represent a complementary aspect of the same physical degrees of freedom. Nevertheless, it seems far from clear as to whether we should think of these degrees of freedom as residing outside of the black hole (e.g., in the thermal atmosphere), on the horizon (e.g., in Chern-Simons states), or inside the black hole (e.g., in degrees of freedom associated with what classically corresponds to the singularity deep within the black hole).

The following puzzle [104] may help bring into focus some of the issues related to the degrees of freedom responsible for black hole entropy and, indeed, the meaning of entropy in quantum gravitational physics. As we have already discussed, one proposal for accounting for black hole entropy is to attribute it to the ordinary entropy of its thermal atmosphere. If one does so, then, as previously mentioned in section 5 above, one has the major puzzle of explaining why the quantum field degrees of freedom near the horizon contribute enormously to entropy, whereas the similar degrees of freedom that are present throughout the universe – and are locally indistinguishable from the thermal atmosphere – are treated as mere "vacuum fluctuations" which do not contribute to entropy. But perhaps an even greater puzzle arises if we assign a negligible entropy to the thermal atmosphere (as compared with the black hole area, A), as would be necessary if we wished to attribute black hole entropy to other degrees of freedom. Consider a black hole enclosed in a reflecting cavity which has come to equilibrium with its Hawking radiation. Surely, far from the black hole, the thermal atmosphere in the cavity must contribute an entropy given by the usual formula for a thermal gas in (nearly) flat spacetime. However, if the thermal atmosphere is to contribute a negligible total entropy (as compared with A), then at some proper distance D from the horizon much greater than the Planck length, the thermal atmosphere must contribute to the entropy an amount that is much less than the usual result ($\propto T^3$) that would be obtained by a naive counting of modes. If that is the case, then consider a box of ordinary thermal matter at infinity whose energy is chosen so that its floating point would be less than this distance D from the horizon. Let us now slowly lower the box to its floating point. By the time it reaches its floating point, the contents of the box are indistinguishable from the thermal atmosphere, so the entropy within the box also must be less than what would be obtained by usual mode counting arguments. It follows that the entropy within the box must have decreased during the lowering process, despite the fact that an observer inside the box still sees it filled with thermal radiation and would view the lowering process as having been adiabatic. Furthermore, suppose one lowers (or, more accurately, pushes) an empty box to the same distance from the black hole. The entropy difference between the empty box and the box filled with radiation should still be given by the usual mode counting formulas. Therefore, the empty box would have to be assigned a negative entropy.

I believe that in order to gain a better understanding of the degrees of freedom responsible for black hole entropy, it will be necessary to achieve a deeper understanding of the notion of entropy itself. Even in flat spacetime, there is far from universal agreement as to the meaning of entropy – particularly in quantum theory – and as to the nature of the second law of thermodynamics. The situation in general relativity is considerably murkier [105], as, for example, there is no unique, rigid notion of "time translations" and classical general relativistic dynamics appears to be incompatible with any notion of "ergodicity". It seems likely that a new conceptual framework will be required in order to have a proper understanding of entropy in quantum gravitational physics.

7 Acknowledgements

This research was supported in part by NSF grant PHY 95-14726 to the University of Chicago.

References

[1] R.M. Wald, *General Relativity*, University of Chicago Press (Chicago, 1984).

[2] S.W. Hawking and G.F.R. Ellis, *The Large Scale Structure of Space-time*, Cambridge University Press (Cambridge, 1973).

[3] P. T. Chrusciel, E. Delay. G.J. Galloway, and R. Howard, "The Area Theorem", gr-qc/0001003.

[4] S.W. Hawking, "Gravitational Radiation from Colliding Black Holes", Phys. Rev. Lett. **26**, 1344-1346 (1971).

[5] J.D. Bekenstein, "Black Holes and Entropy", Phys. Rev. **D7**, 2333-2346 (1973).

[6] J.D. Bekenstein, "Generalized Second Law of Thermodynamics in Black-Hole Physics", Phys. Rev. **D9**, 3292-3300 (1974).

[7] J.M. Bardeen, B. Carter, and S.W. Hawking, "The Four Laws of Black Hole Mechanics" Commun. Math. Phys. **31**, 161-170 (1973).

[8] M. Heusler, *Black Hole Uniqueness Theorems*, Cambridge University Press (Cambridge, 1996).

[9] B. Carter, "Black Hole Equilibrium States" in *Black Holes*, ed. by C. DeWitt and B.S. DeWitt, 57-214, Gordon and Breach (New York, 1973).

[10] H. Friedrich, I. Racz, and R.M. Wald, "On the Rigidity Theorem for Spacetimes with a Stationary Event Horizon or a Compact Cauchy Horizon", Commun. Math. Phys. **204**, 691-707 (1999); gr-qc/9811021.

[11] D. Sudarsky and R.M. Wald, "Extrema of Mass, Stationarity and Staticity, and Solutions to the Einstein-Yang-Mills Equations" Phys. Rev. **D46**, 1453-1474 (1992).

[12] D. Sudarsky and R.M. Wald, "Mass Formulas for Stationary Einstein-Yang-Mills Black Holes and a Simple Proof of Two Staticity Theorems" Phys. Rev. **D47**, R5209-R5213 (1993).

[13] P.T. Chrusciel and R.M. Wald, "Maximal Hypersurfaces in Stationary Asymptotically Flat Spacetimes" Commun. Math Phys. **163**, 561-604 (1994).

[14] I. Racz and R.M. Wald, "Global Extensions of Spacetimes Describing Asymptotic Final States of Black Holes" Class. Quant. Grav. **13**, 539-552 (1996); gr-qc/9507055.

[15] R.M. Wald, *Quantum Field Theory in Curved Spacetime and Black Hole Thermodynamics*, University of Chicago Press (Chicago, 1994).

[16] V. Iyer and R.M. Wald, "Some Properties of Noether Charge and a Proposal for Dynamical Black Hole Entropy", Phys. Rev. **D50**, 846-864 (1994).

[17] R. Sorkin, "Two Topics Concerning Black Holes: Extremality of the Energy, Fractality of the Horizon" in *Proceedings of the Conference on Heat Kernel Techniques and Quantum Gravity*, ed. by S.A. Fulling, 387-407, University of Texas Press, (Austin, 1995); gr-qc/9508002.

[18] W. Israel, "Third Law of Black-Hole Dynamics: a Formulation and Proof", Phys. Rev. Lett. **57**, 397-399 (1986).

[19] M. Aizenman and E.H. Lieb, "The Third Law of Thermodynamics and the Degeneracy of the Ground State for Lattice Systems", J. Stat. Phys. **24**, 279-297 (1981).

[20] R.M. Wald, "'Nernst Theorem' and Black Hole Thermodynamics", Phys. Rev. **D56**, 6467-6474 (1997); gr-qc/9704008.

[21] A. Ashtekar, C. Beetle, O. Dreyer, S. Fairhurst, B. Krishnan, J. Lewandowski, and J. Wisniewski, "Generic Isolated Horizons and Their Applications", gr-qc/0006006.

[22] A. Ashtekar, C. Beetle, and S. Fairhurst, "Isolated Horizons: A Generalization of Black Hole Mechanics", Class. Quant. Grav. **16**, L1-L7 (1999); gr-qc/9812065.

[23] A. Ashtekar, C. Beetle, and S. Fairhurst, "Mechanics of Isolated Horizons", Class. Quant. Grav. **17**, 253-298 (2000); gr-qc/9907068.

[24] A. Ashtekar and A. Corichi, "Laws Governing Isolated Horizons: Inclusion of Dilaton Couplings", Class. Quant. Grav. **17**, 1317-1332 (2000); gr-qc/9910068.

[25] J. Lewandowski, "Spacetimes Admitting Isolated Horizons", Class. Quant. Grav. **17**, L53-L59 (2000); gr-qc/9907058.

[26] A. Ashtekar, S. Fairhurst, and B. Krishnan, "Isolated Horizons: Hamiltonian Evolution and the First Law", gr-qc/0005083.

[27] A. Corichi, U. Nucamendi, and D. Sudarsky, "Einstein-Yang-Mills Isolated Horizons: Phase Space, Mechanics, Hair and Conjectures" Phys. Rev. **D62**, 044046 (19 pages) (2000); gr-qc/0002078.

[28] S.W. Hawking, "Particle Creation by Black Holes", Commun. Math. Phys. **43**, 199-220 (1975).

[29] L. Parker, "Quantized Fields and Particle Creation in Expanding Universes", Phys. Rev. **183**, 1057-1068 (1969).

[30] R.M. Wald, "On Particle Creation by Black Holes", Commun. Math. Phys. **45**, 9-34 (1975).

[31] K. Fredenhagen and R. Haag, "On the Derivation of the Hawking Radiation Associated with the Formation of a Black Hole", Commun. Math. Phys. **127**, 273-284 (1990).

[32] W.G. Unruh, "Experimental Black-Hole Evaporation?", Phys. Rev. Lett. **46**, 1351-1353 (1981).

[33] W.G. Unruh, "Dumb Holes and the Effects of High Frequencies on Black Hole Evaporation" Phys. Rev. **D51**, 2827-2838 (1995); gr-qc/9409008.

[34] R. Brout, S. Massar, R. Parentani, and Ph. Spindel, "Hawking Radiation Without Transplanckian Frequencies", Phys. Rev. **D52**, 4559-4568 (1995); hep-th/9506121.

[35] S. Corley and T. Jacobson, "Hawking Spectrum and High Frequency Dispersion" Phys. Rev. **D54**, 1568-1586 (1996); hep-th/9601073.

[36] T. Jacobson, "On the Origin of the Outgoing Black Hole Modes" Phys. Rev. **D53**, 7082-7088 (1996); hep-th/9601064.

[37] B. Reznik, "Trans-Planckian Tail in a Theory with a Cutoff", Phys. Rev. **D55**, 2152-2158 (1997); gr-qc/9606083.

[38] M. Visser, "Hawking radiation without black hole entropy", Phys. Rev. Lett. **80**, 3436-3439 (1998); gr-qc/9712016.

[39] S. Corley and T. Jacobson, "Lattice Black Holes", Phys. Rev. **D57**, 6269-6279 (1998); hep-th/9709166.

[40] T. Jacobson and D. Mattingly, "Hawking radiation on a falling lattice", Phys. Rev. **D61** 024017 (10 pages) (2000); hep-th/9908099.

[41] W.G. Unruh, "Notes on Black Hole Evaporation", Phys. Rev. **D14**, 870-892 (1976).

[42] B.S. Kay and R.M. Wald, "Theorems on the Uniqueness and Thermal Properties of Stationary, Nonsingular, Quasifree States on Spacetimes with a Bifurcate Killing Horizon", Phys. Rep. **207**, 49-136 (1991).

[43] J.J. Bisognano and E.H. Wichmann, "On the Duality Condition for Quantum Fields", J. Math. Phys. **17**, 303-321 (1976).

[44] J.B. Hartle and S.W. Hawking, "Path Integral Derivation of Black Hole Radiance", Phys. Rev. **D13**, 2188-2203 (1976).

[45] R. Geroch, colloquium given at Princeton University, December, 1971 (unpublished).

[46] J.D. Bekenstein, "Universal Upper Bound on the Entropy-to-Energy Ratio for Bounded Systems", Phys. Rev. **D23**, 287-298 (1981).

[47] W.G. Unruh and R.M. Wald, "Acceleration Radiation and the Generalized Second Law of Thermodynamics", Phys. Rev. **D25**, 942-958 (1982).

[48] W.H. Zurek and K.S. Thorne, "Statistical Mechanical Origin of the Entropy of a Rotating, Charged Black Hole", Phys. Rev. Lett. **54**, 2171-2175 (1986).

[49] K.S. Thorne, W.H. Zurek, and R.H. Price, "The Thermal Atmosphere of a Black Hole", in *Black Holes: The Membrane Paradigm*, ed. by K.S. Thorne, R.H. Price, and D.A. Macdonald, 280-340, Yale University Press (New Haven, 1986).

[50] V.P. Frolov and D.N. Page, "Proof of the Generalized Second Law for Quasistatic, Semiclassical Black Holes", Phys. Rev. Lett. **71**, 3902-3905 (1993).

[51] R.D. Sorkin, "The Statistical Mechanics of Black Hole Thermodynamics", in *Black Holes and Relativistic Stars*, ed. by R.M. Wald, 177-194, University of Chicago Press (Chicago, 1998); gr-qc/9705006.

[52] D.N. Page, "Defining Entropy Bounds", hep-th/0007238.

[53] J.D. Bekenstein, "On Page's Examples Challenging the Entropy Bound", gr-qc/0006003.

[54] D.N. Page, "Huge Violations of Bekenstein's Entropy Bound", gr-qc/0005111.

[55] D.N. Page, "Subsystem Entropy Exceeding Bekenstein's Bound", hep-th/0007237.

[56] J.D. Bekenstein, "Entropy Content and Information Flow in Systems with Limited Energy", Phys. Rev. **D30**, 1669-1679 (1984).

[57] J.D. Bekenstein and M. Schiffer, "Quantum Limitations on the Storage and Transmission of Information", Int. J. Mod. Phys. **C1**, 355 (1990).

[58] C.W. Misner, K.S. Thorne, and J.A. Wheeler, *Gravitation*, Freeman (San Francisco, 1973).

[59] R. Penrose, "Quasi-Local Mass and Angular Momentum ", Proc. Roy. Soc. Lond. **A381**, 53-63 (1982).

[60] J.D. Brown and J.W. York, "Quasilocal Energy and Conserved Charges Derived from the Gravitational Action", Phys. Rev. **D47**, 1407-1419 (1993).

[61] R.D. Sorkin, R.M. Wald, and Z.J. Zhang, "Entropy of Self-Gravitating Radiation", Gen. Rel. Grav. **13**, 1127-1146 (1981).

[62] J.D. Bekenstein, "Entropy Bounds and the Second Law for Black Holes", Phys. Rev. **D27**, 2262-2270 (1983).

[63] J.D. Bekenstein, "Entropy Bounds and Black Hole Remnants", Phys. Rev. **D49**, 1912-1921 (1994).

[64] J.D. Bekenstein, "Non-Archimedian Character of Quantum Buoyancy and the Generalized Second Law of Thermodynamics", Phys. Rev. **D60**, 124010 (9 pages) (1999); gr-qc/9906058.

[65] W.G. Unruh and R.M. Wald, "Entropy Bounds, Acceleration Radiation and the Generalized Second Law", Phys. Rev. **D27**, 2271-2276 (1983).

[66] M.A. Pelath and R.M.Wald, "Comment on Entropy Bounds and the Generalized Second Law", Phys. Rev. **D60**, 104009 (4 pages) (1999); gr-qc/9901032.

[67] E.E. Flanagan, D. Marolf, and R.M. Wald, "Proof of Classical Versions of the Bousso Entropy Bound and of the Generalized Second Law" Phys. Rev. **D62**, 084035 (11 pages) (2000); hep-th/9909373

[68] W. Anderson, "Does the GSL Imply and Entropy Bound?", in *Matters of Gravity*, ed. by J. Pullin, gr-qc/9909022.

[69] G. 't Hooft, "On the Quantization of Space and Time", in *Quantum Gravity*, ed. by M.A. Markov, V.A. Berezin, and V.P. Frolov, 551-567, World Scientific Press (Singapore, 1988).

[70] J.D. Bekenstein, "Do We Understand Black Hole Entropy ?", in *Proceedings of the VII Marcel Grossman Meeting*, 39-58, World Scientific Press (Singapore, 1996); gr-qc/9409015.

[71] L. Susskind, "The World as a Hologram", J. Math. Phys. **36**, 6377-6396 (1995); hep-th/9409089.

[72] R. Bousso, "A Covariant Entropy Conjecture", JHEP **07**, 004 (1999); hep-th/9905177.

[73] R. Bousso, "Holography in General Space-times", JHEP **06**, 028 (1999); hep-th/9906022.

[74] R. Bousso, "The Holographic Principle for General Backgrounds", hep-th/9911002.

[75] G. Gibbons and S.W. Hawking "Action Integrals and Partition Functions in Quantum Gravity", Phys. Rev. **D15**, 2752-2756 (1977).

[76] J.D. Brown and J.W. York, "Microcanonical Functional Integral for the Gravitational Field", Phys. Rev. **D47**, 1420-1431 (1993).

[77] R.M. Wald, "Black Hole Entropy is the Noether Charge", Phys. Rev. **D48**, R3427-R3431 (1993).

[78] V. Iyer and R.M. Wald, "A Comparison of Noether Charge and Euclidean Methods for Computing the Entropy of Stationary Black Holes", Phys. Rev. **D52**, 4430-4439 (1995); gr-qc/9503052.

[79] L. Bombelli, R.K. Koul, J. Lee, and R. Sorkin, "Quantum Source of Entropy for Black Holes" Phys. Rev. **D34**, 373-383 (1986).

[80] C. Callen and F. Wilzcek, "On Geometric Entropy", Phys. Lett **B333**, 55-61 (1994).

[81] C. Holzhey, F. Larsen, and F. Wilzcek, "Geometric and Renormalized Entropy in Conformal Field Theory", Nucl. Phys. **B424**, 443-467 (1994).

[82] L. Susskind and J. Uglam, "Black Hole Entropy in Canonical Quantum Gravity and Superstring Theory", Phys. Rev. **D50**, 2700-2711 (1994).

[83] R. Sorkin, "How Wrinkled is the Surface of a Black Hole?" in *Proceedings of the First Australasian Conference on General Relativity and Gravitation*, ed. by D. Wiltshire, 163-174, University of Adelaide Press, (Adelaide, 1996); gr-qc/9701056.

[84] D. Dou, "Causal Sets, a Possible Interpretation for the Black Hole Entropy, and Related Topics", Ph.D. thesis (SISSA, Trieste, 1999).

[85] G. 't Hooft, "On the Quantum Structure of a Black Hole", Nucl. Phys. **B256**, 727-745 (1985).

[86] S. Mukohyama, "Aspects of Black Hole Entropy", gr-qc/9912103.

[87] V.P. Frolov, D.V. Fursaev and A.I. Zelnikov, "Statistical Origin of Black Hole Entropy in Induced Gravity", Nucl. Phys. **B486**, 339-352 (1997); hep-th/9607104.

[88] V.P. Frolov and D.V. Fursaev, "Mechanism of the Generation of Black Hole Entropy in Sakharov's Induced Gravity", Phys. Rev. **D56**, 2212-2225 (1997); hep-th/9703178.

[89] A. Ashtekar, J. Baez, A. Corichi, and K. Krasnov, "Quantum Geometry and Black Hole Entropy", Phys. Rev. Lett. **80**, 904-907 (1998); gr-qc/9710007.

[90] A. Ashtekar and K. Krasnov, "Quantum Geometry and Black Holes", in *Black Holes, Gravitational Radiation, and the Universe*, ed. by B.R. Iyer and B. Bhawal, 149-170, Kluwer Academic Publishers (Dordrecht, 1999); gr-qc/9804039.

[91] A. Ashtekar, A. Corichi, and K. Krasnov, "Isolated Horizons: the Classical Phase Space", gr-qc/9905089.

[92] D. Marolf, "String/M-branes for Relativists", gr-qc/9908045.

[93] G. Horowitz, "Quantum States of Black Holes", in *Black Holes and Relativistic Stars*, ed. by R.M. Wald, 241-266, University of Chicago Press (Chicago, 1998); gr-qc/9704072.

[94] A. Peet, "TASI Lectures on Black Holes in String Theory", hep-th/0008241.

[95] G.L. Cardoso, B. de Wit, and T. Mohaupt, "Area Law Corrections from State Counting and Supergravity", Class. Quant. Grav. **17** 1007-1015 (2000); hep-th/9910179.

[96] J.M. Maldacena and A. Strominger, "Black Hole Greybody Factors and D-Brane Spectroscopy", Phys.Rev. **D55**, 861-870 (1997); hep-th/9609026.

[97] S. Carlip, "Entropy from Conformal Field Theory at Killing Horizons", Class. Quant. Grav. **16**, 3327-3348 (1999); gr-qc/9906126.

[98] S. Carlip, "Black Hole Entropy from Horizon Conformal Field Theory", gr-qc/9912118.

[99] J. Hartle, "Generalized Quantum Theory in Evaporating Black Hole Spacetimes", in *Black Holes and Relativistic Stars*, ed. by R.M. Wald, 195-219, University of Chicago Press (Chicago, 1998); gr-qc/9705022.

[100] T. Banks, L. Susskind, and M.E. Peskin, "Difficulties for the Evolution of Pure States into Mixed States", Nucl. Phys. **B244**, 125-134 (1984).

[101] J. Ellis, J.S. Hagelin, D.V. Nanopoulos, and M. Srednicki, "Search for Violations of Quantum Mechanics", Nucl. Phys. **B241**, 381-405 (1984).

[102] W.G. Unruh and R.M. Wald, "Evolution Laws Taking Pure States to Mixed States in Quantum Field Theory", Phys. Rev. **D52**, 2176-2182 (1995); hep-th/9503024.

[103] R. Penrose, "Singularities and Time-Asymmetry" in *General Relativity, an Einstein Centennary Survey*, ed. by S.W. Hawking and W. Israel, 581-638, Cambridge University Press (Cambridge, 1979).

[104] R.M. Wald, "Gravitation, Thermodynamics, and Quantum Theory", Class. Quant. Grav. **16**, A177-A190 (1999); gr-qc/9901033.

[105] R.M. Wald, "Black Holes and Thermodynamics", in *Black Holes and Relativistic Stars*, ed. by R.M. Wald, 155-176, University of Chicago Press (Chicago, 1998); gr-qc/9702022.

BLACK HOLES WITH HAIR

ERICK J. WEINBERG
Department of Physics
Columbia University
New York, NY 10027, USA

1. Introduction

Among the most remarkable results of classical general relativity are the black hole uniqueness theorems for pure gravity and for gravity coupled to electromagnetism. The simplicity and elegance of these black holes inspired Chandrasekhar's statement, in the prologue to his treatise [1], that "the black holes of nature are the most perfect macroscopic objects there are in the universe ... and ... they are the simplest objects as well".

These uniqueness theorems, together with related results on black holes coupled to other types of matter and on the behavior of matter as it collapses to form a black hole, led to the widely repeated statement that "black holes have no hair" [2]. This statement had various interpretations. Some took it to mean that the only possible static fields outside a black hole horizon are those required by the conserved long-range charges. A weaker interpretation allowed such "hair", but required that the solution be uniquely determined by its mass, angular momentum, and conserved charges. In either case, there was a question of whether the statement applied to all solutions, or only to stable solutions.

Many in the wider theoretical physics community thought that a general result, restricted perhaps by technical assumptions, had been established. In fact, as was clear to experts in the field, no-hair theorems had only been proven for very specific types of matter, and the more general statement, however interpreted, was only a conjecture.

Over the last decade it has become clear that this conjecture, even in its weaker form, is not in general true. When gravity is coupled to matter theories that have more complex structures — including theories similar to those of the standard model — there are black hole solutions that do, indeed, have hair. These black holes are most naturally subatomic, rather

P.G. Bergmann and V. de Sabbata (eds.),
Advances in the Interplay Between Quantum and Gravity Physics, 523–544.

than astrophysical, in size. Interesting in their own right, they also help clarify which features are general characteristics of classical black holes and which are not, and at the same time lend insight into the quantum mechanical connection between black hole dynamics and thermodynamics.

By now, a variety of solutions with hair are known. In these lectures I will focus on the magnetically charged black holes that arise in spontaneously broken Yang-Mills-Higgs theories and on the properties of the related self-gravitating nonsingular magnetic monopoles. For an extensive review that includes discussions of other types of solutions, see [3].

After a brief general discussion of spherically symmetric black holes, I will review some of the properties of 't Hooft-Polyakov monopoles in flat spacetime. While these are usually understood from a topological point of view, I will present some energetic arguments that are perhaps more helpful in understanding the related black hole solutions. I will then describe the effects of gravity on the singly-charged monopole. These are two-fold. First, there is an upper bound on the mass of a nonsingular monopole, with the monopole going over into an extremal black hole as this limit is reached. Second, it is possible to embed a black hole within the monopole core, thus yielding a black hole with hair. These solutions with hair can be degenerate in mass and charge with pure Reissner-Nordstrom solutions. I will show that in these theories the latter have a classical instability that leads to decay into solutions with hair. In the case of Reissner-Nordstrom black holes with higher magnetic charge, this instability can lead to static black holes without any rotational symmetry. Finally, in the last part of these lectures I will examine in more detail the transition from nonsingular monopole to black hole, focusing on how the "quasi-black holes" that are just short of this transition appear to an external observer. As we will see, these provide interesting insights into the origin of black hole entropy.

2. Spherically symmetric black holes

For the sake of simplicity, in these lectures I will focus for the most part on solutions with static, spherically symmetric metrics. Any such metric can be written in the form

$$ds^2 = -B(r)dt^2 + A(r)dr^2 + R^2(r)(d\theta^2 + \sin^2\theta d\phi^2)\,. \qquad (1)$$

Furthermore, I will use the freedom to redefine coordinates to set $R(r) = r$ and write

$$ds^2 = -B(r)dt^2 + A(r)dr^2 + r^2(d\theta^2 + \sin^2\theta d\phi^2)\,. \qquad (2)$$

A zero of $1/A$ corresponds to a horizon, while a double zero corresponds to an extremal horizon.

The two simplest black holes of this form are the Schwarzschild and the Reissner-Nordstrom solutions. The Schwarzschild black hole is a vacuum solution with

$$B_{\text{Sch}} = A_{\text{Sch}}^{-1} = 1 - \frac{2MG}{r}\,. \tag{3}$$

There is a coordinate singularity at the horizon, $r = 2GM$, and a true curvature singularity at $r = 0$. The maximally extended spacetime contains two exterior regions, each with $2MG < r < \infty$ and $-\infty < t < \infty$; a region, with $0 < r < 2MG$, that lies to the future of the horizon and ends on a spacelike $r = 0$ singularity; and finally a region, also with $0 < r < 2MG$, that lies in the past of the horizon and has an initial spacelike $r = 0$ singularity. It is important to keep in mind that r is actually a timelike coordinate for values less than $2MG$. Hence, it is somewhat misleading to think of the region with $r < 2MG$ as the "interior" of the black hole; one can draw a complete spacelike hypersurface through the spacetime on which r is never less than $2MG$.

The Reissner-Nordstrom solution has Coulomb electric and magnetic fields

$$\mathbf{E} = Q_{\text{E}}\,\frac{\hat{\mathbf{r}}}{r^2} \qquad\qquad \mathbf{B} = Q_{\text{M}}\,\frac{\hat{\mathbf{r}}}{r^2} \tag{4}$$

and a metric

$$B_{\text{RN}} = A_{\text{RN}}^{-1} = 1 - \frac{2MG}{r} + \frac{4\pi G(Q_{\text{E}}^2 + Q_{\text{M}}^2)}{r^2}\,. \tag{5}$$

There are three cases to consider. If

$$M > \sqrt{4\pi(Q_{\text{E}}^2 + Q_{\text{M}}^2)}\, M_{\text{Pl}} \tag{6}$$

(where the Planck mass $M_{\text{Pl}} = G^{-1/2}$ in units where $c = \hbar = 1$) the metric describes a black hole solution with horizons at

$$r_{\pm} = MG \pm \sqrt{M^2G^2 - 4\pi G(Q_{\text{E}}^2 + Q_{\text{M}}^2)} \tag{7}$$

and a timelike curvature singularity at $r = 0$. The maximally extended spacetime contains an infinite sequence of exterior regions. It is possible for a worldline to pass through an infinite sequence of such regions without ever encountering the $r = 0$ singularities.

If

$$M = \sqrt{4\pi(Q_{\text{E}}^2 + Q_{\text{M}}^2)}\, M_{\text{Pl}} \tag{8}$$

there is an extremal black hole, with a horizon at

$$r_0 = MG = \sqrt{4\pi(Q_{\text{E}}^2 + Q_{\text{M}}^2)}\, M_{\text{Pl}}^{-1}\,. \tag{9}$$

As in the previous case, the maximally extended spacetime contains an infinite sequence of exterior regions, and it is possible to avoid the timelike singularity at $r = 0$. On any hypersurface of constant time, the extremal horizon at $r = r_0$ is an infinite proper distance from any point with $r \neq r_0$; nevertheless, a worldline starting at any $r > r_0$ can cross the horizon and reach $r < r_0$ in a finite proper time.

Finally, if

$$M < \sqrt{4\pi(Q_{\rm E}^2 + Q_{\rm M}^2)}\, M_{\rm Pl} \tag{10}$$

there is no horizon, but only a naked singularity at $r = 0$.

The Schwarzschild and Reissner-Nordstrom solutions are the only static black hole solutions in the Einstein-Maxwell theory; if we only require that the solution be stationary, there is also the Kerr-Newman solution, which includes the others as special cases. Thus, these black holes are completely specified by giving their mass, angular momentum, and electric and magnetic charges. This result was the inspiration for the no-hair conjecture. However, although the statement that "black holes have no hair" was widely repeated, this conjecture was actually proven only in a number of very specific contexts.

As an example of these, consider the case of gravity coupled to a scalar field $\phi(x)$ [4]. The dynamics of ϕ are governed by a potential $V(\phi)$ that is assumed to have a single minimum, at $\phi = \phi_0$. To simplify the presentation I will assume spherical symmetry, although the proof is readily extended to the more general case. With a metric of the form given in Eq. (2), and ϕ assumed to depend only on r, the scalar field equations take the form

$$\frac{1}{r^2\sqrt{AB}} \left(\frac{r^2\sqrt{AB}\,\phi'}{A} \right)' = \frac{dV}{d\phi} \tag{11}$$

with primes denoting differentiation with respect to r. Multiplying both sides of this equation by common factors, we obtain

$$(\phi - \phi_0) \left(\frac{r^2\sqrt{AB}\,\phi'}{A} \right)' = (\phi - \phi_0)\, r^2 \sqrt{AB}\, \frac{dV}{d\phi}\,. \tag{12}$$

We now assume that there is a horizon at $r = r_{\rm H}$, and integrate the above equation from $r_{\rm H}$ to infinity. An integration by parts leads to

$$\int_{r_{\rm H}}^{\infty} dr\, \frac{d}{dr} \left[(\phi - \phi_0)\, \frac{r^2\sqrt{AB}\,\phi'}{A} \right] = \int_{r_{\rm H}}^{\infty} dr\, r^2\sqrt{AB} \left[\frac{(\phi')^2}{A} + (\phi - \phi_0)\, \frac{dV}{d\phi} \right]. \tag{13}$$

The left hand side is equal to the sum of surface terms at the horizon and at infinity. The former vanishes because $1/A = 0$ on the horizon. Because ϕ

must approach its vacuum value at $r = \infty$, the decreases in ϕ' and $\phi - \phi_0$ are rapid enough that the surface term at infinity also vanishes. The integral on the right hand side must therefore be equal to zero. The first term in the integrand is manifestly positive (since $A > 0$ outside the horizon), while our assumption that V has a single minimum implies that the second term is also positive. Hence, the only way that the integral can vanish is for $\phi(r)$ to be equal to its vacuum value ϕ_0 everywhere outside the horizon.

This proof relied crucially on the assumed properties of $V(\phi)$. It would have failed if the potential had multiple minima, or if there were additional fields present. Although the proof can be extended to a wider class of scalar field theories [5], this reliance on the details of the theory suggests that it might be possible to construct black holes with hair in a theory with a sufficiently complex structure. As we will see, a natural place to look is the spontaneously broken gauge theories that support magnetic monopole solutions in flat spacetime.

3. Magnetic monopoles in flat spacetime

Consider an SU(2) Yang-Mills theory with a triplet scalar field ϕ^a and a Lagrangian

$$\mathcal{L} = -\frac{1}{4}(F^a_{\mu\nu})^2 + \frac{1}{2}(D_\mu\phi)^2 - V(\phi) \tag{14}$$

where the field strength

$$F^a_{\mu\nu} = \partial_\mu A^a_\nu - \partial_\nu A^a_\mu - e\,\epsilon_{abc} A^b_\mu A^c_\nu\,, \tag{15}$$

the covariant derivative

$$D_\mu\phi^a = \partial_\mu\phi^a - e\,\epsilon_{abc} A^b_\mu \phi^c\,, \tag{16}$$

and the scalar field potential

$$V(\phi) = -\frac{\mu^2}{2}\phi^2 + \frac{\lambda}{2}(\phi^2)^2 \tag{17}$$

with μ^2 and λ both positive.

The potential has a family of gauge-equivalent minima with

$$\phi^2 = v^2 \equiv \frac{\mu^2}{2\lambda} \tag{18}$$

that spontaneously break the SU(2) symmetry down to U(1). Without loss of generality, we can choose the vacuum with $\phi^a = v\delta^{a3}$. The fields corresponding to the physical elementary particles are then the "electromagnetic" U(1) gauge field $\mathcal{A}_\mu = A^3_\mu$, a neutral scalar field $\varphi = \phi^3$, and a complex vector field $W_\mu = (A^1_\mu + iA^2_\mu)/\sqrt{2}$ whose quanta are spin-one particles with electric charge $\pm e$ and mass $m_W = ev$. In terms of these fields, the Lagrangian can be written as

$$\begin{aligned} \mathcal{L} &= -\frac{1}{2}\left|\mathcal{D}_\mu W_\nu - \mathcal{D}_\nu W_\mu\right|^2 - \frac{1}{4}(\mathcal{F}_{\mu\nu})^2 + \frac{1}{2}d_{\mu\nu}\mathcal{F}^{\mu\nu} - \frac{1}{4}(d_{\mu\nu})^2 \\ &\quad + e^2\varphi^2|W_\mu|^2 + \frac{1}{2}(\partial_\mu\varphi)^2 - V(\varphi) \end{aligned} \tag{19}$$

where

$$\mathcal{F}_{\mu\nu} = \partial_\mu \mathcal{A}_\nu - \partial_\nu \mathcal{A}_\mu \tag{20}$$

and

$$\mathcal{D}_\mu = (\partial_\mu - ie\mathcal{A}_\mu) \tag{21}$$

denote the electromagnetic field strength and covariant derivative and

$$d_{\mu\nu} = ie[W^*_\mu W_\nu - W^*_\nu W_\mu] \tag{22}$$

is the magnetic moment density due to the charged vector field.

This theory possesses finite energy magnetic monopole solutions [6, 7]. Their existence is usually motivated by topological arguments. One begins by considering configurations in which the scalar field at spatial infinity has its SU(2) orientation correlated with the direction in space, so that as $r \to \infty$

$$\phi^a \to v\hat{r}^a\,. \tag{23}$$

Because such a configuration cannot be smoothly deformed to the uniform vacuum solution, it should be possible to obtain a static solution by minimizing the energy subject to this boundary condition. In order that the energy be finite, $D_i\phi$ must fall faster than $r^{-3/2}$, which implies a vector potential

$$A^a_i = \epsilon_{iak}\frac{\hat{r}^k}{er} + O(r^{-2}) \tag{24}$$

that gives rise to a Coulomb magnetic field

$$B^a_i = \frac{\hat{r}^a\hat{r}^k}{er^2} + O(r^{-3})\,. \tag{25}$$

Thus, this configuration describes a magnetic monopole with magnetic charge $Q_{\rm M} = 1/e$. Higher charges can be obtained by allowing additional twisting of the asymptotic scalar field, but these must obey the topological quantization condition

$$Q_{\rm M} = \frac{n}{e}\,. \tag{26}$$

One can proceed further by adopting the Ansatz

$$\begin{aligned} \phi^a &= v\,\hat{r}^a\,h(r) \\ A_i^a &= \epsilon_{iak}\,\hat{r}^k \left[\frac{1-u(r)}{er}\right] \end{aligned} \tag{27}$$

with the boundary conditions $h(0) = u(\infty) = 0$ and $u(0) = h(\infty) = 1$. Substituting this Ansatz into the Euler-Lagrange equations of the theory gives a set of coupled ordinary differential equations that can be solved numerically. Their solution is characterized by a core region, of radius $R_{\rm core} \sim (ev)^{-1}$, outside of which u and h approach their asymptotic values exponentially rapidly. The total energy is

$$M_{\rm mon} \sim \frac{Q_{\rm M}^2}{R_{\rm core}} \sim \frac{v}{e}\,. \tag{28}$$

We will find it useful to view this solution from a somewhat different viewpoint [8]. To this end, note that by a singular gauge transformation the fields of Eq. (27) can be brought into the unitary gauge form

$$\begin{aligned} \varphi &= h(r) \\ W_i &= \frac{f_i(\theta,\phi)}{er}\,u(r) \\ \mathcal{A}_i &= \mathcal{A}_i^{\rm Dirac} \end{aligned} \tag{29}$$

where the $f_i(\theta,\phi)$ are complex functions whose explicit form will not be needed and $\mathcal{A}_i^{\rm Dirac}$ is the U(1) Dirac vector potential for a monopole of charge $1/e$. (Because it is only a gauge artifact, the string singularity of $\mathcal{A}_i^{\rm Dirac}$ will be of no concern.) Note that $u(r)$ is directly related to the magnitude of the charged vector meson field.

In this gauge, the structure of the monopole can be understood by making reference to the form of the Lagrangian given in Eq. (19). Thus, we can imagine constructing the monopole solution by beginning with a point Dirac monopole. Because of the $1/r^2$ Coulomb magnetic field, this has a divergent energy density near the origin. However, this divergence can be cancelled by introducing the charged vector field, provided that the magnetic moment of the latter is properly oriented. Indeed, the appearance of a nonzero W field is to be expected whenever the energy gain from the interaction of the magnetic moment with the magnetic field outweighs the cost in mass energy; in the presence of our Coulomb field, this is the case for $r \lesssim (ev)^{-1} \sim R_{\rm core}$. Finally, the vanishing of ϕ at the center of the monopole, which is explained on topological grounds in the usual description, occurs here because it minimizes the contribution of the W mass term to the energy.

The lesson to be drawn from this is that the appearance of a nonzero W field can be understood in terms of "local" physics, without any reference to the topological behavior at spatial infinity. In other words, the value of $W_i(\mathbf{r})$ at a given point is directly related to the value of the magnetic field at that point.

4. Self-gravitating monopoles and magnetically charged black holes with hair

Let us now include gravity in this analysis. One indication of what to expect can be gained by noting that the Schwarzschild radius $2MG \sim v/(eM_{\rm Pl}^2)$ is comparable to the core radius if $v \sim M_{\rm Pl}$. Hence, we might expect the monopole solutions to become black holes when v is greater than some critical value of the order of the Planck mass. (As long as $e \ll 1$, the mass and horizon radius will be much greater than the Planck mass and Planck length, respectively, so that quantum gravity effects should be negligible.) We will also see that these monopoles can have related black hole solutions even when $v \ll M_{\rm Pl}$.

Let us begin by adapting the Ansatz of Eq. (27) to a curved spacetime with a spherically symmetric metric of the form of Eq. (2). The matter field part of the action can then be written in the $(1+1)$-dimensional form

$$S_{\rm matter} = -4\pi \int dt\, dr\, r^2 \sqrt{AB} \left[\frac{K}{A} + U \right] \tag{30}$$

where

$$K = \frac{(u')^2}{e^2 r^2} + \frac{1}{2} v^2 (h')^2 \tag{31}$$

and

$$U = \frac{(1-u^2)^2}{2e^2 r^4} + \frac{u^2 h^2 v^2}{r^2} + \frac{\lambda v^4}{2}(1-h^2)^2 \,. \tag{32}$$

One can view U as being an r-dependent potential for two scalar fields u and h. At large r, its minimum occurs when $u=0$ and $h=1$. Near the origin, it is minimized by $u=1$ and $h=0$. For small scalar self-coupling, $\lambda < e^2$, these are the only minima of U. However, if $\lambda > e^2$ there is an intermediate region of r where the potential has a nontrivial r-dependent minimum that I will denote by $\hat{u}(r)$ and $\hat{h}(r)$.

The matter field equations can be obtained by varying the reduced action of Eq. (30). This gives

$$\frac{1}{\sqrt{AB}} \left(\frac{\sqrt{AB}\, u'}{A} \right)' = \frac{e^2 r^2}{2} \frac{\partial U}{\partial u} \tag{33}$$

$$\frac{1}{r^2\sqrt{AB}}\left(\frac{r^2\sqrt{AB}h'}{A}\right)' = \frac{1}{v^2}\frac{\partial U}{\partial h}\,. \tag{34}$$

These must be supplemented by equations for the metric functions A and B. Einstein's equations reduce to

$$\mathcal{M}' = 4\pi r^2\left(\frac{K}{A}+U\right) \tag{35}$$

$$\frac{(AB)'}{AB} = 16\pi G r K \tag{36}$$

where the mass function $\mathcal{M}(r)$ is defined by

$$\frac{1}{A(r)} = 1 - \frac{2G\mathcal{M}(r)}{r}\,. \tag{37}$$

By substituting Eq. (36) into Eqs. (33) and (34), we can eliminate $B(r)$ and obtain a set of three coupled differential equations for u, h, and A. These are subject to a number of boundary conditions. At spatial infinity, finiteness of the energy requires that $u(\infty) = 0$ and $h(\infty) = 1$. In order that the fields and metric be nonsingular at the origin, we must require that $u(0) = 1$ and $h(0) = \mathcal{M}(0) = 0$. Finally, the coefficients of u'' and h'' in Eqs. (33) and (34) vanish at any zeroes of $1/A$. As a result, these equations give two additional constraints among u, h, u', and h' at every horizon.

In general, a set of one first-order and two second-order equations will allow at most five boundary conditions to be satisfied. Hence, we might hope to find solutions without horizons that are regular at both the origin and infinity (i.e., nonsingular self-gravitating monopoles) or black hole solutions that are finite at spatial infinity and at one horizon, but singular at the origin. Only for special choices of parameters would we expect to be able to have solutions that are regular at two horizons (like the Reissner-Nordstrom metric) or solutions regular at a horizon and at both $r = 0$ and $r = \infty$.

Of course, the presence of the correct number of boundary conditions does not guarantee the existence of a solution. To see whether there actually is a solution, one must resort to numerical techniques [9, 10, 11, 12]. One finds that $1/A$ develops a minimum at a value of r of order v/e. As v is increased this minimum becomes deeper until, at a critical value $v_{\rm cr}$ of order $M_{\rm Pl}$, an extremal horizon appears; this critical value varies with λ/e^2. For $v > v_{\rm cr}$ there are no nonsingular solutions. Later in these lectures I will return to these critical solutions and discuss the approach to the black hole limit in more detail.

There are also solutions with horizons. One type is obtained trivially. Setting $u = 0$ and $h = 1$ everywhere clearly satisfies Eqs. (33) and (34).

Equations (35) and (36) then lead to a Reissner-Nordstrom metric with magnetic charge $1/e$ and arbitrary mass M.

We can also look for solutions with a horizon, but with nontrivial matter fields outside the horizon; i.e., black holes with hair. One can imagine doing this by putting a small Schwarzschild-like black hole in the center of a monopole. In other words, we assume that there is a horizon at $r_{\rm H} = 2G\mathcal{M}_0$, where $\mathcal{M}_0 \ll M_{\rm mon} \sim v/e$. Because such a horizon would correspond to a very light black hole, one would expect that its gravitational effects outside the horizon would be small, and that for $r \gtrsim r_{\rm H}$ the solution would be similar to that for the nonsingular monopole. This expectation is borne out by detailed numerical and analytic investigations [9, 10, 11].

These solutions are possible only in a certain region of parameter space. Thus, consider integrating Eq. (35) to obtain the monotonically increasing mass function

$$\mathcal{M}(r) = \mathcal{M}_0 + 4\pi \int_{r_{\rm H}}^{r} ds\, s^2 \left(\frac{K}{A} + U \right) . \tag{38}$$

There will be a horizon whenever $\mathcal{M}(r)/r = 1/(2G)$. By construction, this occurs at $r = r_{\rm H}$. If v is small, $\mathcal{M}(r)/r$ will initially decrease with increasing r outside the horizon, but will then begin to increase and, when $r \sim R_{\rm core} \sim 1/(ev)$, reach a maximum of order $M_{\rm mon}/R_{\rm core} \sim v^2$, after which it decreases and asymptotically vanishes. As v is increased, the height of the maximum of $\mathcal{M}(r)/r$ will increase until it reaches $1/2G$ at a critical v of order $M_{\rm Pl}$. Since we do not expect to be able to find solutions regular at two horizons and infinity, this sets a maximum value of v for the given $\mathcal{M}_0$.

This analysis assumes that $r_{\rm H}$ is well inside the monopole core; we would not expect to find solutions with hair if $2G\mathcal{M}_0$ were considerably larger than $R_{\rm core}$. This leads to the additional constraint $\mathcal{M}_0 \lesssim M_{\rm Pl}^2/(ev) \sim M_{\rm Pl}^2/(e^2 M_{\rm mon})$. More detailed discussions and numerical analyses of these bounds are given in [9, 10, 11].

It is easy to see that these rough bounds allow the existence of solutions with hair that have masses greater than the extremal Reissner-Nordstrom mass $\sqrt{4\pi} M_{\rm Pl}/e$. This implies that there can be two distinct black hole solutions with the same mass and charge: the solution with hair, and the Reissner-Nordstrom solution. This disproves the weaker form of the no-hair conjecture. It also raises the possibility of a transition from one solution to the other.

5. Instability of the Reissner-Nordstrom solution

To explore this possibility, let us examine the stability under small perturbations of a Reissner-Nordstrom solution with magnetic charge $1/e$ and outer horizon radius $r_{\rm H}$ [13]. For the moment I will consider only spherically symmetric modes and write

$$\begin{aligned} A &= A_{\rm RN}(r) + \delta A(r,t) \\ B &= B_{\rm RN}(r) + \delta B(r,t) \\ h &= 1 + \delta h(r,t) \\ u &= u(r,t) \ . \end{aligned} \tag{39}$$

Linearizing the field equations in the perturbations δA, δB, δh, and u, we find that they separate into a pair of equations involving only δA and δB, another involving only δh, and one involving u. It is clear that the first set give no instability, since otherwise the Reissner-Nordstrom solution would be unstable in the Maxwell-Einstein theory, which we know is not the case. It is also easy to see that δh has no unstable modes. Hence, we need only consider u, which obeys

$$\frac{1}{\sqrt{AB}}\frac{\partial}{\partial t}\left(\frac{\sqrt{AB}}{B}\frac{\partial u}{\partial t}\right) - \frac{1}{\sqrt{AB}}\frac{\partial}{\partial r}\left(\frac{\sqrt{AB}}{A}\frac{\partial u}{\partial r}\right) = \frac{u(1-u^2)}{r^2} - e^2h^2v^2u \ . \tag{40}$$

Using the properties of the unperturbed metric and keeping only terms linear in u, we obtain from this

$$\frac{1}{B_{\rm RN}}\frac{\partial^2 u}{\partial t^2} - \frac{\partial}{\partial r}\left(B_{\rm RN}\frac{\partial u}{\partial r}\right) = -\left(e^2v^2 - \frac{1}{r^2}\right)u \, . \tag{41}$$

An instability would correspond to an exponentially growing solution; i.e., a solution of the form $u = f(r)e^{i\omega t}$ with imaginary frequency ω.

The equation can be recast in a more familiar form by defining a new coordinate x by

$$\frac{dr}{dx} = B_{\rm RN}(r) \, . \tag{42}$$

This maps the exterior region, $r_{\rm H} < r < \infty$, onto the entire real line, $-\infty < x < \infty$, and allows us to rewrite Eq. (41) in the form

$$-\frac{d^2u}{dx^2} + V(x)u = -\frac{d^2u}{dt^2} = \omega^2 u \tag{43}$$

where

$$V(x) = \frac{B_{\rm RN}(r)}{r^2}(e^2v^2r^2 - 1) \tag{44}$$

with r given as a function of x through Eq. (42). The precise shape of V depends on the value of $r_{\rm H}$, but in all cases $V(-\infty) = 0$ and $V(\infty) = e^2v^2$.

Equation (43) is of the form of a non-relativistic Schroedinger equation, and the existence of an instability is equivalent to having a negative energy bound state. This is determined by the value of $r_{\rm H}$. If $r_{\rm H} > 1/(ev)$, then $V(x)$ is everywhere positive and there are no bound states. If instead $r_{\rm H} < 1/(ev)$, there is a range of x where $V(x)$ is negative and a bound state becomes a possibility; numerical analysis shows that this actually happens if

$$r_{\rm H} < \frac{c}{ev} \tag{45}$$

with $c \approx 0.557$ [14]. While linear analysis cannot by itself determine the eventual outcome of the instability, it seems clear that the result is a static magnetically charged black hole with W-boson hair.

Note that this stability analysis did not make use of the full structure of the Yang-Mills theory, but only relied on the existence of a charged vector field with a magnetic moment. Recalling the nontopological analysis of the 't Hooft-Polyakov monopole outlined in Sec. 3, we see that the physical origin of the instability lies in the fact that in a sufficiently strong magnetic field it is energetically favorable to produce a nonzero W field. Making the horizon radius small enough guarantees that there will be a magnetic field of critical strength outside the horizon.

This instability has dramatic consequences for the ultimate fate of a magnetically charged black hole. Because of quantum mechanical effects, black holes emit thermal radiation at a Hawking temperature

$$T_{\rm H} = \frac{\hbar}{4\pi}\left(\frac{B'}{\sqrt{AB}}\right)_{r=r_{\rm H}} . \tag{46}$$

For a Schwarzschild black hole this temperature is inversely proportional to the mass, so that as the black hole radiates its temperature increases without limit, leading to complete evaporation in a finite time. A Reissner-Nordstrom black hole initially follows the same scenario. However, as the mass approaches the extremal value the temperature begins to decrease, with $T = 0$ in the extremal limit. Hence, unless the black hole loses its charge (e.g., by preferential Hawking radiation of particles of one charge over the other), the radiation will eventually cease and the black hole horizon will remain forever.

The instability we have found changes this scenario. Initially, the black hole radiates and loses mass, just as before. This causes the horizon to contract until the inequality (45) is satisfied. At this point, nonzero vector meson fields begin to appear outside the horizon, producing a black hole with hair whose Hawking temperature, like that of a Schwarzschild black hole, never vanishes. Radiation continues unimpeded until the horizon has contracted to a point, leaving behind a nonsingular monopole [15].

6. Static black holes without spherical symmetry

One of the most striking results in classical black hole physics is the fact that in the Einstein-Maxwell theory all static black holes are spherically symmetric [16, 17]. In contrast with electrodynamics, where static solutions corresponding to point multipoles of arbitrary order are possible, higher "mass multipoles" seem to be ruled out. By extending the analysis of the previous section to Reissner-Nordstrom solutions with $q > 1$ units of magnetic charge, we can show that this is not always the case.

It is most convenient to work with the unitary gauge fields φ, $\mathcal{A}_\mu$, and W_μ. The unperturbed solution has $\varphi(\mathbf{r}) = v$ and $W_\mu(\mathbf{r}) = 0$ everywhere, while the metric and electromagnetic field are those of a pure Reissner-Nordstrom solution with magnetic charge q/e. It is natural to expand the perturbations in spherical harmonics of appropriate types. For the scalar field, the electromagnetic field, and the metric, these are the standard scalar, vector, and tensor spherical harmonics. However, the expansion of the charged vector field must be modified. Recall that in the presence of a magnetic charge $Q_{\rm M} = q/e$ a particle carrying electric charge e acquires an additional angular momentum of magnitude $eQ_{\rm M} = q$ directed along the line from the particle to the magnetic charge. Because this is perpendicular to the ordinary orbital angular momentum $\mathbf{r} \times m\mathbf{v}$, the angular momentum of a spinless particle has a lower bound $\mathbf{L}^2 \geq q^2$. Correspondingly, in the expansion of a charged scalar field the usual spherical harmonics $Y_{LM}(\theta, \phi)$ must be replaced by monopole spherical harmonics [18, 19] $\mathcal{Y}_{qLM}$, with $L = q, q+1, \ldots$ and $M = -L, -L+1, \ldots, L$. For a charged vector field (or more precisely, for its spatial components) one must introduce monopole vector harmonics labeled by a total angular momentum J. Since this is the result of adding unit spin angular momentum to the orbital angular momentum L, we can have $J = L-1$, L, or $L+1$. We therefore obtain vector monopole spherical harmonics [20, 21] $\mathbf{C}^{(\lambda)}_{qJM}$, where $J = q-1, q, q+1, \ldots$ and λ distinguishes between different harmonics with the same values of J and M. There are three such harmonics for most values of J, but for $J = q-1$ there is only a single multiplet of vector harmonics.

Note that $J = 0$ can occur only if $q = 1$, so that a spherically symmetric W field is possible only for unit magnetic charge. This explains the old result [22] that no finite energy SU(2) configuration with multiple magnetic charge can be spherically symmetric. It also implies that any instability of the higher-charged Reissner-Nordstrom solutions must lead to a solution with non-spherically symmetric hair.

We now substitute the spherical harmonic expansions of the various fields into the action, keeping only terms quadratic in the perturbation. Because the unperturbed solution is spherically symmetric, the quadratic action splits into a sum of terms with different angular momentum. Each of these, in turn, splits into a part containing the metric and electromagnetic field perturbations, a part containing the scalar field perturbations, and a part involving only the perturbations of the massive vector field. As was noted previously, we know that the first of these cannot give any instability. It is easy to see that the second term is also positive definite. Thus, as with the singly-charged case of Sec. 5, the only possible instability arises from the massive vector modes.

Once again, the presence of an instability is equivalent to the existence of a bound state in a Schroedinger-like problem. However, the analysis is more complicated than previously because there is more than one mode with the same values of J and M for $J \geq q$. Nevertheless, one still finds [14] that for all values of J there is a bound state, and thus an unstable mode, if the horizon radius $r_{\rm H}$ is less than a critical value $r_{\rm cr}(J)$. The largest $r_{\rm cr}(J)$ occurs for the minimum angular momentum, $J = q - 1$. Hence, a Reissner-Nordstrom solution with horizon radius just less than $r_{\rm cr}(q-1)$ has a single multiplet of $2q-1$ normalized negative eigenmodes $\delta W_\mu = \psi^M_\mu$ that obey a differential equation of the form

$$\mathcal{M}_\mu{}^\nu \psi^M_\nu = -\beta^2 m^2 \psi^M_\mu \tag{47}$$

where m is the unperturbed mass of the black hole and β is dimensionless. The solution is therefore classically unstable against decay into a black hole with vector meson hair. Because $J = q - 1 \neq 0$, the latter solution cannot be spherically symmetric. It could, however, be axially symmetric if, e.g., only the mode with $M = 0$ were excited. Other combinations of modes, on the other hand, could lead to solutions with less symmetry, or possibly no rotational symmetry at all.

To see which of these is the case, we must go beyond this linear analysis [23]. If we assume that $r_{\rm H}$ is just below the critical value for instability, so that β is small, we can use a perturbative approach. Let

$$W_\mu = a\, m^{-1/2} \sum_M k_M \psi^M_\mu + \tilde{W}_\mu \equiv V_\mu + \tilde{W}_\mu \tag{48}$$

where $\tilde{W}_\mu$ is orthogonal to the negative modes. The constants k_M determine the angular dependence of the solution; they are assumed to be normalized so that

$$\sum_{M=-q+1}^{q-1} |k_M|^2 = 1\,. \tag{49}$$

We will see that the overall scale a is proportional to β/e.

Substituting Eq. (48) into the W-field equations, one finds that $\tilde{W}_\mu$ is of order $e^2 a^3$. Maxwell's equations show that the perturbation δA_μ of the electromagnetic field is of order ea^2, while from Einstein's equations we find that the metric perturbation $h_{\mu\nu} = O(Gm^2a^2)$. If we assume that $Gm^2 \ll e^2$, the dominant terms in the matter Lagrangian can be written schematically as

$$\mathcal{L}_{\text{matter}} = -V^{\mu *}\mathcal{M}_{\mu\nu}V^{\nu} + -e^2V^4 + e(\delta\mathcal{A})V^2 + (\delta\mathcal{F})^2 + \cdots . \tag{50}$$

The first term is of order $\beta^2 a^2$, the next three are $O(e^2a^4)$, and the omitted terms are suppressed by powers of either a or Gm^2/e^2.

We now integrate Eq. (50) over the region outside the horizon. Extremizing the resulting action with respect to a shows that a is of order β/e. By choosing r_{H} to be sufficiently close to the critical value, we can make a small enough that the omitted terms in Eq. (50) are indeed negligible. We must also minimize with respect to the k_M. For the $q = 2$ doubly-charged black hole, this gives an axially symmetric configuration. This axial symmetry does not survive for larger q. The $q = 3$ and $q = 4$ solutions have tetrahedral and cubic symmetries, respectively. (The somewhat surprising connection between these regular polyhedra and magnetic charges can be understood in terms of the number of zeroes of the $J = q - 1$ vector harmonics [23]. Similar behavior is also found in other contexts [24, 25].) For larger q, there is in general no rotational symmetry at all.

At this point one can go back to the gravitational field equations and determine the metric perturbations. The angular dependence of the matter fields gives rise to higher gravitational multipole moments, with the consequent multipole fields only falling as powers of the distance from the black hole. However, despite the angular dependence, the surface gravity remains constant on the horizon, just as required by the zeroth law of black hole dynamics.

7. The monopole-black hole transition

Let us now return to the case of unit magnetic charge and examine in more detail the transition from a nonsingular monopole to a magnetically-charged solution with a horizon [26]. In this section I will focus on the extremal solutions that form the boundary between these regimes, while in the next I will discuss the solutions that are just short of this critical limit.

Because both $1/A$ and $(1/A)'$ vanish at the extremal horizon $r = r_*$, it is a singular point of Eqs. (33-35), and we can expect to find nonanalytic behavior there. Indeed, since r itself is a singular coordinate at the horizon, in the sense that there is an infinite metric distance from $r = r_*$ to any other value of r, it would not be surprising if the derivatives of fields with respect to r were to diverge at the horizon. Ordinarily, physical considerations would determine the allowable singularities. However, here I am not actually requiring that the extremal solution be physically acceptable, but only that it be the limit of a family of physically acceptable nonsingular solutions. With this in mind, I will allow u' and h' to diverge, and will only require that this divergence be such that $u'/\sqrt{A}$ and $h'/\sqrt{A}$ remain finite. I will also assume that the leading singularities of the matter fields and of the metric functions are of the form $|r - r_*|^\alpha$, with the exponent α possibly being different on the two sides of the horizon.

With these assumptions, Eqs. (33-35) imply that at $r = r_*$ the matter fields must be at a stationary point of the r-dependent potential U, and that

$$1 - 8\pi G r_*^2 U(r_*) = 0 . \tag{51}$$

This allows two possible scenarios: In one, the horizon occurs at the extremal Reissner-Nordstrom value $r_0 = \sqrt{4\pi G/e^2}$, and the matter fields at the horizon are $u_* = 0$ and $h_* = 1$. Since these values are those expected far from the monopole core, where only the Coulomb fields survive, I will refer to this case as having a "Coulomb region horizon". In the other possibility, $r_* < r_0$ and the matter fields have nontrivial values $u_* = \hat{u}(r_*)$ and $h_* = \hat{h}(r_*)$ at the horizon. This gives an extremal solution with hair that I will refer to as having a "core region horizon".

It was argued in Sec. 4 that there were too many boundary conditions for one to expect a solution to be nonsingular at $r = 0$, $r = \infty$, and also at a horizon. The nonanalyticity at an extremal horizon invalidates this argument. One effectively has two independent boundary value problems to solve. Integrating out from the horizon to infinity, there are two free constants in the Taylor expansions of u and h at the horizon that can be adjusted so as to satisfy the two boundary conditions at $r = \infty$. Integrating inward, one must be able to satisfy three conditions at $r = 0$. The Taylor expansions of the matter fields at r_* (which are independent of the expansions on the other side of the horizon) only provide two adjustable constants. To obtain a third, we recall that an extremal solution only arises when v is at a critical value $v_{\rm cr}$; hence, we can think of v as being the third adjustable constant.

Carrying out this analysis in detail, one finds two rather different behaviors. With a Coulomb region horizon, the exterior solution is a pure Reissner-Nordstrom solution with $u = 0$ and $h = 1$. Just inside the horizon, one finds that

$$\begin{aligned} u &= p|x|^{1/2} + C_u|x|^{\gamma_u} + ax + \cdots \\ h &= 1 - C_h|x|^{\gamma_h} + bx + \cdots \\ \frac{1}{A} &= kx^2 + \cdots \end{aligned} \tag{52}$$

where $x \equiv (r - r_*)/r_*$. Here p, k, a, and b are determined in terms of the parameters of the theory, as are the exponents γ_u and γ_h, both of which are greater than 1/2. The constants C_u and C_h can be adjusted so that the boundary conditions at the origin are satisfied. The terms indicated by ellipses are determined by the terms shown explicitly. Note that $k \neq 1$, so that $(1/A)''$ is discontinuous at the horizon. (Solutions with the square root singularity in h rather than u are also possible.)

The solutions with core region horizons behave more smoothly and do not have a square root singularity. Near the horizon,

$$\begin{aligned}
u &= \hat{u}(r_*) + ax + p_1 C_1 |x|^{\gamma_1} + p_2 C_2 |x|^{\gamma_2} + \cdots \\
h &= \hat{h}(r_*) + bx + q_1 C_1 |x|^{\gamma_1} + q_2 C_2 |x|^{\gamma_2} + \cdots \\
\frac{1}{A} &= Fx^2 + \cdots \, . \qquad (53)
\end{aligned}$$

The adjustable coefficients C_1 and C_2 can take on different values inside and outside the horizon, while the other constants are fixed by the parameters of the theory.

Numerical integration of the field equations is needed to determine which type of critical behavior actually happens in a particular case. When $\lambda/e^2 \lesssim 25$, the critical solution has a Coulomb region horizon [26, 27]. For larger values of λ/e^2, a core region horizon is found. The approach to the critical solution is rather curious in this case. Initially, there is a minimum in $1/A$ at $r \approx r_0$ that gets deeper as v approaches $v_{\rm cr}$, just as if a Coulomb region horizon were about to be formed. However, just before $v_{\rm cr}$ is reached a second minimum appears at a smaller value of r; it is this latter minimum that becomes the extremal horizon.

To see what is perhaps the most striking difference between the two cases, we must return to Eq. (36), which we have thus far ignored. Integrating this equation, we find that

$$(AB)_r = (AB)_\infty \exp\left[-16\pi G \int_r^\infty ds\, sK\right] \qquad (54)$$

where K, defined in Eq. (31), contains the gradient terms in the dimensionally reduced matter Lagrangian. This does not lead to anything particularly unusual when there is a core region horizon. For a Coulomb region horizon, on the other hand, the square root singularity in u (or h) leads to a divergence in the integral on the right hand side of this equation. This gives a step-function rise in AB, so that for any two points $r_1 < r_*$ and $r_2 > r_*$

$$\frac{(AB)_{r_1}}{(AB)_{r_2}} = 0 \, . \qquad (55)$$

This behavior at a Coulomb region horizon leads to a phenomenon identified recently by Horowitz and Ross [28, 29]. It is often said that, because the horizon is only a coordinate singularity rather than a true curvature singularity, a freely-falling observer should feel no unusual effects at the time of crossing the horizon. In fact, the acceleration of a radially infalling observer near the horizon can invalidate this conclusion. This can be seen by relating the curvature components in a boosted frame where the observer is instantaneously at rest to the components in a "static" frame where the metric is time-independent. Using orthonormal components in both cases, we have

$$\begin{aligned} R_{t'kt'k} &= R_{tktk} + \sinh^2\alpha(R_{tktk} + R_{rkrk}) \\ R_{r'kr'k} &= R_{rkrk} + \sinh^2\alpha(R_{tktk} + R_{rkrk}) \\ R_{t'kr'k} &= \sinh\alpha\cosh\alpha(R_{tktk} + R_{rkrk}) \\ R_{t'r't'r'} &= R_{trtr} \end{aligned} \tag{56}$$

where primes denote coordinates in the infalling frame, k indicates either transverse angular coordinate, and α is the boost factor. Since α can become large as the observer nears the horizon, it is possible for the curvature components in the infalling frame (i.e., the components actually "felt" by the observer) to be large even though all components in the static frame are small. The fact that this does not happen in the the Schwarzschild and Reissner-Nordstrom metrics is a consequence of the fact that these metrics have the special property that $R_{tktk} + R_{rkrk} = 0$.

In an arbitrary metric of the form of Eq. (2), an infalling particle with an energy to mass ratio E feels a tidal force proportional to

$$R_{t'kt'k} = -\frac{1}{2r}\frac{d}{dr}\left[\frac{E^2}{AB} - \frac{1}{A}\right] . \tag{57}$$

Horowitz and Ross exhibited several examples of dilaton black holes for which this quantity becomes large at exterior points near the horizon. Because this implies that an observer will feel a black hole-induced "singularity" even before crossing the horizon, they termed such solutions "naked black holes".

Comparing Eqs. (55) and (57), we see that the same phenomenon occurs for near-critical monopole solutions that are close to developing a Coulomb region horizon. As the extremal solution is approached, the value of the right hand side of Eq. (57) at the quasi-horizon diverges. Hence, these solutions become "naked black holes" even before they become black holes. It might be tempting to conclude that this singular behavior is a necessary concomitant of the transition from a nonsingular spacetime to one with a horizon. However, the existence of solutions with core region horizons at large λ/e^2 shows that this is not the case.

8. Quasi-black holes and the emergence of black hole entropy

I now turn to solutions that are just short of being black holes; i.e., solutions for which the minimum of $1/A$ has a value ϵ that, while positive, is very close to zero. These are nonsingular and topologically trivial. However, one might expect that as ϵ decreases and the critical solution is approached, it would be harder and harder for an external observer to distinguish these solutions from true black holes. Hence, it seems appropriate to call these "quasi-black holes", and to denote the minimum of $1/A$ at $r = r_*$ a "quasi-horizon".

Let us now consider how these solutions would appear to an observer who remains at a radius $r \gg r_*$ [30]. In order to determine whether or not the solution was actually a black hole, the observer could employ various means to try to probe the region inside the quasi-horizon. One possibility would be send in a particle and wait for it to emerge again. Thus, consider a massive particle moving on a geodesic that starts from an initial radius $r_1 \gg r_*$ at time t, goes in to a minimum radius $r_{\rm min} < r_*$, and then returns again to r_1 at a time $t + \Delta t$. Without loss of generality, we can assume that the geodesic lies in the $\theta = \pi/2$ plane. It will be characterized by the conserved energy per unit mass $E = B(dt/d\tau)$ and angular momentum per unit mass $J = r^2(d\phi/d\tau)$. A standard calculation then gives

$$\frac{dr}{d\tau} = \frac{1}{\sqrt{AB}}\left[E^2 - B\left(\frac{J^2}{r^2} + 1\right)\right]^{1/2} . \tag{58}$$

Integrating $dt/dr = (dt/d\tau)/(dr/d\tau)$ gives the elapsed coordinate time

$$\Delta t = 2\int_{r_{\rm min}}^{r_1} dr\,\frac{A}{\sqrt{AB}}\left[1 - \frac{B}{E^2}\left(\frac{J^2}{r^2} + 1\right)\right]^{-1/2} . \tag{59}$$

For a solution with a core region quasi-horizon, the integral is dominated by the region $r \approx r_*$, and

$$\Delta t \approx k_1 \frac{r_*}{\sqrt{(AB)_{r_*}}}\,\epsilon^{-1/2} \tag{60}$$

where k_1 is a constant of order unity.

In the Coulomb horizon regime, the interior dominates the integral and

$$\Delta t \approx k_2\, r_*\, \epsilon^{-q} \tag{61}$$

where k_2 is also of order unity and $0.7 < q < 1$. Curiously, although the coordinate time needed to traverse the interior diverges in the critical limit, for this case the proper time vanishes as ϵ^q.

One could also probe the quasi-black hole by scattering waves off of it. Consider, for example, a massive scalar field ϕ obeying the curved space Klein-Gordon equation

$$0 = \frac{1}{\sqrt{g}} \partial_\mu [\sqrt{g}\, g^{\mu\nu}\, \partial_\nu \phi] + m^2 \phi \,. \tag{62}$$

By defining $\psi = r\phi$ and introducing a new coordinate y obeying

$$\frac{dr}{dy} = \frac{\sqrt{AB}}{A} \tag{63}$$

we can rewrite this equation as

$$0 = \frac{\partial^2 \psi}{\partial t^2} - \frac{\partial^2 \psi}{\partial y^2} + [U(r) + m^2 B]\psi \tag{64}$$

where the potential

$$\begin{aligned} U &= \frac{1}{2r}\frac{d}{dr}\left[\frac{AB}{A^2}\right] + \frac{J(J+1)}{r^2} B \\ &= \frac{AB}{rA}\left[\frac{8\pi GK}{A} - \frac{d}{dr}\left(\frac{1}{A}\right) + \frac{J(J+1)}{r}\right] . \end{aligned} \tag{65}$$

For either type of critical solution $U(r_*)$ tends to zero as $\epsilon \to 0$. As a result, there is a clear distinction between the reflected wave arising from interaction with the potential in the region $r > r_*$ and that arising from interactions in the region $r < r_*$. As the critical limit is approached, the former becomes indistinguishable from the wave reflected by the corresponding black hole solution. The existence of a reflected wave from the interior region, as well as of a transmitted wave, distinguishes the nonsingular monopole from the black hole. However, both of these suffer a time delay proportional to either $\epsilon^{-1/2}$ or ϵ^{-q}, just as for the particle probe.

An external observer with unlimited time available would be able to use probes such as these to gain information about the interior region of the quasi-black hole. However, any real observer must work on some finite time scale $\Delta\mathcal{T}$. For such an observer, the interior is inaccessible if ϵ is too small [less than $(\Delta\mathcal{T})^{-2}$ or $(\Delta\mathcal{T})^{-1/q}$ for the core- and Coulomb-type solutions, respectively]. The natural way to describe the system would then be by means of a density matrix ρ that was obtained by tracing over the degrees of freedom in the interior. This in turn would give rise to a naturally defined entropy

$$S_{\rm QBH} = -{\rm Tr}\, \rho \ln \rho \tag{66}$$

for the quasi-black hole

An estimate of the value of this entropy can be obtained from a calculation by Srednicki [31]. He considered a free massless scalar field in a flat spacetime. Assuming that the system was in its ground state, by tracing over the degrees of freedom inside an arbitrary spherical region he obtained an entropy

$$S = \kappa M^2 A \tag{67}$$

where κ is a numerical constant of order unity, M is an ultraviolet cutoff, and A is the area of the boundary of the region. In a gravitational context, we expect the Planck mass to provide the ultraviolet cutoff. Hence, it is quite plausible that in the critical limit the entropy of the quasi-black hole will approach the Bekenstein-Hawking black hole entropy $(1/4)M_{\rm Pl}^2 A$. However, in contrast with the black hole case, the quasi-black hole is topologically trivial. Its "interior" is nonsingular and static. Furthermore, this region is unambiguously defined, so that it is at least conceptually clear what it means to trace over the interior degrees of freedom.

9. The third law of black hole dynamics

The third law of thermodynamics has several formulations, one of which states the impossibility of reaching zero temperature in a finite time. Since extremal black holes have zero Hawking temperature, the analogies between thermodynamics and black hole dynamics then suggest that they should be difficult, if not impossible, to create. Indeed, one formulation of the third law of black hole dynamics states that, under certain technical assumptions, a nonextremal black hole cannot be made extremal [32]. The essential difficulty can be understood by recalling that an extremal Reissner-Nordstrom black hole has a mass and a charge that are equal in Planck units, whereas a nonextremal black hole has greater mass than charge. If one tried to make a nonextremal black hole extremal by dropping in matter with more charge than mass, the Coulomb repulsion would tend to overcome the gravitational attraction.

One could also try to produce an extremal black hole by starting with a nonsingular spacetime and allowing the collapse of a shell of matter with a properly adjusted mass to charge ratio. Boulware showed that this could in fact be done [33]. However, this mechanism relies crucially on the shell being infinitely thin; it fails for shells of finite thickness.

The quasi-black holes of the previous section suggest another possibility. Because of the cancellation of the Coulomb energy in the core by the magnetic dipole interaction, these monopoles have greater charge than mass. To bring them to criticality and produce an extremal black hole, it should only be necessary to drop in an appropriate amount of uncharged matter. With a Coulomb region quasi-horizon, one might run into difficulties from the naked black hole behavior. However, these can be avoided by working in the high λ/e^2 regime where the critical solution has a core region horizon. A variation on this process starts with a solution containing a small black hole in the center of the monopole core and an almost critical quasi-horizon further out in the monopole core. Here, the effect of the infalling matter is to replace the initial finite temperature horizon by a zero temperature horizon at a larger value of r. This scenario has been tested by numerical simulations using a massive neutral scalar field as the infalling matter [30]. The results of these are completely consistent with expectations. The possibility of such processes should give us clues for a more precise formulation of the third law of black hole dynamics.

This work was supported in part by the U. S. Department of Energy.

References

1. S. Chandrasekhar, *The Mathematical Theory of Black Holes* (1993) Oxford University Press, New York.
2. See e.g., C. W. Misner, K. S. Thorne, and J. A. Wheeler, *Gravitation* (1973) W. H. Freeman, San Francisco.
3. M. S. Volkov and D. V. Gal'tsov, Phys. Rept. **319** (1999) 1.
4. J. D. Bekenstein, Phys. Rev. D **5** (1972) 1239.
5. J. D. Bekenstein, Phys. Rev. D **51** (1995) 6608.
6. G. 't Hooft, Nucl. Phys. B **79** (1974) 276.
7. A. M. Polyakov, JETP Lett. **20** (1974) 194 [Pisma Zh. Eksp. Teor. Fiz. **20** (1974) 430].
8. K. Lee and E. J. Weinberg, Phys. Rev. Lett. **73** (1994) 1203.
9. K. Lee, V. P. Nair and E. J. Weinberg, Phys. Rev. D **45** (1992) 2751.
10. P. Breitenlohner, P. Forgacs and D. Maison, Nucl. Phys. B **383** (1992) 357.
11. P. Breitenlohner, P. Forgacs and D. Maison, Nucl. Phys. B **442** (1995) 126.
12. M. E. Ortiz, Phys. Rev. D **45** (1992) 2586.
13. K. Lee, V. P. Nair and E. J. Weinberg, Phys. Rev. Lett. **68** (1992) 1100.
14. S. A. Ridgway and E. J. Weinberg, Phys. Rev. D **51** (1995) 638.
15. T. Tamaki and K. Maeda, Phys. Rev. D **62** (2000) 084041.
16. W. Israel, Phys. Rev. **164** (1967) 1776.
17. W. Israel, Commun. Math. Phys. **8** (1968) 245.
18. I. Tamm, Z. Phys **71** (1931) 141.
19. T. T. Wu and C. N. Yang, Nucl. Phys. B **107** (1976) 365.
20. E. J. Weinberg, Phys. Rev. D **49** (1994) 1086.
21. H. A. Olsen, P. Osland and T. T. Wu, Phys. Rev. D **42** (1990) 665.
22. E. J. Weinberg and A. H. Guth, Phys. Rev. D **14** (1976) 1660.
23. S. A. Ridgway and E. J. Weinberg, Phys. Rev. D **52** (1995) 3440.
24. E. Braaten, S. Townsend and L. Carson, Phys. Lett. B **235** (1990) 147.
25. C. J. Houghton and P. M. Sutcliffe, Commun. Math. Phys. **180** (1996) 343.
26. A. Lue and E. J. Weinberg, Phys. Rev. D **60** (1999) 084025.
27. Y. Brihaye, B. Hartmann and J. Kunz, Phys. Rev. D **62** (2000) 044008.
28. G. T. Horowitz and S. F. Ross, Phys. Rev. D **56** (1997) 2180.
29. G. T. Horowitz and S. F. Ross, Phys. Rev. D **57** (1998) 1098.
30. A. Lue and E. J. Weinberg, Phys. Rev. D **61** (2000) 124003.
31. M. Srednicki, Phys. Rev. Lett. **71** (1993) 666.
32. W. Israel, Phys. Rev. Lett. **57** (1986) 397.
33. D. G. Boulware, Phys. Rev. D **8** (1973) 2363.

SUBJECT INDEX